CONTENTS

□ Chapter 3 **87**

□ Chapter 4 **141**

□ Chapter 7 **309**

□ Chapter 8 **363**

□ Chapter 9 **399**

□ Chapter 10 **429**

□ Chapter 11 **483**

□ Chapter 12 **527**

□ Chapter 13 **579**

□ Chapter 14   **617**

PREFACE

The object of this text is to provide an understanding of the fundamentals of logic and computer design for a wide audience of readers. Many of the fundamentals that are presented here have not changed in decades. On the other hand, the advances in underlying technology have had a major effect on the application of these fundamentals and the emphasis to be placed on them. The process of design has been automated by using hardware description languages and logic synthesis, and the quest for speed and low power has changed the fundamentals of computer design.

The content of this third edition continues to focus on fundamentals while at the same time reflecting the relative importance of basic concepts as the technology and the design process evolve. As an illustration, microprogramming, which has declined in use as a primary control unit design method, is treated only as a control unit design technique for implementing complex computer instructions. Also, over time, the fundamental terminology is evolving and, along with it, our perspective of associated concepts. For example, in this edition, sections on NAND circuits and NOR circuits appear in the broader context of technology mapping.

The text continues to provide the option to instructors to provide very basic coverage of either VHDL or Verilog® or omit hardware description language (HDL) coverage entirely. The perspective of the introductory coverage here is the correspondence of the HDL description to the actual hardware it represents. This vital perspective, which is critical in writing HDLs for logic synthesis, can be lost in more detailed treatments focusing on the language and fluency in its use.

In summary, this edition of *Logic and Computer Design Fundamentals* features a strong emphasis on fundamentals underlying contemporary logic design using hardware description languages, synthesis, and verification as well as changes in emphasis in the use of fundamentals of computer design. The focus on basic concepts remains and manual exercises to enhance thorough understanding of these concepts continue as a mainstay.

In order to support the evolving perspective and to deal with growing structural problems, notably chapter length, this edition features a major chapter reorganization. Chapters 1 through 6 of the book treat logic design, and Chapters 7 through 9 deal with digital systems design. Chapters 10 through 14 focus directly on computer design. This arrangement provides solid digital system design fundamentals while accomplishing a gradual, bottom-up development of fundamentals for use in top-down computer design in later chapters. Eleven of the 14 chapters contain new material not present in the second edition, and approximately 50% of the problems are modified or new. There are over a dozen text supplements available on the text website, which represent both new material and deleted material from prior editions. Summaries of the topics covered in each chapter follow.

Chapter 1—Digital Computers and Information introduces computer systems and information representation, including a new section on Gray codes.

Chapter 2—Combinational Logic Circuits deals with the basic theory and concepts for designing and optimizing gate circuits. A new section on multi-level logic optimization appears. In addition to the basic literal count, gate input count is introduced as a more accurate cost criterion for use with multilevel logic circuits.

Chapter 3—Combinational Logic Design provides an overview of the contemporary logic design process and deals with gate characteristics and delay, and technology issues such as use of NAND, NOR, AOI and OAI, and XOR and XNOR gate functions. The details of steps of the design process including problem formulation, logic optimization, technology mapping, and verification are covered for combinational logic. As a part of technology mapping, this chapter contains basic coverage of ROMs, PLAs, and PALs. Coverage of Field Programmable Gate Arrays (FPGAs), with the focus on the parts typically used in student labs, is provided as a website supplement to permit updating as this technology changes during the lifetime of this edition.

Chapter 4—Combinational Functions and Circuits covers the building blocks of combinational design. Remnants of MSI logic have been removed with the focus changed to 1) fundamental combinational functions and their implementations and 2) techniques for utilizing and modifying these functions and their associated implementations. This focus provides fundamentals for a clearer understanding of structured logic design and for visualizing the logic resulting from HDL synthesis. In addition to covering decoding, encoding, code conversion, selecting, and distributing, new functions such as enabling and input-fixing are introduced. Introductory sections on Verilog and VHDL are provided for the various types of functions.

Chapter 5—Arithmetic Functions and Circuits deals with arithmetic functions and their implementations. Beyond number representation, addition, subtraction and multiplication, functions and implementations are introduced for incrementing, decrementing, filling, extension and shifting. Verilog and VHDL descriptions are provided for arithmetic functions.

Chapter 6—Sequential Circuits introduces sequential circuit analysis and design. Latches, master-slave flip-flops, and edge-triggered flip-flops are covered with an emphasis on the D type. Other types of flip-flops (S-R, J-K, and T), which are used less frequently in modern designs, are covered but given less emphasis with in-depth coverage moved to a website supplement. Verilog and VHDL descriptions of flip-flops and sequential circuits are provided.

Chapter 7—Registers and Register Transfers ties together closely the implementation of registers and their applications. Shift register and counter design are based on combining registers with functions and implementations introduced in Chapters 4 and 5. Only the ripple counter is presented as a totally new concept. This approach is in keeping with the reduction of focus on circuits originating as

MSI parts. A new section focuses on register cell design for registers performing multiple operations. Verilog and VHDL descriptions of the various register types are introduced.

Chapter 8—Sequencing and Control covers control unit design. An additional feature added to the Algorithmic State Machine (ASM) representation is a multiway branch that is analogous to the "case" from Verilog and VHDL. Hardware control is emphasized, with reduced emphasis on microprogrammed control.

Chapter 9—Memory Basics covers SRAM, DRAM, and basic memory systems. A new section on synchronous DRAMs treats the basics of these current technologies. Verilog and VHDL memory models are provided on the website.

Chapter 10—Computer Design Basics covers register files, function units, datapaths, and two simple computers. A single-cycle computer and a new multiple-cycle computer are designed in some detail with both employing hard-wired control.

Chapter 11—Instruction Set Architecture introduces many facets of instruction set architecture. It deals with address count, addressing modes, architectures, and the types of instructions. Addressing modes and other aspects of instructions are illustrated with brief segments of instruction code.

Chapter 12—RISC and CISC Central Processing Units introduces pipelined datapaths and control. A Reduced Instruction Set Computer (RISC) design using pipelining is given. A new Complex Instruction Set Computer (CISC) design is presented. This design uses a microprogrammed control unit in conjunction with the RISC foundation to implement complex instructions.

Chapter 13—Input-Output and Communication deals with data transfer between the CPU, input-output interfaces and peripheral devices. Discussions of a keyboard, a CRT display, and a hard disk as peripherals are included, and a keyboard interface is illustrated. Other topics covered range from serial communication, including the Universal Serial Bus (USB) as an illustration, to I/O processors.

Chapter 14—Memory Systems has a particular focus on memory hierarchies. The concept of locality of reference is introduced and illustrated by consideration of the cache/main memory and main memory/hard disk relationships. An overview of cache design parameters is provided. The treatment of memory management focuses on paging and a translation lookaside buffer supporting virtual memory.

In addition to the text itself, there are substantial support features provided, descriptions of which follow.

Companion Website (http://www.prenhall.com/mano) content includes the following material: 1) twelve reading supplements including new material and material deleted from prior editions, 2) VHDL and Verilog source files for all examples, 3) solutions for about one-third of all text Chapter and reading supplement problems, 4) errata, 5) PowerPoint® slides for Chapters 1 through 9, and 6) projection originals for complex figures and tables from the text.

Design Tools packaged with most domestic and international printings of the text consist of the Xilinx® ISE Student Edition software graciously provided without charge by Xilinx, Inc. Download access to the demo XE version of the ModelSim® logic simulator from Model Technology Incorporated is available through Xilinx. These tools can be used to enter schematics and state diagrams, compile and simulate VHDL code, Verilog code, or schematics, synthesize CPLD and FPGA implementations and simulate the resulting implementations. With the purchase of low-cost experimental hardware, these tools provide everything needed for students to perform CPLD-based or FPGA-based experiments.

Instructor's Manual content includes important suggestions for use of the book, information for obtaining alternative CAD tools, and all problem solutions. This manual is available from Prentice Hall to instructors at academic institutions who adopt the book for classroom use.

Because of its broad coverage of both logic and computer design, this book can serve several different objectives in sophomore through junior level courses. Chapters 1 through 11, with selected sections omitted, provide an overview of hardware for computer science, computer engineering, electrical engineering or engineering students in general in a single semester course. Chapters 1 through 8 give a basic introduction to logic design, which can be completed in a single quarter for electrical and computer engineering students. Coverage of Chapters 1 through 10 in a semester, with perhaps some supplementary material, provides a stronger, more contemporary logic design treatment. The entire book, covered in two quarters, provides the basics of logic and computer design for computer engineering and science students. Coverage of the entire book with appropriate supplementary material or a laboratory component can fill a two-semester sequence in logic design and computer architecture. Finally, due to its moderately paced treatment of a wide range of topics, the book is ideal for self-study by engineers and computer scientists.

Among the many contributions to this book, excellent detailed comments and suggestions on drafts of Chapters 1 through 8 were provided by Richard E. Haskell, Oakland University; Eugene Henry, University of Notre Dame; Sung Hu, San Francisco State University; and Walid Hubbi, New Jersey Institute of Technology. Their contributions to text improvements are greatly appreciated. Faculty and students at the University of Wisconsin also contributed to the text. A suggestion from Professor Jim Smith motivated the direction taken in CISC design in Chapter 12, and Professor Leon Shohet suggested specific improvements based on his use of the 2nd Edition. A special thanks goes to Eric Weglarz for his in-depth reviews of new material for both content and clarity. Also thanks to Eric and to Jim Liu for preparing solutions to new and modified problems for the Instructor's Manual. Our special appreciation goes of to all of those at Prentice Hall and elsewhere for their efforts on this edition. Notable are Tom Robbins and Alice Dworkin for their guidance and support, Eric Frank for his contributions during the early stages of this edition, and Daniel Sandin for his very efficient and helpful handling of the production of this edition.

Finally, a very special thanks to Val Kime for her patience and understanding throughout the development of the third edition.

M. MORRIS MANO
CHARLES R. KIME

LOGIC AND COMPUTER DESIGN FUNDAMENTALS

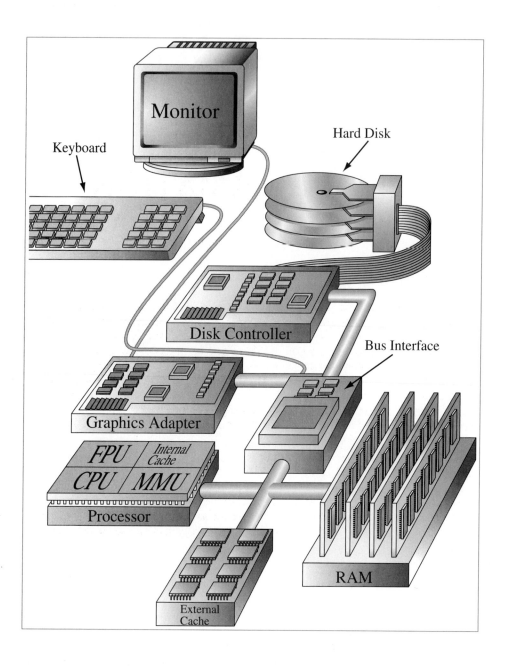

CHAPTER

1

DIGITAL COMPUTERS AND INFORMATION

Logic design fundamentals and computer design fundamentals are the topics of this book. Logic design deals with the basic concepts and tools used to design digital hardware consisting of logic circuits. Computer design deals with the additional concepts and tools used to design computers and other complex digital hardware. Computers and digital hardware in general are referred to as digital systems. Thus, this book is about understanding and designing digital systems. Due to its generality and complexity, the computer provides an ideal vehicle for learning the concepts of, and tools for, digital system design. In addition, due to its widespread use, the computer itself is deserving of study. Hence, the focus in this book is on computers and their design.

The computer will be not only a vehicle, but also a motivator for study. To this end, we use the exploded pictorial diagram of a computer of the class commonly referred to as a PC (personal computer) given on the opposite page. We use this generic computer to highlight the significance of the material covered and its relationship to the overall system. A bit later in the chapter, we will discuss the various major components of the generic computer and see how they relate to a block diagram often used to describe a computer.

1-1 DIGITAL COMPUTERS

Today, digital computers have such a prominent and growing role in modern society that we often say we are in the "information age." Computers are involved in our business transactions, communications, transportation, medical treatment, and entertainment. They monitor our weather and environment. In the industrial world, they are heavily employed in design, manufacturing, distribution, and sales. They have contributed to many scientific discoveries and engineering developments that would have been unattainable otherwise. Notably, the design of a new processor for a modern computer could not be done without the use of many computers!

The most striking property of the digital computer is its generality. It can follow a sequence of instructions, called a program, that operates on given data. The user can specify and change the program or the data according to specific needs. As a result of this flexibility, general-purpose digital computers can perform a variety of information-processing tasks that range over a very wide spectrum of applications. The general-purpose digital computer is the best known example of a *digital system*. Characteristic of a digital system is its manipulation of discrete elements of information. Any set that is restricted to a finite number of elements contains discrete information. Examples of discrete sets are the 10 decimal digits, the 26 letters of the alphabet, the 52 playing cards, and the 64 squares of a chessboard. Early digital computers were used mostly for numeric computations. In this case, the discrete elements used were the digits. From such an application, the term *digital computer* emerged.

Discrete elements of information are represented in a digital system by physical quantities called *signals*. Electrical signals such as voltages and currents are most common. Electronic devices called transistors predominate in the circuitry that implements these signals. The signals in most present-day electronic digital systems use just two discrete values and are therefore said to be *binary*.

We typically represent the two discrete values by ranges of voltage values called HIGH and LOW. Output voltage ranges and input voltage ranges are illustrated in Figure 1-1. The HIGH output voltage value ranges between 4.0 and 5.5 volts, and the LOW output voltage value ranges between −0.5 and 1.0 volt. The high input range allows 3.0 to 5.5 volts to be recognized as a HIGH, and the low input range allows −0.5 to 2.0 volts to be recognized as a LOW. The fact that the input ranges are longer than the output ranges allows the circuits to function correctly in spite of variations in their behavior and undesirable "noise" voltages that may be added to or subtracted from the outputs.

We give the output and input voltage ranges a number of different names. Among these are HIGH(H) and LOW(L), TRUE(T) and FALSE(F), and 1 and 0.

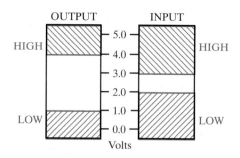

☐ **FIGURE 1-1**
An Example of Voltage Ranges for Binary Signals

It is clear that the higher voltage ranges are associated with HIGH or H, and the lower voltage ranges with LOW or L. We find, however, that for TRUE and 1 and FALSE and 0, there is a choice. TRUE and 1 can be associated with either the higher or lower voltage range and FALSE and 0 with the other range. Unless otherwise indicated, we assume that TRUE and 1 are associated with the higher of the voltage ranges, H, and that FALSE and 0 are associated with the lower of the voltage ranges, L.

Why is binary used? In contrast to the situation in Figure 1-1, consider a system with 10 values representing the decimal digits. In such a system, the voltages available—say, 0 to 5.0 volts—could be divided into 10 ranges, each of length 0.5 volt. A circuit would provide an output voltage within each of these 10 ranges. An input of a circuit would need to determine in which of the 10 ranges an applied voltage lies. If we wish to allow for noise on the voltages, then output voltage might be permitted to range over less than 0.25 volt for a given digit representation, and boundaries between inputs could vary by only less than 0.25 volt. This would require complex and costly electronic circuits and still could be disturbed by small "noise" voltages or small variations in the circuits occurring during their manufacture or use. As a consequence, the use of such multivalued circuits is very limited. Instead, binary circuits are used in which correct circuit operation can be achieved with significant variations in both the two output voltages and the two input ranges. The resulting transistor circuit with an output that is either HIGH or LOW is simple, easy to design, and extremely reliable.

Information Representation

Since 0 and 1 are associated with the binary number system, they are the preferred names for the signal ranges. A binary digit is called a *bit*. Information is represented in digital computers by groups of bits. By using various coding techniques, groups of bits can be made to represent not only binary numbers, but also other groups of discrete symbols. Groups of bits, properly arranged, can even specify to the computer the instructions to be executed and the data to be processed.

Discrete quantities of information either emerge from the nature of the data being processed or may be purposely quantized from continuous values. For example, a payroll schedule is inherently discrete data containing employee names, social security numbers, weekly salaries, income taxes, and so on. An employee's paycheck is processed using discrete data values such as letters of the alphabet (for the employee's name), digits (for the salary), and special symbols such as $. On the other hand, an engineer may measure the speed of rotation of an automobile wheel, which varies continuously with time, but may record only specific values at specific times in tabular form. The engineer is thus quantizing the continuous data, making each number in the table a discrete quantity of information. In a case such as this, if the measurement can be converted to an electronic signal, the quantization of the signal in both value and time can be performed automatically by an analog-to-digital conversion device.

Computer Structure

A block diagram of a digital computer is shown in Figure 1-2. The memory stores programs as well as input, output, and intermediate data. The datapath performs arithmetic and other data-processing operations as specified by the program. The control unit supervises the flow of information between the various units. A datapath, when combined with the control unit, forms a component referred to as a *central processing unit*, or CPU.

The program and data prepared by the user are transferred into memory by means of an input device such as a keyboard. An output device, such as a CRT (cathode-ray tube) monitor, displays the results of the computations and presents them to the user. A digital computer can accommodate many different input and output devices, such as hard disks, floppy disk drives, CD-ROM drives, and scanners. These devices use some digital logic, but often include analog electronic circuits, optical sensors, CRTs or LCDs (liquid crystal displays), and electromechanical components.

The control unit in the CPU retrieves the instructions, one by one, from the program stored in the memory. For each instruction, the control unit manipulates the datapath to execute the operation specified by the instruction. Both program and data are stored in memory. A digital computer is a powerful system. It can perform arithmetic computations, manipulate strings of alphabetic characters, and be programmed to make decisions based on internal and external conditions.

More on the Generic Computer

At this point, we will briefly discuss the generic computer and relate its various parts to the block diagram in Figure 1-2. At the lower left of the diagram at the beginning of this chapter is the heart of the computer, an integrated circuit called the *processor*. Modern processors such as this one are quite complex and consist of millions of transistors. The processor contains four functional modules: the CPU, the FPU, the MMU, and the internal cache.

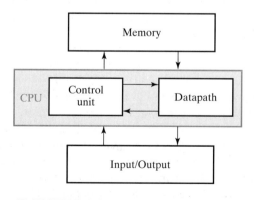

☐ **FIGURE 1-2**
Block Diagram of a Digital Computer

We have already discussed the CPU. The FPU (floating-point unit) is somewhat like the CPU, except that its datapath and control unit are specifically designed to perform floating-point operations. In essence, these operations process information represented in the form of scientific notation (e.g., 1.234×10^7), permitting the generic computer to handle very large and very small numbers. The CPU and the FPU, in relation to Figure 1-2, each contain a datapath and a control unit.

The MMU is the memory management unit. The MMU plus the internal cache and the separate blocks near the bottom of the computer labeled "External Cache" and "RAM" (random access memory) are all part of the memory in Figure 1-2. The two caches are special kinds of memory that allow the CPU and FPU to get at the data to be processed much faster than with RAM alone. RAM is what is most commonly referred to as memory. As its main function, the MMU causes the memory that appears to be available to the CPU to be much, much larger than the actual size of the RAM. This is accomplished by data transfers between the RAM and the hard disk shown at the top of the picture of the generic computer. So the hard disk, which we discuss later as an input/output device, appears conceptually as a part of the memory and input/output.

The connection paths shown between the processor, memory, and external cache are the pathways between integrated circuits. These are typically implemented as fine copper conductors on a printed circuit board. The connection paths below the bus interface are referred to as the processor bus. The connections above the bus interface are referred to as the input/output (I/O) bus. The processor bus and the I/O bus attached to the bus interface carry data having different numbers of bits and have different ways of controlling the movement of data. They may also operate at different speeds. The bus interface hardware handles these differences so that data can be communicated between the two buses.

All of the remaining structures in the generic computer are considered part of I/O in Figure 1-2. In terms of sheer physical volume, these structures dominate. In order to enter information into the computer, a keyboard is provided. In order to view output in the form of text or graphics, a graphics adapter card and CRT monitor are provided. The hard disk, discussed previously, is an electromechanical magnetic storage device. It stores large quantities of information in the form of magnetic flux on spinning disks coated with magnetic materials. In order to control the hard disk and transfer information to and from it, a disk controller is used. The keyboard, graphics adapter card, and disk controller card are all attached to the I/O bus. This allows these devices to communicate through the bus interface with the CPU and other circuitry connected to the processor buses.

The generic computer consists mainly of an interconnection of digital modules. To understand the operation of each module, it is necessary to have a basic knowledge of digital systems and their general behavior. Chapters 1 through 6 of this book deal with logic design of digital circuits in general. Chapters 7 and 8 discuss the primary components of a digital system, their operation, and their design. The operational characteristics of RAM are explained in Chapter 9. Datapath and control for simple computers are introduced in Chapter 10. Chapters 11 through 14 present the basics of computer design. Typical instructions employed in computer instruction set architectures are presented in Chapter 11. The architecture and design of CPUs are examined in Chapter 12. Input and output devices and the various ways that a CPU can communicate with them are discussed in Chapter 13. Finally, memory hierarchy concepts related to the caches and MMU are introduced in Chapter 14.

To guide the reader through this material and to keep in mind the "forest" as we carefully examine many of the "trees," accompanying discussion appears in a blue box at the beginning of each chapter to tie the topics in the chapter to the associated components in the generic computer diagram at the start of this chapter. At the completion of our journey, we will have covered most of the various modules of the computer and will have an understanding of the fundamentals that underlie both its function and design.

Earlier, we mentioned that a digital computer manipulates discrete elements of information and that all information in the computer is represented in binary form. Operands used for calculations may be expressed in the binary number system or in the decimal system by means of a binary code. The letters of the alphabet are also converted into a binary code. The purpose of the remainder of this chapter is to introduce the binary number system, binary arithmetic, and selected binary codes as a basis for further study in the succeeding chapters. In relation to the generic computer, this material is very important and spans all of the components except some in I/O that involve mechanical operations and analog (as contrasted with digital) electronics.

1-2 NUMBER SYSTEMS

The decimal number system is employed in everyday arithmetic to represent numbers by strings of digits. Depending on its position in the string, each digit has an associated value of an integer raised to the power of 10. For example, the decimal number 724.5 is interpreted to represent 7 hundreds plus 2 tens plus 4 units plus 5 tenths. The hundreds, tens, units, and tenths are powers of 10 implied by the position of the digits. The value of the number is computed as follows:

$$724.5 = 7 \times 10^2 + 2 \times 10^1 + 4 \times 10^0 + 5 \times 10^{-1}$$

The convention is to write only the digits and infer the corresponding powers of 10 from their positions. In general, a decimal number with n digits to the left of the decimal point and m digits to the right of the decimal point is represented by a string of coefficients:

$$A_{n-1}A_{n-2}...A_1A_0.A_{-1}A_{-2}...A_{-m+1}A_{-m}$$

Each coefficient A_i is one of 10 digits (0, 1, 2, 3, 4, 5, 6, 7, 8, 9). The subscript value i gives the position of the coefficient and, hence, the weight 10^i by which the coefficient must be multiplied.

The decimal number system is said to be of *base* or *radix* 10, because the coefficients are multiplied by powers of 10 and the system uses 10 distinct digits. In general, a number in base r contains r digits, 0, 1, 2, ..., $r - 1$, and is expressed as a power series in r with the general form

$$A_{n-1}r^{n-1} + A_{n-2}r^{n-2} + ... + A_1r^1 + A_0r^0$$
$$+ A_{-1}r^{-1} + A_{-2}r^{-2} + ... + A_{-m+1}r^{-m+1} + A_{-m}r^{-m}$$

When the number is expressed in positional notation, only the coefficients and the radix point are written down:

$$A_{n-1}A_{n-2}...A_1A_0.A_{-1}A_{-2}...A_{-m+1}A_{-m}$$

In general, the "." is called the *radix point*. A_{n-1} is referred to as the *most significant digit* (msd), and A_{-m} is referred to as the *least significant digit* (lsd) of the number. Note that if $m = 0$, the lsd is $A_{-0} = A_0$. To distinguish between numbers of different bases, it is customary to enclose the coefficients in parentheses and place a subscript after the right parenthesis to indicate the base of the number. However, when the context makes the base obvious, it is not necessary to use parentheses. The following illustrates a base-5 number with $n = 3$ and $m = 1$ and its conversion to decimal:

$$(312.4)_5 = 3 \times 5^2 + 1 \times 5^1 + 2 \times 5^0 + 4 \times 5^{-1}$$
$$= 75 + 5 + 2 + 0.8 = (82.8)_{10}$$

Note that for all the numbers without the base designated, the arithmetic is performed with decimal numbers. Note also that the base-5 system uses only five digits, and, therefore, the values of the coefficients in a number can be only 0, 1, 2, 3, and 4 when expressed in that system.

An alternative method for conversion to base 10 that reduces the number of operations is based on a factored form of the power series:

$$(...((A_{n-1}r + A_{n-2})r + A_{n-3})r + ... + A_1)r + A_0$$
$$+ (A_{-1} + (A_{-2} + (A_{-3} + ... + (A_{-m+2} + (A_{-m+1} + A_{-m}r^{-1})r^{-1})r^{-1}...)r^{-1})r^{-1})r^{-1}$$

For the example above,

$$(312.4)_5 = ((3 \times 5 + 1) \times 5) + 2 + 4 \times 5^{-1}$$
$$= 16 \times 5 + 2 + 0.8 = (82.8)_{10}$$

In addition to decimal, three number systems are used in computer work: binary, octal, and hexadecimal. These are base-2, base-8, and base-16 number systems, respectively.

Binary Numbers

The binary number system is a base-2 system with two digits: 0 and 1. A binary number such as 11010.11 is expressed with a string of 1's and 0's and, possibly, a

☐ **TABLE 1-1**
Powers of Two

n	2^n	n	2^n	n	2^n
0	1	8	256	16	65,536
1	2	9	512	17	131,072
2	4	10	1,024	18	262,144
3	8	11	2,048	19	524,288
4	16	12	4,096	20	1,048,576
5	32	13	8,192	21	2,097,152
6	64	14	16,384	22	4,194,304
7	128	15	32,768	23	8,388,608

binary point. The decimal equivalent of a binary number can be found by expanding the number into a power series with a base of 2. For example,

$$(11010)_2 = 1 \times 2^4 + 1 \times 2^3 + 0 \times 2^2 + 1 \times 2^1 + 0 \times 2^0 = (26)_{10}$$

As noted earlier, the digits in a binary number are called bits. When a bit is equal to 0, it does not contribute to the sum during the conversion. Therefore, the conversion to decimal can be obtained by adding the numbers with powers of two corresponding to the bits that are equal to 1. For example,

$$(110101.11)_2 = 32 + 16 + 4 + 1 + 0.5 + 0.25 = (53.75)_{10}$$

The first 24 numbers obtained from 2 to the power of n are listed in Table 1-1. In computer work, 2^{10} is referred to as K (kilo), 2^{20} as M (mega), and 2^{30} as G (giga). Thus,

$$4K = 2^2 \times 2^{10} = 2^{12} = 4,096 \text{ and } 16M = 2^4 \times 2^{20} = 2^{24} = 16,777,216$$

The conversion of a decimal number to binary can be easily achieved by a method that successively subtracts powers of two from the decimal number. To convert the decimal number N to binary, first find the greatest number that is a power of two (see Table 1-1) and that, subtracted from N, produces a positive difference. Let the difference be designated N_1. Now find the greatest number that is a power of two and that, subtracted from N_1, produces a positive difference N_2. Continue this procedure until the difference is zero. In this way, the decimal number is converted to its powers-of-two components. The equivalent binary number is obtained from the coefficients of a power series that forms the sum of the components. 1's appear in the binary number in the positions for which terms appear in the power series, and 0's appear in all other positions. This method is demonstrated by the conversion of decimal 625 to binary as follows:

$$625 - 512 = 113 = N_1 \qquad 512 = 2^9$$

$$113 - 64 = 49 = N_2 \qquad 64 = 2^6$$

$$49 - 32 = 17 = N_3 \qquad 32 = 2^5$$
$$17 - 16 = 1 = N_4 \qquad 16 = 2^4$$
$$1 - 1 = 0 = N_5 \qquad 1 = 2^0$$
$$(625)_{10} = 2^9 + 2^6 + 2^5 + 2^4 + 2^0 = (1001110001)_2$$

Octal and Hexadecimal Numbers

As previously mentioned, all computers and digital systems use the binary representation. The octal (base-8) and hexadecimal (base-16) systems are useful for representing binary quantities indirectly because they possess the property that their bases are powers of two. Since $2^3 = 8$ and $2^4 = 16$, each octal digit corresponds to three binary digits and each hexadecimal digit corresponds to four binary digits.

The more compact representation of binary numbers in either octal or hexadecimal is much more convenient for people than using bit strings in binary that are three to four times as long. Thus, most computer manuals use either octal or hexadecimal numbers to specify binary quantities. A group of 15 bits, for example, can be represented in the octal system with only five digits. A group of 16 bits can be represented in hexadecimal with four digits. The choice between an octal and a hexadecimal representation of binary numbers is arbitrary, although hexadecimal tends to win out, since bits often appear in a group of size divisible by four.

The octal number system is the base-8 system with digits 0, 1, 2, 3, 4, 5, 6, 7. An example of an octal number is 127.4. To determine its equivalent decimal value, we expand the number in a power series with a base of 8:

$$(127.4)_8 = 1 \times 8^2 + 2 \times 8^1 + 7 \times 8^0 + 4 \times 8^{-1} = (87.5)_{10}$$

Note that the digits 8 and 9 cannot appear in an octal number.

It is customary to use the first r digits from the decimal system, starting with 0, to represent the coefficients in a base-r system when r is less than 10. The letters of the alphabet are used to supplement the digits when r is 10 or more. The hexadecimal number system is a base-16 system with the first 10 digits borrowed from the decimal system and the letters A, B, C, D, E, and F used for the values 10, 11, 12, 13, 14, and 15, respectively. An example of a hexadecimal number is

$$(B65F)_{16} = 11 \times 16^3 + 6 \times 16^2 + 5 \times 16^1 + 15 \times 16^0 = (46687)_{10}$$

The first 16 numbers in the decimal, binary, octal, and hexadecimal number systems are listed in Table 1-2. Note that the sequence of binary numbers follows a prescribed pattern. The least significant bit alternates between 0 and 1, the second significant bit alternates between two 0's and two 1's, the third significant bit alternates between four 0's and four 1's, and the most significant bit alternates between eight 0's and eight 1's.

The conversion from binary to octal is easily accomplished by partitioning the binary number into groups of three bits each, starting from the binary point

☐ **TABLE 1-2**
Numbers with Different Bases

Decimal (base 10)	Binary (base 2)	Octal (base 8)	Hexadecimal (base 16)
00	0000	00	0
01	0001	01	1
02	0010	02	2
03	0011	03	3
04	0100	04	4
05	0101	05	5
06	0110	06	6
07	0111	07	7
08	1000	10	8
09	1001	11	9
10	1010	12	A
11	1011	13	B
12	1100	14	C
13	1101	15	D
14	1110	16	E
15	1111	17	F

and proceeding to the left and to the right. The corresponding octal digit is then assigned to each group. The following example illustrates the procedure:

$$(010\ 110\ 001\ 101\ 011.\ 111\ 100\ 000\ 110)_2 = (26153.7406)_8$$

The corresponding octal digit for each group of three bits is obtained from the first eight entries in Table 1-2. To make the total count of bits a multiple of three, 0's can be added on the left of the string of bits to the left of the binary point. More importantly, 0's must be added on the right of the string of bits to the right of the binary point to make the number of bits a multiple of three and obtain the correct octal result.

Conversion from binary to hexadecimal is similar, except that the binary number is divided into groups of four digits. The previous binary number is converted to hexadecimal as follows:

$$(0010\ 1100\ 0110\ 1011.\ 1111\ 0000\ 0110)_2 = (2C6B.F06)_{16}$$

The corresponding hexadecimal digit for each group of four bits is obtained by reference to Table 1-2.

Conversion from octal or hexadecimal to binary is done by reversing the procedure just performed. Each octal digit is converted to a 3-bit binary equivalent and extra 0's are deleted. Similarly, each hexadecimal digit is converted to its 4-bit binary equivalent. This is illustrated in the following examples:

$$(673.12)_8 = 110\ 111\ 011.\ 001\ 010 = (110111011.00101)_2$$
$$(3A6.C)_{16} = 0011\ 1010\ 0110.\ 1100 = (1110100110.11)_2$$

Number Ranges

In digital computers, the range of numbers that can be represented is based on the number of bits available in the hardware structures that store and process information. The number of bits in these structures is most frequently a power of two, such as 8, 16, 32, and 64. Since the numbers of bits is fixed by the structures, the addition of leading or trailing zeros to represent numbers is necessary, and the range of numbers that can be represented is also fixed.

For example, for a computer processing 16-bit unsigned integers, the number 537 is represented as 0000001000011001. The range of integers that can be handled by this representation is from 0 to $2^{16} - 1$, that is, from 0 to 65,535. If the same computer is processing 16-bit unsigned fractions with the binary point to the left of the most significant digit, then the number 0.375 is represented by 0.0110000000000000. The range of fractions that can be represented is from 0 to $(2^{16} - 1)/2^{16}$, or from 0.0 to 0.9999847412.

In later chapters, we will deal with fixed-bit representations and ranges for binary signed numbers and floating-point numbers. In both of these cases, some bits are used to represent information other than simple integer or fraction values.

1-3 ARITHMETIC OPERATIONS

Arithmetic operations with numbers in base r follow the same rules as for decimal numbers. However, when a base other than the familiar base 10 is used, one must be careful to use only r allowable digits and perform all computations with base-r digits. Examples of the addition of two binary numbers are as follows (note the names of the operands for addition):

Carries:	00000	101100
Augend:	01100	10110
Addend:	+10001	+10111
Sum:	11101	101101

The sum of two binary numbers is calculated following the same rules as for decimal numbers, except that the sum digit in any position can be only 1 or 0. Also, a carry in binary occurs if the sum in any bit position is greater than 1. (A carry in decimal occurs if the sum in any digit position is greater than 9.) Any carry obtained in a given position is added to the bits in the column one significant position higher. In the first example, since all of the carries are 0, the sum bits are simply the sum of the augend and addend bits. In the second example, the sum of the bits in the second column from the right is 2, giving a sum bit of 0 and a carry bit of 1 $(2 = 2 + 0)$. The carry bit is added with the 1's in the third position, giving a sum of 3, which produces a sum bit of 1 and a carry of 1 $(3 = 2 + 1)$.

The following are examples of the subtraction of two binary numbers; as with addition, note the names of the operands:

Borrows:	00000	00110		00110
Minuend:	10110	10110	10011	11110
Subtrahend:	−10010	−10011	−11110	−10011
Difference:	00100	00011		−01011

The rules for subtraction are the same as in decimal, except that a borrow into a given column adds 2 to the minuend bit. (A borrow in the decimal system adds 10 to the minuend digit.) In the first example shown, no borrows occur, so the difference bits are simply the minuend bits minus the subtrahend bits. In the second example, in the right position, the subtrahend bit is 1 with the minuend bit 0, so it is necessary to borrow from the second position as shown. This gives a difference bit in the first position of 1 (2 + 0 − 1 = 1). In the second position, the borrow is subtracted, so a borrow is again necessary. Recall that, in the event that the subtrahend is larger than the minuend, we subtract the minuend from the subtrahend and give the result a minus sign. This is the case in the third example, in which this interchange of the two operands is shown.

The final operation to be illustrated is binary multiplication, which is quite simple. The multiplier digits are always 1 or 0. Therefore, the partial products are equal either to the multiplicand or to 0. Multiplication is illustrated by the following example:

Multiplicand:	1011
Multiplier:	× 101
	1011
	0000
	1011
Product:	110111

Arithmetic operations with octal, hexadecimal, or any other base-r system will normally require the formulation of tables from which one obtains sums and products of two digits in that base. An easier alternative for adding two numbers in base r is to convert each pair of digits in a column to decimal, add the digits in decimal, and then convert the result to the corresponding sum and carry in the base-r system. Since addition is done in decimal, we can rely on our memories for obtaining the entries from the familiar decimal addition table. The sequence of steps for adding the two hexadecimal numbers 59F and E46 is shown in Example 1-1.

EXAMPLE 1-1   Hexadecimal Addition

Perform the addition $(59F)_{16} + (E46)_{16}$:

Hexadecimal		Equivalent Decimal Calculation			
	1 ←		1 ←		
5 9 F	5	Carry	9	15	Carry
E 4 6	14		4	6	
1 3 E 5	1 19 = 16 + 3		14 = E	21 = 16 + 5	

The equivalent decimal calculation columns on the right show the mental reasoning that must be carried out to produce each digit of the hexadecimal sum. Instead of adding F + 6 in hexadecimal, we add the equivalent decimals, 15 + 6 = 21. We then convert back to hexadecimal by noting that 21 = 16 + 5. This gives a sum digit of 5 and a carry of 1 to the next higher order column of digits. The other two columns are added in a similar fashion.   ■

The multiplication of two base-r numbers can be accomplished by doing all the arithmetic operations in decimal and converting intermediate results one at a time. This is illustrated in the multiplication of two octal numbers shown next in Example 1-2.

EXAMPLE 1-2   Octal Multiplication

Perform the multiplication $(762)_8 \times (45)_8$:

Octal	Octal		Decimal		Octal
7 6 2	5×2	=	10 = 8 + 2	=	12
4 5	$5 \times 6 + 1$	=	31 = 24 + 7	=	37
4 6 7 2	$5 \times 7 + 3$	=	38 = 32 + 6	=	46
3 7 1 0	4×2	=	8 = 8 + 0	=	10
4 3 7 7 2	$4 \times 6 + 1$	=	25 = 24 + 1	=	31
	$4 \times 7 + 3$	=	31 = 24 + 7	=	37

The computations on the right show the mental calculations for each pair of octal digits. The octal digits 0 through 7 have the same value as their corresponding decimal digits. The multiplication of two octal digits plus a carry, derived from the calculation on the previous line, is done in decimal, and the result is then converted back to octal. The left digit of the two-digit octal result gives the carry that must be added to the digit product on the next line. The blue digits from the octal results of the decimal calculations are copied to the octal partial products on the left. For example, $(5 \times 2)_8 = (12)_8$. The left digit, 1, is the carry to be added to the product $(5 \times 6)_8$, and the blue least significant digit, 2, is the corresponding digit of the octal partial product. When there is no digit product to which the carry can be added, the carry is written directly into the octal partial product, as in the case of the 4 in 46.   ■

Conversion from Decimal to Other Bases

The conversion of a number in base r to decimal is done by expanding the number in a power series and adding all the terms, as shown previously. We now present a general procedure for the reverse operation of converting a decimal number to a number in base r that is related to the alternative expansion to decimal in Section 1-2. If the number includes a radix point, it is necessary to separate the number into an integer part and a fraction part, since the two parts must be converted differently. The conversion of a decimal integer to a number in base r is done by dividing the number and all successive quotients by r and accumulating the remainders. This procedure is best explained by example.

EXAMPLE 1-3   Conversion of Decimal Integers to Octal

Convert decimal 153 to octal:

The conversion is to base 8. First, 153 is divided by 8 to give a quotient of 19 and a remainder of 1, as shown in blue. Then 19 is divided by 8 to give a quotient of 2 and a remainder of 3. Finally, 2 is divided by 8 to give a quotient of 0 and a remainder of 2. The coefficients of the desired octal number are obtained from the remainders:

$$153/8 = 19 + 1/8 \quad \text{Remainder} = 1 \quad \text{Least significant digit}$$
$$19/8 = 2 \; + 3/8 \qquad\qquad\quad\; = 3$$
$$2/8 = 0 \; + 2/8 \qquad\qquad\quad\; = 2 \quad \text{Most significant digit}$$
$$(153)_{10} = (231)_8$$

Note in Example 1-3 that the remainders are read from last to first, as indicated by the arrow, to obtain the converted number. The quotients are divided by r until the result is 0. We also can use this procedure to convert decimal integers to binary as shown in Example 1-4. In this case, the base of the converted number is 2, and therefore, all the divisions must be done by 2.

EXAMPLE 1-4   Conversion of Decimal Integers to Binary

Convert decimal 41 to binary:

$$41/2 = 20 + 1/2 \quad \text{Remainder} = 1 \quad \text{Least significant digit}$$
$$20/2 = 10 \qquad\qquad\qquad\quad\; = 0$$
$$10/2 = 5 \qquad\qquad\qquad\qquad = 0$$
$$5/2 = 2 + 1/2 \qquad\qquad\quad\; = 1$$
$$2/2 = 1 \qquad\qquad\qquad\qquad = 0$$
$$1/2 = 0 + 1/2 \qquad\qquad\quad\; = 1 \quad \text{Most significant digit}$$
$$(41)_{10} = (101001)_2$$

Of course, the decimal number could be converted by the sum of powers of two:

$$(41)_{10} = 32 + 8 + 1 = (101001)_2$$

The conversion of a decimal fraction to base r is accomplished by a method similar to that used for integers, except that multiplication by r is used instead of division, and integers are accumulated instead of remainders. Again, the method is best explained by example.

EXAMPLE 1-5 Conversion of Decimal Fractions to Binary

Convert decimal 0.6875 to binary:

First, 0.6875 is multiplied by 2 to give an integer and a fraction. The new fraction is multiplied by 2 to give a new integer and a new fraction. This process is continued until the fractional part equals 0 or until there are enough digits to give sufficient accuracy. The coefficients of the binary number are obtained from the integers as follows:

$$0.6875 \times 2 = 1.3750 \qquad \text{Integer} = 1 \quad \text{Most significant digit}$$
$$0.3750 \times 2 = 0.7500 \qquad\qquad\quad = 0$$
$$0.7500 \times 2 = 1.5000 \qquad\qquad\quad = 1$$
$$0.5000 \times 2 = 1.0000 \qquad\qquad\quad = 1 \quad \text{Least significant digit}$$
$$(0.6875)_{10} = (0.1011)_2$$

Note in the foregoing example that the integers are read from first to last, as indicated by the arrow, to obtain the converted number. In the example, a finite number of digits appears in the converted number. The process of multiplying fractions by r does not necessarily end with zero, so we must decide how many digits of the fraction to use from the conversion. Also, remember that the multiplications are by number r. Therefore, to convert a decimal fraction to octal, we must multiply the fractions by 8, as shown in Example 1-6.

EXAMPLE 1-6 Conversion of Decimal Fractions to Octal

Convert decimal 0.513 to a three-digit octal fraction:

$$0.513 \times 8 = 4.104 \qquad \text{Integer} = 4 \quad \text{Most significant digit}$$
$$0.104 \times 8 = 0.832 \qquad\qquad\quad = 0$$
$$0.832 \times 8 = 6.656 \qquad\qquad\quad = 6$$
$$0.656 \times 8 = 5.248 \qquad\qquad\quad = 5 \quad \text{Least significant digit}$$

The answer, to three significant figures, is obtained from the integer digits. Note that the last integer digit, 5, is used for rounding in base 8 of the second-to-the-last digit, 6, to obtain

$$(0.513)_{10} = (0.407)_8.$$

The conversion of decimal numbers with both integer and fractional parts is done by converting each part separately and then combining the two answers. Using the results of Example 1-3 and Example 1-6, we obtain

$$(153.513)_{10} = (231.407)_8$$

1-4 DECIMAL CODES

The binary number system is the most natural system for a computer, but people are accustomed to the decimal system. One way to resolve this difference is to convert decimal numbers to binary, perform all arithmetic calculations in binary, and then convert the binary results back to decimal. This method requires that we store the decimal numbers in the computer in a way that they can be converted to binary. Since the computer can accept only binary values, we must represent the decimal digits by a code that contains 1's and 0's. It is also possible to perform the arithmetic operations directly with decimal numbers when they are stored in the computer in coded form.

An n-bit *binary code* is a group of n bits that assume up to 2^n distinct combinations of 1's and 0's, with each combination representing one element of the set being coded. A set of four elements can be coded with a 2-bit binary code, with each element assigned one of the following bit combinations: 00, 01, 10, 11. A set of 8 elements requires a 3-bit code, and a set of 16 elements requires a 4-bit code. The bit combinations of an n-bit code can be determined from the count in binary from 0 to $2^n - 1$. Each element must be assigned a unique binary bit combination, and no two elements can have the same value; otherwise, the code assignment is ambiguous.

A binary code will have some unassigned bit combinations if the number of elements in the set is not a power of 2. The 10 decimal digits form such a set. A binary code that distinguishes among 10 elements must contain at least four bits, but six out of the 16 possible combinations will remain unassigned. Numerous different binary codes can be obtained by arranging four bits into 10 distinct combinations. The code most commonly used for the decimal digits is the straightforward binary assignment listed in Table 1-2 on page 12. This is called *binary-coded decimal* and is commonly referred to as BCD. Other decimal codes are possible, a few of which are presented in Chapter 3.

Table 1-3 gives a 4-bit code for each decimal digit. A number with n decimal digits will require $4n$ bits in BCD. Thus, decimal 396 is represented in BCD with 12 bits as

$$0011 \quad 1001 \quad 0110$$

with each group of four bits representing one decimal digit. A decimal number in BCD is the same as its equivalent binary number only when the number is between 0 and 9, inclusive. A BCD number greater than 10 has a representation different from its equivalent binary number, even though both contain 1's and 0's.

□ **TABLE 1-3**
Binary-Coded Decimal (BCD)

Decimal Symbol	BCD Digit
0	0000
1	0001
2	0010
3	0011
4	0100
5	0101
6	0110
7	0111
8	1000
9	1001

Moreover, the binary combinations 1010 through 1111 are not used and have no meaning in the BCD code.

Consider decimal 185 and its corresponding value in BCD and binary:

$$(185)_{10} = (0001\ 1000\ 0101)_{BCD} = (10111001)_2$$

The BCD value has 12 bits, but the equivalent binary number needs only 8 bits. It is obvious that a BCD number needs more bits than its equivalent binary value. However, there is an advantage in the use of decimal numbers because computer input and output data are handled by people who use the decimal system. BCD numbers are decimal numbers and not binary numbers, even though they are represented in bits. The only difference between a decimal and a BCD number is that decimals are written with the symbols 0, 1, 2, ..., 9, and BCD numbers use the binary codes 0000, 0001, 0010, ..., 1001.

BCD Addition

Consider the addition of two decimal digits in BCD, together with a possible carry of 1 from a previous less significant pair of digits. Since each digit does not exceed 9, the sum cannot be greater than $9 + 9 + 1 = 19$, the 1 being a carry. Suppose we add the BCD digits as if they were binary numbers. Then the binary sum will produce a result in the range from 0 to 19. In binary, this will be from 0000 to 10011, but in BCD, it should be from 0000 to 1 1001, the first 1 being a carry and the next four bits being the BCD digit sum. When the binary sum is less than 1010 (without a carry), the corresponding BCD digit is correct. But when the binary sum is greater than or equal to 1010, the result is an invalid BCD digit. The addition of binary 6, $(0110)_2$, to the sum converts it to the correct digit and also produces a decimal carry as required. The reason is that the difference between a carry from the most significant bit position of the binary sum and a decimal carry is $16 - 10 = 6$.

Thus, the decimal carry and the correct BCD sum digit are forced by adding 6 in binary. Consider the next three-digit BCD addition example.

EXAMPLE 1-7   BCD Addition

110	BCD carry	1←	1←	
448		0100	0100	1000
+489		+0100	+1000	+1001
937	Binary sum	1001	1101	1 0001
	Add 6		+0110	+0110
	BCD sum		1 0011	1 0111
	BCD result	1001	0011	0111

In each position, the two BCD digits are added as if they were two binary numbers. If the binary sum is greater than 1001, we add 0110 to obtain the correct BCD digit sum and a carry. In the right column, the binary sum is equal to 17. The presence of the carry indicates that the sum is greater than 16 (certainly greater than 9), so a correction is needed. The addition of 0110 produces the correct BCD digit sum, 0111 (7), and a carry of 1. In the next column, the binary sum is 1101 (13), an invalid BCD digit. Addition of 0110 produces the correct BCD digit sum, 0011 (3), and a carry of 1. In the final column, the binary sum is equal to 1001 (9) and is the correct BCD digit. ■

Parity Bit

To detect errors in data communication and processing, an additional bit is sometimes added to a binary code word to define its parity. A *parity bit* is the extra bit included to make the total number of 1's in the resulting code word either even or odd. Consider the following two characters and their even and odd parity:

	With Even Parity	With Odd Parity
1000001	01000001	11000001
1010100	11010100	01010100

In each case, we use the extra bit in the leftmost position of the code to produce an even number of 1's in the character for even parity or an odd number of 1's in the character for odd parity. In general, one parity or the other is adopted, with even parity being more common. Parity may be used with binary numbers as well as with codes, including ASCII for characters, and the parity bit may be placed in any fixed position in the code.

The parity bit is helpful in detecting errors during the transmission of information from one location to another. Assuming that even parity is used, the simplest case is handled as follows: An even (or odd) parity bit is generated at the sending end for all 7-bit ASCII characters; the 8-bit characters that include parity bits are transmitted to their destination. The parity of each character is then checked at the receiving end; if the parity of the received character is not even (odd), it means that at least one bit has changed its value during the transmission. This method detects one, three, or any odd number of errors in each character transmitted. An even number of errors is undetected. Other error-detection codes, some of which are based on additional parity bits, may be needed to take care of an even number of errors. What is done after an error is detected depends on the particular application. One possibility is to request retransmission of the message on the assumption that the error was random and will not occur again. Thus, if the receiver detects a parity error, it sends back a NAK (negative acknowledge) control character consisting of the even-parity eight bits, 10010101, from Table 1-5 on page 25. If no error is detected, the receiver sends back an ACK (acknowledge) control character, 00000110. The sending end will respond to a NAK by transmitting the message again, until the correct parity is received. If, after a number of attempts, the transmission is still in error, an indication of a malfunction in the transmission path is given.

1-5 GRAY CODES

As we count up or down using binary codes, the number of bits that change from one binary value to the next varies. This is illustrated by the binary code for the octal digits on the left in Table 1-4. As we count from 000 up to 111 and "roll over" to 000, the number of bits that change between the binary values ranges from 1 to 3.

□ **TABLE 1-4**
 Gray Code

Binary Code	Bit Changes	Gray Code	Bit Changes
000		000	
	1		1
001		001	
	2		1
010		011	
	1		1
011		010	
	3		1
100		110	
	1		1
101		111	
	2		1
110		101	
	1		1
111		100	
	3		1
000		000	

For many applications, multiple bit changes as the circuit counts is not a problem. There are applications, however, in which a change of more than one bit when counting up or down can cause serious problems. One such problem is illustrated by an optical shaft angle encoder shown in Figure 1-3(a). The encoder is a disk attached to a rotating shaft for measurement of the rotational position of the shaft. The disk contains areas that are clear for binary 1 and opaque for binary 0. An illumination source is placed on one side of the disk, and optical sensors, one for each of the bits to be encoded, are placed on the other side of the disk. When a clear region lies between the source and a sensor, the sensor responses to the light with a binary 1 output. When an opaque region lies between the source and the sensor, the sensor responds to the dark with a binary 0.

The rotating shaft, however, can be in any angular position. For example, suppose that the shaft and disk are positioned so that the sensors lie right at the boundary between 011 and 100. In this case, sensors in positions B_2, B_1 and B_0 have the light partially blocked. In such a situation, it is unclear whether the three sensors will see light or dark. As a consequence, each sensor may produce either a 1 or a 0. Thus, the resulting encoded binary number for a value between 3 and 4 may be 000, 001, 010, 011, 100, 101, 110, or 111. Either 011 or 100 will be satisfactory in this case, but the other six values are clearly erroneous!

The solution to this problem becomes apparent by noting that in those cases in which only a single bit changes when going from one value to the next or previous value, this problem cannot occur. For example, if the sensors lie on the boundary between 2 and 3, the resulting code is either 010 or 011, either of which is satisfactory. If we change the encoding of the values 0 through 7 such that only one bit value changes as we count up or down (including rollover from 7 to 0), then the encoding will be satisfactory for all positions. A code having the property that only one bit at a time changes between codes during counting is a *Gray code*. There are multiple Gray codes for any set of *n* consecutive integers, with *n* even.

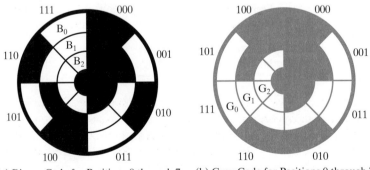

(a) Binary Code for Positions 0 through 7   (b) Gray Code for Positions 0 through 7

□ **FIGURE 1-3**
Optical Shaft Angle Encoder

A specific Gray code for the octal digits, called a *binary reflected Gray code*, appears on the right in Table 1-4. Note that the counting order for binary codes is now 000, 001, 011, 010, 110, 111, 101, 100, and 000. If we want binary codes for processing, then we can build a digital circuit or use software that converts these codes to binary before they are used in further processing of the information.

Figure 1-3(b) shows the optical shaft angle encoder using the Gray code from Table 1-4. Note that any two segments on the disk adjacent to each other have only one region that is clear for one and opaque for the other. The Gray code is named for Frank Gray who patented its use for shaft encoders in 1953.

The optical shaft encoder illustrates one use of the Gray code concept. There are many other similar uses in which a physical variable, such as position or voltage, has a continuous range of values that is converted to a digital representation. A quite different use of Gray codes appears in low-power CMOS (Complementary Metal Oxide Semiconductor) logic circuits that count up or down. In CMOS, power is consumed only when a bit changes. For the example codes given in Table 1-4 with continuous counting (either up or down), there are 14 bit changes for binary counting for every eight bit changes for Gray code counting. Thus, the power consumed at the counter outputs for the Gray code counter is only 57% of that consumed at the binary counter outputs.

A Gray code for a counting sequence of n binary code words (n must be even) can be constructed by replacing each of the first $n/2$ numbers in the sequence with a code word consisting of 0 followed by the even parity for each bit of the binary code word and the bit to its left. For example, for the binary code word 0100, the Gray code word is 0, parity(0,1), parity(1,0), parity(0,0) = 0110. Next, take the sequence of numbers formed and copy it in reverse order with the leftmost 0 replaced by a 1. This new sequence provides the Gray code words for the second $n/2$ of the original n code words. For example, for BCD codes, the first five Gray code words are 0000, 0001, 0011, 0010, and 0110. Reversing the order of these codes and replacing the leftmost 0 with a 1, we obtain 1110, 1010, 1011, 1001, and 1000 for the last five Gray codes.

For the special cases in which the original binary codes are 0 through $2^n - 1$, each Gray code word may be formed directly from the corresponding binary code word by copying its leftmost bit and then replacing each of the remaining bits with the even parity or the bit of the number and the bit to its left.

1-6 ALPHANUMERIC CODES

Many applications of digital computers require the handling of data consisting not only of numbers, but also of letters. For instance, an insurance company with thousands of policyholders uses a computer to process its files. To represent the names and other pertinent information, it is necessary to formulate a binary code for the letters of the alphabet. In addition, the same binary code must represent numerals and special characters such as $. Any alphanumeric character set for English is a set of elements that includes the 10 decimal digits, the 26 letters of the alphabet, and several (more than three) special characters. If only capital letters are included, we need a binary code of at least six bits, and if both uppercase letters and lowercase letters are included, we need a binary code of at least seven bits. Binary codes play an important role in digital computers. The codes must be in binary because computers can handle only 1's and 0's. Note that binary encoding merely changes the symbols, not the meaning of the elements of information being encoded.

ASCII Character Code

The standard binary code for the alphanumeric characters is called ASCII (American Standard Code for Information Interchange). It uses seven bits to code 128 characters, as shown in Table 1-5. The seven bits of the code are designated by B_1 through B_7, with B_7 being the most significant bit. Note that the most significant three bits of the code determine the column of the table and the least significant four bits the row of the table. The letter A, for example, is represented in ASCII as 1000001 (column 100, row 0001). The ASCII code contains 94 characters that can be printed and 34 nonprinting characters used for various control functions. The printing characters consist of the 26 uppercase letters, the 26 lowercase letters, the 10 numerals, and 32 special printable characters such as %, @, and $.

The 34 control characters are designated in the ASCII table with abbreviated names. They are listed again below the table with their full functional names. The control characters are used for routing data and arranging the printed text into a prescribed format. There are three types of control characters: format effectors, information separators, and communication control characters. Format effectors are characters that control the layout of printing. They include the familiar typewriter controls such as backspace (BS), horizontal tabulation (HT), and carriage return (CR). Information separators are used to separate the data into divisions—for example, paragraphs and pages. They include characters such as record separator (RS) and file separator (FS). The communication control characters are used during the transmission of text from one location to the other. Examples of communication control characters are STX (start of text) and ETX (end of text), which are used to frame a text message transmitted via communication wires.

ASCII is a 7-bit code, but most computers manipulate an 8-bit quantity as a single unit called a *byte*. Therefore, ASCII characters most often are stored one per byte, with the most significant bit set to 0. The extra bit is sometimes used for specific purposes, depending on the application. For example, some printers recognize an additional 128 8-bit characters, with the most significant bit set to 1. These characters enable the printer to produce additional symbols, such as those from the Greek alphabet or characters with accent marks as used in languages other than English.

 UNICODE This supplement on Unicode, a 16-bit standard code for representing the symbols and ideographs for the world's languages, is available on the Companion Website (http://www.prenhall.com/mano) for the text.

□ **TABLE 1-5**
American Standard Code for Information Interchange (ASCII)

$B_4B_3B_2B_1$	$B_7B_6B_5$							
	000	001	010	011	100	101	110	111
0000	NULL	DLE	SP	0	@	P	`	p
0001	SOH	DC1	!	1	A	Q	a	q
0010	STX	DC2	"	2	B	R	b	r
0011	ETX	DC3	#	3	C	S	c	s
0100	EOT	DC4	$	4	D	T	d	t
0101	ENQ	NAK	%	5	E	U	e	u
0110	ACK	SYN	&	6	F	V	f	v
0111	BEL	ETB	'	7	G	W	g	w
1000	BS	CAN	(	8	H	X	h	x
1001	HT	EM	)	9	I	Y	i	y
1010	LF	SUB	*	:	J	Z	j	z
1011	VT	ESC	+	;	K	[	k	{
1100	FF	FS	,	<	L	\	l		
1101	CR	GS	-	=	M	]	m	}
1110	SO	RS	.	>	N	^	n	~
1111	SI	US	/	?	O	_	o	DEL

Control Characters:

NULL	NULL	DLE	Data link escape
SOH	Start of heading	DC1	Device control 1
STX	Start of text	DC2	Device control 2
ETX	End of text	DC3	Device control 3
EOT	End of transmission	DC4	Device control 4
ENQ	Enquiry	NAK	Negative acknowledge
ACK	Acknowledge	SYN	Synchronous idle
BEL	Bell	ETB	End of transmission block
BS	Backspace	CAN	Cancel
HT	Horizontal tab	EM	End of medium
LF	Line feed	SUB	Substitute
VT	Vertical tab	ESC	Escape
FF	Form feed	FS	File separator
CR	Carriage return	GS	Group separator
SO	Shift out	RS	Record separator
SI	Shift in	US	Unit separator
SP	Space	DEL	Delete

1-7 CHAPTER SUMMARY

In this chapter, we have introduced digital systems and digital computers and have shown why such systems use signals having only two values. We have briefly introduced computer structure with a block diagram and discussion of the nature of the blocks. Number system concepts, including base (radix) and radix point, were presented. Because of their correspondence to two-valued signals, binary numbers were discussed in detail. Octal (base 8) and hexadecimal (base 16) were also emphasized, since they are useful as shorthand notation for binary. Arithmetic operations in bases other than base 10 and the conversion of numbers from one base to another were covered. Because of the predominance of decimal in normal use, Binary Coded Decimal (BCD) was treated. The parity bit was presented as a technique for error detection, and the Gray code, which is critical to selected applications, was defined. Finally, the representation of information in the form of characters instead of numbers by means of the ASCII code for the English alphabet was presented.

In subsequent chapters, we will treat the representation of signed numbers and floating-point numbers. We will also introduce additional codes for the decimal digits. Although these topics fit well with the topics in this chapter, they are difficult to motivate without associating them with the hardware used to implement the operations they denote. Thus, we delay their presentation until we examine the associated hardware.

REFERENCES

1. GRAY, F. *Pulse Code Communication.* U. S. Patent 2 632 058, March 17, 1953.

2. MANO, M. M. *Computer Engineering: Hardware Design.* Englewood Cliffs, NJ: Prentice Hall, 1988.

3. MANO, M. M. *Digital Design*, 3rd ed. Englewood Cliffs, NJ: Prentice Hall, 2002.

4. MANO, M. M. *Computer System Architecture,* 3rd ed. Englewood Cliffs, NJ: Prentice Hall, 1993.

5. PATTERSON, D. A., AND HENNESSY, J. L. *Computer Organization and Design: The Hardware/Software Interface.* 2nd ed. San Mateo, CA: Morgan Kaufmann, 1998.

6. TANENBAUM, A. S. *Structured Computer Organization,* 4th ed. Upper Saddle River, NJ: Prentice Hall, 1999.

7. WHITE, R. *How Computers Work,* Emeryville, CA: Ziff-Davis Press, 1993.

8. WILLIAMS, M. R. *A History of Computing Technology.* Englewood Cliffs, NJ: Prentice-Hall, 1985.

PROBLEMS

The plus (+) indicates a more advanced problem and the asterisk (*) indicates a solution is available on the Companion Website for the text.

1–1. *List the binary, octal, and hexadecimal numbers from 16 to 31.

1–2. What is the exact number of bits in a memory that contains **(a)** 48K bits; **(b)** 384M bits; **(c)** 8G bits?

1–3. What is the decimal equivalent of the largest binary integer that can be obtained with **(a)** 12 bits and **(b)** 24 bits?

1–4. *Convert the following binary numbers to decimal: 1001101, 1010011.101, and 10101110.1001.

1–5. Convert the following decimal numbers to binary: 125, 610, 2003, and 18944.

1–6. Each of the following five numbers has a different base: $(11100111)_2$, $(22120)_3$, $(3113)_4$, $(4110)_5$, and $(343)_8$. Which of the five numbers have the same value in decimal?

1–7. *Convert the following numbers from the given base to the other three bases listed in the table:

Decimal	Binary	Octal	Hexadecimal
369.3125	?	?	?
?	10111101.101	?	?
?	?	326.5	?
?	?	?	F3C7.A

1–8. *Convert the following decimal numbers to the indicated bases using the methods of Examples 1-3 on page 16 and 1-6 on page 17:
(a) 7562.45 to octal   **(b)** 1938.257 to hexadecimal   **(c)** 175.175 to binary.

1–9. *Perform the following conversion by using base 2 instead of base 10 as the intermediate base for the conversion:
(a) $(673.6)_8$ to hexadecimal   **(b)** $(E7C.B)_{16}$ to octal   **(c)** $(310.2)_4$ to octal

1–10. Perform the following binary multiplications:
(a) 1101×1001 **(b)** 0101×1011 **(c)** 100101×0110110

1–11. +Division is composed of multiplications and subtractions. Perform the binary division $1011110 \div 101$ to obtain a quotient and remainder.

1–12. There is considerable evidence to suggest that base 20 has historically been used for number systems in a number of cultures.
(a) Write the digits for a base-20 system, using an extension of the same digit representation scheme employed for hexadecimal.
(b) Convert $(2003)_{10}$ to base 20.   **(c)** Convert $(BCH.G)_{20}$ to decimal.

1–13. *In each of the following cases, determine the radix r:
(a) $(BEE)_r = (2699)_{10}$ **(b)** $(365)_r = (194)_{10}$

1–14. The following calculation was performed by a particular breed of unusually intelligent chicken. If the radix r used by the chicken corresponds to its total number of toes, how many toes does the chicken have on each foot?
$$((35)_r + (24)_r) \times (21)_r = (1501)_r$$

1–15. *Represent the decimal numbers 694 and 835 in BCD, and then show the steps necessary to form their sum.

1–16. *Find the binary representations for each of the following BCD numbers:
 (a) 0100 1000 0110 0111   **(b)** 0011 0111 1000.0111 0101

1–17. List the 5-bit binary number equivalents for 16 through 31 with a parity bit added in the rightmost position giving odd parity to the overall 6-bit numbers. Repeat for even parity.

1–18. By using the procedure given in Section 1-5, find the Gray code for the hexadecimal digits.

1–19. + What is the percentage of power consumed for continuous counting (either up or down but not both) at the outputs of a binary Gray code counter compared to a binary counter as a function of the number of bits, n, in the two counters?

1–20. What bit position in an ASCII code must be complemented to change the ASCII letter represented from uppercase to lowercase and vice versa?

1–21. Write your full name in ASCII, using an 8-bit code **(a)** with the leftmost bit always 0 and **(b)** with the leftmost bit selected to produce even parity. Include a space between names and a period after the middle initial.

1–22. Decode the following ASCII code: 1001010 1101111 1101000 1101110 0100000 1000100 1101111 1100101.

1–23. *Show the bit configuration that represents the decimal number 365 in **(a)** binary, **(b)** BCD, **(c)** ASCII.

1–24. A computer represents information in groups of 32 bits. How many different integers can be represented in **(a)** binary, **(b)** BCD, and **(c)** 8-bit ASCII, all using 32 bits?

COMBINATIONAL LOGIC CIRCUITS

I n this chapter, we will learn about gates, the most primitive logic elements used in digital systems. In addition, we will learn the mathematical techniques used in designing circuits from these gates and learn how to design cost-effective circuits. These techniques, which are fundamental to the design of almost all digital circuits, are based on Boolean algebra. One aspect of design is to avoid unnecessary circuitry and excess cost, a goal accomplished by a technique called optimization. Karnaugh maps provide a graphical method for enhancing understanding of optimization and solving small optimization problems for "two-level" logic circuits. More general optimization methods for circuits with more than two levels are introduced. Types of logic gates characteristic of contemporary integrated circuit implementation are discussed. Exclusive OR and Exclusive NOR gates are introduced, along with associated algebraic techniques.

In terms of the diagram at the beginning of Chapter 1, concepts from this chapter apply to most of the generic computer. Exceptions are circuits that are largely memory, such as caches and RAM, and analog electronic circuits in the monitor and hard disk controller. Nevertheless, with its use throughout the design of most of the computer, what we study in this chapter is fundamental to an in-depth understanding of computers and digital systems and how they are designed.

2-1 BINARY LOGIC AND GATES

Digital circuits are hardware components that manipulate binary information. The circuits are implemented using transistors and interconnections in complex semi-conductor devices called *integrated circuits*. Each basic circuit is referred to as a *logic gate*. For simplicity in design, we model the transistor-based electronic circuits as logic gates. Thus, the designer need not be concerned with the internal electronics of the individual gates, but only with their external logic properties. Each gate performs a specific logical operation. The outputs of gates are applied to the inputs of other gates to form a digital circuit.

In order to describe the operational properties of digital circuits, it is necessary to introduce a mathematical notation that specifies the operation of each gate and that can be used to analyze and design circuits. This binary logic system is one of a class of mathematical systems referred to generally as *Boolean algebras*. The name is in honor of the English mathematician George Boole, who in 1854 published a book introducing the mathematical theory of logic. The specific Boolean algebra we will study is used to describe the interconnection of digital gates and to design logic circuits through the manipulation of Boolean expressions. We first introduce the concept of binary logic and show its relationship to digital gates and binary signals. We then present the properties of the Boolean algebra, together with other concepts and methods useful in designing logic circuits.

Binary Logic

Binary logic deals with binary variables, which take on two discrete values, and with the operations of mathematical logic applied to these variables. The two values the variables take may be called by different names, as mentioned in Section 1-1, but for our purpose, it is convenient to think in terms of binary values and assign 1 or 0 to each variable. In the first part of this book, variables are designated by letters of the alphabet, such as A, B, C, X, Y, and Z. Later this notation will be expanded to include strings of letters, numbers, and special characters. Associated with the binary variables are three basic logical operations called AND, OR, and NOT:

1. AND. This operation is represented by a dot or by the absence of an operator. For example, $Z = X \cdot Y$ or $Z = XY$ is read "Z is equal to X AND Y." The logical operation AND is interpreted to mean that $Z = 1$ if and only if $X = 1$ and $Y = 1$; otherwise $Z = 0$. (Remember that X, Y, and Z are binary variables and can be equal to only 1 or 0.)

2. OR. This operation is represented by a plus symbol. For example, $Z = X + Y$ is read "Z is equal to X OR Y," meaning that $Z = 1$ if $X = 1$ or if $Y = 1$, or if both $X = 1$ and $Y = 1$. $Z = 0$ if and only if $X = 0$ and $Y = 0$.

3. NOT. This operation is represented by a bar over the variable. For example, $Z = \overline{X}$ is read "Z is equal to NOT X," meaning that Z is what X is not. In other words, if $X = 1$, then $Z = 0$; but if $X = 0$, then $Z = 1$. The NOT operation is also referred to as the *complement* operation, since it changes a 1 to 0 and a 0 to 1.

Binary logic resembles binary arithmetic, and the operations AND and OR have similarities to multiplication and addition, respectively. This is why the symbols used for AND and OR are the same as those used for multiplication and addition. However, binary logic should not be confused with binary arithmetic. One should realize that an arithmetic variable designates a number that may consist of many digits, whereas a logic variable is always either a 1 or a 0. The following equations define the logical OR operation:

$$0 + 0 = 0$$
$$0 + 1 = 1$$
$$1 + 0 = 1$$
$$1 + 1 = 1$$

These resemble binary addition, except for the last operation. In binary logic, we have $1 + 1 = 1$ (read "one OR one is equal to one"), but in binary arithmetic, we have $1 + 1 = 10$ (read "one plus one is equal to two"). To avoid ambiguity, the symbol $\vee$ is sometimes used for the OR operation instead of the $+$ symbol. But as long as arithmetic and logic operations are not mixed, each can use the $+$ symbol with its own independent meaning.

The next equations define the logical AND operation:

$$0 \cdot 0 = 0$$
$$0 \cdot 1 = 0$$
$$1 \cdot 0 = 0$$
$$1 \cdot 1 = 1$$

This operation is identical to binary multiplication, provided that we use only a single bit. Alternative symbols to the $\cdot$ for AND and $+$ for OR, are symbols $\wedge$ and $\vee$, respectively, that represent conjunctive and disjunctive operations in propositional calculus.

For each combination of the values of binary variables such as X and Y, there is a value of Z specified by the definition of the logical operation. The definitions may be listed in compact form in a truth table. A *truth table* for an operation is a table of combinations of the binary variables showing the relationship between the values that the variables take on and the values of the result of the operation. The truth tables for the operations AND, OR, and NOT are shown in Table 2-1. The tables list all possible combinations of values for two variables and the results of the operation. They clearly demonstrate the definition of the three operations.

□ **TABLE 2-1**
Truth Tables for the Three Basic Logical Operations

AND			OR			NOT	
X	**Y**	**Z = X·Y**	**X**	**Y**	**Z = X + Y**	**X**	**Z = X̄**
0	0	0	0	0	0	0	1
0	1	0	0	1	1	1	0
1	0	0	1	0	1		
1	1	1	1	1	1		

Logic Gates

Logic gates are electronic circuits that operate on one or more input signals to produce an output signal. Electrical signals such as voltages or currents exist throughout a digital system in either of two recognizable values. Voltage-operated circuits respond to two separate voltage ranges that represent a binary variable equal to logic 1 or logic 0, as illustrated in Figure 1-1. The input terminals of logic gates accept binary signals within the allowable range and respond at the output terminals with binary signals that fall within a specified range. The intermediate regions between the allowed ranges in the figure are crossed only during changes from 1 to 0 or from 0 to 1. These changes are called *transitions*, and the intermediate regions are called the *transition regions*.

The graphics symbols used to designate the three types of gates—AND, OR, and NOT—are shown in Figure 2-1(a). The gates are electronic circuits that produce the equivalents of logic-1 and logic-0 output signals in accordance with their respective truth tables if the equivalents of logic-1 and logic-0 input signals are applied. The two input signals X and Y to the AND and OR gates take on one of four possible combinations: 00, 01, 10, or 11. These input signals are shown as timing diagrams in Figure 2-1(b), together with the timing diagrams for the corresponding output signal for each type of gate. The horizontal axis of a *timing diagram* represents time, and the vertical axis shows a signal as it changes between the two possible voltage levels. The low level represents logic 0 and the high level represents logic 1. The AND gate responds with a logic-1 output signal when both input signals are logic 1. The OR gate responds with a logic-1 output signal if either

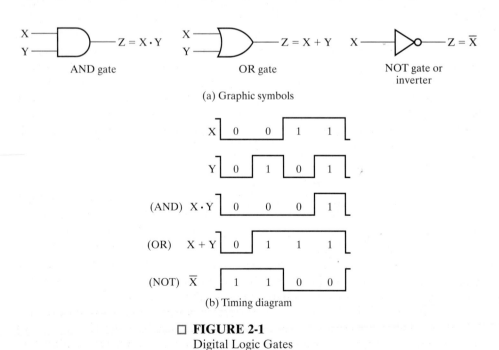

(a) Graphic symbols

(b) Timing diagram

☐ **FIGURE 2-1**
Digital Logic Gates

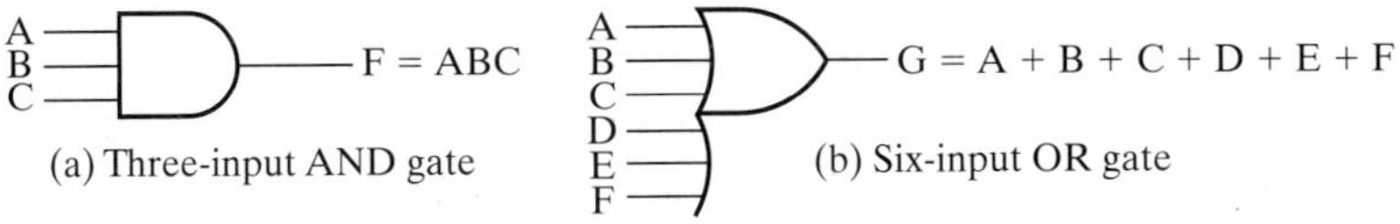

A —⟍
B —⟍ F = ABC
C —⟍   (a) Three-input AND gate

A —⟍
B —⟍
C —⟍
D —⟍ G = A + B + C + D + E + F
E —⟍
F —⟍   (b) Six-input OR gate

□ **FIGURE 2-2**
Gates with More than Two Inputs

input signal is logic 1. The NOT gate is more commonly referred to as an *inverter*. The reason for this name is apparent from the response in the timing diagram. The output logic signal is an inverted version of input logic signal X.

AND and OR gates may have more than two inputs. An AND gate with three inputs and an OR gate with six inputs are shown in Figure 2-2. The three-input AND gate responds with a logic-1 output if all three inputs are logic 1. The output is logic 0 if any input is logic 0. The six-input OR gate responds with a logic 1 if any input is logic 1; its output becomes a logic 0 only when all inputs are logic 0.

2-2 BOOLEAN ALGEBRA

The Boolean algebra we present is an algebra dealing with binary variables and logic operations. The variables are designated by letters of the alphabet, and the three basic logic operations are AND, OR, and NOT (complementation). A *Boolean expression* is an algebraic expression formed by using binary variables, the constants 0 and 1, the logic operation symbols, and parentheses. A Boolean function can be described by a Boolean equation consisting of a binary variable identifying the function followed by an equal sign and a Boolean expression. Optionally, the function identifier is followed by parentheses enclosing a list of the function variables separated by commas. A *single-output Boolean function* is a mapping from each of the possible combinations of values 0 and 1 on the function variables to value 0 or 1. A *multiple-output Boolean function* is a mapping from each of the possible combinations of values 0 and 1 on the function variables to combinations of 0 and 1 on the function outputs. Consider an example Boolean equation representing function F:

$$F(X, Y, Z) = X + \overline{Y}Z$$

The two parts of the expression, X and $\overline{Y}Z$, are called *terms* of the expression for F. The function F is equal to 1 if term X is equal to 1 or if $\overline{Y}Z$ term is equal to 1 (i.e., both $\overline{Y}$ and Z are equal to 1). Otherwise, F is equal to 0. The complement operation dictates that if $\overline{Y} = 1$, then Y must equal 0. Therefore, we can say that $F = 1$ if $X = 1$, or if $Y = 0$ and $Z = 1$. A Boolean equation expresses the logical relationship between binary variables. It is evaluated by determining the binary value of the expression for all possible combinations of values for the variables.

A Boolean function can be represented by a truth table. A *truth table* for a function is a list of all combinations of 1's and 0's that can be assigned to the binary variables and a list that shows the value of the function for each binary combination.

☐ **TABLE 2-2**
Truth Table
for the Function $F = X + \overline{Y}Z$

X	Y	Z	F
0	0	0	0
0	0	1	1
0	1	0	0
0	1	1	0
1	0	0	1
1	0	1	1
1	1	0	1
1	1	1	1

The truth tables for the logic operations given in Table 2-1 are special cases of truth tables for functions. The number of rows in a truth table is 2^n, where n is the number of variables in the function. The binary combinations for the truth table are the n-bit binary numbers that correspond to counting in decimal from 0 through $2^n - 1$. Table 2-2 shows the truth table for the function $F = X + \overline{Y}Z$. There are eight possible binary combinations that assign bits to the three variables X, Y, and Z. The column labeled F contains either 0 or 1 for each of these combinations. The table shows that the function is equal to 1 if $X = 1$ and if $Y = 0$ and $Z = 1$. Otherwise, the function is equal to 0.

An algebraic expression for a Boolean function can be transformed into a circuit diagram composed of logic gates that implements the function. The logic circuit diagram for function F is shown in Figure 2-3. An inverter on input Y generates the complement, $\overline{Y}$. An AND gate operates on $\overline{Y}$ and Z, and an OR gate combines X and $\overline{Y}Z$. In logic circuit diagrams, the variables of the function F are taken as the inputs of the circuit, and the binary variable F is taken as the output of the circuit. If the circuit has a single output, F is a single output function. If the circuit has multiple outputs, function F is a multiple output function with multiple equations required to represent its outputs. Circuit gates are interconnected by wires that carry logic signals. Logic circuits of this type are called *combinational logic circuits*, since the variables are "combined" by the logical operations. This is in contrast to the sequential logic to be treated in Chapter 6, in which variables are stored over time as well as being combined.

☐ **FIGURE 2-3**
Logic Circuit Diagram for $F = X + \overline{Y}Z$

There is only one way that a Boolean function can be represented in a truth table. However, when the function is in algebraic equation form, it can be expressed in a variety of ways. The particular expression used to represent the function dictates the interconnection of gates in the logic circuit diagram. By manipulating a Boolean expression according to Boolean algebraic rules, it is often possible to obtain a simpler expression for the same function. This simpler expression reduces both the number of gates in the circuit and the numbers of inputs to the gates. To see how this is done, it is necessary first to study the basic rules of Boolean algebra.

Basic Identities of Boolean Algebra

Table 2-3 lists the most basic identities of Boolean algebra. The notation is simplified by omitting the symbol for AND whenever doing so does not lead to confusion. The first nine identities show the relationship between a single variable X, its complement $\overline{X}$, and the binary constants 0 and 1. The next five identities, 10 through 14, have counterparts in ordinary algebra. The last three, 15 through 17, do not apply in ordinary algebra, but are useful in manipulating Boolean expressions.

The basic rules listed in the table have been arranged into two columns that demonstrate the property of duality of Boolean algebra. The *dual* of an algebraic expression is obtained by interchanging OR and AND operations and replacing 1's by 0's and 0's by 1's. An equation in one column of the table can be obtained from the corresponding equation in the other column by taking the dual of the expressions on both sides of the equals sign. For example, relation 2 is the dual of relation 1 because the OR has been replaced by an AND and the 0 by 1. It is important to note that most of the time the dual of an expression is not equal to the original expression, so that an expression usually cannot be replaced by its dual.

The nine identities involving a single variable can be easily verified by substituting each of the two possible values for X. For example, to show that $X + 0 = X$, let $X = 0$ to obtain $0 + 0 = 0$, and then let $X = 1$ to obtain $1 + 0 = 1$. Both

□ **TABLE 2-3**
   Basic Identities of Boolean Algebra

1.	$X + 0 = X$		2.	$X \cdot 1 = X$	
3.	$X + 1 = 1$		4.	$X \cdot 0 = 0$	
5.	$X + X = X$		6.	$X \cdot X = X$	
7.	$X + \overline{X} = 1$		8.	$X \cdot \overline{X} = 0$	
9.	$\overline{\overline{X}} = X$				
10.	$X + Y = Y + X$		11.	$XY = YX$	Commutative
12.	$X + (Y + Z) = (X + Y) + Z$		13.	$X(YZ) = (XY)Z$	Associative
14.	$X(Y + Z) = XY + XZ$		15.	$X + YZ = (X + Y)(X + Z)$	Distributive
16.	$\overline{X + Y} = \overline{X} \cdot \overline{Y}$		17.	$\overline{X \cdot Y} = \overline{X} + \overline{Y}$	DeMorgan's

equations are true according to the definition of the OR logic operation. Any expression can be substituted for the variable X in all the Boolean equations listed in the table. Thus, by identity 3 and with $X = AB + C$, we obtain

$$AB + C + 1 = 1$$

Note that identity 9 states that double complementation restores the variable to its original value. Thus, if $X = 0$, then $\overline{X} = 1$ and $\overline{\overline{X}} = 0 = X$.

Identities 10 and 11, the commutative laws, state that the order in which the variables are written will not affect the result when using the OR and AND operations. Identities 12 and 13, the associative laws, state that the result of applying an operation over three variables is independent of the order that is taken, and therefore, the parentheses can be removed altogether as follows:

$$X + (Y + Z) = (X + Y) + Z = X + Y + Z$$

$$X(YZ) = (XY)Z = XYZ$$

These two laws and the first distributive law, identity 14, are well known from ordinary algebra, so they should not impose any difficulty. The second distributive law, given by identity 15, is the dual of the ordinary distributive law and does not hold in ordinary algebra. As illustrated previously, each variable in an identity can be replaced by a Boolean expression, and the identity still holds. Thus, consider the expression $(A + B)(A + CD)$. Letting $X = A, Y = B$, and $Z = CD$, and applying the second distributive law, we obtain

$$(A + B)(A + CD) = A + BCD$$

The last two identities in Table 2-3,

$$\overline{X + Y} = \overline{X} \cdot \overline{Y} \text{ and } \overline{X \cdot Y} = \overline{X} + \overline{Y}$$

are referred to as DeMorgan's theorem. This is a very important theorem and is used to obtain the complement of an expression and of the corresponding function. DeMorgan's theorem can be illustrated by means of truth tables that assign all the possible binary values to X and Y. Table 2-4 shows two truth tables that verify the first part of DeMorgan's theorem. In A, we evaluate $\overline{X + Y}$ for all possible values of X and Y. This is done by first evaluating $X + Y$ and then taking its complement. In B, we evaluate $\overline{X}$ and $\overline{Y}$ and then AND them together. The result is the same

□ **TABLE 2-4**
Truth Tables to Verify DeMorgan's Theorem

A) X	Y	X + Y	$\overline{X + Y}$		B) X	Y	$\overline{X}$	$\overline{Y}$	$\overline{X} \cdot \overline{Y}$
0	0	0	1		0	0	1	1	1
0	1	1	0		0	1	1	0	0
1	0	1	0		1	0	0	1	0
1	1	1	0		1	1	0	0	0

for the four binary combinations of X and Y, which verifies the identity of the equation.

Note the order in which the operations are performed when evaluating an expression. In part B of the table, the complement over a single variable is evaluated first, followed by the AND operation, just as in ordinary algebra with multiplication and addition. In part A, the OR operation is evaluated first. Then, noting that the complement over an expression such as $X + Y$ is considered as specifying NOT $(X + Y)$, we evaluate the expression within the parentheses and take the complement of the result. It is customary to exclude the parentheses when complementing an expression, since a bar over the entire expression joins it together. Thus, $\overline{(X + Y)}$ is expressed as $\overline{X + Y}$ when designating the complement of $X + Y$.

DeMorgan's theorem can be extended to three or more variables. The general DeMorgan's theorem can be expressed as

$$\overline{X_1 + X_2 + \ldots + X_n} = \overline{X}_1 \overline{X}_2 \ldots \overline{X}_n$$

$$\overline{X_1 X_2 \ldots X_n} = \overline{X}_1 + \overline{X}_2 + \ldots + \overline{X}_n$$

Observe that the logic operation changes from OR to AND or from AND to OR. In addition, the complement is removed from the entire expression and placed instead over each variable. For example,

$$\overline{A + B + C + D} = \overline{A}\,\overline{B}\,\overline{C}\,\overline{D}$$

Algebraic Manipulation

Boolean algebra is a useful tool for simplifying digital circuits. Consider, for example, the Boolean function represented by

$$F = \overline{X}YZ + \overline{X}Y\overline{Z} + XZ$$

The implementation of this equation with logic gates is shown in Figure 2-4(a). Input variables X and Z are complemented with inverters to obtain $\overline{X}$ and $\overline{Z}$. The three terms in the expression are implemented with three AND gates. The OR gate forms the logical OR of the terms. Now consider a simplification of the expression for F by applying some of the identities listed in Table 2-3:

$$
\begin{aligned}
F &= \overline{X}YZ + \overline{X}Y\overline{Z} + XZ \\
&= \overline{X}Y(Z + \overline{Z}) + XZ && \text{by identity 14} \\
&= \overline{X}Y \cdot 1 + XZ && \text{by identity 7} \\
&= \overline{X}Y + XZ && \text{by identity 2}
\end{aligned}
$$

The expression is reduced to only two terms and can be implemented with gates as shown in Figure 2-4(b). It is obvious that the circuit in (b) is simpler than the one in (a), yet, both implement the same function. It is possible to use a truth table to verify that the two implementations are equivalent. This is shown in Table 2-5. As expressed in Figure 2-4(a), the function is equal to 1 if $X = 0, Y = 1$, and $Z = 1$; if $X = 0, Y = 1$, and $Z = 0$; or if X and Z are both 1. This produces the

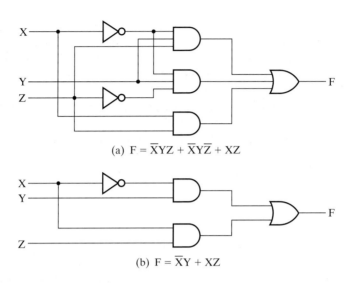

(a) $F = \overline{X}YZ + \overline{X}Y\overline{Z} + XZ$

(b) $F = \overline{X}Y + XZ$

☐ **FIGURE 2-4**
Implementation of Boolean Function with Gates

☐ **TABLE 2-5**
Truth Table for Boolean Function

X	Y	Z	(a) F	(b) F
0	0	0	0	0
0	0	1	0	0
0	1	0	1	1
0	1	1	1	1
1	0	0	0	0
1	0	1	1	1
1	1	0	0	0
1	1	1	1	1

four 1's for F in part (a) of the table. As expressed in Figure 2-4(b), the function is equal to 1 if $X = 0$ and $Y = 1$ or if $X = 1$ and $Z = 1$. This produces the same four 1's in part (b) of the table. Since both expressions produce the same truth table, they are equivalent. Therefore, the two circuits have the same output for all possible binary combinations of the three input variables. Each circuit implements the same function, but the one with fewer gates is preferable because it requires fewer components.

When a Boolean equation is implemented with logic gates, each term requires a gate, and each variable within the term designates an input to the gate. We define a *literal* as a single variable within a term that may or may not be complemented. The expression for the function in Figure 2-4(a) has three terms and eight literals; the one in Figure 2-4(b) has two terms and four literals. By reducing the number of terms, the number of literals, or both in a Boolean expression, it is often possible to obtain a simpler circuit. Boolean algebra is applied to reduce an expression for the purpose of obtaining a simpler circuit. For highly complex functions, finding the best expression based on counts of terms and literals is very difficult, even by the use of computer programs. Certain methods, however, for reducing expressions are often included in computer tools for synthesizing logic circuits. These methods can obtain good, if not the best, solutions. The only manual method for the general case is a cut-and-try procedure employing the basic relations and other manipulations that become familiar with use. The following examples use identities from Table 2-3 to illustrate a few of the possibilities:

1. $X + XY = X(1 + Y) = X$
2. $XY + X\overline{Y} = X(Y + \overline{Y}) = X$
3. $X + \overline{X}Y = (X + \overline{X})(X + Y) = X + Y$

Note that the intermediate step $X = X \cdot 1$ has been omitted when X is factored out in equation 1. The relationship $1 + Y = 1$ is useful for eliminating redundant terms, as is done with the term XY in this same equation. The relation $Y + \overline{Y} = 1$ is useful for combining two terms, as is done in equation 2. The two terms being combined must be identical except for one variable, and that variable must be complemented in one term and not complemented in the other. Equation 3 is simplified by means of the second distributive law (identity 15 in Table 2-3). The following are three more examples of simplifying Boolean expressions:

4. $X(X + Y) = X + XY = X$
5. $(X + Y)(X + \overline{Y}) = X + Y\overline{Y} = X$
6. $X(\overline{X} + Y) = X\overline{X} + XY = XY$

Note that the intermediate steps $XX = X = X \cdot 1$ have been omitted during the manipulation of equation 4. The expression in equation 5 is simplified by means of the second distributive law. Here again, we omit the intermediate steps $Y\overline{Y} = 0$ and $X + 0 = X$.

Equations 4 through 6 are the duals of equations 1 through 3. Remember that the dual of an expression is obtained by changing AND to OR and OR to AND throughout (and 1's to 0's and 0's to 1's if they appear in the expression). The *duality principle* of Boolean algebra states that a Boolean equation remains valid if we take the dual of the expressions on both sides of the equals sign. Therefore, equations 4, 5, and 6 can be obtained by taking the dual of equations 1, 2, and 3, respectively.

Along with the results just given in equations 1 through 6, the following *consensus theorem* is useful when simplifying Boolean expressions:

$$XY + \overline{X}Z + YZ = XY + \overline{X}Z$$

The theorem shows that the third term, YZ, is redundant and can be eliminated. Note that Y and Z are associated with X and $\overline{X}$ in the first two terms and appear together in the term that is eliminated. The proof of the consensus theorem is obtained by first ANDing YZ with $(X + \overline{X}) = 1$ and proceeds as follows:

$$XY + \overline{X}Z + YZ \;=\; XY + \overline{X}Z + YZ(X + \overline{X})$$

$$=\; XY + \overline{X}Z + XYZ + \overline{X}YZ$$

$$=\; XY + XYZ + \overline{X}Z + \overline{X}YZ$$

$$=\; XY(1 + Z) + \overline{X}Z(1 + Y)$$

$$=\; XY + \overline{X}Z$$

The dual of the consensus theorem is

$$(X + Y)(\overline{X} + Z)(Y + Z) = (X + Y)(\overline{X} + Z)$$

The following example shows how the consensus theorem can be applied in manipulating a Boolean expression:

$$(A + B)(\overline{A} + C) \;=\; A\overline{A} + AC + \overline{A}B + BC$$

$$=\; AC + \overline{A}B + BC$$

$$=\; AC + \overline{A}B$$

Note that $A\overline{A} = 0$ and $0 + AC = AC$. The redundant term eliminated in the last step by the consensus theorem is BC.

Complement of a Function

The complement representation for a function F, $\overline{F}$, is obtained from an interchange of 1's to 0's and 0's to 1's for the values of F in the truth table. The complement of a function can be derived algebraically by applying DeMorgan's theorem. The generalized form of this theorem states that the complement of an expression is obtained by interchanging AND and OR operations and complementing each variable and constant, as shown in Example 2-1.

EXAMPLE 2-1   Complementing Functions

Find the complement of each of the functions represented by the equations $F_1 = \overline{X}Y\overline{Z} + \overline{X}\,\overline{Y}Z$ and $F_2 = X(\overline{Y}\overline{Z} + YZ)$. Applying DeMorgan's theorem as many times as necessary, we obtain the complements as follows:

$$\overline{F_1} \;=\; \overline{\overline{X}Y\overline{Z} + \overline{X}\,\overline{Y}Z} = (\overline{\overline{X}Y\overline{Z}}) \cdot (\overline{\overline{X}\,\overline{Y}Z})$$

$$=\; (X + \overline{Y} + Z)(X + Y + \overline{Z})$$

$$\overline{F_2} \;=\; \overline{X(\overline{Y}\overline{Z} + YZ)} = \overline{X} + (\overline{\overline{Y}\overline{Z} + YZ})$$

$$= \quad \overline{X} + (\overline{\overline{YZ}} \cdot \overline{YZ})$$

$$= \quad \overline{X} + (Y + Z)(\overline{Y} + \overline{Z}) \qquad \blacksquare$$

A simpler method for deriving the complement of a function is to take the dual of the function equation and complement each literal. This method follows from the generalization of DeMorgan's theorem. Remember that the dual of an expression is obtained by interchanging AND and OR operations and 1's and 0's. To avoid confusion in handling complex functions, adding parentheses around terms before taking the dual is helpful, as illustrated in the next example.

EXAMPLE 2-2   Complementing Functions by Using Duals

Find the complements of the functions in Example 2-1 by taking the duals of their equations and complementing each literal.

We begin with

$$F_1 = \overline{X}Y\overline{Z} + \overline{X}\,\overline{Y}Z \; = (\overline{X}Y\overline{Z}) + (\overline{X}\,\overline{Y}Z)$$

The dual of F_1 is

$$(\overline{X} + Y + \overline{Z})(\overline{X} + \overline{Y} + Z)$$

Complementing each literal, we have

$$(X + \overline{Y} + Z)(X + Y + \overline{Z}) = \overline{F}_1$$

Now,

$$F_2 = X(\overline{Y}\,\overline{Z} + YZ) \; = X((\overline{Y}\,\overline{Z}) + (YZ))$$

The dual of F_2 is

$$X + (\overline{Y} + \overline{Z})(Y + Z)$$

Complementing each literal yields

$$\overline{X} + (Y + Z)(\overline{Y} + \overline{Z}) = \overline{F}_2. \qquad \blacksquare$$

2-3 STANDARD FORMS

A Boolean function expressed algebraically can be written in a variety of ways. There are, however, specific ways of writing algebraic equations that are considered to be standard forms. The standard forms facilitate the simplification procedures for Boolean expressions and frequently result in more desirable logic circuits.

The standard forms contain *product terms* and *sum terms*. An example of a product term is $X\overline{Y}Z$. This is a logical product consisting of an AND operation among three literals. An example of a sum term is $X + Y + \overline{Z}$. This is a logical sum consisting of an OR operation among the literals. It must be realized that the words "product" and "sum" do not imply arithmetic operations in Boolean algebra; instead, they specify the logical operations AND and OR, respectively.

Minterms and Maxterms

It has been shown that a truth table defines a Boolean function. An algebraic expression representing the function is derived from the table by finding the logical sum of all product terms for which the function assumes the binary value 1. A product term in which all the variables appear exactly once, either complemented or uncomplemented, is called a *minterm*. Its characteristic property is that it represents exactly one combination of the binary variables in a truth table. It has the value 1 for that combination and 0 for all others. There are 2^n distinct minterms for n variables. The four minterms for the two variables X and Y are $\overline{X}\,\overline{Y}$, $\overline{X}Y$, $X\overline{Y}$, and XY. The eight minterms for the three variables X, Y, and Z are listed in Table 2-6. The binary numbers from 000 to 111 are listed under the variables. For each binary combination, there is a related minterm. Each minterm is a product term of exactly three literals. A literal is a complemented variable if the corresponding bit of the related binary combination is 0 and is an uncomplemented variable if it is 1. A symbol m_j for each minterm is also shown in the table, where the subscript j denotes the decimal equivalent of the binary combination for which the minterm has the value 1. This list of minterms for any given n variables can be formed in a similar manner from a list of the binary numbers from 0 through $2^n - 1$. In addition, the truth table for each minterm is given in the right half of the table. These truth tables clearly show that each minterm is 1 for the corresponding binary combination and 0 for all other combinations. Such truth tables will be helpful later in using minterms to form Boolean expressions.

A sum term that contains all the variables in complemented or uncomplemented form is called a *maxterm*. Again, it is possible to formulate 2^n maxterms with n variables. The eight maxterms for three variables are listed in Table 2-7. Each maxterm is a logical sum of the three variables, with each variable being complemented if the corresponding bit of the binary number is 1 and uncomplemented

□ **TABLE 2-6**
Minterms for Three Variables

X	Y	Z	Product Term	Symbol	m_0	m_1	m_2	m_3	m_4	m_5	m_6	m_7
0	0	0	$\overline{X}\,\overline{Y}\,\overline{Z}$	m_0	1	0	0	0	0	0	0	0
0	0	1	$\overline{X}\,\overline{Y}Z$	m_1	0	1	0	0	0	0	0	0
0	1	0	$\overline{X}Y\overline{Z}$	m_2	0	0	1	0	0	0	0	0
0	1	1	$\overline{X}YZ$	m_3	0	0	0	1	0	0	0	0
1	0	0	$X\overline{Y}\,\overline{Z}$	m_4	0	0	0	0	1	0	0	0
1	0	1	$X\overline{Y}Z$	m_5	0	0	0	0	0	1	0	0
1	1	0	$XY\overline{Z}$	m_6	0	0	0	0	0	0	1	0
1	1	1	XYZ	m_7	0	0	0	0	0	0	0	1

□ **TABLE 2-7**
Maxterms for Three Variables

X	Y	Z	Sum Term	Symbol	M_0	M_1	M_2	M_3	M_4	M_5	M_6	M_7
0	0	0	$X + Y + Z$	M_0	0	1	1	1	1	1	1	1
0	0	1	$X + Y + \overline{Z}$	M_1	1	0	1	1	1	1	1	1
0	1	0	$X + \overline{Y} + Z$	M_2	1	1	0	1	1	1	1	1
0	1	1	$X + \overline{Y} + \overline{Z}$	M_3	1	1	1	0	1	1	1	1
1	0	0	$\overline{X} + Y + Z$	M_4	1	1	1	1	0	1	1	1
1	0	1	$\overline{X} + Y + \overline{Z}$	M_5	1	1	1	1	1	0	1	1
1	1	0	$\overline{X} + \overline{Y} + Z$	M_6	1	1	1	1	1	1	0	1
1	1	1	$\overline{X} + \overline{Y} + \overline{Z}$	M_7	1	1	1	1	1	1	1	0

if it is 0. The symbol for a maxterm is M_j, where j denotes the decimal equivalent of the binary combination for which the maxterm has the value 0. In the right half of the table, the truth table for each maxterm is given. Note that the value of the maxterm is 0 for the corresponding combination and 1 for all other combinations. It is now clear where the terms "minterm" and "maxterm" come from: a minterm is a function, not equal to 0, having the minimum number of 1's in its truth table; a maxterm is a function, not equal to 1, having the maximum of 1's in its truth table. Note from Table 2-6 and Table 2-7 that a minterm and maxterm with the same subscript are the complements of each other; that is, $M_j = \overline{m_j}$. For example, for $j = 3$, we have

$$\overline{m_3} = \overline{\overline{X}YZ} = X + \overline{Y} + \overline{Z} = M_3$$

A Boolean function can be represented algebraically from a given truth table by forming the logical sum of all the minterms that produce a 1 in the function. This expression is called a *sum of minterms*. Consider the Boolean function F in Table 2-8(a). The function is equal to 1 for each of the following binary combinations of the variables $X, Y,$ and Z: 000, 010, 101 and 111. These combinations correspond to minterms 0, 2, 5, and 7. By examining Table 2-8 and the truth tables for these minterms in Table 2-6, it is evident that the function F can be expressed algebraically as the logical sum of the stated minterms:

$$F = \overline{X}\,\overline{Y}\overline{Z} + \overline{X}Y\overline{Z} + X\overline{Y}Z + XYZ = m_0 + m_2 + m_5 + m_7$$

This can be further abbreviated by listing only the decimal subscripts of the minterms:

$$F(X, Y, Z) = \Sigma m(0, 2, 5, 7)$$

The symbol Σ stands for the logical sum (Boolean OR) of the minterms. The numbers following it represent the minterms of the function. The letters in parentheses following F form a list of the variables in the order taken when the minterms are converted to product terms.

☐ **TABLE 2-8**
Boolean Functions of Three Variables

(a) X	Y	Z	F	$\overline{F}$	(b) X	Y	Z	E
0	0	0	1	0	0	0	0	1
0	0	1	0	1	0	0	1	1
0	1	0	1	0	0	1	0	1
0	1	1	0	1	0	1	1	0
1	0	0	0	1	1	0	0	1
1	0	1	1	0	1	0	1	1
1	1	0	0	1	1	1	0	0
1	1	1	1	0	1	1	1	0

Now consider the complement of a Boolean function. The binary values of $\overline{F}$ in Table 2-8(a) are obtained by changing 1's to 0's and 0's to 1's in the values of F. Taking the logical sum of minterms of $\overline{F}$, we obtain

$$\overline{F} = \overline{X}\overline{Y}Z + \overline{X}YZ + X\overline{Y}\overline{Z} + XY\overline{Z} = m_1 + m_3 + m_4 + m_6$$

or, in abbreviated form,

$$\overline{F}(X,Y,Z) = \Sigma m(1,3,4,6)$$

Note that the minterm numbers for $\overline{F}$ are the ones missing from the list of the minterm numbers of F. We now take the complement of $\overline{F}$ to obtain F:

$$F = \overline{m_1 + m_3 + m_4 + m_6} = \overline{m_1} \cdot \overline{m_3} \cdot \overline{m_4} \cdot \overline{m_6}$$

$$= M_1 \cdot M_3 \cdot M_4 \cdot M_6 \ (\text{Since } \overline{m_j} = M_j)$$

$$= (X + Y + \overline{Z})(X + \overline{Y} + \overline{Z})(\overline{X} + Y + Z)(\overline{X} + \overline{Y} + Z)$$

This shows the procedure for expressing a Boolean function as a *product of maxterms*. The abbreviated form for this product is

$$F(X,Y,Z) = \Pi M(1,3,4,6)$$

where symbol Π denotes the logical product (Boolean AND) of the maxterms whose numbers are listed in parentheses. Note that the decimal numbers included in the product of maxterms will always be the same as the minterm list of the complemented function, such as $(1, 3, 4, 6)$ in the foregoing example. Maxterms are seldom used directly when dealing with Boolean functions, as we can always replace them with the minterm list of $\overline{F}$.

The following is a summary of the most important properties of minterms:

1. There are 2^n minterms for n Boolean variables. These minterms can be evaluated from the binary numbers from 0 to $2^n - 1$.

2. Any Boolean function can be expressed as a logical sum of minterms.

3. The complement of a function contains those minterms not included in the original function.

4. A function that includes all the 2^n minterms is equal to logic 1.

A function that is not in the sum-of-minterms form can be converted to that form by means of a truth table, since the truth table always specifies the minterms of the function. Consider, for example, the Boolean function

$$E = \overline{Y} + \overline{X}\overline{Z}$$

The expression is not in sum-of-minterms form, because each term does not contain all three variables X, Y, and Z. The truth table for this function is listed in Table 2-8(b). From the table, we obtain the minterms of the function:

$$E(X, Y, Z) = \Sigma m(0, 1, 2, 4, 5)$$

The minterms for the complement of E are given by

$$\overline{E}(X, Y, Z) = \Sigma m(3, 6, 7)$$

Note that the total number of minterms in E and $\overline{E}$ is equal to eight, since the function has three variables, and three variables produce a total of eight minterms. With four variables, there will be a total of 16 minterms, and for two variables, there will be 4 minterms. An example of a function that includes all the minterms is

$$G(X, Y) = \Sigma m(0, 1, 2, 3) = 1$$

Since G is a function of two variables and contains all four minterms, it is always equal to logic 1.

Sum of Products

The sum-of-minterms form is a standard algebraic expression that is obtained directly from a truth table. The expression so obtained contains the maximum number of literals in each term and usually has more product terms than necessary. This is because, by definition, each minterm must include all the variables of the function, complemented or uncomplemented. Once the sum of minterms is obtained from the truth table, the next step is to try to simplify the expression to see whether it is possible to reduce the number of product terms and the number of literals in the terms. The result is a simplified expression in *sum-of-products* form. This is an alternative standard form of expression that contains product terms with one, two, or any number of literals. An example of a Boolean function expressed as a sum of products is

$$F = \overline{Y} + \overline{X}Y\overline{Z} + XY$$

The expression has three product terms, the first with one literal, the second with three literals, and the third with two literals.

The logic diagram for a sum-of-products form consists of a group of AND gates followed by a single OR gate, as shown in Figure 2-5. Each product term

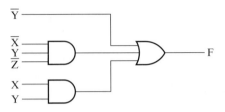

☐ **FIGURE 2-5**
Sum-of-Products Implementation

requires an AND gate, except for a term with a single literal. The logical sum is formed with an OR gate that has single literals and the outputs of the AND gates as inputs. It is assumed that the input variables are directly available in their complemented and uncomplemented forms, so inverters are not included in the diagram. The AND gates followed by the OR gate form a circuit configuration referred to as a *two-level implementation* or *two-level circuit.*

If an expression is not in sum-of-products-form, it can be converted to the standard form by means of the distributive laws. Consider the expression

$$F = AB + C(D + E)$$

This is not in sum-of-products form, because the term D + E is part of a product, but is not a single literal. The expression can be converted to a sum of products by applying the appropriate distributive law as follows:

$$F = AB + C(D + E) = AB + CD + CE$$

The function F is implemented in a nonstandard form in Figure 2-6(a). This requires two AND gates and two OR gates. There are three levels of gating in the circuit. F is implemented in sum-of-products form in Figure 2-6(b). This circuit requires three AND gates and an OR gate and uses two levels of gating. The decision as to whether to use a two-level or multiple-level (three levels or more) implementation is complex. Among the issues involved are the number of gates, number of gate inputs, and the amount of delay between the time the input values are set and the time the resulting output values appear. Two-level implementations are the natural form for certain implementation technologies, as we will see in Chapter 4.

Product of Sums

Another standard form of expressing Boolean functions algebraically is the *product of sums.* This form is obtained by forming a logical product of sum terms. Each logical sum term may have any number of distinct literals. An example of a function expressed in product-of-sums form is

$$F = X(\overline{Y} + Z)(X + Y + \overline{Z})$$

This expression has sum terms of one, two, and three literals. The sum terms perform an OR operation, and the product is an AND operation.

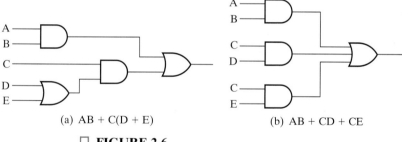

(a) AB + C(D + E)   (b) AB + CD + CE

□ **FIGURE 2-6**
Three-Level and Two-Level Implementation

The gate structure of the product-of-sums expression consists of a group of OR gates for the sum terms (except for a single literal term), followed by an AND gate. This is shown in Figure 2-7 for the preceding function *F*. As with the sum of products, this standard type of expression results in a two-level gating structure.

2-4 TWO-LEVEL CIRCUIT OPTIMIZATION

The complexity of the digital logic gates that implement a Boolean function is directly related to the algebraic expression from which the function is implemented. Although the truth table representation of a function is unique, when expressed algebraically, the function appears in many different forms. Boolean expressions may be simplified by algebraic manipulation as discussed in Section 2-2. However, this procedure of simplification is awkward because it lacks specific rules to predict each succeeding step in the manipulative process and it is difficult to determine whether the simplest expression has been achieved. By contrast, the map method provides a straightforward procedure for optimizing Boolean functions of up to four variables.   Maps for five and six variables can be drawn as well, but are more cumbersome to use. The map is also known as the *Karnaugh map*, or *K-map*. The map is a diagram made up of squares, with each square representing one minterm of the function. Since any Boolean function can be expressed as a sum of minterms, it follows that a Boolean function is recognized graphically in the map by those squares whose minterms are included in the function. In fact, the map presents a visual diagram of all possible ways a function may be expressed in a standard form. By recognizing various patterns, the user can derive alternative algebraic

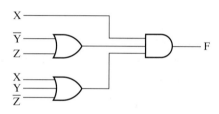

□ **FIGURE 2-7**
Product-of-Sums Implementation

expressions for the same function, from which the simplest can be selected. The optimized expressions produced by the map are always in sum-of-products or product-of-sums form. Thus, maps handle optimization for two-level implementations, but do not apply directly to possible simpler implementations for the general case with three or more levels. Initially, this section covers sum-of-products optimization and, later, applies it to performing product-of-sums optimization.

Cost Criteria

In the prior section, counting literals and terms was mentioned as a way of measuring the simplicity of a logic circuit. We introduce two cost criteria to formalize this concept.

The first criterion is *literal cost*, the number of literal appearances in a Boolean expression corresponding exactly to the logic diagram. For example, for the circuits in Figure 2-6, the corresponding Boolean expressions are

$$F = AB + C(D + E) \quad \text{and} \quad F = AB + CD + CE$$

There are five literal appearances in the first equations and six literal appearances in the second equation, so the first equation is the simplest in terms of literal cost. Literal cost has the advantage that it is very simple to evaluate by counting literal appearances. It does not, however, represent circuit complexity accurately in all cases, even for the comparison of different implementations of the same logic function. The following Boolean equations, both for function G, illustrate this situation:

$$G = ABCD + \overline{A}\,\overline{B}\,\overline{C}\,\overline{D} \quad \text{and} \quad G = (\overline{A} + B)(\overline{B} + C)(\overline{C} + D)(\overline{D} + A)$$

The implementations represented by these equations both have a literal cost of eight. But, the first equation has two terms and the second equation has four terms. This suggests that the first equation has a lower cost than the second equation.

To capture the difference illustrated, we define *gate input cost* as the number of inputs to the gates in the implementation corresponding exactly to the given equation or equations. This cost can be determined easily from the logic diagram by simply counting the total number of inputs to the gates in the logic diagram. For sum-of-products or product-of-sums equations, it can be found from the equation by finding the sum of

(1) all literal appearances,
(2) the number of terms excluding terms that consist only of a single literal, and, optionally,
(3) the number of distinct complemented single literals.

In (1), all gate inputs from outside the circuit are represented. In (2), all gate inputs within the circuit, except for those to inverters are represented and in (3), inverters needed to complement the input variables are counted in the event that complemented input variables are not provided. For the two preceding equations, excluding the count from (3), the respective gate input counts are $8 + 2 = 10$ and $8 + 4 = 12$.

Including the count from (3), that of input inverters, the respective counts are 14 and 16. So the first equation for G has a lower gate input cost even though the literal costs are equal.

Gate input cost is currently a good measure for contemporary logic implementations since it is proportional to the number of transistors and wires used in implementing a logic circuit. Representation of gate inputs becomes particularly important in measuring cost for circuits with more than two levels. Typically, as the number of levels increases, literal cost represents a smaller proportion of the actual circuit cost since more and more gates have no inputs from outside the circuit itself. Later, in Figure 2-29, we introduce complex gate types for which evaluation of the gate input cost from an equation is invalid, since the correspondence between the AND, OR and NOT operations in the equation and the gates in the circuit can no longer be established. In such cases, as well as for equation forms more complex than sum-of-products and product-of-sums, the gate input count must be determined directly from the implementation.

Regardless of the cost criteria used, we see later that the simplest expression is not necessarily unique. It is sometimes possible to find two or more expressions that satisfy the cost criterion applied. In that case, either solution is satisfactory from the cost standpoint.

Two-Variable Map

There are four minterms for a Boolean function with two variables. Hence, the two-variable map consists of four squares, one for each minterm, as shown in Figure 2-8(a). The map is redrawn in Figure 2-8(b) to show the relationship between the squares and the two variables X and Y. The 0 and 1 marked on the left side and the top of the map designate the values of the variables. The variable X appears complemented in row 0 and uncomplemented in row 1. Similarly, Y appears complemented in column 0 and uncomplemented in column 1. Note that the four combinations of these binary values correspond to the truth table rows associated with the four minterms.

A function of two variables can be represented in a map by marking the squares that correspond to the minterms of the function. As an example, the function XY is shown in Figure 2-9(a). Since XY is equal to minterm m_3, a 1 is placed

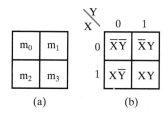

□ **FIGURE 2-8**
Two-Variable Map

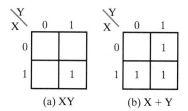

□ **FIGURE 2-9**
Representation of Functions in the Map

inside the square that belongs to m_3. Figure 2-9(b) shows the map for the logical sum of three minterms:

$$m_1 + m_2 + m_3 = \overline{X}Y + X\overline{Y} + XY = X + Y$$

The optimized expression $X + Y$ is determined from the two-square area for the variable X in the second row and the two-square area for Y in the second column. Together, these two areas enclose the three squares belonging to X or Y. This simplification can be justified by algebraic manipulation:

$$\overline{X}Y + X\overline{Y} + XY = \overline{X}Y + X(\overline{Y} + Y) = (\overline{X} + X)(Y + X) = X + Y$$

The exact procedure for combining squares in the map will be clarified in the examples that follow.

Three-Variable Map

There are eight minterms for three binary variables. Therefore, a three-variable map consists of eight squares, as shown in Figure 2-10. The map drawn in part (b) is marked with binary numbers for each row and each column to show the binary values of the minterms. Note that the numbers along the columns do not follow the binary count sequence. The characteristic of the listed sequence is that only one bit changes in value from one adjacent column to the next, which corresponds to the Gray code introduced in Chapter 1.

A minterm square can be located in the map in two ways. First, we can memorize the numbers listed in Figure 2-10(a) for each minterm location, or we can refer to the binary numbers along the rows and columns in Figure 2-10(b). For example, the square assigned to m_5 corresponds to row 1 and column 01. When these two numbers are combined, they give the binary number 101, whose decimal equivalent is 5.

Another way of looking at square $m_5 = X\overline{Y}Z$ is to consider it to be in the row marked X and the column belonging to $\overline{Y}Z$ (column 01). Note that there are four squares where each variable is equal to 1 and four where each is equal to 0.

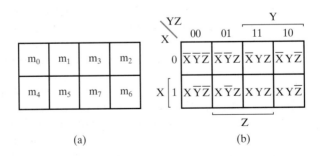

(a)   (b)

☐ **FIGURE 2-10**
Three-Variable Map

The variable appears uncomplemented in the four squares where it is equal to 1 and complemented in the four squares where it is equal to 0. For convenience, we write the variable name along the four squares where it is uncomplemented. After one becomes familiar with maps, the use of the variable names alone is sufficient to label the map regions. To this end, it is important to note the location of these labels to obtain all minterms on the map.

In the two-variable map, the function XY demonstrated that a function or a term for a function can consist of a single square of the map. But to achieve simplification, we need to consider multiple squares corresponding to product terms. To understand how combining squares simplifies Boolean functions, we must recognize the basic property possessed by adjacent squares: Any two adjacent squares placed horizontally or vertically (but not diagonally) to form a rectangle correspond to minterms that differ in only a single variable. The single variable appears uncomplemented in one square and complemented in the other. For example, m_5 and m_7 lie in two adjacent squares. Variable Y is complemented in m_5 and uncomplemented in m_7, while the other two variables match in both squares. The logical sum of two such adjacent minterms can be simplified into a single product term of two variables:

$$m_5 + m_7 = X\overline{Y}Z + XYZ = XZ(\overline{Y} + Y) = XZ$$

Here the two squares differ in the variable Y, which can be removed when the logical sum (OR) of the two minterms is formed. Thus, on a 3-variable map, any two minterms in adjacent squares that are ORed together produce a product term of two variables. This is shown in Example 2-3.

EXAMPLE 2-3   Simplifying a Boolean Function Using a Map

Simplify the Boolean function

$$F(X, Y, Z) = \Sigma m(2, 3, 4, 5)$$

First, a 1 is marked in each minterm that represents the function. This is shown in Figure 2-11, where the squares for minterms 010, 011, 100, and 101 are marked with 1's. For convenience, all of the remaining squares for which the function has value 0 are left blank rather than entering the 0's. The next step is to explore collections of squares on the map representing product terms to be considered for the simplified expression. We call such objects *rectangles*, since their shape is that of a rectangle (including, of course, a square). Rectangles that correspond to product terms are restricted, however, to contain numbers of squares that are powers of 2, such as 1, 2, 4, 8, ... So our goal is to find the fewest such rectangles that include all of the minterms marked with 1's. This will give the fewest product terms. In the map in the figure, two rectangles enclose all four squares containing 1's. The upper right rectangle represents the product term $\overline{X}Y$. This is determined by observing that the rectangle is in row 0, corresponding to $\overline{X}$, and the last two columns, corresponding to Y. Similarly, the lower left rectangle represents the product term $X\overline{Y}$. (The second row represents X and the two left columns represent $\overline{Y}$.) Since these

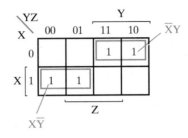

☐ **FIGURE 2-11**
Map for Example 2-3: $F(X, Y, Z) = \Sigma m(2,3,4,5) = \overline{X}Y + X\overline{Y}$

two rectangles include all of the 1's in the map, the logical sum of the corresponding two product terms gives the optimized expression for F:

$$F = \overline{X}Y + X\overline{Y}$$ ■

In some cases, two squares in the map are adjacent and form a rectangle of size two, even though they do not touch each other. For example, in Figure 2-10, m_0 is adjacent to m_2 and m_4 is adjacent to m_6 because the minterms differ by one variable. This can be readily verified algebraically:

$$m_0 + m_2 = \overline{X}\,\overline{Y}\,\overline{Z} + \overline{X}Y\overline{Z} = \overline{X}\,\overline{Z}(\overline{Y} + Y) = \overline{X}\,\overline{Z}$$
$$m_4 + m_6 = X\overline{Y}\,\overline{Z} + XY\overline{Z} = X\overline{Z}(\overline{Y} + Y) = X\overline{Z}$$

The rectangles corresponding to these two product terms, $\overline{X}\,\overline{Z}$ and $X\overline{Z}$, are shown on the map in Figure 2-12(a). Based on the location of these rectangles, we must modify the definition of adjacent squares to include this and other, similar cases. We do so by considering the map as being drawn on a cylinder, as shown in Figure 2-12(b), where the right and left edges touch each other to correctly establish minterm adjacencies and form the rectangles. In the maps in Figure 2-12, we have simply used numbers rather than m's to represent the minterms. Both of these notations will be used freely.

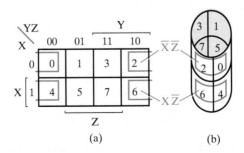

☐ **FIGURE 2-12**
Three-Variable Map: Flat and on a Cylinder to Show Adjacent Squares

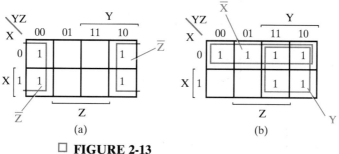

□ **FIGURE 2-13**
Product Terms Using Four Minterms

A four-square rectangle represents a product term that is the logical sum of four minterms. For the three-variable case, such a product term is only one literal. As an example, the logical sum of the four adjacent minterms 0, 2, 4, and 6 reduces to a single literal term $\overline{Z}$:

$$m_0 + m_2 + m_4 + m_6 = \overline{X}\,\overline{Y}\,\overline{Z} + \overline{X}Y\overline{Z} + X\overline{Y}\,\overline{Z} + XY\overline{Z}$$

$$= \overline{X}\,\overline{Z}(\overline{Y} + Y) + X\overline{Z}(\overline{Y} + Y)$$

$$= \overline{X}\,\overline{Z} + X\overline{Z} = \overline{Z}(\overline{X} + X) = \overline{Z}$$

The rectangle for this product term is shown in Figure 2-13(a). Note that the product term $\overline{Z}$ uses the fact that the left and right edges of the map are adjacent in order to form the rectangle. Two other examples of rectangles corresponding to product terms derived from four minterms are shown in Figure 2-13(b).

In general, as more squares are combined, we obtain a product term with fewer literals. Three-variable maps exhibit the following characteristics:

One square represents a minterm of three literals.

A rectangle of two squares represents a product term of two literals.

A rectangle of four squares represents a product term of one literal.

A rectangle of eight squares encompasses the entire map and produces a function that is always equal to logic 1.

These characteristics are illustrated in Example 2-4.

EXAMPLE 2-4   Simplifying Three-Variable Functions with Maps

Simplify the following two Boolean functions:

$$F_1(X, Y, Z) = \Sigma m(3, 4, 6, 7)$$

$$F_2(X, Y, Z) = \Sigma m(0, 2, 4, 5, 6)$$

The map for F_1 is shown in Figure 2-14(a). There are four squares marked with 1's, one for each minterm of the function. Two adjacent squares are combined

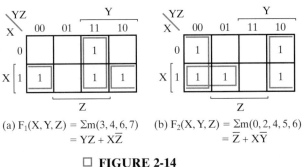

(a) $F_1(X, Y, Z) = \Sigma m(3, 4, 6, 7)$
$= YZ + X\overline{Z}$

(b) $F_2(X, Y, Z) = \Sigma m(0, 2, 4, 5, 6)$
$= \overline{Z} + X\overline{Y}$

☐ **FIGURE 2-14**
Maps for Example 2-4

in the third column to give a two-literal term YZ. The remaining two squares with 1's are also adjacent by the cylinder-based definition and are shown in the diagram with their values enclosed in half rectangles. When combined, these two squares give the two-literal term $X\overline{Z}$. The optimized function thus becomes

$$F_1 = YZ + X\overline{Z}$$

The map for F_2 is shown in Figure 2-14(b). First, we immediately combine the four adjacent squares in the first and last columns based on what we learned from Figure 2-13, to give the single literal term $\overline{Z}$. The remaining single square representing minterm 5 is combined with an adjacent square that already is being used once. This is not only permissible, but desirable, since the two adjacent squares give the two-literal term $X\overline{Y}$, while the single square represents the three-literal minterm $X\overline{Y}Z$. The optimized function is

$$F_2 = \overline{Z} + X\overline{Y}$$

On occasion there are alternative ways of combining squares to produce equally optimized expressions. An example of this is demonstrated in the map of Figure 2-15. Minterms 1 and 3 are combined to give the term $\overline{X}Z$, and minterms 4 and 6 produce the term $X\overline{Z}$. However, there are two ways that the square of minterm 5 can be combined with another adjacent square to produce a third two-literal term. Combining it with minterm 4 gives the term $X\overline{Y}$; combining it instead with minterm 1 gives the term $\overline{Y}Z$. Each of the two possible optimized expressions listed in Figure 2-15 has three terms of two literals each, so there are two possible optimized solutions for this function.

If a function is not expressed as a sum of minterms, we can use the map to obtain the minterms of the function and then simplify the function. It is necessary, however, to have the algebraic expression in sum-of-products form, from which each product term is plotted in the map. The minterms of the function are then read directly from the map. As an example, consider the Boolean function

$$F = \overline{X}Z + \overline{X}Y + X\overline{Y}Z + YZ$$

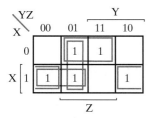

□ **FIGURE 2-15**

$F(X,Y,Z) = \Sigma m(1,3,4,5,6)$
$= \overline{X}Z + X\overline{Z} + X\overline{Y}$
$= \overline{X}Z + X\overline{Z} + \overline{Y}Z$

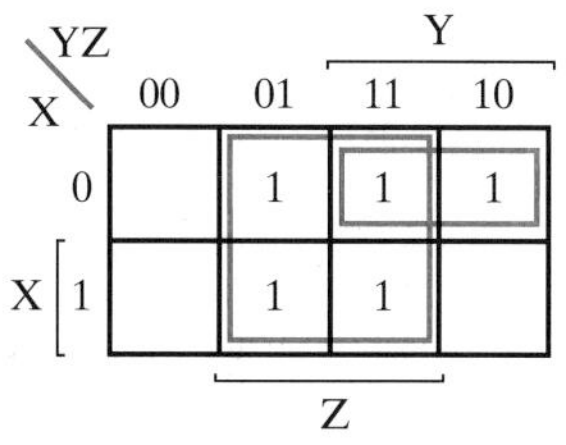

□ **FIGURE 2-16**

$F(X,Y,Z) = \Sigma m(1,2,3,5,7) = Z + \overline{X}Y$

Three product terms in the expression have two literals and are represented in a three-variable map by two squares each. The two squares corresponding to the first term, $\overline{X}Z$, are found in Figure 2-16 from the coincidence of $\overline{X}$ (first row) and Z (two middle columns), to give 1's in squares 001 and 011. Note that when marking 1's in the squares, it is possible to find a 1 already placed there from a preceding term. This happens with the second term, $\overline{X}Y$, which has 1's in squares 011 and 010; but square 011 is common with the first term, $\overline{X}Z$, so only one 1 is marked in it. Continuing in this fashion, we find that the function has five minterms, as indicated by the five 1's in the figure. The minterms are read directly from the map to be 1, 2, 3, 5, and 7. The function as originally given has four product terms. It can be optimized on the map to only two simple terms as

$$F = Z + \overline{X}Y$$

giving a significant reduction in the cost of implementation.

Four-Variable Map

There are 16 minterms for four binary variables, and therefore, a four-variable map consists of 16 squares, as shown in Figure 2-17. The minterm assignment in each square is indicated in part (a) of the diagram. The map is redrawn in (b) to show the relationship of the four variables. The rows and columns are numbered so that only one bit of the binary number changes in value between any two adjacent columns or rows, guaranteeing the same property for adjacent squares. The row and column number corresponds to a two-bit Gray code, as introduced in Chapter 1. The minterms corresponding to each square can be obtained by combining the row number with the column number. For example, when combined, the numbers in the third row (11) and the second column (01) give the binary number 1101, the binary equivalent of 13. Thus, the square in the third row and second column represents minterm m_{13}. In addition, each variable is marked on the map to show the eight squares in which it appears uncomplemented. The other eight squares, in which no label is indicated, correspond to the variable being complemented. Thus, W appears complemented in the first two rows and uncomplemented in the second two rows.

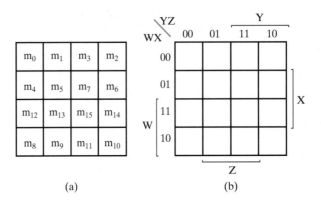

(a)   (b)

□ **FIGURE 2-17**
Four-Variable Map

The method used to simplify four-variable functions is similar to that used to simplify three-variable functions. Adjacent squares are defined to be squares next to each other, as for two- and three-variable maps. To show adjacencies between squares, the map of Figure 2-18(a) is drawn as a torus in Figure 2-18(b), with the top and bottom edges, as well as the right and left edges, touching each other to show adjacent squares. For example, m_0 and m_2 are two adjacent squares, as are m_0 and m_8. The combinations of squares that can be chosen during the optimization process in the four-variable map are as follows:

One square represents a minterm of four literals.

A rectangle of 2 squares represents a product term of three literals.

A rectangle of 4 squares represents a product term of two literals.

A rectangle of 8 squares represents a product term of one literal.

A rectangle of 16 squares produces a function that is always equal to logic 1.

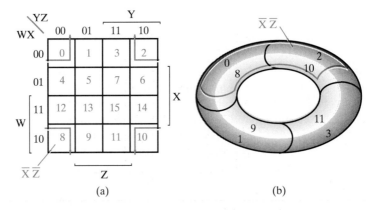

(a)   (b)

□ **FIGURE 2-18**
Four-Variable Map: Flat and on a Torus to Show Adjacencies

No other combination of squares can be used. An interesting product term of two literals $\overline{X}\,\overline{Z}$, is shown in Figure 2-18. In (b), when the map is viewed as a torus, the adjacencies of the squares that represent this product term are clear, but in (a) these squares are on the four corners of the map and appear quite removed from each other. This product term is important to recall, since it is often missed. It also serves as a reminder that the left edge and the right edge of the map are adjacent, as are the top edge and the bottom edge. Thus, in general, rectangles on a map cross the left and right edges, top and bottom edges, or both.

The next examples show the procedure for simplifying four-variable Boolean functions.

EXAMPLE 2-5 Simplifying a 4-Variable Function with a Map
Simplify the Boolean function

$$F(W, X, Y, Z) = \Sigma m(0, 1, 2, 4, 5, 6, 8, 9, 12, 13, 14)$$

The minterms of the function are marked with 1's in the map of Figure 2-19. Eight squares in the two left columns are combined to form a rectangle for the one literal term, $\overline{Y}$. The remaining three 1's cannot be combined to give a simplified term; rather, they must be combined as two- or four-square rectangles. The top two 1's on the right are combined with the top two 1's on the left to give the term $\overline{W}\,\overline{Z}$. Note again that it is permissible to use the same square more than once. We are now left with a square marked with a 1 in the third row and fourth column (minterm 1110). Instead of taking this square alone, which will give a term of four literals, we combine it with squares already used to form a rectangle of four squares in the two middle rows and the two end columns, giving the term $X\overline{Z}$. The optimized expression is the logical sum of the three terms:

$$F = \overline{Y} + \overline{W}\,\overline{Z} + X\overline{Z}$$

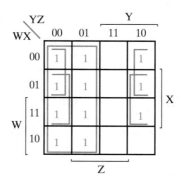

□ **FIGURE 2-19**
Map for Example 2-5: $F = \overline{Y} + \overline{W}\,\overline{Z} + X\overline{Z}$

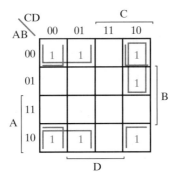

□ **FIGURE 2-20**
Map for Example 2-6: $F = \overline{B}\,\overline{D} + \overline{B}\,\overline{C} + \overline{A}C\overline{D}$

EXAMPLE 2-6   Simplifying a 4-Variable Function with a Map

Simplify the Boolean function

$$F = \overline{A}\,\overline{B}\,\overline{C} + \overline{B}C\overline{D} + A\overline{B}\,\overline{C} + \overline{A}BC\overline{D}$$

This function has four variables: A, B, C, and D. It is expressed in sum-of-products form with three terms of three literals each and one term of four literals. The area in the map covered by the function is shown in Figure 2-20. Each term of three literals is represented in the map by two squares. $\overline{A}\,\overline{B}\,\overline{C}$ is represented by squares 0000 and 0001, $\overline{B}C\overline{D}$ by squares 0010 and 1010, and $A\overline{B}\,\overline{C}$ by squares 1000 and 1001. The term with four literals is minterm 0110. The function is simplified on the map by taking the 1's in the four corners, to give the term $\overline{B}\,\overline{D}$. This product term is in the same map locations as $\overline{X}\,\overline{Z}$ in Figure 2-18. The two 1's in the top row are combined with the two 1's in the bottom row to give the term $\overline{B}\,\overline{C}$. The remaining 1, in square 0110, is combined with its adjacent square, 0010, to give the term $\overline{A}C\overline{D}$. The optimized function is thus

$$F = \overline{B}\,\overline{D} + \overline{B}\,\overline{C} + \overline{A}C\overline{D} \qquad\qquad ■$$

2-5   MAP MANIPULATION

When combining squares in a map, it is necessary to ensure that all the minterms of the function are included. At the same time, it is necessary to minimize the number of terms in the optimized function by avoiding any redundant terms whose minterms are already included in other terms. In this section, we consider a procedure that assists in the recognition of useful patterns in the map. Other topics to be covered are the optimization of products of sums and the optimization of incompletely specified functions.

Essential Prime Implicants

The procedure for combining squares in a map may be made more systematic if we introduce the terms "implicant," "prime implicant," and "essential prime implicant." A product term is an *implicant* of a function if the function has the value 1 for all minterms of the product term. Clearly, all rectangles on a map made up of squares containing 1's correspond to implicants. If the removal of any literal from an implicant P results in a product term that is not an implicant of the function, then P is a *prime implicant*. On a map for an n-variable function, the set of prime implicants corresponds to the set of all rectangles made up of 2^m squares containing 1's ($m = 0, 1, \ldots n$), with each rectangle containing as many squares as possible.

If a minterm of a function is included in only one prime implicant, that prime implicant is said to be *essential*. Thus, if a square containing a 1 is in only one rectangle representing a prime implicant, then that prime implicant is essential. In Figure 2-15 on page 55, the terms $\overline{X}Z$ and $X\overline{Z}$ are essential prime implicants, and the terms $X\overline{Y}$ and $\overline{Y}Z$ are nonessential prime implicants.

The prime implicants of a function can be obtained from a map of the function as all possible maximum collections of 2^m squares containing 1's ($m = 0, 1, \ldots n$) that constitute rectangles. This means that a single 1 on a map represents a prime implicant if it is not adjacent to any other 1's. Two adjacent 1's form a rectangle representing a prime implicant, provided that they are not within a rectangle of four or more squares containing 1's. Four 1's form a rectangle representing a prime implicant if they are not within a rectangle of eight or more squares containing 1's, and so on. Each essential prime implicant contains at least one square that is not contained in any other prime implicant.

The systematic procedure for finding the optimized expression from the map requires that we first determine all prime implicants. Then, the optimized expression is obtained from the logical sum of all the essential prime implicants, plus other prime implicants needed to include remaining minterms not included in the essential prime implicants. This procedure will be clarified by examples.

EXAMPLE 2-7   Simplification Using Prime Implicants

Consider the map of Figure 2-21. There are three ways that we can combine four squares into rectangles. The product terms obtained from these combinations are the prime implicants of the function, $\overline{A}D, B\overline{D}$ and $\overline{A}B$. The terms $\overline{A}D$ and $B\overline{D}$ are essential prime implicants, but $\overline{A}B$ is not essential. This is because minterms 1 and 3 can be included only in the term $\overline{A}D$, and minterms 12 and 14 can be included only in the term $B\overline{D}$. But minterms 4, 5, 6, and 7 are each included in two prime implicants, one of which is $\overline{A}B$, so the term $\overline{A}B$ is not an essential prime implicant. In fact, once the essential prime implicants are chosen, the third term is not needed because all the minterms are already included in the essential prime implicants. The optimized expression for the function of Figure 2-21 is

$$F = \overline{A}D + B\overline{D}$$

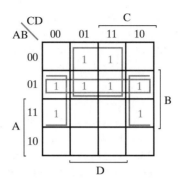

☐ **FIGURE 2-21**
Prime Implicants for Example 2-7: $\overline{A}D$, $B\overline{D}$, and $\overline{A}B$

EXAMPLE 2-8   Simplification Via Essential and Nonessential Prime Implicants
A second example is shown in Figure 2-22. The function plotted in part (a) has seven minterms. If we try to combine squares, we will find that there are six prime implicants. In order to obtain a minimum number of terms for the function, we must first determine the prime implicants that are essential. As shown in part (b) of the figure, the function has four essential prime implicants. The product term $\overline{A}\,\overline{B}\,\overline{C}\,\overline{D}$ is essential because it is the only prime implicant that includes minterm 0. Similarly, the product terms $B\overline{C}D$, $AB\overline{C}$, and $A\overline{B}C$ are essential prime implicants because they are the only prime implicants that include minterms 5, 12, and 10, respectively. Minterm 15 is included in two nonessential prime implicants. The optimized expression for the function consists of the logical sum of the four essential prime implicants and one prime implicant that includes minterm 15:

$$F = \overline{A}\,\overline{B}\,\overline{C}\,\overline{D} + B\overline{C}D + AB\overline{C} + A\overline{B}C + \begin{cases} ACD \\ \text{or} \\ ABD \end{cases}$$

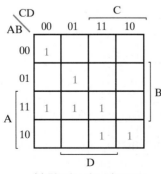

(a) Plotting the minterms

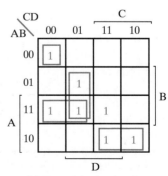

(b) Essential prime implicants

☐ **FIGURE 2-22**
Simplification with Prime Implicants in Example 2-8

The identification of essential prime implicants in the map provides an additional tool which shows the terms that must absolutely appear in every sum-of-products expression for a function and provides a partial structure for a more systematic method for choosing patterns of squares.

Nonessential Prime Implicants

Beyond using all essential prime implicants, the following rule can be applied to include the remaining minterms of the function in nonessential prime implicants:

Selection Rule: Minimize the overlap among prime implicants as much as possible. In particular, in the final solution, make sure that each prime implicant selected includes at least one minterm not included in any other prime implicant selected.

In most cases, this results in a simplified, although not necessarily minimum cost, sum-of-products expression. The use of the selection rule is illustrated in the next example.

EXAMPLE 2-9   Simplifying a Function Using the Selection Rule

Find a simplified sum-of-products form for $F(A, B, C, D) = \Sigma m$ (0, 1, 2, 4, 5, 10, 11, 13, 15).

The map for F is given in Figure 2-23, with all prime implicants shown. $\overline{A}\,\overline{C}$ is the only essential prime implicant. Using the preceding selection rule, we can choose the remaining prime implicants for the sum-of-products form in the order indicated by the numbers. Note how the prime implicants 1 and 2 are selected in order to include minterms without overlapping. Prime implicant 3 $(\overline{A}\,\overline{B}\,D)$ and prime implicant $\overline{B}C\overline{D}$ both include the one remaining minterm 0010, and prime implicant 3 is arbitrarily selected to include the minterm and complete the sum-of-products expression:

$$F(A, B, C, D) = \overline{A}\,\overline{C} + ABD + A\overline{B}C + \overline{A}\,\overline{B}\,\overline{D}$$

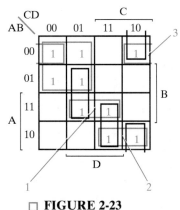

□ **FIGURE 2-23**
Map for Example 2-9

Product-of-Sums Optimization

The optimized Boolean functions derived from the maps in all of the previous examples were expressed in sum-of-products form. With only minor modification, the product-of-sums form can be obtained.

The procedure for obtaining an optimized expression in product-of-sums form follows from the basic properties of Boolean functions. The 1's placed in the squares of the map represent the minterms of the function. The minterms not included in the function belong to the complement of the function. From this, we see that the complement of a function is represented in the map by the squares not marked by 1's. If we mark the empty squares with 0's and combine them into valid rectangles, we obtain an optimized expression of the complement of the function. We then take the complement of $\overline{F}$ to obtain the function F as a product of sums. This is done by taking the dual and complementing each literal, as described in Example 2-2 on page 41.

EXAMPLE 2-10   Simplifying a Product-of-Sums Form

Simplify the following Boolean function in product-of-sums form:

$$F(A,B,C,D) = \Sigma m(0,1,2,5,8,9,10)$$

The 1's marked in the map of Figure 2-24 represent the minterms of the function. The squares marked with 0's represent the minterms not included in F and therefore denote the complement of F. Combining the squares marked with 0's, we obtain the optimized complemented function

$$\overline{F} = AB + CD + B\overline{D}$$

Taking the dual and complementing each literal gives the complement of $\overline{F}$. This is F in product-of-sums form:

$$F = (\overline{A} + \overline{B})(\overline{C} + \overline{D})(\overline{B} + D) \qquad\qquad ■$$

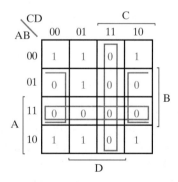

☐ **FIGURE 2-24**
Map for Example 2-10: $F = (\overline{A} + \overline{B})(\overline{C} + \overline{D})(\overline{B} + D)$

The previous example shows the procedure for obtaining the product-of-sums optimization when the function is originally expressed as a sum of minterms. The procedure is also valid when the function is originally expressed as a product of maxterms or a product of sums. Remember that the maxterm numbers are the same as the minterm numbers of the complemented function, so 0's are entered in the map for the maxterms or for the complement of the function. To enter a function expressed as a product of sums into the map, we take the complement of the function and, from it, find the squares to be marked with 0's. For example, the function

$$F = (\overline{A} + \overline{B} + C)(B + D)$$

can be plotted in the map by first obtaining its complement,

$$\overline{F} = AB\overline{C} + \overline{B}\,\overline{D}$$

and then marking 0's in the squares representing the minterms of $\overline{F}$. The remaining squares are marked with 1's. Then, combining the 1's gives the optimized expression in sum-of-products form. Combining the 0's and then complementing gives the optimized expression in product-of-sums form. Thus, for any function plotted on the map, we can derive the optimized function in either one of the two standard forms.

Don't-Care Conditions

The minterms of a Boolean function specify all combinations of variable values for which the function is equal to 1. The function is assumed to be equal to 0 for the rest of the minterms. This assumption, however, is not always valid, since there are applications in which the function is not specified for certain variable value combinations. There are two cases in which this occurs. In the first case, the input combinations never occur. As an example, the four-bit binary code for the decimal digits has six combinations that are not used and not expected to occur. In the second case, the input combinations are expected to occur, but we do not care what the outputs are in response to these combinations. In both cases, the outputs are said to be unspecified for the input combinations. Functions that have unspecified outputs for some input combinations are called *incompletely specified functions*. In most applications, we simply do not care what value is assumed by the function for the unspecified minterms. For this reason, it is customary to call the unspecified minterms of a function *don't-care conditions*. These conditions can be used on a map to provide further simplification of the function.

It should be realized that a don't-care minterm cannot be marked with a 1 on the map, because that would require that the function always be a 1 for such a minterm. Likewise, putting a 0 in the square requires the function to be 0. To distinguish the don't-care condition from 1's and 0's, an X is used. Thus, an X inside a square in the map indicates that we do not care whether the value of 0 or 1 is assigned to the function for the particular minterm.

In choosing adjacent squares to simplify the function in a map, the don't-care minterms may be used. When simplifying function F, using the 1's we can choose to include those don't-care minterms that give the simplest prime implicants for F. When simplifying function $\overline{F}$ using the 0's, we can choose to include those don't-care minterms that give the simplest prime implicants for $\overline{F}$, irrespective of those included in the prime implicants for F. In both cases, whether or not the don't care minterms are included in the terms in the final expression is irrelevant. The handling of don't care conditions is illustrated in the next example.

EXAMPLE 2-11 Simplification with Don't Care Conditions

To clarify the procedure for handling the don't-care conditions, consider the following incompletely specified function F that has three don't-care minterms d:

$$F(A,B,C,D) = \Sigma m(1,3,7,11,15)$$

$$d(A,B,C,D) = \Sigma m(0,2,5)$$

The minterms of F are the variable combinations that make the function equal to 1. The minterms of d are the don't-care minterms. The map optimization is shown in Figure 2-25. The minterms of F are marked by 1's, those of d are marked by X's, and the remaining squares are filled with 0's. To get the simplified function in sum-of-products form, we must include all five 1's in the map, but we may or may not include any of the X's, depending on what yields the simplest expression for the function. The term CD includes the four minterms in the third column. The remaining minterm in square 0001 can be combined with square 0011 to give a three-literal term. However, by including one or two adjacent X's, we can combine four squares into a rectangle to give a two-literal term. In part (a) of the figure, don't-care minterms 0 and 2 are included with the 1's, which results in the simplified function

$$F = CD + \overline{A}\,\overline{B}$$

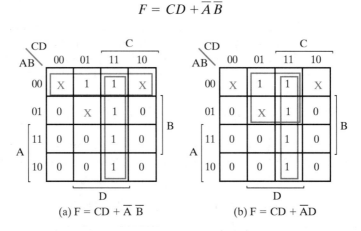

(a) $F = CD + \overline{A}\ \overline{B}$ (b) $F = CD + \overline{A}D$

☐ **FIGURE 2-25**
Example with Don't-Care Conditions

In part (b), don't-care minterm 5 is included with the 1's, and the simplified function now is

$$F = CD + \overline{A}D$$

The two expressions represent two functions that are algebraically unequal. Both include the specified minterms of the original incompletely specified function, but each includes different don't-care minterms. As far as the incompletely specified function is concerned, both expressions are acceptable. The only difference is in the value of F for the unspecified minterms.

It is also possible to obtain an optimized product-of-sums expression for the function of Figure 2-25. In this case, the way to combine the 0's is to include don't-care minterms 0 and 2 with the 0's giving the optimized complemented function

$$\overline{F} = \overline{D} + A\overline{C}$$

Taking the complement of $\overline{F}$ gives the optimized expression in product-of-sums form:

$$F = D(\overline{A} + C)$$

The foregoing example shows that the don't-care minterms in the map are initially considered as representing both 0 and 1. The 0 or 1 value eventually assigned depends on the optimization process. Due to this process, the optimized function will have a 0 or 1 value for each minterm of the original function, including those that were initially don't cares. Thus, although the outputs in the initial specification may contain X's, the outputs in a particular implementation of the specification are only 0's and 1's.

2-6 MULTIPLE-LEVEL CIRCUIT OPTIMIZATION

Although we have found that two-level circuit optimization can reduce the cost of combinational logic circuits, there are often additional cost savings available by using circuits with more than two levels. Such circuits are referred to as multiple-level circuits. These savings are illustrated by the implementation of the function

$$G = ABC + ABD + E + ACF + ADF$$

Figure 2-26(a) gives the two-level implementation of G which has a gate-input cost of 17. Now suppose that we apply the distributive law of Boolean algebra to G to give

$$G = AB(C + D) + E + A(C + D)F$$

This equation gives the multiple-level implementation of G in Figure 2-26(b) which has a gate input cost of 13, an improvement of 4 gate inputs. In Figure 2-26(b),

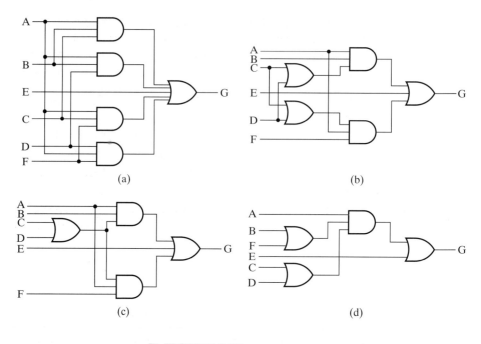

(a)

(b)

(c)

(d)

☐ **FIGURE 2-26**
Multiple-Level Circuit Example

$C + D$ is implemented twice. Instead, one implementation of this subfunction can be shared to give the circuit in Figure 2-26(c) with a gate-input cost of 11, an improvement of 2. This common use of $(C + D)$ suggests that G can be written as

$$G = (AB + AF)(C + D) + E$$

This increases the cost to 12. But by factoring out A from $AB + AF$, we obtain

$$G = A(B + F)(C + D) + E$$

Figure 2-26(d) gives the multi-level implementation of G using this equation which has a gate input cost of only nine, is slightly more than one-half of the original cost.

This reduction was achieved by a sequence of applications of algebraic identities at each step observing the effect on the gate input cost. Just as with the use of Boolean algebra to obtain simplified two-level circuits, the procedure used here is not particularly systematic. Further, an algorithmic procedure corresponding to the use of Karnaugh maps for two-level circuit optimization that gives an optimum circuit cost does not exist due to the broader range of possible actions and the number of solutions possible. So multiple-level optimization is based on the use of a set of transformations that are applied in conjunction with cost evaluation to find a good, but not necessarily optimum solution. In the remainder of this section, we consider such transformations and illustrate their application in reducing circuit cost. The transformations, to be illustrated by the next example, are defined as follows:

1. *Factoring* is finding a factored form from either a sum-of-products expression or product-of-sums expression for a function.

2. *Decomposition* is the expression of a function as a set of new functions.

3. *Extraction* is the expression of multiple functions as a set of new functions.

4. *Substitution* of a function G into a function F is expressing F as a function of G and some or all of the original variables of F.

5. *Elimination* is the inverse of substitution in which function G in an expression for function F is replaced by the expression for G. Elimination is also called *flattening* or *collapsing*.

EXAMPLE 2-12 Multilevel Optimization Transformations

The following functions will be used in illustrating the transformations:

$$G = A \overline{C} E + A \overline{C} F + A \overline{D} E + A \overline{D} F + B C D \overline{E} \overline{F}$$
$$H = \overline{A} B C D + A C E + A C F + B C E + B C F$$

The first transformation to be illustrated is factoring by using function G. Initially, we will look at *algebraic factoring,* which avoids axioms that are unique to Boolean algebra, such as those involving the complement and idempotence. Factors can be found not only for the entire expression for G, but also for its subexpressions. For example, since the first four terms of G all contain variable A, it can be factored out of these terms giving:

$$G = A \, (\overline{C} \, E + \overline{C} \, F + \overline{D} \, E + \overline{D} \, F) + B \, C \, D \, \overline{E} \, \overline{F}$$

In this case, note that A and $\overline{C} E + \overline{C} F + \overline{D} E + \overline{D} F$ are factors, and $B C D \overline{E} \overline{F}$ is not involved in the factoring operation. By factoring out $\overline{C}$ and $\overline{D}$, $\overline{C} E + \overline{C} F + \overline{D} E + \overline{D} F$ can be written as $\overline{C} (E + F) + \overline{D} (E + F)$ which be rewritten as $(\overline{C} + \overline{D})(E + F)$. Placing this expression in G gives:

$$G = A(\overline{C} + \overline{D})(E + F) + B \, C \, D \, \overline{E} \, \overline{F}$$

The term $B C D \overline{E} \overline{F}$ could be factored into product terms, but such factoring will not reduce the gate input count and so is not considered. The gate input count for the original sum-of-products expression for G is 26 and for the factored form of G is 18, for a saving of 8 gate inputs. Due to the factoring, there are more gates in series from inputs to outputs, a maximum of four levels instead of three levels including input inverters. This may result in an increase in the delay through the circuit after technology mapping has been applied.

The second transformation to be illustrated is decomposition which allows operations beyond algebraic factoring. The factored form of G can be written as a decomposition as follows:

$$G = A \ (\overline{C} + \overline{D}) \ X_2 + B \ X_1 \ \overline{E} \ \overline{F}$$
$$X_1 = C \ D$$
$$X_2 = E + F$$

Once X_1 and X_2 have been defined, they can be complemented, and the complements can replace $\overline{C} + \overline{D}$ and $\overline{E} \ \overline{F}$, respectively, in G. An illustration of the substitution transformation is

$$G = A \ \overline{X}_1 \ X_2 + B \ X_1 \ \overline{X}_2$$
$$X_1 = C \ D$$
$$X_2 = E + F$$

The gate input count for this decomposition is 14, for a saving of 12 gate inputs from the original sum-of-products expression for G, and of 4 gate inputs from the factored form of G.

In order to illustrate extraction, we need to perform decomposition on H and extract common subexpressions in G and H. Factoring out B from H, we have

$$H = B(\overline{A} \ C \ D + A \ E + A + C \ E + C \ F)$$

Determining additional factors in H, we can write

$$H = B(\overline{A} \ (C \ D) + (A + C)(E + F))$$

Factors X_1, X_2, and X_3 can now be extracted to obtain

$$X_1 = C \ D$$
$$X_2 = E + F$$
$$X_3 = A + C$$

and factors X_1 and X_2 can be shared between G and H. Performing substitution, we can write G and H as

$$G = A \ \overline{X}_1 \ X_2 + B \ X_1 \ \overline{X}_2$$
$$H = B(\overline{A} \ X_1 + X_3 \ X_2)$$

A logic diagram is given for the original sum-of-products in Figure 2-27(a) and for the extracted form in Figure 2-27(b). The gate input cost for the original G and H without shared terms, except for input inverters, is 48. For decomposed G and H without shared terms between G and H, it is 31. With shared terms, it is 25, cutting the gate input cost in half. ■

This example illustrates the value of the transformations in reducing input count cost. In general, due to the wide range of alternative solutions and the complexity in

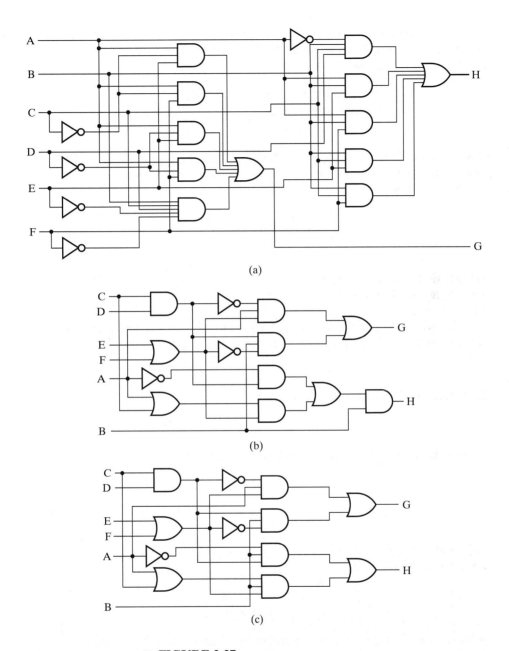

☐ **FIGURE 2-27**
Multiple-Level Circuit Optimization Example

determining the divisors to use in decomposition and extraction, obtaining truly optimum solutions in terms of gate input count is usually not feasible, so only good solutions are sought. The key to successful transformations is the determination of the factors to be used in decomposition or extraction and choice of the transformation sequence to apply. These decisions are complex and beyond the scope of our study here, but are regularly incorporated into logic synthesis tools.

Our discussion thus far has dealt only with multilevel optimization in terms of reducing gate input count. In a large proportion of designs, the length of the longest path or paths through the circuit is often constrained due to the path delay, the length of time it takes for a change in a signal to propagate down a path through the gates. In such cases, the number of gates in series may need to be reduced. Such a reduction using the final transformation, elimination, is illustrated in the following example.

EXAMPLE 2-13   Example of Transformation for Delay Reduction

In the circuit in Figure 2-27(b), the paths from C, D, E, F and A to H all pass through four 2-input gates. Assuming that all multi-input gates contribute the same delay to the path, a delay greater than that contributed by an inverter, these are the longest delay paths in the circuit. Due to a specification on maximum path delay for the circuit, these paths must be shortened to at most three multi-input gates or their equivalent in multi-input gates and inverter delays. This path shortening should be done with a minimum increase in gate input count.

The elimination transform which replaces intermediate variables, X_i, with the expressions on their right hand sides or removes other factoring such as that of variable B is the mechanism for reducing the number of gates in series. To determine which factor or combination of factors should be eliminated, we need to look at the effect on gate input count. The increase in gate input count for the combinations of eliminations that reduce the problem path lengths by at least one gate are of interest. There are only three such combinations: elimination of the factoring of B, elimination of intermediate variables X_1, X_2, and X_3, and elimination of the factor B and the three intermediate variables X_1, X_2, and X_3. The respective gate input count increases for these actions are 0, 12, and 12, respectively. Clearly, the removal of the factor B is the best choice since the gate input count does not increase. This also demonstrates that, due to the additional decomposition of H, the gate input cost gain of 3 that occurred by factoring out B at the beginning has disappeared. The logic diagram resulting from elimination of the factor B is given in Figure 2-27(c).   ▨

While the necessary delay reduction was obtained by using elimination to reduce the number of gates along the paths in Example 2-13, in general, such a gate reduction may not reduce delay, or may even increase it due to differences in the delay characteristics of the gates to be discussed further in Chapter 3.

2-7   OTHER GATE TYPES

Since Boolean functions are expressed in terms of AND, OR, and NOT operations, it is a straightforward procedure to implement a Boolean function with AND, OR, and NOT gates. We find, however, that the possibility of considering gates with other logic operations is of considerable practical interest. Factors to be taken into consideration when constructing other types of gates are the feasibility and economy of implementing the gate with electronic components, the ability of the gate to implement Boolean functions alone or in conjunction with other gates, and the convenience of representing gate functions that are frequently used. In this section, we introduce these other gate types which are used throughout the rest of the text. Specific techniques for incorporating these gate types in circuits are given in section 3-5.

The graphics symbols and truth tables of six logic gate types are shown in Figure 2-28, with six additional gate types given in Figure 2-29. The gates in Figure 2-28 are referred to as *primitive* gates, and those in Figure 2-29 are referred to as *complex* gates.

Although the gates in Figure 2-28 are shown with just two binary input variables, X and Y, and one output binary variable, F, with the exception of the inverter and the buffer, all may have more than two inputs. The distinctively shaped symbols shown, as well as rectangular symbols not shown, are specified in detail in the Institute of Electrical and Electronics Engineers' (IEEE) *Standard Graphic Symbols for Logic Functions* (IEEE Standard 91–1984). The AND, OR, and NOT gates were defined previously. The NOT circuit inverts the logic sense of a binary signal to produce the complement operation. Recall that this circuit is typically called an *inverter* rather than a NOT gate. The small circle at the output of the graphic symbol of an inverter is formally called a *negation indicator* and designates the logical complement. We informally refer to the negation indicator as a "bubble." The triangle symbol by itself designates a buffer circuit. A *buffer* produces the logical function $Z = X$, since the binary value of the output is equal to the binary value of the input. This circuit is used primarily to amplify an electrical signal to permit more gates to be attached to the output or to decrease the time it takes for signals to propagate through the circuit.

The 3-state buffer is unique in that outputs of 3-state buffers can be connected together provided that only one of the signals on their E inputs is 1 at any given time. This type of buffer and its basic use are discussed in detail later in this section.

The NAND gate represents the complement of the AND operation, and the NOR gate represents the complement of the OR operation. Their respective names are abbreviations of NOT-AND and NOT-OR, respectively. The graphics symbols for the NAND gate and NOR gate consist of an AND symbol and an OR symbol, respectively, with a bubble on the output, denoting the complement operation. In contemporary integrated circuit technology, NAND and NOR gates are the natural primitive gate functions for the simplest and fastest electronic circuits. If we consider the inverter as a degenerate version of NAND and NOR gates with just one input, NAND gates alone or NOR gates alone can implement any Boolean function. Thus, these gate types are much more widely used than AND and OR gates in actual logic circuits. As a consequence, actual circuit implementations are often done in terms of these gate types.

Graphics Symbols

Name	Distinctive shape	Algebraic equation	Truth table
AND	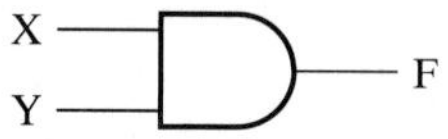	$F = XY$	X Y | F 0 0 | 0 0 1 | 0 1 0 | 0 1 1 | 1
OR	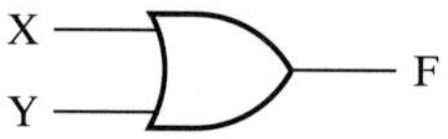	$F = X + Y$	X Y | F 0 0 | 0 0 1 | 1 1 0 | 1 1 1 | 1
NOT (inverter)	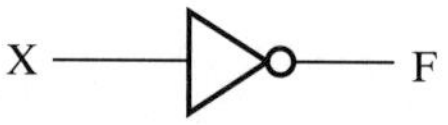	$F = \overline{X}$	X | F 0 | 1 1 | 0
Buffer	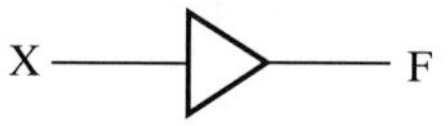	$F = X$	X | F 0 | 0 1 | 1
3-State Buffer	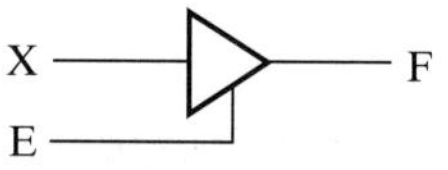		E X | F 0 0 | Hi-Z 0 1 | Hi-Z 1 0 | 0 1 1 | 1
NAND		$F = \overline{X \cdot Y}$	X Y | F 0 0 | 1 0 1 | 1 1 0 | 1 1 1 | 0
NOR	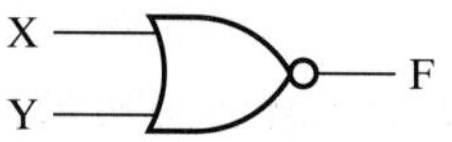	$F = \overline{X + Y}$	X Y | F 0 0 | 1 0 1 | 0 1 0 | 0 1 1 | 0

☐ **FIGURE 2-28**
Primitive Digital Logic Gates

Graphics Symbols

Name	Distinctive shape symbol	Algebraic equation	Truth table
Exclusive–OR (XOR)	 X, Y → F	$F = X\overline{Y} + \overline{X}Y$ $= X \oplus Y$	X Y │ F 0 0 │ 0 0 1 │ 1 1 0 │ 1 1 1 │ 0
Exclusive–NOR (XNOR)	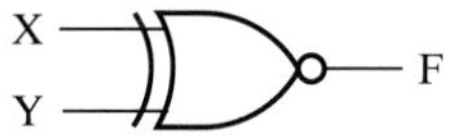 X, Y → F	$F = XY + \overline{X}\,\overline{Y}$ $= \overline{X \oplus Y}$	X Y │ F 0 0 │ 1 0 1 │ 0 1 0 │ 0 1 1 │ 1
AND-OR-INVERT (AOI)	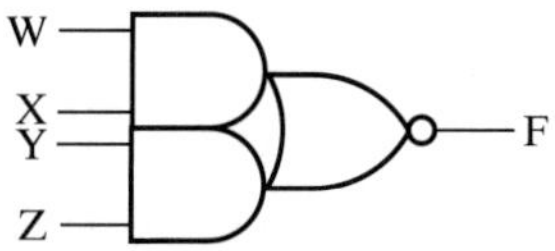 W, X, Y, Z → F	$F = \overline{WX + YZ}$	
OR-AND -INVERT (OAI)	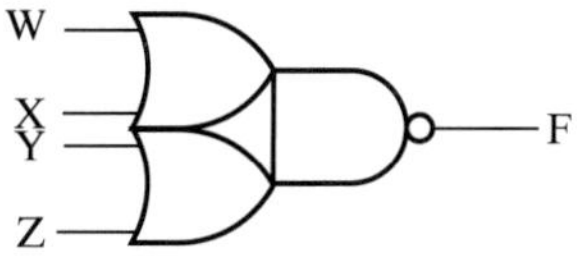 W, X, Y, Z → F	$F = \overline{(W + X)(Y + Z)}$	
AND-OR (AO)	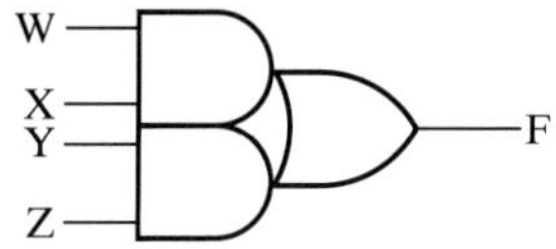 W, X, Y, Z → F	$F = WX + YZ$	
OR-AND (OA)	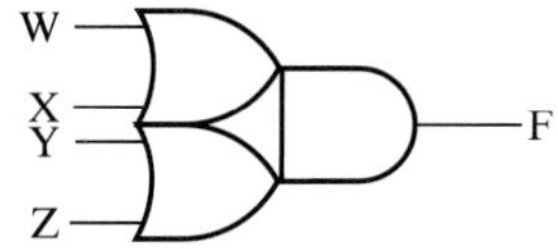 W, X, Y, Z → F	$F = (W + X)(Y + Z)$	

☐ **FIGURE 2-29**
Complex Digital Logic Gates

A gate type that alone can be used to implement all Boolean functions is called a *universal gate*. To show the NAND gate is a universal gate, we need only show that the logical operations of AND, OR, and NOT can be obtained with NAND gates only. This is done in Figure 2-30. The complement operation is obtained from a one-input NAND gate corresponds to a NOT gate. In fact, the one-input NAND is an invalid symbol and is replaced by the NOT symbol, as shown in the figure. The AND operation requires a NAND gate followed by a NOT gate. The NOT inverts the output of the NAND giving an AND operation as

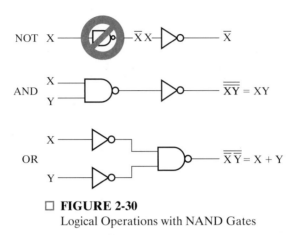

☐ **FIGURE 2-30**
Logical Operations with NAND Gates

the result. The OR operation is achieved using a NAND gate with NOTs on each input. When DeMorgan's theorem is applied as shown in Figure 2-30, the inversions cancel and an OR function results.

The exclusive-OR (XOR) gate shown in Figure 2-29 is similar to the OR gate, but excludes (has the value 0 for) the combination with both X and Y equal to 1. The graphics symbol for the XOR gate is similar to that for the OR gate, except for the additional curved line on the inputs. The exclusive-OR has the special symbol ⊕ to designate its operation. The exclusive-NOR is the complement of the exclusive-OR, as indicated by the bubble at the output of its graphics symbol.

The AND-OR-INVERT (AOI) gate forms the complement of a sum-of-products. The are many different AND-OR-INVERT gates depending on the number of AND gates and the numbers of inputs to each AND and directly to the OR gate. For example, suppose that the function implemented by an AOI is

$$F = \overline{XY + Z}$$

This AOI is referred to as a 2-1 AOI since it consists of a 2-input AND and 1 input directly to the OR gate. If the function implemented is

$$F = \overline{TUV + WX + YZ}$$

then the AOI is called a 3-2-2 AOI. The OR-AND-INVERT (OAI) is the dual of the AOI and implements the complement of a product-of-sums form. The AND-OR (AO) and OR-AND (OA) are versions of the AOI and OAI without the complement.

In general, complex gates are used to reduce the circuit complexity needed for implementing specific Boolean functions in order to reduce integrated circuit cost. In addition, they reduce the time required for signals to propagate through a circuit.

CMOS Circuits   This supplement, which discusses the implementation of both primitive and complex gates with CMOS technology, is available on the Companion Website for the text.

2-8 EXCLUSIVE-OR OPERATOR AND GATES

In addition to the exclusive-OR gate shown in Figure 2-29, there is an exclusive-OR operator with its own algebraic identities. The exclusive-OR (XOR), denoted by $\oplus$, is a logical operation that performs the function

$$X \oplus Y = X\overline{Y} + \overline{X}Y$$

It is equal to 1 if exactly one input variable is equal to 1. The exclusive-NOR, also known as the *equivalence*, is the complement of the exclusive-OR and is expressed by the function

$$\overline{X \oplus Y} = XY + \overline{X}\,\overline{Y}$$

It is equal to 1 if both X and Y are equal to 1 or if both are equal to 0. The two functions can be shown to be the complement of each other, either by means of a truth table or, as follows, by algebraic manipulation:

$$\overline{X \oplus Y} = \overline{X\overline{Y} + \overline{X}Y} = (\overline{X} + Y)(X + \overline{Y}) = XY + \overline{X}\,\overline{Y}$$

The following identities apply to the exclusive-OR operation:

$$X \oplus 0 = X \qquad\qquad X \oplus 1 = \overline{X}$$

$$X \oplus X = 0 \qquad\qquad X \oplus \overline{X} = 1$$

$$X \oplus \overline{Y} = \overline{X \oplus Y} \qquad \overline{X} \oplus Y = \overline{X \oplus Y}$$

Any of these identities can be verified by using a truth table or by replacing the $\oplus$ operation by its equivalent Boolean expression. It can also be shown that the exclusive-OR operation is both commutative and associative; that is,

$$A \oplus B = B \oplus A$$

$$(A \oplus B) \oplus C = A \oplus (B \oplus C) = A \oplus B \oplus C$$

This means that the two inputs to an exclusive-OR gate can be interchanged without affecting the operation. It also means that we can evaluate a three-variable exclusive-OR operation in any order, and for this reason, exclusive-ORs with three or more variables can be expressed without parentheses.

A two-input exclusive-OR function may be constructed with conventional gates. Two NOT gates, two AND gates, and an OR gate are used. The associativity of the exclusive-OR operator suggests the possibility of exclusive-OR gates with more than two inputs. The exclusive-OR concept for more than two variables, however, is replaced by the odd function to be discussed next. Thus, there is no symbol for exclusive-OR for more than two inputs. By duality, the exclusive-NOR is replaced by the even function and has no symbol for more than two inputs.

Odd Function

The exclusive-OR operation with three or more variables can be converted into an ordinary Boolean function by replacing the ⊕ symbol with its equivalent Boolean expression. In particular, the three-variable case can be converted to a Boolean expression as follows:

$$X \oplus Y \oplus Z = (X\overline{Y} + \overline{X}Y)\overline{Z} + (XY + \overline{X}\overline{Y})Z$$

$$= X\overline{Y}\overline{Z} + \overline{X}Y\overline{Z} + \overline{X}\overline{Y}Z + XYZ$$

The Boolean expression clearly indicates that the three-variable exclusive-OR is equal to 1 if only one variable is equal to 1 or if all three variables are equal to 1. Hence, whereas in the two-variable function only one variable need be equal to 1, with three or more variables an odd number of variables must be equal to 1. As a consequence, the multiple-variable exclusive-OR operation is defined as the *odd function*. In fact, strictly speaking, this is the correct name for the ⊕ operation with three or more variables; the name "exclusive-OR" is applicable to the case with only two variables.

The definition of the odd function can be clarified by plotting the function on a map. Figure 2-31(a) shows the map for the three-variable odd function. The four minterms of the function differ from each other in at least two literals and hence cannot be adjacent on the map. These minterms are said to be *distance two* from each other. The odd function is identified from the four minterms whose binary values have an odd number of 1's. The four-variable case is shown in Figure 2-31(b). The eight minterms marked with 1's in the map constitute the odd function. Note the characteristic pattern of the distance between the 1's in the map. It should be mentioned that the minterms not marked with 1's in the map have an even number of 1's and constitute the complement of the odd function, called the *even function*. The odd function is implemented by means of two-input exclusive-OR gates, as

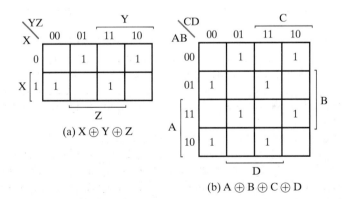

(a) $X \oplus Y \oplus Z$

(b) $A \oplus B \oplus C \oplus D$

☐ **FIGURE 2-31**
Maps for Multiple-Variable Odd Functions

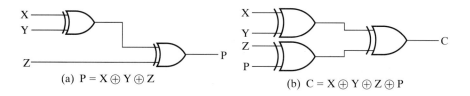

(a) $P = X \oplus Y \oplus Z$

(b) $C = X \oplus Y \oplus Z \oplus P$

□ **FIGURE 2-32**
Multiple-Input Odd Functions

shown in Figure 2-32. The even function is obtained by replacing the output gate with an exclusive-NOR gate.

2-9 HIGH-IMPEDANCE OUTPUTS

Thus far, we have considered gates that have only output values logic 0 and logic 1. In this section, we introduce two important structures, three-state buffers and transmission gates, that provide a third output value referred to as the *high-impedance state* and denoted by Hi-Z or just plain Z or z. The Hi-Z value behaves as an open circuit, which means that, looking back into the circuit, we find that the output appears to be disconnected. High impedance outputs may appear on any gate, but here we restrict consideration to two gate structures with single data inputs. Gates with Hi-Z output values can have their outputs connected together, provided that no two gates drive the line at the same time to opposite 0 and 1 values. In contrast, gates with only logic 0 and logic 1 outputs cannot have their outputs connected together.

Three-State Buffers The 3-state buffer was introduced earlier as one of the primitive gates. As the name implies, a three-state logic output exhibits three distinct states. Two of the "states" are the logic 1 and logic 0 of conventional logic. The third "state" is the *Hi-Z* value, that, for three-state logic, is referred to as the *Hi-Z state.*

The graphic symbol and truth table for a 3-state buffer are given in Figure 2-33. The symbol in Figure 2-33(a) is distinguished from the symbol for a normal buffer by the enable input, *EN*, entering the bottom of the buffer symbol. From the truth table in Figure 2-33(b), if $EN = 1$, *OUT* is equal to *IN*, behaving like a normal buffer. But for $EN = 0$, the output value is high impedance (*Hi-Z*), regardless of the value of *IN*.

Three-state buffer outputs can be connected together to form a multiplexed output line. Figure 2-34(a) shows two 3-state buffers with their outputs connected to form output line *OL*. We are interested in the output of this structure in terms of the four inputs *EN1*, *EN0*, *IN1*, and *IN0*. The output behavior is given by the truth table in Figure 2-34(b). For *EN1* and *EN0* equal to 0, both buffer outputs are *Hi-Z*. Since both appear as open circuits, *OL* is also an open circuit, represented by a *Hi-Z* value. For $EN1 = 0$ and $EN0 = 1$, the output of the top buffer is *IN0*

EN	IN	OUT
0	X	Hi-Z
1	0	0
1	1	1

IN ─▷─ OUT
EN ─

(a) Logic symbol   (b) Truth table

☐ **FIGURE 2-33**
Three-state Buffer

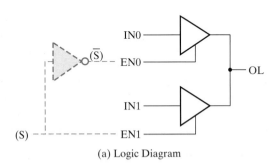

(a) Logic Diagram

EN1	EN0	IN1	IN0	OL
0	0	X	X	Hi-Z
(S)0	($\overline{S}$)1	X	0	0
0	1	X	1	1
1	0	0	X	0
1	0	1	X	1
1	1	0	0	0
1	1	1	1	1
1	1	0	1	🐛🐛
1	1	1	0	🐛🐛

(b) Truth table

☐ **FIGURE 2-34**
Three-state Buffers Forming a Multiplexed Line OL

and the output of bottom buffer is *Hi-Z*. Since the value of *IN*0 combined with an open circuit is just *IN*0, *OL* has value *IN*0, giving the second and third rows of the truth table. A corresponding, but opposite, case occurs for *EN*1 = 1 and *EN*0 = 0, so *OL* has value *IN*1, giving the fourth and fifth rows of the truth table. For *EN*1 and *EN*0 both 1, the situation is more complicated. If *IN*1 = *IN*0, then their mutual value appears at *OL*. But if *IN*1 ≠ *IN*0, then their values conflict at the output. The conflict results in an electrical current flowing from the buffer output that is at 1 into the buffer output that is at 0. This current is often large enough to cause heating and may even destroy the circuit, as symbolized by the "smoke" icons in the truth table. Clearly, such a situation must be avoided. The designer must ensure that *EN0* and *EN1* never equal 1 at the same time. In the general case, for *n* 3-state buffers attached to a bus line, *EN* can equal 1 for only one of the buffers and must be 0 for the rest. One way to ensure this is to use a decoder to generate the *EN* signals. For the two-buffer case, the decoder is just an inverter with select input *S*, as shown in dotted lines in Figure 2-34(a). It is interesting to examine the truth table with the inverter in place. It consists of the shaded area of the table in Figure 2-34(b). Clearly, the value on *S* selects between inputs *IN0* and *IN1*. Further, the circuit output OL is never in the Hi-Z state.

Transmission Gates   In integrated circuit logic, there is a CMOS transistor circuit logic that is important enough to be separately represented at the gate level. This circuit, called a *transmission gate* (TG), is one form of an electronic switch for connecting and disconnecting two points in a circuit. Figure 2-35(a) shows the IEEE symbol for the transmission gate. It has four external connections or ports. *C* and $\overline{C}$ are the control inputs and *X* and *Y* are the signals to be connected or disconnected by the TG. In Figure 2-35(*b*) and (*c*), the switch model for the transmission gate appears. If *C* = 1 and $\overline{C}$ = 0, *X* and *Y* are connected as represented in the model by a "closed" switch and signals ·can pass from *X* to *Y* or from *Y* to *X*. If *C* = 0 and $\overline{C}$ = 1, *X* and *Y* are disconnected as represented in the model by an "open" switch and signals cannot pass between *X* and *Y*. In normal use, the control inputs are connected by an inverter as shown in Figure 2-35(d), so that *C* and $\overline{C}$ are the complements of each other.

To illustrate the use of a transmission gate, an exclusive-OR gate constructed from two transmission gates and two inverters is shown in Figure 2-36(a). Input *C* controls the paths through the transmission gates, and input *A* provides the output for *F*. If input *C* is equal to 1, a path exists through transmission gate TG1 connecting *F* to $\overline{A}$, and no path exists through TG0. If input *C* is equal to 0, a path exists through TG0 connecting *F* to *A*, and no path exists through TG1. Thus, the output *F* is connected to *A*. This results in the exclusive-OR truth table, as indicated in Figure 2-36(b).

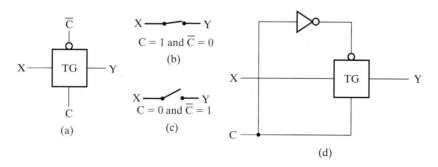

□ **FIGURE 2-35**
Transmission Gate (TG)

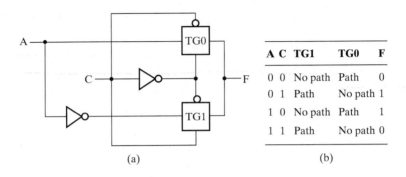

A C	TG1	TG0	F
0 0	No path	Path	0
0 1	Path	No path	1
1 0	No path	Path	1
1 1	Path	No path	0

(a)   (b)

☐ **FIGURE 2-36**
Transmission Gate Exclusive OR

2-10 CHAPTER SUMMARY

The primitive logic operations AND, OR, and NOT define three primitive logic components called gates, from which digital systems are implemented. A Boolean algebra defined in terms of these operations provides a tool for manipulating Boolean functions in designing digital logic circuits. Minterm and maxterm standard forms correspond directly to truth tables for functions. These standard forms can be manipulated into sum-of-products and product-of-sums forms, which correspond to two-level gate circuits. Two cost measures to be minimized in optimizing a circuit are the number of input literals to the circuit and the total number of inputs to the gates in the circuit. K-maps with two to four variables are an effective alternative to algebraic manipulation in optimizing small circuits. These maps can be used to optimize sum-of-products forms, product-of-sums forms, and incompletely specified functions with don't-care conditions. Transforms for optimizing multiple level circuits with three or more levels of gating are illustrated.

The primitive operations AND and OR are not directly implemented by primitive logic elements in the most popular logic family. Thus, NAND and NOR primitives as well as complex gates that implement these families are introduced. A more complex primitive, the exclusive-OR, as well as its complement, the exclusive-NOR, are presented along with their mathematical properties.

REFERENCES

1. BOOLE, G. *An Investigation of the Laws of Thought.* New York: Dover, 1854.

2. KARNAUGH, M. "A Map Method for Synthesis of Combinational Logic Circuits," *Transactions of AIEE, Communication and Electronics*, 72, part I (Nov. 1953), 593-99.

3. DIETMEYER, D. L. *Logic Design of Digital Systems,* 3rd ed. Boston: Allyn Bacon, 1988.

4. MANO, M. M. *Digital Design,* 3rd ed. Upper Saddle River, NJ: Prentice Hall, 2002.

5. ROTH, C. H. *Fundamentals of Logic Design*, 4th ed. St. Paul: West, 1992.

6. HAYES, J. P. *Introduction to Digital Logic Design.* Reading, MA: Addison-Wesley, 1993.

7. WAKERLY, J. F. *Digital Design: Principles and Practices,* 3rd ed. Upper Saddle River, NJ: Prentice Hall, 2000.

8. GAJSKI, D. D. *Principles of Digital Design.* Upper Saddle River, NJ: Prentice Hall, 1997.

9. *IEEE Standard Graphic Symbols for Logic Functions.* (Includes IEEE Std 91a-1991 Supplement and IEEE Std 91-1984.) New York: The Institute of Electrical and Electronics Engineers, 1991.

PROBLEMS

The plus (+) indicates a more advanced problem and the asterisk (*) indicates a solution is available on the Companion Website for the text.

2–1. *Demonstrate by means of truth tables the validity of the following identities:

(a) DeMorgan's theorem for three variables: $\overline{XYZ} = \overline{X} + \overline{Y} + \overline{Z}$

(b) The second distributive law: $X + YZ = (X + Y)(X + Z)$

(c) $\overline{X}Y + \overline{Y}Z + X\overline{Z} = X\overline{Y} + Y\overline{Z} + \overline{X}Z$

2–2. *Prove the identity of each of the following Boolean equations, using algebraic manipulation:

(a) $\overline{X}\overline{Y} + \overline{X}Y + XY = \overline{X} + Y$

(b) $\overline{A}B + \overline{B}\overline{C} + AB + \overline{B}C = 1$

(c) $Y + \overline{X}Z + X\overline{Y} = X + Y + Z$

(d) $\overline{X}\overline{Y} + \overline{Y}Z + XZ + XY + Y\overline{Z} = \overline{X}\overline{Y} + XZ + Y\overline{Z}$

2–3. +Prove the identity of each of the following Boolean equations, using algebraic manipulation:

(a) $AB + B\overline{C}\overline{D} + \overline{A}BC + \overline{C}D = B + \overline{C}D$

(b) $WY + \overline{W}Y\overline{Z} + WXZ + \overline{W}X\overline{Y} = WY + \overline{W}X\overline{Z} + \overline{X}Y\overline{Z} + X\overline{Y}Z$

(c) $A\overline{C} + \overline{A}B + \overline{B}C + \overline{D} = (\overline{A} + \overline{B} + \overline{C} + \overline{D})(A + B + C + \overline{D})$

2–4. +Given that $A \cdot B = 0$ and $A + B = 1$, use algebraic manipulation to prove that

$$(A + C) \cdot (\overline{A} + B) \cdot (B + C) = B \cdot C$$

2–5. +A specific Boolean algebra with just two elements 0 and 1 has been used in this chapter. Other Boolean algebras can be defined with more than two elements by using elements that correspond to binary strings. These algebras form the mathematical foundation for bitwise logical operations that we will study in Chapter 7. Suppose that the strings are each a nibble (half of a byte) of four bits. Then there are 2^4, or 16, elements in the algebra, where an element I is the 4-bit nibble in binary corresponding to I in decimal. Based on bitwise application of the two-element Boolean algebra, define each of the following for the new algebra so that the Boolean identities hold:

(a) The OR operation $A + B$ for any two elements A and B
(b) The AND operation $A \cdot B$ for any two elements A and B
(c) The element that acts as the 0 for the algebra
(d) The element that acts as the 1 for the algebra
(e) For any element A, the element $\overline{A}$.

2–6. Simplify the following Boolean expressions to expressions containing a minimum number of literals:

(a) $\overline{A}\,\overline{C} + \overline{A}BC + B\overline{C}$
(b) $(A + B)(\overline{A} + \overline{B})$
(c) $ABC + \overline{A}C$
(d) $BC + B(AD + \overline{C}D)$
(e) $(B + \overline{C} + B\overline{C})(BC + A\overline{B} + AC)$

2–7. *Reduce the following Boolean expressions to the indicated number of literals:

(a) $\overline{X}\,\overline{Y} + XYZ + \overline{X}Y$ to three literals
(b) $X + Y(Z + \overline{X + Z})$ to two literals
(c) $\overline{W}X(\overline{Z} + \overline{Y}Z) + X(W + \overline{W}YZ)$ to one literal
(d) $(AB + \overline{A}\,\overline{B})(\overline{C}\,\overline{D} + CD) + \overline{A}\,\overline{C}$ to four literals

2–8. Using DeMorgan's theorem, express the function

$$F = \overline{A}BC + \overline{B}\,\overline{C} + A\overline{B}$$

(a) with only OR and complement operations.
(b) with only AND and complement operations.

2–9. *Find the complement of the following expressions:

(a) $A\overline{B} + \overline{A}B$
(b) $(\overline{V}W + X)Y + \overline{Z}$
(c) $WX(\overline{Y}Z + Y\overline{Z}) + \overline{W}\,\overline{X}(\overline{Y} + Z)(Y + \overline{Z})$
(d) $(A + \overline{B} + C)(\overline{A}\,\overline{B} + C)(A + \overline{B}\,\overline{C})$

2–10. *Obtain the truth table of the following functions, and express each function in sum-of-minterms and product-of-maxterms form:

(a) $(XY + Z)(Y + XZ)$
(b) $(\overline{A} + B)(\overline{B} + C)$
(c) $WX\overline{Y} + WX\overline{Z} + WXZ + Y\overline{Z}$

2–11. For the Boolean functions E and F, as given in the following truth table:

X	Y	Z	E	F
0	0	0	1	0
0	0	1	0	0
0	1	0	1	1
0	1	1	0	0
1	0	0	0	1
1	0	1	1	0
1	1	0	1	1
1	1	1	0	1

(a) List the minterms and maxterms of each function.

(b) List the minterms of $\overline{E}$ and $\overline{F}$.

(c) List the minterms of $E + F$ and $E \cdot F$.

(d) Express E and F in sum-of-minterms algebraic form.

(e) Simplify E and F to expressions with a minimum of literals.

2–12. *Convert the following expressions into sum-of-products and product-of-sums forms:

(a) $(AB + C)(B + \overline{C}D)$

(b) $\overline{X} + X(X + \overline{Y})(Y + \overline{Z})$

(c) $(A + B\overline{C} + CD)(\overline{B} + EF)$

2–13. Draw the logic diagram for the following Boolean expressions. The diagram should correspond exactly to the equation. Assume that the complements of the inputs are not available.

(a) $W\overline{X}\overline{Y} + \overline{W}Z + YZ$

(b) $A(B\overline{D} + \overline{B}D) + D(BC + \overline{B}\overline{C})$

(c) $W\overline{Y}(X + Z) + \overline{X}Z(W + Y) + W\overline{X}(Y + Z)$

2–14. Optimize the following Boolean functions by means of a three-variable map:

(a) $F(X, Y, Z) = \Sigma m(1, 3, 6, 7)$

(b) $F(X, Y, Z) = \Sigma m(3, 5, 6, 7)$

(c) $F(A, B, C) = \Sigma m(0, 1, 2, 4, 6)$

(d) $F(A, B, C) = \Sigma m(0, 3, 4, 5, 7)$

2–15. *Optimize the following Boolean expressions using a map:

(a) $\overline{X}\overline{Z} + Y\overline{Z} + XYZ$

(b) $\overline{A}B + \overline{B}C + \overline{A}\,\overline{B}\,\overline{C}$

(c) $\overline{A}\overline{B} + A\overline{C} + \overline{B}C + \overline{A}B\overline{C}$

2–16. Optimize the following Boolean functions by means of a four-variable map:

(a) $F(A,B,C,D) = \Sigma m(2,3,8,9,10,12,13,14)$

(b) $F(W,X,Y,Z) = \Sigma m(0,2,5,6,8,10,13,14,15)$

(c) $F(A,B,C,D) = \Sigma m(0,2,3,7,8,10,12,13)$

2–17. Optimize the following Boolean functions, using a map:

(a) $F(W,X,Y,Z) = \Sigma m(0,2,5,8,9,11,12,13)$

(b) $F(A,B,C,D) = \Sigma m(3,4,6,7,9,12,13,14,15)$

2–18. *Find the minterms of the following expressions by first plotting each expression on a map:

(a) $XY + XZ + \overline{X}YZ$

(b) $XZ + \overline{W}X\overline{Y} + WXY + \overline{W}YZ + W\overline{Y}Z$

(c) $\overline{B}\overline{D} + ABD + \overline{A}BC$

2–19. *Find all the prime implicants for the following Boolean functions, and determine which are essential:

(a) $F(W,X,Y,Z) = \Sigma m(0,2,5,7,8,10,12,13,14,15)$

(b) $F(A,B,C,D) = \Sigma m(0,2,3,5,7,8,10,11,14,15)$

(c) $F(A,B,C,D) = \Sigma m(1,3,4,5,9,10,11,12,13,14,15)$

2–20. Optimize the following Boolean functions by finding all prime implicants and essential prime implicants and applying the selection rule:

(a) $F(W,X,Y,Z) = \Sigma m(0,1,4,5,7,8,9,12,14,15)$

(b) $F(A,B,C,D) = \Sigma m(1,5,6,7,11,12,13,15)$

(c) $F(W,X,Y,Z) = \Sigma m(0,2,3,4,5,7,8,10,11,12,13,15)$

2–21. Optimize the following Boolean functions in product-of-sums form:

(a) $F(W,X,Y,Z) = \Sigma m(0,2,3,4,8,10,11,15)$

(b) $F(A,B,C,D) = \Pi M(0,2,4,5,8,10,11,12,13,14)$

2–22. *Optimize the following expressions in (1) sum-of-products and (2) product-of-sums forms:

(a) $A\overline{C} + \overline{B}D + \overline{A}CD + ABCD$

(b) $(\overline{A} + \overline{B} + \overline{D})(A + \overline{B} + \overline{C})(\overline{A} + B + \overline{D})(B + \overline{C} + \overline{D})$

(c) $(\overline{A} + \overline{B} + D)(\overline{A} + \overline{D})(A + B + \overline{D})(A + \overline{B} + C + D)$

2–23. Optimize the following functions into (1) sum-of-products and (2) product-of-sums forms:

(a) $F(A,B,C,D) = \Sigma m(2,3,5,7,8,10,12,13)$

(b) $F(W,X,Y,Z) = \Pi M(2,10,13)$

2–24. Optimize the following Boolean functions F together with the don't-care conditions d:

(a) $F(A,B,C,D) = \Sigma m(0,3,5,7,11,13)$, $d(A,B,C,D) = \Sigma m(4,6,14,15)$

(b) $F(W,X,Y,Z) = \Sigma m(0,6,8,13,14)$, $d(W,X,Y,Z) = \Sigma m(2,4,7,10,12)$

(c) $F(A,B,C) = \Sigma m(0,1,2,4,5)$, $d(A,B,C) = \Sigma m(3,6,7)$

2–25. *Optimize the following Boolean functions F together with the don't-care conditions d. Find all prime implicants and essential prime implicants, and apply the selection rule.

(a) $F(A,B,C) = \Sigma m(3,5,6)$, $d(A,B,C) = \Sigma m(0,7)$

(b) $F(W,X,Y,Z) = \Sigma m(0,2,4,5,8,14,15)$, $d(W,X,Y,Z) = \Sigma m(7,10,13)$

(c) $F(A,B,C,D) = \Sigma m(4,6,7,8,12,15)$,
$d(A,B,C,D) = \Sigma m(2,3,5,10,11,14)$

2–26. Optimize the following Boolean functions F together with the don't-care conditions d in (1) sum-of-products and (2) product-of-sums form:

(a) $F(A,B,C,D) = \Pi M(1,3,4,6,9,11)$,
$d(A,B,C,D) = \Sigma m(0,2,5,10,12,14)$

(b) $F(W,X,Y,Z) = \Sigma m(3,4,9,15)$, $d(W,X,Y,Z) = \Sigma m(0,2,5,10,12,14)$

2–27. Use decomposition to find minimum gate input count, multiple-level implementations for the functions using AND and OR gates and inverters.

(a) $F(A,B,C,D) = A\overline{B}C + \overline{A}BC + A\overline{B}D + \overline{A}BD$

(b) $F(W,X,Y,Z) = WY + XY + \overline{W}XZ + W\overline{X}Z$

2–28. Use extraction to find a shared, minimum gate input count, multiple-level implementation for the pair of functions given using AND and OR gates and inverters.

(a) $F(A,B,C,D) = \Sigma m(0,5,11,14,15), d(A,B,C,D) = \Sigma m(10)$

(b) $G(A,B,C,D) = \Sigma m(2,7,10,11,14), d(A,B,C,D) = \Sigma m(15)$

2–29. Use elimination to flatten each of the function sets given into a two-level sum-of-products form.

(a) $F(A,B,G,H) = AB\overline{G} + \overline{B}G + \overline{A}H$, $G(C,D) = C\overline{D} + \overline{C}D$,
$H(B,C,D) = B + CD$

(b) $T(U,V,Y,Z) = YZU + \overline{Y}\overline{Z}V$, $U(W,X) = W + \overline{X}$,
$V(W,X,Y) = \overline{W}Y + X$

2–30. *Prove that the dual of the exclusive-OR is also its complement.

2–31. Implement the following Boolean function with exclusive-OR and AND gates, using a minimum number of gate inputs:

$$F(A,B,C,D) = AB\overline{C}D + A\overline{D} + \overline{A}D$$

2–32. **(a)** Implement function $H = \overline{X}Y + XZ$ using two three-state buffers and an inverter.
   (b) Construct an exclusive-OR gate by interconnecting two three-state buffers and two inverters.

2–33. **(a)** Connect the outputs of three 3-state buffers together, and add additional logic to implement the function

$$F = \overline{A}\,B\,C + A\,B\,D + A\,\overline{B}\,\overline{D}$$

   Assume that C, D, and $\overline{D}$ are data inputs to the buffers and A and B pass through logic that generates the enable inputs.

   (b) Is your design in part (a) free of three-state output conflicts? If not, change the design if necessary to be free of such conflicts.

2–34. Use only transmission gates and inverters to implement the function in problem 2-32.

2–35. Depending on the design and the overall logic family being used, it is usually not a good idea to leave the output of a three-state or transmission gate circuit in the high impedance (Hi-Z) state.
   (a) For the transmission gate circuit designed in problem 2-33, give all input combinations for which the output F is the high-impedance state.
   (b) Modify the enable logic driving the enable inputs so that the output is either a 0 or a 1 (instead of Hi-Z).

COMBINATIONAL LOGIC DESIGN

I n this chapter, we learn about the design of combinational circuits. We introduce the use of a hierarchy and top-down design, both of which are essential to the design of digital circuits. Further, computer-aided design is briefly discussed, including hardware description languages (HDLs) and logic synthesis, two concepts with crucial roles in the efficient design of modern, complex circuits.

Concepts related to the underlying technology for digital circuit implementation are covered in the design space section. The properties of logic gates, including integration levels, logic families, and parameters for logic technologies, are presented. Fan-in, fan-out, and propagation delay for gates are defined, and the positive and negative logic concepts are introduced. Finally, trade-offs between dimensions of the design space, such as cost and performance, are touched upon.

A design procedure with five major steps is presented. The first three steps, specification, formulation, and optimization are illustrated by examples. Fixed implementation technologies are introduced, and technology mapping for these technologies, the next step of the design procedure, is defined and illustrated. The final step of the design procedure, verification, is illustrated by an example using both a manual method and logic simulation. The chapter concludes with an introduction to programmable implementation technologies.

The various concepts in this chapter are pervasive in the design of the generic computer in the diagram at the beginning of Chapter 1. Concepts from this chapter apply across all of the digital components of the generic computer including memories.

3-1 DESIGN CONCEPTS AND AUTOMATION

In Chapter 1, we learned about binary numbers and binary codes that represent discrete quantities of information. In Chapter 2, we introduced the various logic gates and learned how to optimize Boolean expressions and equations in order to achieve economical gate implementations. The purpose of this chapter is to use the knowledge acquired in the previous chapters to formulate systematic design procedures for combinational circuits. In addition, the design steps will be related to the use of computer-aided design tools. The various examples introduced provide design practice. Modern digital design involves a number of techniques and tools that are essential to the design of complex circuits and systems. Design hierarchy, top-down design, computer-aided design tools, hardware description languages, and logic synthesis, are among the most important tools for effective and efficient digital design.

Logic circuits for digital systems may be combinational or sequential. A combinational circuit consists of logic gates whose outputs at any time are determined by logic operations on the input values. A combinational circuit performs an operation that can be specified logically by a set of Boolean equations. In contrast, sequential circuits employ elements that store bit values. Sequential circuit outputs are functions of the inputs and the bit values in the storage elements, which, in turn, are a function of previously applied inputs and stored values. As a consequence, the outputs of a sequential circuit depend not only on the current input values, but also on past inputs. The behavior of the circuit must be specified by a sequence in time of inputs and internal stored bit values. Sequential circuits are presented in Chapter 6.

A combinational circuit consists of input variables, output variables, logic gates, and interconnections. The interconnected logic gates accept signals from the inputs and generate signals at the outputs. A block diagram of a combinational circuit is shown in Figure 3-1. The n input variables come from the environment of the circuit, and the m output variables are available for use by the environment. Each input and output variable exists physically as a binary signal that takes on values logic 1 or logic 0.

For n input variables, there are 2^n possible binary input combinations. For each binary combination of the input variables, there is one possible binary value on each output. Thus, a combinational circuit can be specified by a truth table that lists the output values for each combination of the input variables. A combinational circuit can also be described by m Boolean functions, one for each output variable. Each such function is expressed as a function of the n input variables. Before defining the design process, we introduce two fundamental concepts related to design: design hierarchy and top-down design.

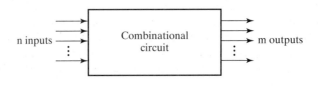

☐ **FIGURE 3-1**
Block Diagram of a Combinational Circuit

Design Hierarchy

A circuit may be specified by a symbol showing its inputs and outputs and a description defining exactly how it operates. In terms of implementation, however, a circuit is composed of logic gates or other logic structures that are interconnected. A complex digital system may contain millions of interconnected gates. In fact, a single very large-scale integrated (VLSI) processor circuit often contains several tens of millions of gates. With such complexity, the interconnected gates appear to be an incomprehensible maze. Thus, no complex system or circuit can be designed simply by interconnecting gates one at a time. In order to deal with such circuit complexity, a "divide and conquer" approach is used. The circuit is broken up into pieces we call *blocks*. The blocks are interconnected to form the circuit. The functions of these blocks and their interfaces are carefully defined, so that the circuit formed by interconnecting the blocks obeys the circuit specification. If a block is still too large and complex to be designed as a single entity, it can be broken into smaller blocks. This process can be repeated as necessary. Note that since we are working primarily with logic circuits, we use the term "circuit" in this discussion, but the ideas apply equally well to the "systems" covered in later chapters.

The "divide and conquer" approach is illustrated by a combinational circuit in Figure 3-2. The circuit implements a 9-input odd function. A symbol for this overall circuit is shown in part (a) of the figure. In part (b), a logic diagram, or *schematic*, is given for the circuit represented by the symbol in part (a). In this schematic, the designer has chosen to break up the circuit into four identical blocks, each of which is a 3-input odd function. The symbol for the 3-input odd function is repeated four times. The four symbols are interconnected to form the 9-input odd-function circuit. In part (c), a 3-input odd function block is shown to consist of two interconnected exclusive-OR gates. Finally, in part (d), the exclusive-OR is implemented using NAND gates. Note that in each case, as we move downward from the top level, symbols are replaced by schematics that represent the implementation of the symbol.

This design approach is referred to as *hierarchical design*, and the resulting related symbols and schematics constitute a *hierarchy* representing the circuit designed. The structure of the hierarchy can be represented without the interconnections by starting with the top block and connecting below each block those blocks from which it is made. Using this representation, the hierarchy for the 9-input odd-function circuit is shown in Figure 3-3(a) using this representation. Note that the resulting structure has the form of a tree with the root at the top. The "leaves" of the tree are the NAND gates, in this case 32 of them. In order to provide a more compact representation of the hierarchy, we can reuse blocks as shown in Figure 3-3(b). This diagram corresponds to blocks used in Figure 3-2 with only one copy of each distinct block shown. These diagrams and the circuits in Figure 3-2 are helpful in illustrating a number of useful concepts associated with hierarchies and hierarchical blocks.

First of all, a hierarchy reduces the complexity required to represent the schematic diagram of a circuit. For example, in Figure 3-3(a), 32 NAND blocks

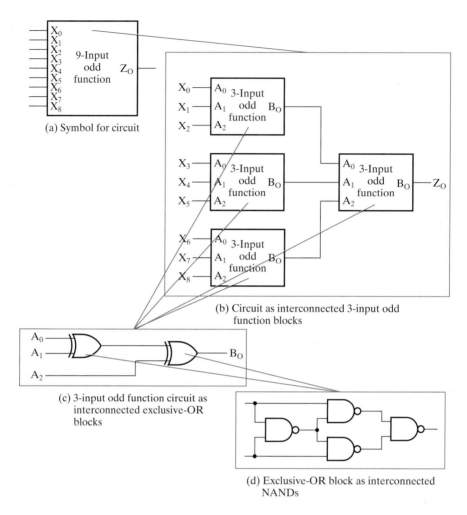

(a) Symbol for circuit

(b) Circuit as interconnected 3-input odd function blocks

(c) 3-input odd function circuit as interconnected exclusive-OR blocks

(d) Exclusive-OR block as interconnected NANDs

☐ **FIGURE 3-2**
Example of Design Hierarchy and Reusable Blocks

appear. This means that if a 9-input odd-function circuit was designed directly in terms of NAND gates, the schematic for the circuit would consist of 32 interconnected NAND gate symbols, in contrast to just 10 symbols used to describe the circuit implementation as a hierarchy in Figure 3-2. Thus, a hierarchy gives a simplified representation of a complex circuit.

Second, the hierarchy ends at a set of "leaves" in Figure 3-3. In this case, the leaves consist of NAND gates. Since the NAND gates are electronic circuits, and we are interested here only in designing the logic, the NAND gates are commonly called *primitive blocks*. These are rudimentary blocks, such as gates, that have a symbol, but no logic schematic. Primitive blocks are a rudimentary type of *predefined blocks*. In general, more complex structures that likewise have symbols,

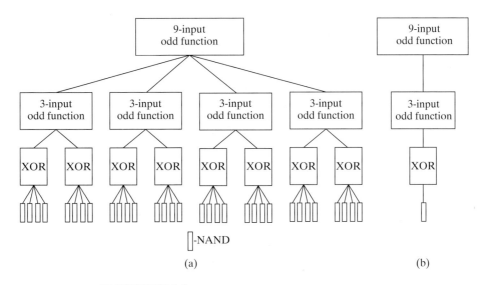

□ **FIGURE 3-3**
Diagrams Representing the Hierarchy for Figure 3-2

but no logic schematics, are also predefined blocks. Instead of schematics, their function can be defined by a program or description that can serve as a model. For example, in the hierarchy depicted in Figure 3-3, the exclusive-OR gates could have been considered as predefined blocks. In such a case, the diagram describing the exclusive-OR blocks in Figure 3-2(d) would not be necessary. The hierarchical representations in Figure 3-3 would then end with the exclusive-OR blocks. In any hierarchy, the "leaves" consist of predefined blocks, some of which may be primitives.

A third very important property that results from hierarchical design is the reuse of blocks as illustrated in Figures 3-3(a) and (b). In part (a), there are four copies of the 3-input odd-function block and eight copies of the exclusive-OR block. In part (b), there is only one copy of the 3-input odd-function block and one copy of the exclusive-OR block. This represents the fact that the designer has to design only one 3-input odd-function block and one exclusive-OR block and can use these blocks four times and eight times, respectively, in the 9-input odd-function circuit. In general, suppose that at various levels of the hierarchy, the blocks used are carefully defined in such a manner that many of them are identical. For these repeated blocks, only one design is necessary. This design can be used everywhere the block is required. The appearance of a block within a design is called an *instance* of the block and its use is called an *instantiation*. The block is *reusable* in the sense that it can be used in multiple places in the circuit design and, possibly, in the design of other circuits as well. This concept greatly reduces the design effort required for complex circuits. Note that, in the implementation of the circuit, sepa-rate hardware has to be provided for each instance of the block as represented in Figure 3-3(a). The reuse, as represented in Figure 3-3(b), is confined to the schemat-ics, not to the actual hardware implementation.

After completing a discussion of the design process, in Chapters 4 and 5, we focus on predefined, reusable blocks that typically lie at the lower levels of logic design hierarchies. These are blocks that provide basic functions used in digital design. They allow designers to do much of the design process above the primitive block level. We refer to these particular blocks as *functional blocks*. Thus, a functional block is a predefined collection of interconnected gates. Many of these functional blocks have been available for decades as medium-scale integrated (MSI) circuits that were interconnected to form larger circuits or systems. Similar blocks are now in computer-aided design tool libraries used for designing larger integrated circuits. These functional blocks provide a catalog of elementary digital components that are widely used in the design and implementation of integrated circuits for computers and digital systems.

Top-Down Design

Ideally, the design process is performed *top down*. This means that the circuit function is specified typically by text or a hardware description language (HDL), plus constraints on cost, performance, and reliability. At high levels of the design, the circuit is then repeatedly divided into blocks as necessary until the blocks are small enough to perform logic design. For manual logic design, the blocks may need to be further divided. In automated synthesis, the HDL description is converted to logic automatically. Then, for both manual design and automated synthesis, the logic is optimized and then mapped to the available primitive elements. In fact, reality departs significantly from this ideal view, particularly at the higher levels of the design. In order to obtain reusability and to make maximum use of predefined modules, it is often necessary to perform portions of the design *bottom up*. In addition, a particular circuit design obtained during the design process may violate one of the constraints in the initial specification. In this case, it is necessary to backtrack upward through the hierarchy until a level is reached at which the violation can be eliminated. A portion of the design is then revised at that level and the revisions are carried back downward through the hierarchy.

In this text, since reader familiarity with logic and computer design is probably limited, we need to build a ready set of functional blocks to provide direction in top-down design. Likewise, a sense of how to break up a circuit into blocks that can serve to guide the top-down approach also must be mastered. So the focus in much of the text will be on bottom-up rather than top-down design. To begin building the basis for top-down design, in Chapters 4 and 5, we focus our efforts on the design of frequently used functional blocks. In Chapters 7 and 10, we illustrate how larger circuits and systems are broken down into blocks and how these blocks are implemented with functional blocks. Finally, beginning with Chapter 11, we apply these ideas to look at design from more of a top-down perspective.

Computer-Aided Design

Designing complex systems and integrated circuits would not be feasible without the use of *computer-aided design* (*CAD*) tools. *Schematic capture* tools support the drawing of blocks and interconnections at all levels of the hierarchy. At the level of primitives and functional blocks, *libraries* of graphics symbols are provided. Schematic capture tools support the construction of a hierarchy by permitting the generation of symbols for hierarchical blocks and the replication of symbols for reuse.

The primitive blocks and the functional block symbols from libraries have associated models that allow the behavior and the timing of the hierarchical blocks and the entire circuit to be verified. This verification is performed by applying inputs to the blocks or circuit and using a *logic simulator* to determine the outputs. We will be illustrating logic simulation in a number of examples.

The primitive blocks from libraries can also have associated data, such as physical area information and delay parameters, that can be used by *logic synthesizers* to optimize designs being generated automatically from hardware description language specifications.

Hardware Description Languages

Thus far, we have mentioned hardware description languages only casually. In modern design, however, such languages have become crucial to the design process. Initially, we justify such languages by describing their uses. We will then briefly discuss VDHL and Verilog®, the most popular of these languages. Beginning in Chapter 4, we introduce both of these languages in detail, although, in any given course, we expect that only one of them will be covered.

Hardware description languages resemble programming languages, but are specifically oriented to describing hardware structures and behavior. They differ markedly from typical programming languages in that they represent extensive parallel operation whereas most programming languages represent serial operation. An obvious use for a hardware description language is to provide an alternative to schematics. When a language is used in this fashion, it is referred to as a *structural description* in which the language describes an interconnection of components. Such a structural description, referred to as a *netlist*, can be used as input to logic simulation just as a schematic is used. For this application, models for each of the primitive blocks are required. If an HDL is used, then these models can also be written in the HDL providing a more uniform, portable representation for simulation input.

The power of an HDL becomes more apparent, however, when it is used to represent more than just schematic information. It can represent Boolean equations, truth tables, and complex operations such as arithmetic. Thus, in top-down design, a very high-level description of an entire system can be precisely specified using an HDL. As a part of the design process, this high-level description can then be refined and partitioned into lower-level descriptions. Ultimately, a final description in terms of primitive components and functional blocks can be obtained as the result of the design process. Note that all of these descriptions can be simulated. Since they represent the same system in terms of function, but not necessarily timing, they should respond by giving the same logic values for the same applied inputs. This vital simulation property supports design verification and is one of the principal reasons for the use of HDLs.

A final major reason for the growth of the use of HDLs is logic synthesis. An HDL description of a system can be written at an intermediate level referred to as a register transfer language (RTL) level. A logic synthesis tool with an accompanying library of components can convert such a description into an interconnection of primitive components that implements the circuit. This replacement of the manual logic design process makes the design of complex logic much more efficient.

Currently, there are two HDLs, VHDL and Verilog, that are widely-used, standard hardware design languages. The language standards are defined, approved, and published by the Institute of Electrical and Electronics Engineers (IEEE). All implementations of these languages must obey their respective standard. This standardization gives HDLs another advantage over schematics. HDLs are portable across computer-aided design tools whereas schematic capture tools are typically unique to a particular vendor. In addition to the standard languages, a number of major companies have their own internal languages, often developed long before the standard languages and incorporating features unique to their particular products.

VHDL stands for VHSIC Hardware Description Language. VHDL was developed under contract for the U. S. Department of Defense as a part of the Very-High-Speed Integrated Circuits (VHSIC) program and subsequently became an IEEE standard language. Verilog® was developed by a company, Gateway Design Automation, which was bought by Cadence® Design Systems, Inc. For a while, Verilog was a proprietary language, but eventually became an IEEE standard language. In this text, we present brief introductions to both VHDL and Verilog. These portions of the text are optional and permit your instructor to cover one of the two languages or neither.

Regardless of the HDL, there is a typical procedure used in employing an HDL description as simulation input. These procedure steps are analysis, elaboration, and initialization, followed finally by the simulation. Analysis and elaboration are typically performed by a compiler similar to those for programming languages. *Analysis* checks the description for violations of the syntax and semantic rules for the HDL and produces an intermediate representation of the design. *Elaboration* traverses the design hierarchy represented by the description; in this process, the design hierarchy is flattened to an interconnection of modules that are described only by their behaviors. The end result of the analysis and elaboration performed by the compiler is a simulation model of the original HDL description. This model is then passed to the simulator for execution. *Initialization* sets all of the variables in the simulation model to specified or default values. *Simulation* executes the simulation model in either batch or interactive mode with inputs specified by the user.

Because of the ability to describe fairly complex hardware efficiently in an HDL, a special HDL structure called a *testbench* may be used. The testbench is a description that includes the design to be tested, typically referred to as the device under test (DUT). The testbench describes a collection of hardware and software functions that apply inputs to the DUT and analyze the outputs for correctness. This approach bypasses the need to provide separate inputs to the simulator and to analyze, often manually, the simulator outputs. Construction of a testbench provides a uniform verification mechanism that can be used at multiple levels in the top-down design process for verification of correct function of the design.

Logic Synthesis

As indicated earlier, the availability of logic synthesis tools is one of the driving forces behind the growing use of HDLs. Logic synthesis transforms an RTL description of a circuit in an HDL into an optimized netlist representing storage elements and combinational logic. Subsequently, this netlist may be transformed by using physical design tools into an actual integrated circuit layout. This layout serves as the basis for integrated circuit manufacture. The logic synthesis tool takes care of a large portion of the details of a design and allows exploration of the cost/performance trade-offs essential to advanced designs.

　　Figure 3-4 gives a simple high-level flow of the steps involved in logic synthesis. The user provides an HDL description of the circuit to be designed as well as various constraints or bounds on the design. Electrical constraints include allowable gate fanouts and output loading restrictions. Area and speed constraints direct the optimization steps of the synthesis. Area constraints typically give the maximum permissible area that a circuit is allowed to occupy within the integrated circuit. Alternatively, a general directive may be given which specifies that area is to be minimized. Speed constraints are typically maximum allowable values for propagation delay on various paths in the circuit. Alternatively, a general directive may be given to maximize speed. Area and speed both translate into the cost of a circuit. A

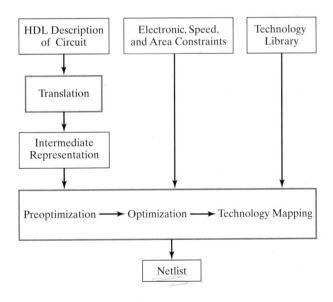

□ **FIGURE 3-4**
High-Level Flow for Logic Synthesis Tool

fast circuit will typically have larger area and thus cost more to manufacture. A circuit that need not operate fast can be optimized for area, and, relatively speaking, costs less to manufacture. In some sophisticated synthesis tools, power consumption can also be used as a constraint. Additional information used by a synthesis tool is a technology library that describes the logic elements available for use in the netlist as well as their delay and loading properties. The latter information is essential in meeting constraints and performing optimization.

The first major step in the synthesis process in Figure 3-4 is a translation of the HDL description into an intermediate form. The information in this representation may be an interconnection of generic gates and storage elements, not taken from a library of primitive blocks, called a *technology library*. It may also be in an alternate form that represents clusters of logic and the interconnections between the clusters.

The second major step in the synthesis process is optimization. A preoptimization step may be used to simplify the intermediate form. For example, logic that is identical in the intermediate form may be shared. Next is the optimization in which the intermediate form is processed to attempt to meet the constraints specified. Typically, two-level and multiple-level optimization are performed. Optimization is followed by *technology mapping* which replaces AND gates, OR gates, and inverters with gates from the technology library. In order to evaluate area and speed parameters associated with these gates, additional information from the technology library is used. In sophisticated synthesis tools, further optimization may be applied during technology mapping in order to improve the likelihood of meeting the constraints on the design. Optimization can be a very complex, time consuming process for large circuits. Many optimization passes may be necessary to achieve the desired results or to demonstrate that constraints are difficult, if not impossible, to meet. The designer may need to modify the constraints or the HDL in order to achieve a satisfactory design. Modification of the HDL may include manual design of some portions of the logic in order to achieve the design goals.

The output of the optimization/technology mapping processes is typically a netlist corresponding to a schematic diagram made up of storage elements, gates, and other combinational logic functional blocks. This output serves as input to physical design tools that physically place the logic elements and route the interconnections between them to produce the layout of the circuit for manufacture. In the case of programmable parts such as field-programmable gate arrays as discussed in Section 3-6, an analog to the physical design tools produces the binary information used to program the logic within the parts.

3-2 THE DESIGN SPACE

For a given design, there is typically a target implementation technology that specifies the primitive elements available and the properties of those elements. In addition, there is a set of constraints that applies to the design. This section deals with the potential primitive gate functions and properties and briefly discusses design constraints and the trade-offs that must be considered while attempting to meet the constraints.

Gate Properties

Digital circuits are constructed with integrated circuits. An integrated circuit (abbreviated IC) is a silicon semiconductor crystal, informally called a chip, containing the electronic components for the digital gates and storage elements. The various components are interconnected on the chip. The chip is mounted in a ceramic or plastic container, and connections are welded from the chip to the external pins to form the integrated circuit. The number of pins may range from 14 on a small IC package to several hundred on a large package. Each IC has a numeric designation printed on the surface of the package for identification. Each vendor publishes datasheets or a catalog that contains the description and all the necessary information about the ICs that it manufactures. Typically, this information is available on vendor websites.

Levels of Integration

As the technology of ICs has improved, the number of gates present in a single silicon chip has increased considerably. Customary reference to a package as being either a small-, medium-, large-, or very large-scale integrated device is used to differentiate between chips with just a few internal gates and those with thousands to tens of millions of gates.

    Small-scale integrated (SSI) devices contain several independent primitive gates in a single package. The inputs and outputs of the gates are connected directly to the pins in the package. The number of gates is usually less than 10 and is limited by the number of pins available on the IC.

    Medium-scale integrated (MSI) devices have approximately 10 to 100 gates in a single package. They usually perform specific elementary digital functions, such as the addition of four bits. MSI digital functions are similar to the functional blocks described in Chapters 4 and Chapter 5.

    Large-scale integrated (LSI) devices contain between 100 and a few thousand gates in a single package. They include digital systems such as small processors, small memories, and programmable modules.

    Very large-scale integrated (VLSI) devices contain several thousand to tens of millions of gates in a single package. Examples are complex microprocessor and digital signal processing chips. Because of their small transistor dimensions, high density, and comparatively low cost, VLSI devices have revolutionized digital system and computer design. VLSI technology gives designers the capability to create complex structures that previously were not economical to manufacture.

Circuit Technologies

Digital integrated circuits are classified not only by their function, but also by their specific implementation technology. Each technology has its own basic electronic device and circuit structures upon which more complex digital circuits and functions are developed. The specific electronic devices used in the construction of the basic circuits provide the name for the technology. Currently, silicon-based

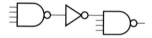

☐ **FIGURE 3-5**
Implementation of a 7-input NAND Gate
using NAND Gates with 4 or Fewer Inputs

Complementary Metal Oxide Semiconductor (CMOS) technology dominates due to its high circuit density, high performance, and low power consumption. Alternative technologies based on Gallium Arsenide (GaAs) and Silicon Germanium (SiGe) are used selectively for very high speed circuits.

Technology Parameters

For each specific implementation technology, there are details that differ in their electronic circuit design and circuit parameters. The most important parameters used to characterize an implementation technology follow:

Fan-in specifies the number of inputs available on a gate.

Fan-out specifies the number of standard loads driven by a gate output. *Maximum Fan-out* for an output specifies the fan-out that the output can drive without impairing gate performance. Standard loads may be defined in a variety of ways depending upon the technology.

Noise margin is the maximum external noise voltage superimposed on a normal input value that will not cause an undesirable change in the circuit output.

Cost for a gate specifies a measure of its contribution to the cost of the integrated circuit containing it.

Propagation delay is the time required for a change in value of a signal to propagate from input to output. The operating speed of a circuit is inversely related to the longest propagation delays through the gates of the circuit.

Power dissipation is the power drawn from the power supply and consumed by the gate. The power consumed is dissipated as heat, so the power dissipation must be considered in relation to the operating temperature and cooling requirements of the chip.

Although all of these parameters are important to the designer, further details on only selected parameters are provided next.

FAN-IN For high-speed technologies, *fan-in*, the number of inputs to a gate, is often restricted on gate primitives to no more than four or five. This is primarily due to electronic considerations related to gate speed. To build gates with larger fan-in, interconnected gates with lower fan-in are used during technology mapping. A mapping for a 7-input NAND gate illustrated in Figure 3-5 is made up of two 4-input NANDs and an inverter.

PROPAGATION DELAY The determination of propagation delay is illustrated in Figure 3-6. Three propagation delay parameters are defined. The *high-to-low*

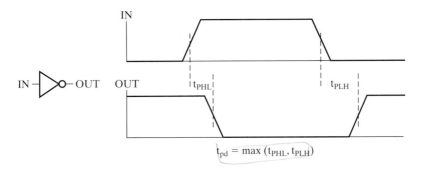

□ **FIGURE 3-6**
Propagation Delay for an Inverter

propagation time t_{PHL} is the delay measured from the reference voltage on the input IN to the reference voltage on the output OUT, with the output voltage going from H to L. The reference voltage we are using is the 50% point, halfway between the minimum and the maximum values of the voltage signals; other reference voltages may be used, depending on the logic family. The *low-to-high propagation time* t_{PLH} is the delay measured from the reference voltage on the input voltage IN to the reference voltage on the output voltage OUT, with the output voltage going from L to H. We define the *propagation delay* t_{pd} as the maximum of these two delays. The reason we have chosen the maximum value is that we will be most concerned with finding the longest time for a signal to propagate from inputs to outputs. Otherwise, the definitions given for t_{pd} may be inconsistent, depending on the use of the data. Manufacturers usually specify the maximum and typical values for both t_{PHL} and t_{PLH} or for t_{pd} for their products.

Two different models, transport delay and inertial delay, are employed in modeling gates during simulation. For *transport delay*, the change in an output in response to the change of an input occurs after a specified propagation delay. *Inertial delay* is similar to transport delay, except that if the input changes cause the output to change twice in an interval less than the *rejection time*, then the first of the two output changes does not occur. The rejection time is a specified value no larger than the propagation delay and is often equal to the propagation delay. An AND gate modeled with both a transport delay and an inertial delay is illustrated in Figure 3-7. To help visualize the delay behavior, we have also given the AND output with no delay. A colored bar on this waveform shows a 2 ns propagation delay time after each input change and a smaller black bar shows a rejection time of 1 ns. The output modeled with the transport delay is identical to that for no delay, except that it is shifted to the right by 2 ns. For the inertial delay, the waveform is likewise shifted. To define the waveform for the delayed output, we will call each change in a waveform an *edge*. To determine whether a particular edge appears in the ID output, it must be determined whether a second edge occurs in the ND output before the end of the rejection time for the edge in question, and whether the edge will result in a change in the ID output. Since edge b occurs before the end of the rejection time for edge a in the ND output, edge a does not

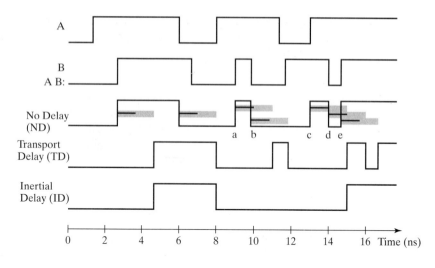

☐ **FIGURE 3-7**
Examples of Behavior of Transport and Inertial Delays

appear in the ID output. Since edge b does not change the state of ID, it is ignored. Since edge d occurs at the rejection time after edge c in the ND output, edge c does appear. Edge e, however, occurs within the rejection time after edge d, so edge d does not appear. Since edge c appeared and edge d did not appear, edge e does not cause a change.

FAN-OUT   One approach to measuring fan-out is the use of a *standard load*. Each input on a driven gate provides a load on the output of the driving gate which is measured in standard load units. For example, the input to a specific inverter can have load equal to 1.0 standard load. If a gate drives six such inverters, then the fan-out is equal to 6.0 standard loads. In addition, the output of a gate has a maximum load that it can drive called its maximum fan-out. The determination of the maximum fan-out is a function of the particular logic family. Our discussion will be restricted to CMOS, currently the most popular logic family. For CMOS gates, the loading of a gate output by the integrated circuit wiring and the inputs of other gates is modeled as a capacitance. This capacitive loading has no effect on the logic levels as loading often does for other families. Instead, the load on the output of a gate determines the time required for the output of the gate to change from L to H and from H to L. If the load on the output is increased, then this time, called the *transition time*, increases. Thus, the maximum fan-out for a gate is the number of standard loads of capacitance that can be driven with the transition time no greater than its maximum allowable value. For example, a gate with a maximum fan-out of 8 standard loads could drive up to 8 inverters that present 1.0 standard load on each on their inputs.

Because it represents capacitance driven, the actual fan-out of the gate, in terms of standard loads, also affects the propagation delays of the gate. Thus, a simple expression for propagation delay can be given by a formula or table that considers a fixed delay plus a delay per standard load times the number of standard loads driven, as shown in the next example.

EXAMPLE 3-1   Calculation of Gate Delay Based on Fan-Out

A 4-input NAND gate output is attached to the inputs of the following gates with the given number of standard loads representing their inputs:

> 4-input NOR gate - 0.80 standard load
> 3-input NAND gate - 1.00 standard load, and
> inverter - 1.00 standard load.

The formula for the delay of the 4-input NAND gate is

$$t_{pd} = 0.07 + 0.021 \times SL \text{ ns}$$

where SL is the sum of the standard loads driven by the gate.

Ignoring the wiring delay, the delay projected for the NAND gate as loaded is

$$t_{pd} = 0.07 + 0.021 \times (0.80 + 1.00 + 1.00) = 0.129 \text{ ns}$$

In modern high-speed circuits, the portion of the gate delay due to wiring capacitance is often significant. While ignoring such delay is unwise, it is difficult to evaluate since it depends on the layout of the wires in the integrated circuit. Nevertheless, since we do not have this information or a method to obtain a good estimate of it, we ignore this delay component here.   ■

Both fan-in and fan-out must be dealt with in the technology mapping step of the design process. Gates with fan-ins larger than those available for technology mapping can be implemented with multiple gates. Gates with fan-outs that either exceed their maximum allowable fan-out or have too high a delay need to be replaced with multiple gates or have buffers added at their outputs.

COST   For integrated circuits, the cost of a primitive gate is usually based on the area occupied by the layout cell for the circuit. The layout cell area is proportional to the size of the transistors and the wiring in the gate layout. Ignoring the wiring area, the area of the gate is proportional to the number of transistors in the gate, which in turn is usually proportional to the gate input count. If the actual area of the layout is known, then a normalized value of this area provides a more accurate estimation of cost than gate input count.

Positive and Negative Logic

Excluding transitions, the binary signals at the inputs and outputs of any gate have one of two values: H or L. One value represents logic 1 and the other logic 0. There are two different assignments of signal levels to logic values, as shown in Figure 3-8. Choosing the high level H to represent logic 1 defines a *positive-logic* system. Choosing the low level L to represent logic 1 defines a *negative-logic* system. The terms "positive" and "negative" are somewhat misleading, since both signals may be positive voltages or both may be negative voltages. It is not the actual signal values

Signal value	Logic value	Signal value	Logic value
H	1	H	0
L	0	L	1
(a) Positive logic		(b) Negative logic	

□ **FIGURE 3-8**

Signal Assignment and Logic Polarity

that determine the type of logic, but rather the assignment of logic values to the *relative amplitudes* of the two signal ranges.

Integrated circuit data sheets define digital gates in terms of both logic values and signal values H and L. If H and L are used, it is up to the user to decide on a positive or negative logic assignment. Consider, for example, the truth table in Figure 3-9(a). This table is given in a data book for the CMOS gate shown in Figure 3-9(b). The table specifies the physical behavior of the gate when H is 5 volts and L is 0 volts. The truth table of Figure 3-9(c) assumes positive logic, with 1 assigned to H and 0 assigned to L. The table is the same as the truth table for the AND operation. The graphics symbol for a positive-logic AND gate is shown in Figure 3-9(d).

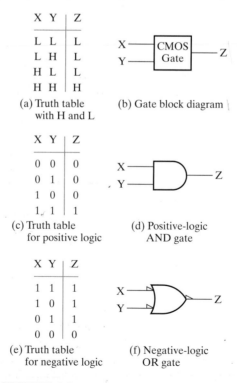

X	Y	Z
L	L	L
L	H	L
H	L	L
H	H	H

(a) Truth table with H and L

(b) Gate block diagram

X	Y	Z
0	0	0
0	1	0
1	0	0
1	1	1

(c) Truth table for positive logic

(d) Positive-logic AND gate

X	Y	Z
1	1	1
1	0	1
0	1	1
0	0	0

(e) Truth table for negative logic

(f) Negative-logic OR gate

□ **FIGURE 3-9**

Demonstration of Positive and Negative Logic

Now consider the negative logic assignment for the same physical gate, with 1 assigned to L and 0 assigned to H. The result is the truth table of Figure 3-9(e). This table represents the OR operation with the table rows reversed. The graphic symbol for the negative-logic OR gate is shown in Figure 3-9(f). The small triangles on the inputs and output are *polarity indicators*. The presence of a polarity indicator on an input or output signifies that negative logic is assumed for the corresponding signal. Thus, the same physical gate can operate as either a positive-logic AND gate or as a negative-logic OR gate.

The conversion from positive logic to negative logic and vice versa is an operation that changes 1's to 0's and 0's to 1's at both the inputs and the output of a gate. Since this interchange of 1's and 0's is a part of taking the dual, the conversion operation produces the dual of the gate function. Thus, the change of all gate inputs and outputs from one polarity to the other results in taking the dual of the function, with all AND operations (or graphics symbols) converted to OR operations (or graphics symbols) and vice versa. In addition, one must not forget to include the polarity indicator in the graphics symbols when negative logic is assumed, and one must also recognize that the polarity definitions for circuit inputs and circuit outputs have been changed. In this book we do not use negative logic, but assume that all gates operate with a positive-logic assignment.

Design Trade-Offs

Previously, it has been shown that there is a relationship between the actual fan-out of a gate and its propagation delay. Higher fan-out increases the propagation delay. For example, a circuit has a gate G having a fan-out equal to 16.00 standard loads. The delay through this gate, including the delay component based on the standard loads driven, is 0.406 ns. To reduce this delay, a buffer is added to the output of gate G and the 16.00 standard loads are connected to the buffer output. The output of gate G now drives just the buffer. The delay for this combination of the gate and the buffer in series is only 0.323 ns, giving a delay reduction of over 20%. Gate G alone has a cost of 2.0 while gate G plus the buffer have a cost of 3.0. These two circuits illustrate a cost/performance trade-off, the most common of the trade-offs with which a designer must deal. While this example uses two simple circuits, cost/performance trade-offs can be made at much higher levels in a system design. These trade-offs may influence functional specifications for the systems and the implementation approach used for system functions.

Continuing with the simple example, the designer has two choices. If gate G alone is fast enough, it should be selected because of its lower cost. On the other hand, if it is not fast enough, then gate G plus the buffer should be selected. In order to make this trade-off, we need to have one or more constraints on the design of the circuit. Suppose that the constraint is that the maximum input-to-output delay $t_{pd,max} = 0.35$ ns. Since gate G alone does not meet this constraint, the buffer must be added. Alternatively, suppose that the constraint is that the maximum number of area units for the circuit is 2.5. Since gate G plus the buffer do not meet this constraint, gate G alone must be chosen. Finally, suppose that both the delay and area constraints given are expected to be met. Then neither solution is satisfactory. A new design must be found that meets the constraints or the constraints must be relaxed so that one of the two circuits meets them.

A sample list of possible circuit constraints is as follows:

Maximum input-to-output delay
Maximum area units
Maximum power dissipation
Maximum standard loads presented to circuits driving the circuit inputs
Minimum standard load and drive provided by the circuit outputs

Typically, not all constraints on this list are specified for a given circuit. For example, instead of having both delay and cost constraints, the delay constraint may be fixed with the cost to be minimized while meeting all other constraints.

3-3 DESIGN PROCEDURE

The design of a combinational circuit starts from the specification of the problem and culminates in a logic diagram or netlist that describes a logic diagram. The procedure involves the following steps:

1. **Specification:** Write a specification for the circuit if one is not already available.
2. **Formulation:** Derive the truth table or initial Boolean equations that define the required relationships between inputs and outputs.
3. **Optimization:** Apply two-level and multiple-level optimization. Draw a logic diagram or provide a netlist for the resulting circuit using ANDs, ORs, and inverters.
4. **Technology Mapping:** Transform the logic diagram or netlist to a new diagram or netlist using the available implementation technology.
5. **Verification:** Verify the correctness of the final design.

The specification can take a variety of forms such as text or an HDL description and should include the respective symbols or names for the inputs and outputs. Formulation converts the specification into forms that can be optimized. These forms are typically truth tables or Boolean expressions. It is important that verbal specifications be interpreted correctly when formulating truth tables or expressions. Often the specifications are incomplete, and any wrong interpretation may result in an incorrect truth table or expression. Optimization can be performed by any of a number available methods, such as algebraic manipulation, the K-map method, or computer-based simplification programs. In a particular application, specific criteria serve as a guide for choosing the optimization method. A practical design must consider constraints such as the cost of the gates used, maximum allowable propagation time of a signal through the circuit, and limitations on the fan-out of each gate. This is complicated by the fact that gate costs, gate delays, and fan-out limits are not known until the technology mapping stage. As a consequence, it is difficult to make a general statement about what constitutes an acceptable end result for optimization. In many cases, the optimization begins by satisfying an elementary objective, such as producing the simplified Boolean expressions in a standard form for each output. The next step is multiple level opti-mization with terms shared between multiple outputs. In more sophisticated syn-thesis tools, optimization and technology mapping may be interspersed to improve the likelihood of meeting constraints. It may be necessary to repeat optimization and technology mapping multiple times to meet the specified constraints.

The remainder of this chapter illustrates the design procedure by using three examples. In the rest of this section, we perform the first three steps of design, specification, formulation, and optimization. We then consider implementation technologies and the final two steps in separate sections.

The first two example specifications are for a class of circuits called *code con-verters,* which translate information from one binary code to another. The inputs to the circuit are the bit combinations specified by the first code, and the outputs gener-ate the corresponding bit combination of the second code. The combinational circuit performs the transformation from one code to the other. The first code converter example converts the BCD code to the excess-3 code for the decimal digits. The other converts the BCD code to the seven signals required to drive a seven-segment light-emitting diode (LED) display. The third example is the design of a 4-bit equal-ity comparator that represents a circuit having a large number of inputs.

EXAMPLE 3-2   Design of a BCD–to–Excess-3 Code Converter

□ **TABLE 3-1**
Truth Table for Code Converter Example

Decimal Digit	Input BCD				Output Excess-3			
	A	B	C	D	W	X	Y	Z
0	0	0	0	0	0	0	1	1
1	0	0	0	1	0	1	0	0
2	0	0	1	0	0	1	0	1
3	0	0	1	1	0	1	1	0
4	0	1	0	0	0	1	1	1
5	0	1	0	1	1	0	0	0
6	0	1	1	0	1	0	0	1
7	0	1	1	1	1	0	1	0
8	1	0	0	0	1	0	1	1
9	1	0	0	1	1	1	0	0

SPECIFICATION: The *excess-3 code* for a decimal digit is the binary combination corresponding to the decimal digit plus 3. For example, the excess-3 code for deci-mal digit 5 is the binary combination for $5 + 3 = 8$, which is 1000. The excess-3 code has desirable properties with respect to implementing decimal subtraction. Each BCD digit is four bits with the bits, from most significant to least significant, labeled A, B, C, D. Each excess-3 digit is four bits, with the bits, from most signifi-cant to least significant, labeled W, X, Y, Z.

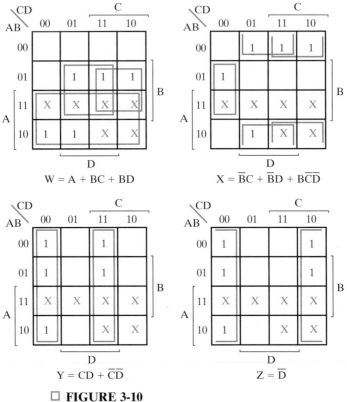

☐ **FIGURE 3-10**
Maps for BCD–to–Excess-3 Code Converter

FORMULATION: The excess-3 code word is easily obtained from a BCD code word by adding binary 0011 (3) to it. The resulting truth table relating the input and output variables is shown in Table 3-1. Note that the four BCD input variables may have 16 bit combinations, but only 10 are listed in the truth table. The six combinations 1010 through 1111 are not listed under the inputs, since these combinations have no meaning in the BCD code, and we can assume that they will never occur. Hence, for these input combinations, it does not matter what binary values we assign to the excess-3 outputs, and therefore, we can treat them as don't-care conditions.

OPTIMIZATION: Since this is a four-variable function, we use the K-maps in Figure 3-10 for the initial optimization of the four output functions. The maps are plotted to obtain simplified sum-of-products Boolean expressions for the outputs. Each of the four maps represents one of the outputs of the circuit as a function of the four inputs. The 1's in the maps are obtained directly from the truth table output columns. For example, the column under output W has 1's for minterms 5, 6, 7, 8, and 9. Therefore, the map for W must have 1's in the squares corresponding to these minterms. The six don't-care minterms, 10 through 15, are each marked with an X in all the maps. The optimized functions are listed in sum-of-products form under the map for each output variable.

The two-level AND-OR logic diagram for the circuit can be obtained directly from the Boolean expressions derived from the maps. We apply multiple-level optimization as a second optimization step to determine if the gate input count, which is currently 26 (including inverters), can be reduced. In this optimization, we consider sharing subexpressions between the four output expressions. The following manipulation illustrates optimization with multiple-output circuits implemented with three levels of gates:

$$T_1 = C + D$$

$$W = A + BC + BD = A + BT_1$$

$$X = \overline{B}C + \overline{B}D + B\overline{C}\,\overline{D} = \overline{B}T_1 + B\overline{C}\overline{D}$$

$$Y = CD + \overline{C}\,\overline{D}$$

$$Z = \overline{D}$$

The manipulation allows the gate producing $C + D$ to be shared by the logic for W and X, and reduced the gate input count from 26 to 22. This optimized result is viewed as being adequate and gives the logic diagram in Figure 3-11. ▪

EXAMPLE 3-3   Design of a BCD–to–Seven-Segment Decoder

SPECIFICATION: Digital readouts found in many consumer electronic products such as alarm clocks often use Light Emitting Diodes (LEDs). Each digit of the readout

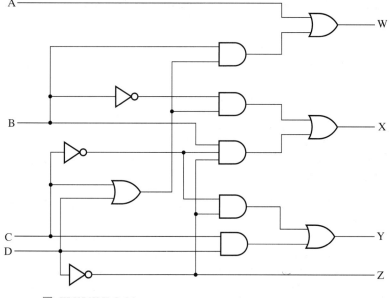

☐ **FIGURE 3-11**
Logic Diagram of BCD–to–Excess-3 Code Converter

(a) Segment designation   (b) Numeric designation for display

☐ **FIGURE 3-12**
Seven-Segment Display

is formed from seven LED segments, Each segment can be illuminated by a digital signal. A BCD–to–seven-segment decoder is a combinational circuit that accepts a decimal digit in BCD and generates the appropriate outputs for the segments of the display for that decimal digit. The seven outputs of the decoder (a, b, c, d, e, f, g) select the corresponding segments in the display, as shown in Figure 3-12(a). The numeric designations chosen to represent the decimal digits are shown in Figure 3-12(b). The BCD–to–seven-segment decoder has four inputs, A, B, C, and D for the BCD digit and seven outputs, a through g, for controlling the segments.

FORMULATION: The truth table of the combinational circuit is listed in Table 3-2. On the basis of Figure 3-12(b), each BCD digit illuminates the proper segments for the decimal display. For example, BCD 0011 corresponds to decimal 3, which is displayed as segments a, b, c, d, and g. The truth table assumes that a logic 1 signal illuminates the segment and a logic 0 signal turns the segment off. Some seven-segment displays operate in reverse fashion and are illuminated by a logic 0 signal. For these displays, the seven outputs must be complemented. The six binary combinations 1010 through 1111 have no meaning in BCD. In the previous example, we assigned these combinations to don't-care conditions. If we do the same here, the design will

☐ **TABLE 3-2**
Truth Table for BCD–to–Seven-Segment Decoder

BCD Input				Seven-Segment Decoder						
A	B	C	D	a	b	c	d	e	f	g
0	0	0	0	1	1	1	1	1	1	0
0	0	0	1	0	1	1	0	0	0	0
0	0	1	0	1	1	0	1	1	0	1
0	0	1	1	1	1	1	1	0	0	1
0	1	0	0	0	1	1	0	0	1	1
0	1	0	1	1	0	1	1	0	1	1
0	1	1	0	1	0	1	1	1	1	1
0	1	1	1	1	1	1	0	0	0	0
1	0	0	0	1	1	1	1	1	1	1
1	0	0	1	1	1	1	1	0	1	1
All other inputs				0	0	0	0	0	0	0

most likely produce some arbitrary and meaningless displays for the unused combinations. As long as these combinations do not occur, we can use that approach to reduce the complexity of the converter. A safer choice, turning off all the segments when any one of the unused input combinations occurs, avoids any spurious displays if any of the combinations occurs, but increases the converter complexity. This choice can be accomplished by assigning all 0's to minterms 10 through 15.

OPTIMIZATION: The information from the truth table can be transferred into seven K-maps from which the initial optimized output functions can be derived. The plotting of the seven functions in map form is left as an exercise. One possible way of simplifying the seven functions results in the following Boolean functions:

$$a = \overline{A}C + \overline{A}BD + \overline{B}\,\overline{C}D + A\overline{B}\,\overline{C}$$

$$b = \overline{A}\,\overline{B} + \overline{A}\,\overline{C}\,\overline{D} + \overline{A}CD + A\overline{B}\,\overline{C}$$

$$c = \overline{A}B + \overline{A}D + \overline{B}\,\overline{C}D + A\overline{B}\,\overline{C}$$

$$d = \overline{A}C\overline{D} + \overline{A}\,\overline{B}C + \overline{B}\,\overline{C}\overline{D} + A\overline{B}\,\overline{C} + \overline{A}BC\overline{D}$$

$$e = \overline{A}C\overline{D} + \overline{B}\,\overline{C}\overline{D}$$

$$f = \overline{A}B\overline{C} + \overline{A}\,\overline{C}\,\overline{D} + \overline{A}B\overline{D} + A\overline{B}\,\overline{C}$$

$$g = \overline{A}C\overline{D} + \overline{A}\,\overline{B}C + \overline{A}B\overline{C} + A\overline{B}\,\overline{C}$$

Independent implementation of these seven functions requires 27 AND gates and 7 OR gates. However, by sharing the six product terms common to the different output expressions, the number of AND gates can be reduced to 14 along with a substantial savings in gate input count. For example, the term $\overline{B}\,\overline{C}\,\overline{D}$ occurs in a, c, d, and e. The output of the AND gate that implements this product term goes directly to the inputs of the OR gates in all four functions. For this function, we stop optimization with the two-level circuit and shared AND gates, realizing that it might be possible to reduce the gate input count even further by applying multiple level optimization. ■

The BCD–to–seven-segment decoder is called a decoder by most manufacturers of integrated circuits because it decodes a binary code for a decimal digit. However, it is actually a code converter that converts a four-bit decimal code to a seven-bit code. The word "decoder" is usually reserved for another type of circuit, presented in the next chapter.

In general, the total number of gates can be reduced in a multiple-output combinational circuit by using common terms of the output functions. The maps of the output functions may help find the common terms by finding identical implicants from two or more maps. Some of the common terms may not be prime implicants of the individual functions. The designer must be inventive and combine squares in the maps in such a way as to create common terms. This can be done more formally by using a procedure for simplifying multiple-output functions. The prime implicants are defined not only for each individual function, but also for all possible combinations of the output functions. These prime implicants are formed by using the AND operator on every possible nonempty subset of the output functions and finding the prime implicants of each of the results. Using this entire set of prime implicants, a formal selection process can be used to find the optimum two-level multiple output circuit. Such a procedure is implemented in various forms in logic simplification software within logic synthesis tools and is the method used to obtain the equations in Example 3-3.

EXAMPLE 3-4   Design of a 4-bit Equality Comparator

SPECIFICATION: The inputs to the circuit consist of two vectors: A(3:0) and B(3:0). Vector A consists of four bits, A(3), A(2), A(1), and A(0), with A(3) as the most significant bit. Vector B has a similar description with B replaced by A. The output of the circuit is a single bit variable E. Output E is equal to 1 if A and B are equal and equal to 0 if A and B are unequal.

FORMULATION: Since this circuit has eight inputs, use of a truth table for formulation is impractical. In order for A and B to be equal, the bit values in each of the respective positions, 3 down to 0, of A and B must be equal. If all of the bit positions for A and B contain equal values in every position, then E = 1; otherwise, E = 0.

OPTIMIZATION: For this circuit, we use intuition to immediately develop a multiple level circuit using hierarchy. Since comparison of a bit from A and the corresponding bit from B must be done in each of the bit positions, we can decompose the problem into four 1-bit comparison circuits and an additional circuit that combines the four comparison circuit outputs to obtain E. For bit position i, we define the circuit output E_i to be 0 if A_i and B_i have the same values and $E_i = 1$ if A_i and B_i have different values. This circuit can be described by the equation

$$E_i = \overline{A}_i B_i + A_i \overline{B}_i$$

which has the circuit diagram shown in Figure 3-13(a). By using hierarchy and reuse, we can employ four copies of this circuit, one for each of the four bits of A and B. Output $E = 1$ only if all of the E_i values are 0. This can be described by the equation

$$E = \overline{E_0 + E_1 + E_2 + E_3}$$

and has the diagram given in Figure 3-13(b). Both of the circuits given are optimum two-level circuits. The overall circuit can be described hierarchically by the diagram in Figure 3-13(c). ∎

3-4   TECHNOLOGY MAPPING

There are three primary ways of designing VLSI circuits. In *full custom design*, an entire design of the chip, down to the smallest detail of the layout, is performed.

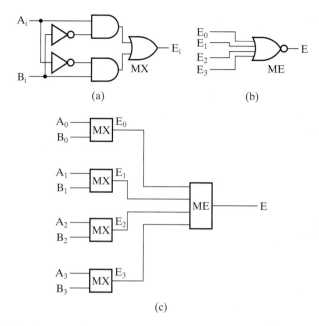

□ **FIGURE 3-13**
Hierarchical Diagram for a 4-bit Equality Comparator

Since this process is very expensive, custom design can be justified only for dense, fast ICs that are likely to be sold in sizable quantities.

A closely related technique is *standard cell design*, in which large parts of the design have been performed ahead of time or, possibly, used in previous designs. The predesigned parts are connected to form the IC design. This intermediate-cost methodology gives lower density and lower performance than full custom design.

The third approach to VLSI design is the use of a *gate array*. A gate array uses a rectangular pattern of gates fabricated in silicon. This pattern is repeated thousands of times, so that the entire chip contains identical gates. Depending on the technology used, pattern arrays of 1000 to millions of gates can be fabricated within a single IC. The application of a gate array requires that the design specify how the gates are interconnected and how the interconnections are routed. Many steps of the fabrication process are common and independent of the final logic function. These steps are economical, since they can be used for numerous different designs. In order to customize the gate array to the particular design, additional fabrication steps are required to interconnect the gates. Due to the commonality of fabrication steps and ability to share the results of these steps with many different designs, this is the lowest cost method among the fixed implementation technologies.

For standard cell and gate array technologies, circuits are constructed by interconnecting cells. The collection of cells available for a given implementation technology is called a *cell library*. In order to design in terms of a cell library, it is necessary to characterize each of the cells (i.e., provide a detailed specification of the cells for use by the designer). A library of characterized cells provides a foundation for the technology mapping of circuits. Coupled with the library is a technology mapping procedure. In this section, we consider technology mapping procedures for cell libraries consisting of (1) single gate types such as NAND gates, and (2) multiple gate types. Technology mapping may focus on a number of the dimensions of the design space, particularly on cost and performance. For simplicity, our procedures focus only on optimizing cost. Further, these procedures are rudimentary versions of technology mapping algorithms used in computer-aided design tools and are suitable for manual application to only the simplest of circuits. Nevertheless, they give us some insight into how a design using AND gates, OR gates, and inverters can be transformed into cost-effective designs using cell types supported by available implementation technologies.

Cell Specification

Specifications for cells used in standard cell and gate array designs typically have many components. Typical components include the following:

1. A schematic or logic diagram for the function of the cell.
2. A specification of the area the cell occupies, often normalized to the area of a small cell such as that of a minimum area inverter.
3. The input loading, in standard loads, that each input of a cell presents to the output driving it.
4. Delays from each input of a cell to each output of a cell (if a path from the input to output exists), including the effect of the number of standard loads driven by the output.
5. One or more templates for the cell for use in performing technology mapping.
6. One or more HDL models for the cell.

If the tools used provide automated layout, then the following additional components are also included in the specification:

7. An integrated circuit layout for the cell.
8. A floorplan layout showing the locations of the inputs, outputs and power and ground connections for the cell for use during the cell interconnection process.

The first five components listed are included in a simple technology library of cells in the next subsection. Some of these components are discussed in more detail.

Libraries

The cells for a particular design technology are organized into one or more libraries. A *library* is a collection of cell specifications. A circuit that initially consists of

AND, OR and NOT gates is converted by technology mapping to one that uses only cells from the applicable libraries. A very small technology library is described in Table 3-3. This library contains primitive inverting gates with fan-ins up to four and a single AOI circuit.

The first column of the table contains a descriptive name for the cell and the second column contains the cell schematic. The third column contains the area of the cell normalized to the area of a minimum inverter. Area can be used as a very simple measure of the cost of the cell. The next column gives the typical load that a cell input places on the gate driving it. The load values are normalized to a quantity called a standard load which in this case is the capacitance presented to the driving circuit by the input of an inverter. In the case of the cells given, the input loads are almost all the same. The fifth column gives a simple linear equation for calculating the typical input-to-output delay for the cell. The variable SL is the sum of all of the standard loads presented by the inputs of cells driven by the cell output. It may also contain an estimate, in standard loads, of the capacitance of the wiring connecting the cell output to the inputs of other cells. This equation illustrates the notion that cell delays consist of some fixed delay, plus a delay that is dependent upon the capacitance loading of the cell as represented by SL. Cell delay calculation is illustrated in Example 3-5.

EXAMPLE 3-5 Calculation of Cell Delay

This example illustrates the effect of loading on cell delay. A 2NAND output drives the following cells: an inverter, a 4NAND, and a 4NOR. The sum of the standard loads in this case is

$$SL = 1.00 + 0.95 + 0.80 = 2.75$$

With this value, the delay of the 2NAND driving the cells specified is

$$t_p = 0.05 + 0.014 \times 2.75 = 0.089 \text{ ns}$$

The final column of the table gives templates for the cell function that use only basic functions as components. In this case, the basic functions are a 2-input NAND gate and an inverter. Use of these basic function templates provides a way of representing each cell function in a "standard" form. As illustrated by the 4-input NAND and NOR cells, the basic function template for a cell is not necessarily unique. It should be noted that these diagrams represent only a netlist, not actual location, orientation, or interconnect layout. For example, consider the template for the 3NAND. If the left NAND and the following inverter were connected to the top input of the right NAND, instead of to its bottom input, the template would be unchanged. The value of these templates will become apparent in the next section on mapping techniques.

Mapping Techniques

In this subsection, we consider the mapping process for fixed cell technologies. A convenient way to implement a Boolean function with NAND gates is to obtain the optimized Boolean function in terms of the Boolean operators AND, OR, and

☐ **TABLE 3-3**
Example Cell Library for Technology Mapping

Cell Name	Cell Schematic	Normalized Area	Typical Input Load	Typical Input-to-Output Delay	Basic Function Templates
Inverter		1.00	1.00	$0.04 + 0.012 \times SL$	
2NAND		1.25	1.00	$0.05 + 0.014 \times SL$	
3NAND		1.50	1.00	$0.06 + 0.017 \times SL$	
4NAND		2.00	0.95	$0.07 + 0.021 \times SL$	
2NOR		1.25	1.00	$0.06 + 0.018 \times SL$	
3NOR		2.00	0.95	$0.15 + 0.012 \times SL$	
4NOR		3.25	0.80	$0.17 + 0.012 \times SL$	
2-2 AOI		2.25	0.95	$0.07 + 0.019 \times SL$	

NOT and then map the function to NAND logic. The conversion of an algebraic expression from AND, OR, and NOT to NAND can be done by a simple procedure that changes AND-OR logic diagrams to NAND logic diagrams. The same conversion applies for NOR gates.

Given an optimized circuit that consists of AND gates, OR gates, and inverters, the following procedure produces a circuit using either NAND (or NOR) gates with unrestricted gate fan-in:

1. Replace each AND and OR gate with the NAND (NOR) gate and inverter equivalent circuits shown in Figure 3-14 (a) and (b).
2. Cancel all inverter pairs.
3. Without changing the logic function, (a) "push" all inverters lying between (i) either a circuit input or a driving NAND(NOR) gate output and (ii) the

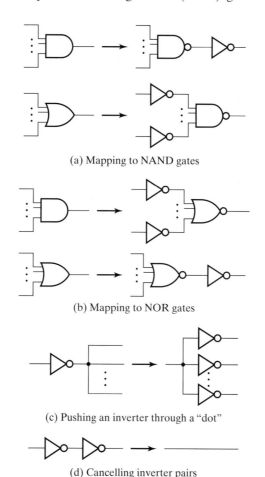

(a) Mapping to NAND gates

(b) Mapping to NOR gates

(c) Pushing an inverter through a "dot"

(d) Cancelling inverter pairs

□ **FIGURE 3-14**
Mapping of AND Gates, OR Gates and Inverters to
NAND gates, NOR gates, and Inverters

driven NAND(NOR) gate inputs toward the driven NAND(NOR) gate inputs. Cancel pairs of inverters in series whenever possible during this step. (b) Replace inverters in parallel with a single inverter that drives all of the outputs of the parallel inverters. (c) Repeat (a) and (b) until there is at most one inverter between the circuit input or driving NAND(NOR) gate output and the attached NAND(NOR) gate inputs.

In Figure 3-14 (c), the rule for pushing an inverter through a "dot" is given. The inverter on the input line to the dot is replaced with inverters on each of the output lines from the dot. The cancellation of pairs of inverters in Figure 3-14(d) is based on the Boolean algebraic identity

$$\overline{\overline{X}} = X$$

The next example illustrates this procedure for NAND gates.

EXAMPLE 3-6   Implementation with NAND Gates

Implement the following optimized function with NAND gates:

$$F = AB + \overline{(AB)}C + (\overline{AB})\overline{D} + E$$

The AND, OR, inverter implementation is given in Figure 3-15(a). In Figure 3-15 (b), step 1 of the procedure has been applied replacing each AND gate and OR gate with its equivalent circuit using NAND gates and inverters from Figure 3-14(a). Labels appear on dots and inverters to assist in the explanation. In step 2, the

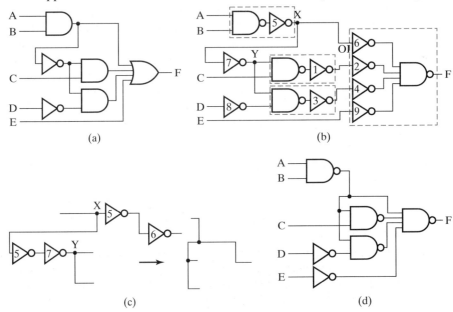

(a)

(b)

(c)

(d)

☐ **FIGURE 3-15**
Solution to Example 3-6

inverter pairs (1, 2) and (3, 4), cancel, giving direct connections between the corresponding NAND gates in Figure 3-15(d). As shown in Figure 3-15(c), inverter 5 is pushed through X and cancels with inverters 6 and 7, respectively. This gives direct connections between the corresponding NAND gates in Figure 3-15(d). No further steps can be applied since inverters 8 and 9 cannot be paired with other inverters, and must remain in the final mapped circuit in Figure 3-15(d). The next example illustrates this procedure for NOR gates. ■

EXAMPLE 3-7   Implementation with NOR Gates

Implement the same optimized Boolean function used in Example 3-7 with NOR gates:

$$F = AB + (\overline{AB})C + (\overline{AB})\overline{D} + E$$

The AND, OR, inverter implementation is given in Figure 3-16(a). In Figure 3-16(b), step 1 of the procedure has been applied replacing each AND gate and OR gate with its equivalent circuit using NOR gates and inverters from Figure 3-14(b). Labels appear on dots and inverters to assist in the explanation. In step 2, inverter 1 can be pushed through dot X to cancel with inverters 2 and 3, respectively. The pair of inverters on the D input line cancel as well. The single inverters on input lines A, B, and C and output line F must remain, giving the final mapped circuit that appears in Figure 3-16(c). ■

In Example 3-6, the gate input cost of the mapped circuit is 12, and, in Example 3-7, the gate input cost is 14, so the NAND implementation is less costly. Also, the NAND implementation involves a maximum of three gates in series while the NOR implementation has a maximum of five gates in series. Due to the longer series of gates in the NOR circuit, the maximum delay from an input change to a corresponding output change is likely to be longer.

In the preceding procedure and examples, the mapping target consisted of a single gate type, either NAND gates or NOR gates. The following procedure handles multiple gate types:

1. Replace each AND and OR gate with an optimum equivalent circuit consisting only of 2-input NAND gates and inverters.

2. In each line in the circuit attached to a circuit input, a NAND gate input, a NAND gate output, or a circuit output in which no inverter appears, insert a serial pair of inverters.

3. Perform a replacement of connections of NAND gates and inverters by the available library cells such that the gate input cost which results within fan-out free subcircuits is optimized. A *fan-out free subcircuit* is a circuit in which each gate output drives a single gate input. (This step is not covered here in detail due to its complexity, but is provided on the course website with an example. The templates shown in the right column of Table 3-3 are used to match connections of NAND gates and inverters to available library cells.)

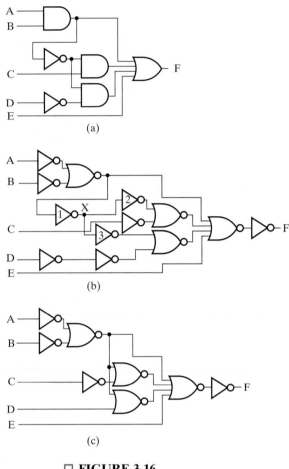

☐ **FIGURE 3-16**
Solution to Example 3-7

4. (a) Without changing the logic function, "push" all inverters, lying between (i) a circuit input or a driving gate output and (ii) the driven gate inputs, toward the driven gate inputs. Cancel pairs of inverters in series whenever possible during this step. (b) Replace inverters in parallel with a single inverter that drives all of the outputs of the parallel inverters. (c) Repeat (a) and (b) until there is at most one inverter between the circuit input or driving gate output and the attached driven gate inputs.

This procedure is one of the foundations for technology mapping in commercial synthesis tools. The intermediate replacement of the initial circuit gates with only 2-input NAND gates and inverters breaks the circuit up into small pieces in order to provide the maximum flexibility in mapping cells to achieve an optimized result. Example 3-8 shows an implementation approach using a small cell library.

EXAMPLE 3-8 Implementation with a Small Cell Library

Implement the same optimized Boolean function used in Examples 3-6 and 3-7

$$F = AB + (\overline{AB})C + (\overline{AB})\overline{D} + E$$

with a cell library containing a 2-input NAND gate, 3-input NAND gate, a 2-input NOR gate, and an inverter. The AND, OR, inverter implementation is given in Figure 3-17(a). In Figure 3-17(b), steps 1 and 2 of the procedure have been applied. Each AND gate and each OR gate has been replaced with its equivalent circuit made up of 2-input NAND gates and inverters. Pairs of inverters have been added to the internal lines without inverters. Due to lack of space, the pairs of inverters on the inputs and outputs are not shown. Application of step 3 results in the mapping to the cells from the cell library shown in Figure 3-17(c). The blue outlines enclose connections of NAND gates and inverters, each of which is to be replaced by an available cell using the templates in Table 3-3. In this case, all of the available cells have been used at least once. Application of step 4 cancels out three of the inverters, giving the final mapped circuit in Figure 3-17(d). ■

The solution for Example 3-8 has a gate input cost of 12, compared to costs of 12 and 14 for Examples 3-6 and 3-7, respectively. Although the cost in Examples 3-6 and 3-8 are identical, it should be noted that the cell libraries differ. In particular, Example 3-6 benefits from the use of a 4-input NAND gate that is unavailable in Example 3-8. Without this cell, the solution would cost two additional gate inputs. Thus, the use of a more diverse cell library has provided a cost benefit.

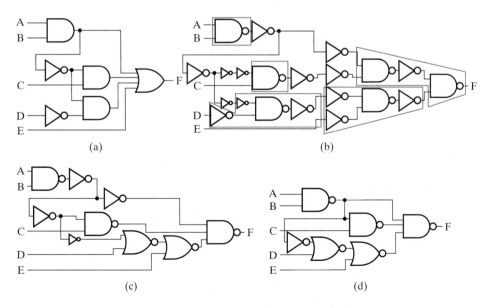

□ **FIGURE 3-17**
Solution to Example 3-8

To provide continuity with examples in earlier sections of this chapter, the following example shows the mapping of the BCD–to–Excess-3 Code Converter for an expanded cell library.

EXAMPLE 3-9   Technology Mapping for BCD–to–Excess-3 Code Converter

The final result of the technology mapping for the BCD–to–Excess-3 Code Converter is given in Figure 3-18. The original AND, OR, inverter logic diagram appears in Figure 3-11, and the cell library used is given in Table 3-3. The optimization has resulted in the use of the following cells from that library: inverters, 2-input NANDs, a 2-input NOR, and a 2-2 AOI. ■

The gate input cost of the mapped circuit in Example 3-9 is 22, with a gate input cost of 22 for the original AND, OR, NOT circuit. The optimization procedure, aside from locally minimizing inverters, works separately on the various parts of the circuit. These parts are separated by gate fan-outs in the original AND-OR circuit. The selection of these points in the optimization can effect the optimality of the final result. In the case of this circuit, a different original circuit may yield a better optimization. In general, this problem of separate optimization and mapping is handled by using combined optimization steps and mapping steps in commercial logic optimization tools.

ADVANCED TECHNOLOGY MAPPING This supplement on technology mapping, including detailed examples illustrating the mapping procedure for general cell libraries, is available on the Companion Website for the text.

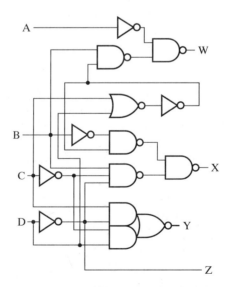

☐ **FIGURE 3-18**
Technology Mapping Example: BCD–to–Excess-3 Code Converter

3-5   VERIFICATION

In this section, manual logic analysis and computer simulation-based logic analysis are considered, both with the goal of verification of circuit function (i.e., determination of whether or not a given circuit implements its specified function). If the circuit does not meet its specification, then it is incorrect. As a consequence, verification plays a vital role in preventing incorrect circuit designs from being manufactured and used. Logic analysis also can be used for other purposes, including redesign of a circuit and determination of the function of a circuit.

In order to verify a combinational circuit, it is essential that the specification be unambiguous and correct. As a consequence, specifications such as truth tables, Boolean equations, and HDL code are most useful. Initially, we examine manual verification by continuing with the design examples we introduced in this chapter.

Manual Logic Analysis

Manual logic analysis consists of finding Boolean equations for the circuit outputs or, additionally, finding the truth table for the circuit. The approach taken here emphasizes finding the equations and then using those equations to find the truth table. In finding the equation for a circuit, it is often convenient to break up the circuit into subcircuits by defining intermediate variables at selected points in the circuit. Points typically selected are those at which a gate output drives two or more gate inputs. Such points are often referred to as *fan-out points*. Fan-out points from a single inverter on an input typically would not be selected. The determination of logic equations for a circuit is illustrated using the BCD–to–Excess-3 Code Converter circuit designed in previous sections.

EXAMPLE 3-10   Manual Verification of BCD–to–Excess-3 Code Converter

Figure 3-19 shows (a) the original truth table specification, (b) the final circuit implementation, and (c) an incomplete truth table to be completed from the implementation and then compared to the original truth table. The truth table values are to be determined from Boolean equations for $W, X, Y,$ and Z derived from the circuit. The point $T1$ is selected as an intermediate variable to simplify the analysis:

$$T1 = \overline{\overline{C + D}} = C + D$$

$$W = \overline{\overline{A} \cdot (T1 \cdot B)} = A + B \cdot T1$$

$$X = \overline{(\overline{B} \cdot T1) \cdot (B \cdot \overline{C} \cdot \overline{D})} = \overline{B} \cdot T1 + B\overline{C} \cdot \overline{D}$$

$$Y = \overline{\overline{CD} + \overline{\overline{C}D}} = CD + \overline{C}\overline{D}$$

$$Z = \overline{D}$$

Input BCD				Output Excess-3			
A	B	C	D	W	X	Y	Z
0	0	0	0	0	0	1	1
0	0	0	1	0	1	0	0
0	0	1	0	0	1	0	1
0	0	1	1	0	1	1	0
0	1	0	0	0	1	1	1
0	1	0	1	1	0	0	0
0	1	1	0	1	0	0	1
0	1	1	1	1	0	1	0
1	0	0	0	1	0	1	1
1	0	0	1	1	1	0	0

(a)

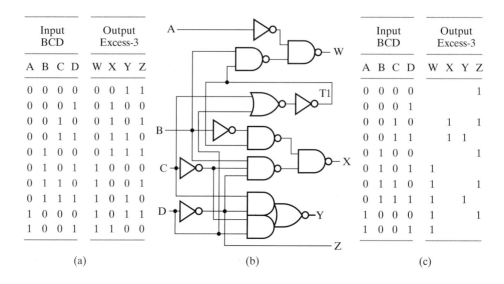

(b)

Input BCD				Output Excess-3			
A	B	C	D	W	X	Y	Z
0	0	0	0				1
0	0	0	1				
0	0	1	0		1		1
0	0	1	1		1	1	
0	1	0	0				1
0	1	0	1	1			
0	1	1	0	1			1
0	1	1	1	1		1	
1	0	0	0	1			1
1	0	0	1	1			

(c)

☐ **FIGURE 3-19**
Verification: BCD–to–Excess-3 Code Converter

Substituting the expression for *T1* in the equations for *W* and *X*, we have

$$W = A + B\,(C + D) = A + B\,C + B\,D$$

$$X = \overline{B}\,(C + D) + B\,\overline{C}\,\overline{D} = \overline{B}\,C + \overline{B}\,D + B\,\overline{C}\,\overline{D}$$

Each of the product terms in the four output equations can be mapped to 1's in the truth table in Figure 3-19 (c). The mappings of the 1's for A, $\overline{B}\,C$, BD, CD, and $\overline{D}$ are shown. After the remaining product terms are mapped to 1's, the blank entries are filled with 0's. The new truth table in this case will match the original one, verifying that the circuit is correct. ■

LOGIC ANALYSIS   This supplement, including additional logic analysis techniques and examples, is available on the Companion Website for the text.

Simulation

An alternative to manual verification is the use of computer simulation for verification. Using a computer permits truth table verification to be done for a significantly larger number of variables and greatly reduces the tedious analysis effort required. Since simulation uses applied values, if possible, it is desirable for thorough verification to apply all possible input combinations. The next example illustrates the use of Xilinx ISE4.2i FPGA development tools and XE II Modelsim simulator to verify the BCD–to–Excess-3 Code Converter using all possible input combinations from the truth table.

EXAMPLE 3-11   Simulation-based Verification of BCD–to–Excess-3 Code Converter

Figure 3-19 shows (a) the original truth table specification, and (b) the final circuit implementation of the BCD–to–Excess 3 Code Converter. The circuit implementation has been entered into Xilinx ISE 4.2i as the schematic shown in Figure 3-20. Since there are no AOIs available in the symbol library, the AOI has been modeled with available symbols. In addition to entering the schematic, the input combinations given in Figure 3-19(a) have also been entered as a waveform. These input waveforms are given in the INPUTS section of the simulation output shown in Figure 3-21. The simulation of the input waveforms applied to the circuit produces the output waveforms given in the OUTPUT section. Examining each input combination and the corresponding output combination represented by the waveforms, we can manually verify whether the outputs match the original truth table. Beginning with $(A,B,C,D) = (0,0,0,0)$ in the input waveform, we find that the corresponding output waveform represents $(W,X,Y,Z) = (0,0,1,1)$. Continuing, for $(A,B,C,D) = (0,0,0,1)$, the values for the output waveforms are $(W,X,Y,Z) = (0,1,0,0)$. In both cases, the values are correct. This process of checking the waveform values against the specification can be continued for the remaining eight input combinations to complete the verification.   ▪

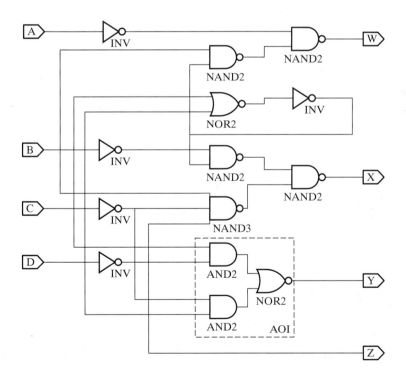

□ **FIGURE 3-20**
Schematic for Simulation of BCD–to–Excess-3 code Converter

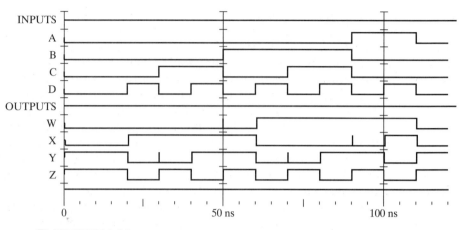

☐ **FIGURE 3-21**
Example 3-10: Simulation Results—BCD–to–Excess-3 Code Converter

 ADVANCED VERIFICATION This supplement, containing additional verification techniques and examples, is available on the Companion Website for the text.

3-6 PROGRAMMABLE IMPLEMENTATION TECHNOLOGIES

Thus far, we have introduced implementation technologies that are fixed in the sense that they are fabricated as integrated circuits or by connecting together integrated circuits. In contrast, programmable logic devices (PLDs) are fabricated with structures that implement logic functions and structures that are used to control connections or to store information specifying the actual logic functions implemented. These latter structures require *programming*, a hardware procedure that determines which functions are implemented. The next three subsections deal with three types of simple programmable logic devices (PLDs): the read-only memory (ROM), the programmable logic array (PLA), and the programmable array logic (PAL®) device. In a Companion Website supplement, the more complex field-programmable gate array (FPGA) is discussed and illustrated. Before treating PLDs, we deal with the supporting programming technologies. In PLDs, programming technologies are applied to (1) establish or break interconnections, (2) build lookup tables, and (3) control transistor switching. We will relate the technologies to these three applications.

The oldest of programming technologies for controlling connections is the use of *fuses*. Each of the programmable points in the PLD consists of a connection formed by a fuse. When a voltage considerably higher than the normal power supply voltage is applied across the fuse, the high current breaks the connection by "blowing out" the fuse. The two connection states, CLOSED and OPEN, are represented by an intact and blown fuse, respectively.

A second programming technology for controlling connections is *mask programming*, which is done by the semiconductor manufacturer during the last steps of the chip fabrication process. Connections are made or not made in the metal layers serving as conductors in the chip. Depending on the desired function for the chip, the structure of these layers is determined by the fabrication process. The procedure is costly because masks for generating the layers and custom fabrication are required for each customer. For this reason, mask programming is economical only if a large quantity of the same PLD configuration is ordered.

A third programming technology for controlling connections is the use of *antifuse.* As the name suggests, the antifuse is just the opposite of a fuse. In contrast to a fuse, an antifuse consists of a small area in which two conductors are separated by a material having a high resistance. The antifuse acts as an OPEN path before programming. By applying a voltage somewhat higher than the normal power supply voltage across the two conductors, the material separating the two conductors is melted or otherwise changed to a low resistance. The low-resistance material conducts, causing a connection (i.e., a CLOSED path) to be formed.

All three of the preceding connection technologies are permanent. The devices cannot be reprogrammed, because irreversible physical changes have occurred as a result of device programming. Thus, if the programming is incorrect or needs to be changed, the device must be discarded.

The final programming technology that can be applied for controlling connections is a single-bit storage element driving the gate of an MOS n-channel transistor at the programming point. If the stored bit value is a 1, then the transistor is turned ON, and the connection between its source and drain forms a CLOSED path. For the stored bit value equal to 0, the transistor is OFF, and the connection between its source and drain is an OPEN path. Since storage element content can be changed electronically, the device can be easily reprogrammed. But in order to store values, the power supply must be available. Thus, the storage element technology is volatile; that is, the programmed logic is lost in the absence of the power supply voltage.

The second application of programming technologies is building lookup tables. In addition to controlling connections, storage elements are ideal for building these tables. In this case, the input combination for the truth table is used to select a storage element containing the corresponding output value for the truth table and provide it as the logic function output. The hardware consists of (1) the storage elements, (2) the hardware to program values into the storage elements, and (3) the logic that selects the storage element contents to be applied to the logic function output. Because storage elements are being selected by the input values, the storage elements combined with the hardware in (3) resemble a memory in which stored data values are selected to appear on the memory output by using an address applied to the inputs. Thus, the logic can be implemented simply by storing the truth table in the memory—hence the term *lookup table.*

The third application of programming technologies is control of transistor switching. The most popular technology is based on storing charge on a floating gate. The latter is located below the regular gate within an MOS transistor and is completely isolated by an insulating dielectric. Stored negative charge on the floating gate makes the transistor impossible to turn ON. The absence of stored negative charge makes it possible for the transistor to turn ON if a HIGH is applied to its regular gate. Since it is possible to add or remove the stored charge, these technologies can permit erasure and reprogramming.

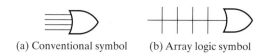

(a) Conventional symbol   (b) Array logic symbol

□ **FIGURE 3-22**
Conventional and Array Logic Symbols for OR Gate

Two technologies using control of transistor switching are called *erasable* and *electrically erasable.* Programming applies combinations of voltage higher than normal power supply voltages to the transistor. Erasure uses exposure to a strong ultraviolet light source for a specified amount of time. Once this type of chip has been erased, it can be reprogrammed. An electrically erasable device can be erased by a process somewhat similar to the programming process, using voltages higher than the normal power supply value. Since transistor control prevents or allows a connection to be established between the source and the drain, it is really a form of connection control, giving a choice between (1) always OPEN or (2) OPEN or CLOSED, depending on an applied HIGH or LOW, respectively, on the regular transistor gate. A third technology based on control of transistor switching is *flash* technology, which is very widely used in *flash memories.* Flash technology is a form of electrically-erasable technology that has a variety of erase options including the erase of stored charge from individual floating gates, all of the floating gates, or specific subsets of floating gates.

A typical PLD may have hundreds to millions of gates. Some, but not all, programmable logic technologies have high fan-in gates. In order to show the internal logic diagram for such technologies in a concise form, it is necessary to employ a special gate symbology applicable to array logic. Figure 3-22 shows the conventional and array logic symbols for a multiple-input OR gate. Instead of having multiple input lines to the gate, we draw a single line to the gate. The input lines are drawn perpendicular to this line and are selectively connected to the gate. If an x is present at the intersection of two lines, there is a connection. If an x is not present, then there is no connection. In a similar fashion, we can draw the array logic for an AND gate. Since this was first done for a fuse-based technology, the graphics representation, when marked with the selected connections, is referred to as a *fuse map.* We will use the same graphics representation and terminology even when the programming technology is not fuses. This type of graphics representation for the inputs of gates will be used subsequently in drawing logic diagrams.

We next consider three distinct programmable device structures. We will describe each of the structures and indicate which of the technologies is typically used in its implementation. These types of PLDs differ in the placement of programmable connections in the AND-OR array. Figure 3-23 shows the locations of the connections for the three types. Programmable read-only memory (PROM) as well as flash memory has a fixed AND array constructed as a decoder and programmable connections for the output OR gates. The PROM implements Boolean functions in sum-of-minterms form. The programmable array logic (PAL®) device has a programmable connection AND array and a fixed OR array. The AND gates are programmed to provide the product terms for the Boolean functions, which are

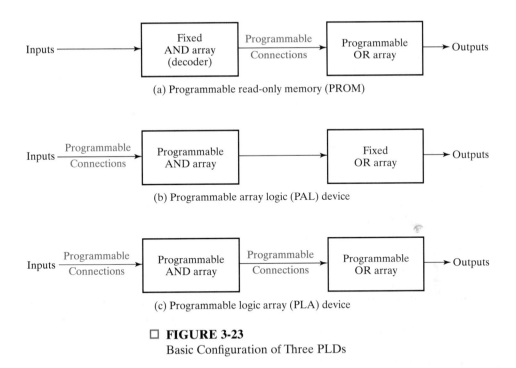

(a) Programmable read-only memory (PROM)

(b) Programmable array logic (PAL) device

(c) Programmable logic array (PLA) device

☐ **FIGURE 3-23**
Basic Configuration of Three PLDs

logically summed in each OR gate. The most flexible of the three types of PLD is the programmable logic array (PLA), which has programmable connections for both AND and OR arrays. The product terms in the AND array may be shared by any OR gate to provide the required sum-of-products implementation. The names PLA and PAL® emerged for devices from different vendors during the development of PLDs.

Read-Only Memory

A read-only memory (ROM) is essentially a device in which "permanent" binary information is stored. The information must be specified by the designer and is then embedded into the ROM to form the required interconnection or electronic device pattern. Once the pattern is established, it stays within the ROM even when power is turned off and on again; that is, ROM is nonvolatile.

A block diagram of a ROM device is shown in Figure 3-24. There are k inputs and n outputs. The inputs provide the address for the memory, and the outputs give

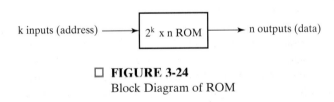

☐ **FIGURE 3-24**
Block Diagram of ROM

the data bits of the stored word that is selected by the address. The number of words in a ROM device is determined from the fact that k address input lines can specify 2^k words. Note that ROM does not have data inputs, because it does not have a write operation. Integrated circuit ROM chips have one or more enable inputs and come with three-state outputs to facilitate the construction of large arrays of ROM.

Consider, for example, a 32×8 ROM. The unit consists of 32 words of 8 bits each. There are five input lines that form the binary numbers from 0 through 31 for the address. Figure 3-25 shows the internal logic construction of this ROM. The five inputs are decoded into 32 distinct outputs by means of a 5–to–32-line decoder. Each output of the decoder represents a memory address. The 32 outputs are connected through programmable connections to each of the eight OR gates. The diagram uses the array logic convention used in complex circuits. (See Figure 3-22.) Each OR gate must be considered as having 32 inputs. Each output of the decoder is connected through a fuse to one of the inputs of each OR gate. Since each OR gate has 32 internal programmable connections, and since there are eight OR gates, the ROM contains $32 \times 8 = 256$ programmable connections. In general, a $2^k \times n$ ROM will have an internal k–to–2^k line decoder and n OR gates. Each OR gate has 2^k inputs, which are connected through programmable connections to each of the outputs of the decoder.

Four technologies are used for ROM programming. If mask programming is used, then the ROM is called simply a ROM. If fuses are used, the ROM can be programmed by the user having the proper programming equipment. In this case, the ROM is referred to as a *programmable ROM*, or PROM. If the ROM uses the erasable floating-gate technology, then the ROM is referred to as an *erasable, programmable ROM*, or EPROM. Finally, if the electrically erasable technology is used, the

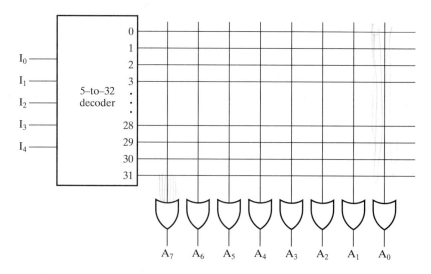

◻ **FIGURE 3-25**
Internal Logic of a 32×8 ROM

ROM is referred to an *electrically erasable, programmable ROM*, or EEPROM or E^2PROM. As discussed previously, flash memory is a modified version of E^2PROM. The choice of programming technology depends on many factors, including the number of identical ROMs to be produced, the desired permanence of the programming, the desire for reprogrammability, and the desired performance in terms of delay.

Programmable Logic Array

The programmable logic array (PLA) is similar in concept to the PROM, except that the PLA does not provide full decoding of the variables and does not generate all the minterms. The decoder is replaced by an array of AND gates that can be programmed to generate product terms of the input variables. The product terms are then selectively connected to OR gates to provide the sum of products for the required Boolean functions.

The internal logic of a PLA with three inputs and two outputs is shown in Figure 3-26. Such a circuit is too small to be cost effective, but is presented here to demonstrate the typical logic configuration of a PLA. The diagram uses the array logic graphics symbols for complex circuits. Each input goes through a buffer and an inverter, represented in the diagram by a composite graphics symbol that has both the true and the complement outputs. Programmable connections run from

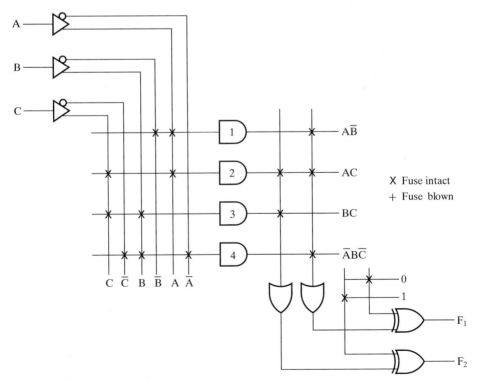

□ **FIGURE 3-26**
PLA with Three Inputs, Four Product Terms, and Two Outputs

each input and its complement to the inputs of each AND gate, as indicated by the intersections between the vertical and horizontal lines. The outputs of the AND gates have programmable connections to the inputs of each OR gate. The output of the OR gate goes to an XOR gate, where the other input can be programmed to receive a signal equal to either logic 1 or logic 0. The output is inverted when the XOR input is connected to 1 (since $X \oplus 1 = \overline{X}$). The output does not change when the XOR input is connected to 0 (since $X \oplus 0 = X$). The particular Boolean functions implemented in the PLA of the figure are

$$F_1 = A\overline{B} + AC + \overline{A}B\overline{C}$$

$$F_2 = \overline{AC + BC}$$

The product terms generated in each AND gate are listed by the output of the gate in the diagram. The product term is determined from the inputs with CLOSED circuit connections. The output of an OR gate gives the logic sum of the selected product terms. The output may be complemented or left in its true form, depending on the programming of the connection associated with the XOR gate.

The size of a PLA is specified by the number of inputs, the number of product terms, and the number of outputs. A typical PLA has 16 inputs, 48 product terms, and eight outputs. For n inputs, k product terms, and m outputs, the internal logic of the PLA consists of n buffer-inverter gates, k AND gates, m OR gates, and m XOR gates. There are $2n \times k$ programmable connections between the inputs and the AND array, $k \times m$ programmable connections between the AND and OR arrays, and m programmable connections associated with the XOR gates.

As with a ROM, the PLA may be mask programmable or field programmable. With mask programming, the customer submits a PLA program table to the manufacturer. The table is used by the vendor to produce a custom-made PLA that has the internal logic specified by the customer. Field programming uses a PLA called a *field-programmable logic array*, or FPLA. This device can be programmed by the user by means of a commercial hardware programming unit.

Programmable Array Logic Devices

The programmable array logic (PAL®) device is a PLD with a fixed OR array and a programmable AND array. Because only the AND gates are programmable, the PAL device is easier to program than, but is not as flexible as, the PLA. Figure 3-27 presents the logic configuration of a typical programmable array logic device. The particular device shown has four inputs and four outputs. Each input has a buffer-inverter gate, and each output is generated by a fixed OR gate. The device has four sections, each composed of a three-wide AND-OR array, meaning that there are three programmable AND gates in each section. Each AND gate has 10 programmable input connections, indicated in the diagram by 10 vertical lines intersecting each horizontal line. The horizontal line symbolizes the multiple-input configuration of an AND gate. One of the outputs shown is connected to a buffer-inverter gate and then fed back into the inputs of the AND gates through programmed connections. This is often done with all device outputs.

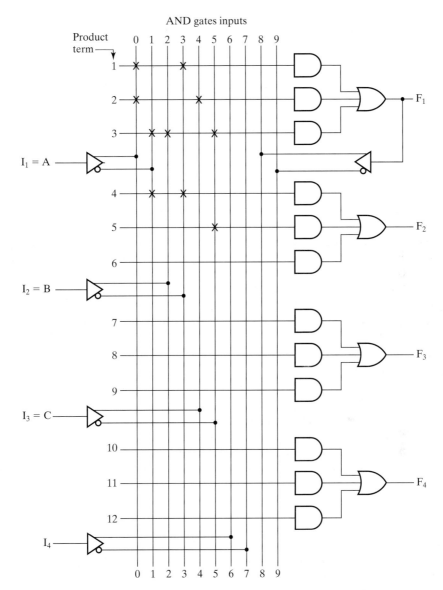

☐ **FIGURE 3-27**
PAL® Device with Four Inputs, Four Outputs, and a Three-wide AND-OR Structure

The particular Boolean functions implemented in the PAL in Figure 3-27 are

$$F_1 = A\overline{B} + AC + \overline{A}B\overline{C}$$

$$F_2 = \overline{AC + BC} = \overline{A}\overline{B} + \overline{C}$$

These functions are the same as those implemented using the PLA. Since the output complement is not available, F_2 is converted to a sum of products form.

Commercial PAL devices contain more gates than the one shown in Figure 3-27. A small PAL integrated circuit may have up to eight inputs, eight outputs, and eight sections, each consisting of an eight-wide AND-OR array. Each PAL device output is driven by a three-state buffer and also serves as an input. These input/outputs can be programmed to be an input only, an output only, or bidirectional with a variable signal driving the three-state buffer enable signal. Flip-flops are often included in a PAL device between the array and the three-state buffer at the outputs. Since each output is fed back as an input through a buffer-inverter gate into the AND programmed array, a sequential circuit can be easily implemented.

VLSI Programmable Logic Devices   This supplement, which covers the basics of two typical Field Programmable Gate Arrays (FPGAs) used in course laboratories is available on the Companion Website for the text. The supplement uses multiplexers, adders, flip-flops, latches, and SRAMs. An appendix to the supplement provides a brief introduction to these components.

3-7   Chapter Summary

This chapter began with the introduction of two important design concepts, design hierarchy and top-down design, that are used throughout the remainder of the book. Computer-aided design was briefly introduced, with a focus on hardware description languages (HDLs) and logic synthesis.

In Section 3-2, properties of the underlying gate technology were introduced. Two component types with an added output value called high-impedance (Hi-Z), three-state buffers, and transmission gates were described. Important technology parameters, including fan-in, fan-out, and propagation delay, were defined and illustrated. Positive and negative logic conventions describe two different relationships between voltage levels and logical values.

The core of this chapter is a five-step design procedure described in Section 3-3. These steps apply to both manual and computer-aided design. The design begins with a defining specification and proceeds through a formulation step in which the specification is converted to a table or equations. The optimization step applies two-level and multiple-level optimization to obtain a circuit composed of AND gates, OR gates, and inverters. Technology mapping converts this circuit into one that efficiently uses the gates in the available implementation technology. Finally, verification is applied to assure that the final circuit satisfies the initial specification. Three examples illustrate the first three of these steps.

In order to discuss technology mapping, fixed implementation technologies including full custom, standard cell, and gate array approaches were introduced. Cell specification and cell libraries also were introduced, and mapping techniques resembling those used in CAD tools were presented and illustrated for single cell types and multiple cell types.

The final section of the chapter dealt with programmable logic technologies. Three basic technologies—read-only memory, programmable logic arrays, and programmable array logic devices—provide technology mapping alternatives.

REFERENCES

1. HACTEL, G., AND F. SOMENZI. *Logic Synthesis and Verification Algorithms.* Boston: Kluwer Academic Publishers, 1996.

2. DE MICHELI, G. *Synthesis and Optimization of Digital Circuits.* New York: McGraw-Hill, Inc., 1994.

3. KNAPP, S. *Frequently-Asked Questions (FAQ) About Programmable Logic* (http://www.optimagic.com/faq.html). OptiMagicTM, Inc., ©1997-2001.

4. LATTICE SEMICONDUCTOR CORPORATION. *Lattice GALs(R)*(http://www.latticesemi.com/products/spld/GAL/index.cfm). Lattice Semiconductor Corporation, © 1995-2002.

5. TRIMBERGER, S. M., ED. *Field-Programmable Gate Array Technology.* Boston: Kluwer Academic Publishers, 1994.

6. XILINX, INC., *Xilinx Spartan™-IIE Data Sheet* (http://direct.xilinx.com/bvdocs/publications/ds077_2.pdf). Xilinx, Inc. © 1994-2002.

7. ALTERA(R) CORPORATION, Altera FLEX 10KE Embedded *Programmable Logic Device Family Data Sheet* ver. 2.4 (http://www.altera.com/literature/ds/dsf10ke.pdf). Altera Corporation, © 1995 - 2002.

PROBLEMS

The plus (+) indicates a more advanced problem and the asterisk (*) indicates a solution is available on the text website.

3–1. Design a circuit to implement the following pair of Boolean equations:

$$F = A (C \overline{E} + D E) + \overline{A} \, D$$
$$F = B (C \overline{E} + D E) + \overline{B} \, C$$

To simplify drawing the schematic, the circuit is to use a hierarchy based on the factoring shown in the equation. Three instances (copies) of a single hierarchical circuit component made up of two AND gates, an OR gate, and an inverter are to be used. Draw the logic diagram for the hierarchical component and for the overall circuit diagram using a symbol for the hierarchical component.

3–2. A hierarchical component with the function

$$H = \overline{X} \, Y + X \, Z$$

is to be used along with inverters to implement the following equation:

$$G = \overline{A} \, \overline{B} \, C + \overline{A} \, B D + A \, \overline{B} \, \overline{C} + A B \overline{D}$$

The overall circuit can be obtained by using Shannon's expansion theorem,

$$F = \overline{X} \cdot F_0(X) + X \cdot F_1(X)$$

where $F_0(X)$ is F evaluated with variable $X = 0$ and $F_1(X)$ is F evaluated with variable $X = 1$. This expansion F can be implemented with function H by letting $Y = F_0$ and $Z = F_1$. The expansion theorem can then be applied to each of F_0 and F_1 using a variable in each, preferably one that appears in both true and complemented form. The process can then be repeated until all F_i's are single literals or constants. For G, use $X = A$ to find G_0 and G_1 and then use $X = B$ for G_0 and G_1. Draw the top level diagram for G using H as a hierarchical component.

3–3. An integrated circuit logic family has NAND gates with a fan-out of 8 standard loads and buffers with a fan-out of 16 standard loads. Sketch a schematic showing how the output signal of a single NAND gate can be applied to 38 other gate inputs using as few buffers as possible. Assume that each input is one standard load.

3–4. *The NOR gates in Figure 3-28 have propagation delay $t_{pd} = 0.078$ ns and the inverter has a propagation delay $t_{pd} = 0.052$ ns. What is the propagation delay of the longest path through the circuit?

3–5. The waveform in Figure 3-29 is applied to an inverter. Find the output of the inverter, assuming that

(a) it has no delay.
(b) it has a transport delay of 0.06 ns.
(c) it has an inertial delay of 0.06 ns with a rejection time of 0.06 ns.

3–6. Assume that t_{pd} is the average of t_{PHL} and t_{PLH}. Find the delay from each input to the output in Figure 3-30 by

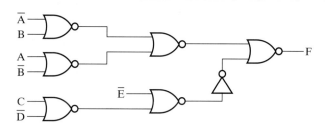

☐ **FIGURE 3-28**
Circuit for Problem 3-4

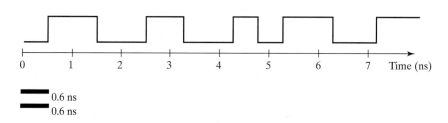

0.6 ns
0.6 ns

☐ **FIGURE 3-29**
Waveform for Problem 3-5

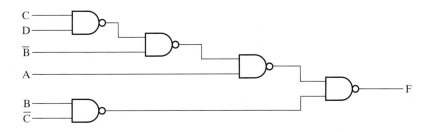

☐ **FIGURE 3-30**
Circuit for Problem 3–6

(a) Finding t_{PHL} and t_{PLH} for each path assuming $t_{PHL} = 0.30$ ns and $t_{PLH} = 0.50$ ns for each gate. From these values, find t_{pd} for each path.

(b) Using $t_{pd} = 0.40$ ns for each gate.

(c) Compare your answers in part (a) and (b) and discuss any differences.

3–7. +The rejection time for inertial delays is required to be less than or equal to the propagation delay. In terms of the discussion of the example in Figure 3-7, why is this condition necessary to determine the delayed output?

3–8. +For a given gate, $t_{PHL} = 0.05$ ns and $t_{PLH} = 0.10$ ns. Suppose that an inertial delay model is to be developed from this information for typical gate delay behavior.

(a) Assuming a positive output pulse (L H L), what would the propagation delay and rejection time be?

(b) Discuss the applicability of the parameters in (a) assuming a negative output pulse (H L H).

3–9. *Show that a positive logic NAND gate is a negative logic NOR gate and vice versa.

3–10. A majority function has an output value of 1 if there are more 1's than 0's on its inputs. The output is 0 otherwise. Design a three-input majority function.

3–11. *Find a function to detect an error in the representation of a decimal digit in BCD. In other words, write an equation with value 1 when the inputs are any one of the six unused bit combinations in the BCD code, and value 0, otherwise.

3–12. Design an Excess-3–to–BCD code converter that gives output code 0000 for all invalid input combinations.

3–13. **(a)** A low-voltage lighting system is to use a binary logic control for a particular light. This light lies at the intersection point of a T-shaped hallway. There is a switch for this light at each of the three end points of the T. These switches have binary outputs 0 and 1 depending on their position and are named X_1, X_2, and X_3. The light is controlled by an amplifier driving a thyristor. When Z, the input to the amplifier, is 1, the light is ON and when Z is 0, the light is OFF. You are to find a function $Z = F(X_1, X_2, X_3)$ so that if any one of the switches is changed, the value of Z changes turning the light ON or OFF.

(b) The function for Z is not unique. How many different functions for Z are there?

3–14. +A traffic light control at a simple intersection uses a binary counter to produce the following sequence of combinations on lines A, B, C, and D: 0000, 0001, 0011, 0010, 0110, 0111, 0101, 0100, 1100, 1101, 1111, 1110, 1010, 1011, 1001, 1000. After 1000, the sequence repeats, beginning again with 0000, forever. Each combination is present for 5 seconds before the next one appears. These lines drive combinational logic with outputs to lamps RNS (Red - North/South), YNS (Yellow - North/South), GNS (Green - North/South), REW (Red - East/West), YEW (Yellow - East/West), and GEW (Green - East/West). The lamp controlled by each output is ON for a 1 applied and OFF for a 0 applied. For a given direction, assume that green is on for 30 seconds, yellow is on for 5 seconds, and red is on for 45 seconds. (The red intervals overlap for 5 seconds.) Divide up the 80 seconds available for the cycle through the 16 combinations into 16 intervals and determine which lamps should be lit in each interval based on expected driver behavior. Assume that, for interval 0000, a change has just occurred and that GNS = 1, REW = 1, and all other outputs are 0. Design the logic to produce the six outputs using AND and OR gates and inverters.

3–15. Design a combinational circuit that accepts a 3-bit number and generates a 6-bit binary number output equal to the square of the input number.

3–16. +Design a combinational circuit that accepts a 4-bit number and generates a 3-bit binary number output that approximates the square root of the number. For example, if the square root is 3.5 or larger, give a result of 4. If the square root is < 3.5 and ≥ 2.5, give a result of 3.

3–17. Design a circuit with a 4-bit BCD input A, B, C, D that produces an output W, X, Y, Z that is equal to the input + 6 in binary. For example, 9 (1001) + 6 (0110) = 15 (1111).

3–18. A traffic metering system for controlling the release of traffic from an entrance ramp onto a superhighway has the following specifications for a part of its controller. There are three parallel metering lanes, each with its own stop (red)–go (green) light. One of these lanes, the car pool lane, is given priority for a green light over the other two lanes. Otherwise, a "round robin" scheme in which the green lights alternate is used for the other two (left and right) lanes. The part of the controller that determines which light is to be green (rather than red) is to be designed. The specifications for the controller follow:

Inputs:

PS - Car pool lane sensor (car present - 1; car absent - 0)
LS - Left lane sensor (car present - 1; car absent - 0)
RS - Right lane sensor (car present - 1; car absent - 0)
RR - Round robin signal (select left - 1; select right - 0)

Outputs:

PL - Car pool lane light (green - 1; red - 0)
LL - Left lane light (green - 1; red - 0)
RL - Right lane light (green - 1; red - 0)

Operation:

1. If there is a car in the car pool lane, PL is 1.
2. If there are no cars in the car pool lane and the right lane, and there is a car in the left lane, LL is 1.
3. If there are no cars in the car pool lane and in the left lane, and there is a car in the right lane, RL is 1.
4. If there is no car in the car pool lane, there are cars in both the left and right lanes, and RR is 1, then LL = 1.
5. If there is no car in the car pool lane, there are cars in both the left and right lanes, and RR is 0, then RL = 1.
6. If any PL, LL, or RL is not specified to be 1 above, then it has value 0.

(a) Find the truth table for the controller part.

(b) Find a minimum multiple-level gate implementation with minimum gate input count using AND gates, OR gates and inverters.

3–19. Complete the design of the BCD–to–seven-segment decoder by performing the following steps:

(a) Plot the seven maps for each of the outputs for the BCD–to–seven-segment decoder specified in Table 3-2.

(b Simplify the seven output functions in sum–of–products form, and determine the total number of gate inputs that will be needed to implement the decoder.

(c) Verify that the seven output functions listed in the text give a valid simplification. Compare the number of gate inputs with that obtained in part (b) and explain the difference.

3–20. +A NAND gate with eight inputs is required. For each of the following cases, minimize the number of gates used in the multiple-level result:

(a) Design the 8-input NAND gate using 2-input NAND gates and NOT gates.

(b) Design the 8-input NAND gate using 2-input NAND gates, 2-input NOR gates, and NOT gates only if needed.

(c) Compare the number of gates used in (a) and (b).

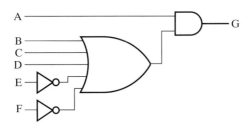

☐ **FIGURE 3-31**
Circuit for Problem 3–21

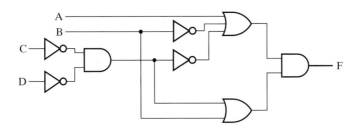

☐ **FIGURE 3-32**
Circuit for Problem 3–22

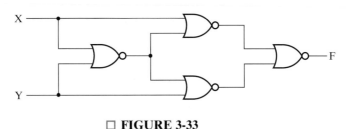

☐ **FIGURE 3-33**
Circuit for Problem 3–23

3–21. Perform a minimum-cost (use minimum total normalized area as cost) technology mapping using NAND cells and inverters from Table 3-3 for the circuit shown in Figure 3-31.

3–22. Perform a low-cost (use minimum total normalized area as cost) technology mapping using all cells from Table 3-3 for the circuit shown in Figure 3-32.

3–23. By using manual methods, verify that the circuit of Figure 3-33 generates the exclusive-NOR function.

3–24. *Manually verify that the functions for the outputs F and G of the hierarchical circuit in Figure 3-34 are

$$F = \overline{X}\,Y + X\,\overline{Y}\,Z + X\,Y\,\overline{Z}$$
$$G = X\,Z + \overline{X}\,Y\,Z + \overline{X}\,\overline{Y}\,\overline{Z}$$

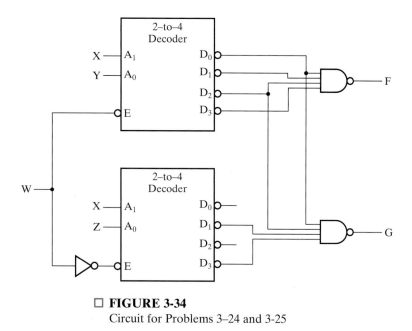

☐ **FIGURE 3-34**
Circuit for Problems 3–24 and 3-25

3–25. Manually verify that the truth table for the outputs F and G of the hierarchical circuit in Figure 3-34 is as follows:

W	X	Y	Z	F	G
0	0	0	0	1	1
0	0	0	1	1	1
0	0	1	0	1	0
0	0	1	1	1	0
0	1	0	0	1	1
0	1	0	1	1	1
0	1	1	0	1	0
0	1	1	1	1	0
1	0	0	0	0	0
1	0	0	1	0	1
1	0	1	0	0	0
1	0	1	1	0	1
1	1	0	0	0	0
1	1	0	1	0	1
1	1	1	0	0	0
1	1	1	1	0	1

3–26. The logic diagram for a 74HC138 MSI CMOS circuit is given in Figure 3-35. Find the Boolean function for each of the outputs. Describe the circuit function carefully.

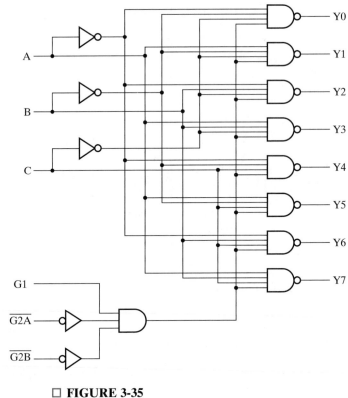

☐ **FIGURE 3-35**
Circuit for Problem 3–26 and Problem 3–27

3–27. Do Problem 3–26 by using logic simulation to find the output waveforms of the circuit or a partial truth table listing, rather than finding Boolean functions.

3–28. In Figure 3-21, simulation results are given for the BCD–to–Excess-3 Code Converter for the BCD inputs for 0 through 9. Perform a similar logic simulation to determine the results for BCD inputs 10 through 15.

CHAPTER

4

COMBINATIONAL FUNCTIONS AND CIRCUITS

I n this chapter, we will learn about a number of functions and the corresponding fundamental circuits that are very useful in designing larger digital circuits. The fundamental, reusable circuits, which we call functional blocks, implement functions of a single variable, decoders, encoders, code converters, multiplexers, and programmable logic. Aside from being important building blocks for larger circuits and systems, many of these functions are strongly tied to the various components of hardware description languages and serve as a vehicle for HDL presentation. As an alternative to truth tables, equations, and schematics, VHDL and Verilog hardware description languages are introduced.

In the generic computer diagram at the beginning of Chapter 1, multiplexers are very important for selecting data in the processor, in memory, and on I/O boards. Decoders are used for selecting boards attached to the input-output bus and to decode instructions to determine the operations performed in the processor. Encoders are used in a number of components, such as the keyboard. Programmable logic is used to handle complex instructions within processors as well as in many other components of the computer. Overall, functional blocks are widely used, so concepts from this chapter apply across most components of the generic computer, including memories.

4-1 COMBINATIONAL CIRCUITS

In Chapter 3, we defined and illustrated combinational circuits and their design. In this section, we define specific combinational functions and corresponding

□ **141**

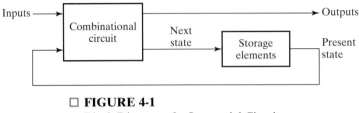

☐ **FIGURE 4-1**
Block Diagram of a Sequential Circuit

combinational circuits, referred to as *functional blocks.* In some cases, we will go through the design process for obtaining a circuit from the function, while in other cases, we will simply present the function and an implementation of it. These functions have special importance in digital design. In the past, the functional blocks were manufactured as small and medium scale integrated circuits. Today, in very large scale integrated (VLSI) circuits, functional blocks are used to design circuits with many such blocks. Combinational functions and their implementations are fundamental to the understanding of VLSI circuits. By using a hierarchy, we typically construct circuits as instances of these functions or the associated functional blocks.

Large-scale and very large-scale integrated circuits are almost always sequential circuits as described in Section 3.1 and detailed beginning in Chapter 6. The functions and functional blocks discussed in this chapter, are combinational. However, they are often combined with storage elements to form sequential circuits as shown in Figure 4-1. Inputs to the combinational circuit can come from both the external environment and from the storage elements. Outputs from the combinational circuit go to both the external environment and to the storage elements. In later chapters, we use the combinational functions and blocks defined here, in Chapter 5, and in Chapter 6 with storage elements to form sequential circuits that perform very useful functions. Further, the functions and blocks defined here and in Chapter 5 serve as a basis for describing and understanding both combinational and sequential circuits using hardware description languages in this and subsequent chapters.

4-2 RUDIMENTARY LOGIC FUNCTIONS

Value-fixing, transferring, inverting, and enabling are among the most elementary of combinational logic functions. The first two operations, value-fixing and transferring, do not involve any Boolean operators. They use only variables and constants. As a consequence, logic gates are not involved in the implementation of these operations. Inverting involves only one logic gate per variable, and enabling involves one or two logic gates per variable.

Value-Fixing, Transferring, and Inverting

If a single-bit function depends on a single variable X, at most, four different functions are possible. Table 4-1 gives the truth tables for these functions. The first and last columns of the table assign either constant value 0 or constant value 1 to the

□ **TABLE 4-1**
Functions of One Variable

X	F = 0	F = X	F = $\overline{X}$	F = 1
0	0	0	1	1
1	0	1	0	1

function, thus performing *value-fixing*. In the second column, the function is simply the input variable X, thus *transferring* X from input to output. In the third column, the function is $\overline{X}$, thus *inverting* input X to become output $\overline{X}$.

The implementations for these four functions are given in Figure 4-2. Value fixing is implemented by connecting a constant 0 or constant 1 to output F as shown in Figure 4-2(a). Figure 4-2(b) shows alternative representations used in logic schematics. For the positive logic convention, constant 0 is represented by the electrical ground symbol and constant 1 by a power supply voltage symbol. The latter symbol is labeled with either V_{CC} or V_{DD}. Transferring is implemented by a simple wire connecting X to F as in Figure 4-2(c). Finally, inverting is represented by an inverter which forms $F = \overline{X}$ from input X as shown in Figure 4-2(d).

Multiple-Bit Functions

The functions defined so far can be applied to multiple bits on a bitwise basis. We can think of these multiple-bit functions as vectors of single bit functions. For example, suppose that we have four functions, F_3, F_2, F_1, and F_0 that make up a four-bit function F. We can order the four functions with F_3 as the most significant bit and F_0 the least significant bit, providing the vector $F = (F_3, F_2, F_1, F_0)$. Suppose that F consists of rudimentary functions $F_3 = 0$, $F_2 = 1$, $F_1 = A$ and $F_0 = \overline{A}$. Then we can write F as the vector $(0, 1, A, \overline{A})$. For $A = 0$, $F = (0, 1, 0, 1)$ and for $A = 1$, $F = (0, 1, 1, 0)$. This multiple-bit function can be referred to as F(3:0) or simply as F and is implemented in Figure 4-3(a). For convenience in schematics, we often represent a set of multiple, related wires by using a single line of greater thickness with a slash across the line. An integer giving the number of wires represented accompanies the slash as shown in Figure 4-3(b). In order to connect the values 0, 1, X, and $\overline{X}$ to the

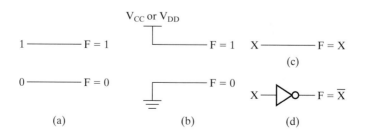

□ **FIGURE 4-2**
Implementation of Functions of a Single Variable X

☐ **FIGURE 4-3**
Implementation of Multi-bit Rudimentary Functions

appropriate bits of F, we break F up into four wires with each wire labeled with the bit of F. Also, in the process of transferring, we may wish to use only a subset of the elements in F, for example, F_2 and F_1. The notation for the bits of F can be used for this purpose as shown in Figure 4-3(c). In Figure 4-3(d), a more complex case illustrates the use of F_3, F_1, F_0 at a destination. Note that since F_3, F_1, and F_0 are not all together, we cannot use the range notation $F(3:0)$ to denote this subvector. Instead, we must use a combination of two subvectors, $F(3), F(1:0)$ denoted by subscripts 3, 1:0. The actual notation used for vectors and subvectors varies among the schematic drawing tools or HDL tools available. Figure 4-3 illustrates just one approach. For a specific tool, the documentation should be consulted.

Value-fixing, transferring and inverting have a variety of applications in logic design. Value-fixing involves replacing one or more variables with constant values 1 and 0. Value fixing may be permanent or temporary. In permanent value-fixing, the value can never be changed. In temporary value-fixing, the values can be changed, often by mechanisms that are somewhat different than those employed in ordinary logical operation. A major application of fixed and temporary value-fixing is in programmable logic devices. Any logic function that is within the capacity of the programmable device can be implemented by fixing a set of values, as illustrated in the next example.

EXAMPLE 4-1   Value-Fixing for Implementing a Function

Consider the truth table shown in Figure 4-4(a). A and B are the input variables and I_0 through I_3 are also variables. Values 0 and 1 can be assigned to I_0 through I_3 depending upon the desired function. Note that Y is actually a function of six variables giving a fully-expanded truth table containing 64 rows and seven columns. But, by putting I_0 through I_3 in the output column, we considerably reduce the size of the table. The equation for the output Y for this truth table is

$$Y(A, B, I_0, I_1, I_2, I_3) = \overline{A} \ \overline{B} \ I_0 + \overline{A} \ B I_1 + A \ \overline{B} \ I_2 + A B I_3$$

The implementation of this equation is given in Figure 4-4(b). By fixing the values of I_0 through I_3, we can implement any function $Y(A, B)$. As shown in Table 4-2, we can implement $Y = A + B$ by using $I_0 = 0, I_1 = 1, I_2 = 1$, and $I_3 = 1$. Or, we can implement $Y = A \ \overline{B} + \overline{A} \ B$ by using $I_0 = 0, I_1 = 1, I_2 = 1$, and $I_3 = 0$. Either of these functions can

A B	Y
0 0	I_0
0 1	I_1
1 0	I_2
1 1	I_3

(a)

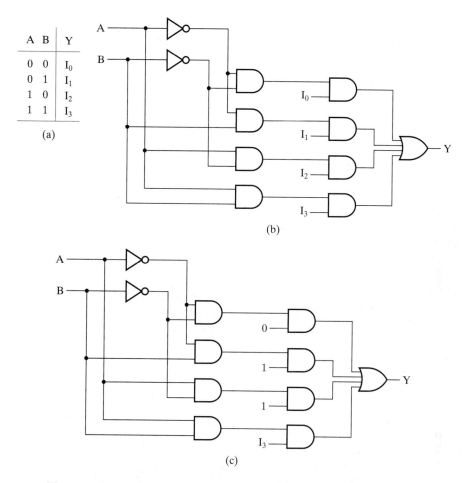

(b)

(c)

□ **FIGURE 4-4**
Implementation of Two Functions by Using Value-Fixing

□ **TABLE 4-2**
Function Implementation by Value-Fixing

A B	$Y = A + B$	$Y = A\bar{B} + \bar{A}B$	$Y = A + B$ ($I_3 = 1$) or $Y = A\bar{B} + \bar{A}B$ ($I_3 = 0$)
0 0	0	0	0
0 1	1	1	1
1 0	1	1	1
1 1	1	0	I_3

be implemented permanently, or can be implemented temporarily by fixing $I_0 = 0, I_1 = 1, I_2 = 1$, and using I_3 as a variable with $I_3 = 1$ for $A + B$ and $I_3 = 0$ for $A\bar{B} + \bar{A}B$. The final circuit is shown in Figure 4-4(c). ■

Enabling

The concept of enabling a signal first appeared in Section 2-9 where Hi-Z outputs and three-state buffers were introduced. In general, enabling permits an input signal to pass through to an output. In addition to replacing the input signal with the Hi-Z state, disabling also can replace the input signal with a fixed output value, either 0 or 1. The additional input signal, often called *ENABLE* or *EN*, is required to determine whether the output is enabled or not. For example, if *EN* is 1, then the input *X* reaches the output (enabled) but if *EN* is 0, the output is fixed at 0 (disabled). For this case, with the disabled value at 0, the input signal is ANDed with the *EN* signal as shown in Figure 4-5(a). If the disabled value is 1, then the input signal *X* is ORed with the complement of the *EN* signal as shown in Figure 4-5(b). Alternatively, the signal to the output may be enabled with *EN* = 0 instead of 1 and the EN signal denoted as $\overline{EN}$ as *EN* is inverted as illustrated in Figure 4-5(b).

EXAMPLE 4-2 Enabling Application

In most automobiles, the lights, radio, and power windows operate only if the ignition switch is turned on. In this case, the ignition switch acts as an "enabling" signal. Suppose that we model this automotive subsystem using the following variables and definitions:

> Ignition switch *IG* - Value 0 if off and value 1 if on
> Light Switch *LS* - Value 0 if off and value 1 if on
> Radio Switch *RS* - Value 0 if off and value 1 if on
> Power Window Switch *WS* - Value 0 if off and value 1 if on
> Lights *L* - Value 0 if off and value 1 if on
> Radio *R* - Value 0 if off and value 1 if on
> Power Windows *W* - Value 0 if off and value 1 if on

Table 4-3 contains the condensed truth table for the operation of this automobile subsystem. Note that when the ignition switch *IS* is off (0), all of the controlled accessories are off (0) regardless of their switch settings. This is indicated by the first row of the table. With the use of X's, this condensed truth table with just nine

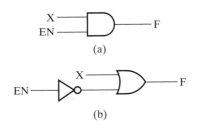

(a)

(b)

☐ **FIGURE 4-5**
Enabling Circuits

□ **TABLE 4-3**
Truth Table for Enabling Application

Input Switches				Accessory Control		
IS	LS	RS	WS	L	R	W
0	X	X	X	0	0	0
1	0	0	0	0	0	0
1	0	0	1	0	0	1
1	0	1	0	0	1	0
1	0	1	1	0	1	1
1	1	0	0	1	0	0
1	1	0	1	1	0	1
1	1	1	0	1	1	0
1	1	1	1	1	1	1

rows represents the same information as the usual 16-row truth table. Whereas X's in output columns represent don't-care conditions, X's in input columns are used to represent product terms that are not minterms. For example, 0XXX represents the product term $\overline{IS}$. Just as with minterms, each variable is complemented if the corresponding bit in the input combination from the table is 0 and is not complemented if the bit is 1. If the corresponding bit in the input combination is an X, then the variable does not appear in the product term. When the ignition switch IS is on (1), then the accessories are controlled by their respective switches. When IS is off (0), all accessories are off. So IS replaces the normal values of the outputs L, R, and W with a fixed value 0 and meets the definition of an $ENABLE$ signal. ■

4-3 DECODING

In digital computers, discrete quantities of information are represented by binary codes. An n-bit binary code is capable of representing up to 2^n distinct elements of coded information. Decoding is the conversion of an n-bit input code to an m-bit output code with $n \leq m \leq 2^n$ such that each valid input code word produces a unique output code. Decoding is performed by a *decoder*, a combinational circuit with an *n*-bit binary code applied to its inputs and an *m*-bit binary code appearing at the outputs. The decoder may have unused bit combinations on its inputs for which no corresponding m-bit code appears at the outputs. Among all of the specialized functions defined here, decoding is the most important since this function and the corresponding functional blocks are incorporated into many of the other functions and functional blocks defined here.

In this section, the functional blocks that implement decoding are called n–to–m-line decoders, where $m \leq 2^n$. Their purpose is to generate the 2^n (or fewer) minterms from the n input variables. For $n = 1$ and $m = 2$, we obtain the 1-to-2-line decoding function with input A and outputs D_0 and D_1. The truth table for this

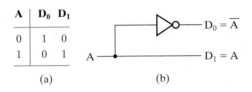

(a) (b)

□ **FIGURE 4-6**
A 1-to-2-Line Decoder

decoding function is given in Figure 4-6(a). If $A = 0$, then $D_0 = 1$ and $D_1 = 0$. If $A = 1$, then $D_0 = 0$ and $D_1 = 1$. From this truth table, $D_0 = \overline{A}$ and $D_1 = A$ giving the circuit shown in Figure 4-6(b).

A second decoding function for $n = 2$ and $m = 4$ with the truth table given in Figure 4-7(a) better illustrates the general nature of decoders. This table has 2-variable minterms as its outputs, with each row containing one output value equal to 1 and three output values equal to 0. Output D_i is equal to 1 whenever the two input values on A_1 and A_0 are the binary code for the number i. As a consequence, the circuit implements the four possible minterms of two variables, one minterm for each output. In the logic diagram in Figure 4-7(b), each minterm is implemented by a 2-input AND gate. These AND gates are connected to two 1-to-2-line decoders, one for each of the lines driving the AND gate inputs.

Decoder Expansion

Large decoders can be constructed by simply implementing each minterm function using a single AND gate with more inputs. Unfortunately, as decoders become

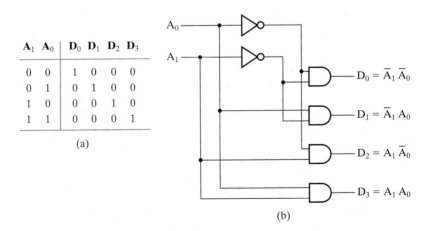

(a) (b)

□ **FIGURE 4-7**
A 2-to-4-Line Decoder

larger, this approach gives a high gate input count. In this section, we give a procedure that uses design hierarchy and collections of AND gates to construct any decoder with n inputs and 2^n outputs. The resulting decoder has the same or a lower gate input count than the one constructed by simply enlarging each AND gate.

To construct a 3–to–8-line decoder ($n = 3$), we can use a 2-to-4-line decoder and a 1-to-2-line decoder feeding eight 2-input AND gates to form the minterms. Hierarchically, the 2-to-4-line decoder can be implemented using two 1-to-2-line decoders feeding four 2-input AND gates as observed in Figure 4-7. The resulting structure is shown in Figure 4-8.

The general procedure is as follows:

1. Let $k = n$.
2. If k is even, divide k by 2 to obtain $k/2$. Use 2^k AND gates driven by two decoders of output size $2^{k/2}$. If is k is odd, obtain $(k + 1)/2$ and $(k - 1)/2$. Use 2^k AND gates driven by a decoder of output size $2^{(k + 1)/2}$ and a decoder of output size $2^{(k - 1)/2}$.
3. For each decoder resulting from step 2, repeat step 2 with k equal to the values obtained in step 2 until $k = 1$. For $k = 1$, use a 1-to-2 decoder.

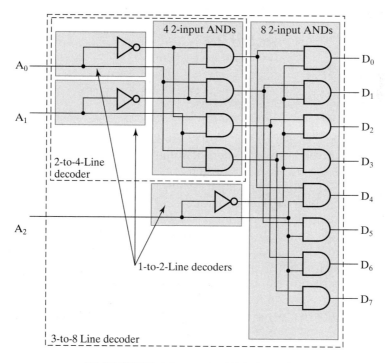

□ **FIGURE 4-8**
A 3-to-8-Line Decoder

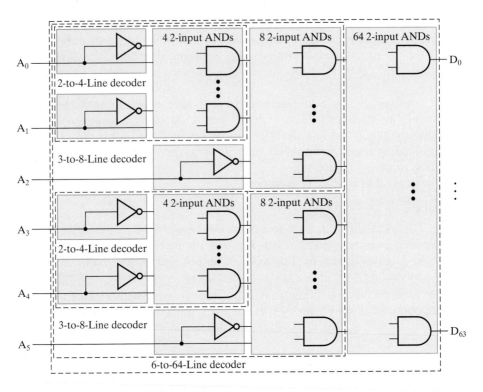

□ **FIGURE 4-9**
A 6-to-64-Line Decoder

EXAMPLE 4-3   6–to–64-Line Decoder

For a 6–to–64-line decoder ($k = n = 6$), in the first execution of step 2, 64 2-input AND gates are driven by two decoders of output size $2^3 = 8$ (i.e., by two 3-to-8-line decoders). In the second execution of step 2, $k = 3$. Since k is odd, the result is ($k + 1)/2 = 2$ and $(k - 1)/2 = 1$. Eight 2-input AND gates are driven by a decoder of output size $2^2 = 4$ and by a decoder of output size $2^1 = 2$ (i.e., by a 2-to-4-line decoder and by a 1–to–2-line decoder, respectively). Finally, on the next execution of step 2, $k = 2$, giving four 2-input AND gates driven by two decoders with output size 2 (i.e., by two 1-to-2-line decoders). Since all decoders have been expanded, the algorithm terminates with step 3 at this point. The resulting structure is shown in Figure 4-9. This structure has a gate input count of $6 + 2 (2 \times 4) + 2 (2 \times 8) + 2 \times 64 = 182$. If a single AND gate for each minterm were used, the resulting gate input count would be $6 + (6 \times 64) = 390$, so a substantial gate input count reduction has been achieved. ■

As an alternative expansion situation, suppose that multiple decoders are needed and that the decoders have common input variables. In this case, instead of implementing separate decoders, parts of the decoders can be shared. For example, suppose that three decoders d_a, d_b, and d_c are functions of input variables as follows:

$$d_a (A, B, C, D)$$
$$d_b (A, B, C, E)$$
$$d_c (C, D, E, F)$$

A 3-to-8-line decoder for A, B, and C can be shared between d_a and d_b. A 2-to-4-line decoder for C and D can be shared between d_a and d_c. A 2-to-4-line decoder for C and E can be shared between d_b and d_c. If we implement all of this sharing, we would have C entering three different decoders and the circuit would be redundant. To use C just once in shared decoders larger than 1 to 2, we can consider the following distinct cases:

1. (A, B) shared by d_a and d_b, and (C, D) shared by d_a and d_c
2. (A, B) shared by d_a and d_b, and (C, E) shared by d_b and d_c, or
3. (A, B, C) shared by d_a and d_b.

Since cases 1 and 2 will clearly have the same costs, we will compare the cost of cases 1 and 3. For case 1, the costs of functions d_a, d_b, and d_c will be reduced by the cost of two 2-to-4-line decoders (exclusive of inverters) or 16 gate inputs. For case 3, the costs for functions d_a and d_b are reduced by one 3-to-8-line decoder (exclusive of inverters) or 24 gate inputs. So case 3 should be implemented. Formalization of this procedure into an algorithm is beyond our current scope, so only this illustration of the approach is given.

Decoder and Enabling Combinations

The function, n-to-m-line decoding with enabling, can be implemented by attaching m enabling circuits to the decoder outputs. Then, m copies of the enabling signal EN are attached to the enable control inputs of the enabling circuits. For $n = 2$ and $m = 4$, the resulting 2-to-4-line decoder with enable is shown in Figure 4-10, along with its truth table. For $EN = 0$, all of the outputs of the decoder are 0. For $EN = 1$, one of the outputs of the decode, determined by the value on (A_1, A_0), is 1 and all others are 0. If the decoder is controlling a set of lights, then with $EN = 0$, all lights are off, and with $EN = 1$, exactly one light is on, with the other three off. For large decoders $(n \geq 4)$, the gate input count can be reduced by placing the enable circuits on the inputs to the decoder and their complements rather than on each of the decoder outputs.

In Section 4-5, selection using multiplexers will be covered. The inverse of selection is *distribution* in which information received from a single line is transmitted to one of 2^n possible output lines. The circuit which implements such distribution is called a *demultiplexer*. The specific output to which the input signal is transmitted is controlled by the bit combination on n selection lines. The 2-to-4-line decoder with enable in Figure 4-10 is an implementation of a 1-to-4-line demultiplexer. For the demultiplexer, input EN provides the data, while the other

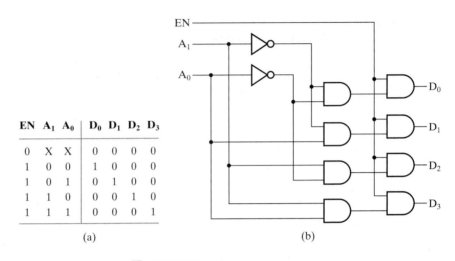

EN	A_1	A_0	D_0	D_1	D_2	D_3
0	X	X	0	0	0	0
1	0	0	1	0	0	0
1	0	1	0	1	0	0
1	1	0	0	0	1	0
1	1	1	0	0	0	1

(a)   (b)

□ **FIGURE 4-10**
A 2-to-4-Line Decoder with Enable

inputs act as the selection variables. Although the two circuits have different applications, their logic diagrams are exactly the same. For this reason, a decoder with enable input is referred to as a decoder/demultiplexer. The data input EN has a path to all four outputs, but the input information is directed to only one of the outputs, as specified by the two selection lines A_1 and A_0. For example, if $(A_1, A_0) = 10$, output D_2 has the value applied to input EN, while all other outputs remain inactive at logic 0. If the decoder is controlling a set of four lights, with $(A_1, A_0) = 10$ and EN periodically changing between 1 and 0, the light controlled by D_2 flashes on and off and all other lights are off.

4-4 ENCODING

An encoder is a digital function that performs the inverse operation of a decoder. An encoder has 2^n (or fewer) input lines and n output lines. The output lines generate the binary code corresponding to the input value. An example of an encoder is the octal-to-binary encoder whose truth table is given in Table 4-4. This encoder has eight inputs, one for each of the octal digits, and three outputs that generate the corresponding binary number. It is assumed that only one input has a value of 1 at any given time, so that the table has only eight rows with specified output values. For the remaining 56 rows, all of the outputs are don't cares. From the truth table, we can observe that A_i is 1 for the columns in which D_j is 1 only if subscript j has a binary representation with a 1 in the ith position. For example, output $A_0 = 1$ if the input is 1 or 3 or 5 or 7. Since all of these values are odd, they have a 1 in the 0 position of their binary representation. This approach can be used to find the truth table. From the table, the encoder can be implemented with n OR gates, one for each output variable A_i. Each OR gate

□ **TABLE 4-4**
   Truth Table for Octal-to-Binary Encoder

Inputs								Outputs		
D_7	D_6	D_5	D_4	D_3	D_2	D_1	D_0	A_2	A_1	A_0
0	0	0	0	0	0	0	1	0	0	0
0	0	0	0	0	0	1	0	0	0	1
0	0	0	0	0	1	0	0	0	1	0
0	0	0	0	1	0	0	0	0	1	1
0	0	0	1	0	0	0	0	1	0	0
0	0	1	0	0	0	0	0	1	0	1
0	1	0	0	0	0	0	0	1	1	0
1	0	0	0	0	0	0	0	1	1	1

combines the input variables D_j having a 1 in the rows for which A_i has value 1. For the 8-to-3-line encoder, the resulting output equations are

$$A_0 = D_1 + D_3 + D_5 + D_7$$

$$A_1 = D_2 + D_3 + D_6 + D_7$$

$$A_2 = D_4 + D_5 + D_6 + D_7$$

which can be implemented with three 4-input OR gates.

The encoder just defined has the limitation that only one input can be active at any given time: if two inputs are active simultaneously, the output produces an incorrect combination. For example, if D_3 and D_6 are 1 simultaneously, the output of the encoder will be 111 because all the three outputs are equal to 1. This represents neither a binary 3 nor a binary 6. To resolve this ambiguity, some encoder circuits must establish an input priority to ensure that only one input is encoded. If we establish a higher priority for inputs with higher subscript numbers, and if both D_3 and D_6 are 1 at the same time, the output will be 110 because D_6 has higher priority than D_3. Another ambiguity in the octal-to-binary encoder is that an output of all 0's is generated when all the inputs are 0, but this output is the same as when D_0 is equal to 1. This discrepancy can be resolved by providing a separate output to indicate that at least one input is equal to 1.

Priority Encoder

A priority encoder is a combinational circuit that implements a priority function. As mentioned in the preceding paragraph, the operation of the priority encoder is such that if two or more inputs are equal to 1 at the same time, the input having the highest priority takes precedence. The truth table for a four-input priority encoder is given in Table 4-5. With the use of X's, this condensed truth table with just five rows represents the same information as the usual 16-row truth table. Whereas X's in output columns represent don't-care conditions, X's in input

☐ **TABLE 4-5**
Truth Table of Priority Encoder

Inputs				Outputs		
D_3	D_2	D_1	D_0	A_1	A_0	V
0	0	0	0	X	X	0
0	0	0	1	0	0	1
0	0	1	X	0	1	1
0	1	X	X	1	0	1
1	X	X	X	1	1	1

columns are used to represent product terms that are not minterms. For example, 001X represents the product term $\overline{D}_3\overline{D}_2D_1$. Just as with minterms, each variable is complemented if the corresponding bit in the input combination from the table is 0 and is not complemented if the bit is 1. If the corresponding bit in the input combination is an X, then the variable does not appear in the product term. Thus, for 001X, the variable D_0, corresponding to the position of the X, does not appear in $\overline{D}_3\overline{D}_2D_1$.

The number of rows of a full truth table represented by a row in the condensed table is 2^p, where p is the number of X's in the row. For example, in Table 4-5, 1XXX represents $2^3 = 8$ truth table rows, all having the same value for all outputs. In forming a condensed truth table, we must include each minterm in at least one of the rows in the sense that minterm can be obtained by filling in 1's and 0's for the X's. Also, a minterm must never be included in more than one row such that the rows in which it appears have one or more conflicting output values.

We form Table 4-5 as follows: Input D_3 has the highest priority; so, regardless of the values of the other inputs, when this input is 1, the output for A_1A_0 is 11 (binary 3). From this we obtain the last row of the table. D_2 has the next priority level. The output is 10 if $D_2 = 1$, provided that $D_3 = 0$, regardless of the values of the lower priority inputs. From this, we obtain the fourth row of the table. The output for D_1 is generated only if all inputs with higher priority are 0, and so on down the priority levels. From this, we obtain the remaining rows of the table. The valid output designated by V is set to 1 only when one or more of the inputs are equal to 1. If all inputs are 0, V is equal to 0, and the other two outputs of the circuit are not used and are specified as don't-care conditions in the output part of the table.

The maps for simplifying outputs A_1 and A_0 are shown in Figure 4-11. The minterms for the two functions are derived from Table 4-5. The output values in the table can be transferred directly to the maps by placing them in the squares covered by the corresponding product term represented in the table. The optimized equation for each function is listed under the map for the function. The

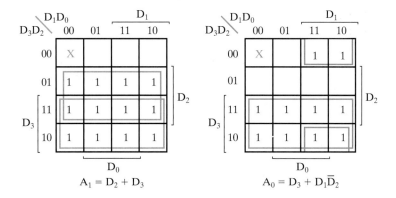

□ **FIGURE 4-11**
Maps for Priority Encoder

equation for output *V* is an OR function of all the input variables. The priority encoder is implemented in Figure 4-12 according to the following Boolean functions:

$$A_0 = D_3 + D_1\overline{D}_2$$

$$A_1 = D_2 + D_3$$

$$V = D_0 + D_1 + D_2 + D_3$$

Encoder Expansion

Thus far, we have considered only small encoders. Encoders can be expanded to larger numbers of inputs by expanding OR gates. In the implementation of

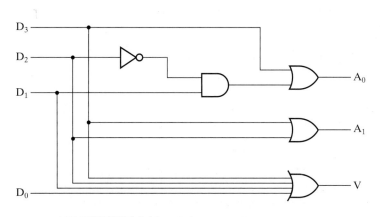

□ **FIGURE 4-12**
Logic Diagram of a 4-Input Priority Encoder

decoders, the use of multiple-level circuits with OR gates beyond the output levels shared in implementing the more significant bits in the output codes reduces the gate input count as n increases for $n \geq 5$. For $n \geq 3$, multiple-level circuits result from technology mapping anyway due to limited gate fan-in. Designing multiple-level circuits with shared gates reduces the cost of the encoders after technology mapping.

4-5 SELECTING

Selection of information to be used in a computer is a very important function, not only in communicating between the parts of the system, but also within the parts as well. Among other uses, selecting in combination with value fixing can implement combinational functions. Circuits that perform selection typically have a set of inputs from which selections are made, a single output, and a set of control lines for making the selection. First, we consider selection using multiplexers; then we briefly examine selection circuits implemented by using three-state drivers or transmission gates.

Multiplexers

A multiplexer is a combinational circuit that selects binary information from one of many input lines and directs the information to a single output line. The selection of a particular input line is controlled by a set of input variables, called *selection inputs*. Normally, there are 2^n input lines and n selection inputs whose bit combinations determine which input is selected. We begin with $n = 1$, a 2-to-1-line multiplexer. This function has two information inputs, I_0 and I_1, and a single select input S. The truth table for the circuit is given in Table 4-6. Examining the table, if the select input $S = 0$, the output of the multiplexer takes on the values of I_0, and, if input $S = 1$, the output of the multiplexer takes on the values of I_1. Thus, S selects either input I_0 or input I_1 to appear at output Y. From this discussion, we can see that the equation for the 2-to-1-line multiplexer output Y is

$$Y = \bar{S}\, I_0 + S\, I_1$$

□ **TABLE 4-6**
Truth Table for 2-to-1-Line Multiplexer

S	I_0	I_1	Y
0	0	0	0
0	0	1	0
0	1	0	1
0	1	1	1
1	0	0	0
1	0	1	1
1	1	0	0
1	1	1	1

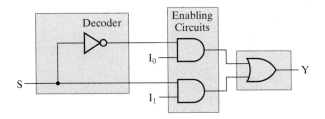

☐ **FIGURE 4-13**
A single bit 2-to-1-Line Multiplexer

   This same equation can be obtained by using a 3-variable K-map. As shown in Figure 4-13, the implementation of the preceding equation can be decomposed into a 1-to-2-line decoder, two enabling circuits and a 2-input OR gate.
   Suppose that we wish to design a 4-to-1-line multiplexer. In this case, the function Y depends on four inputs I_0, I_1, I_2, and I_3 and two select inputs S_1 and S_0. By placing the values of I_0 through I_3 in the Y column, we can form Table 4-7, a condensed truth table for this multiplexer. In this table, the information variables do not appear as input columns of the table but appear in the output column. Each row represents multiple rows of the full truth table. In Table 4-7, the row 00 I_0 represents all rows in which $(S_1, S_0) = 00$ and, for $I_0 = 1$, gives $Y = 1$ and, for $I_0 = 0$, gives $Y = 0$. Since there are six variables, and only S_1 and S_0 are fixed, this single row represents 16 rows of the corresponding full truth table. From the table, we can write the equation for Y as

$$Y = \overline{S}_1 \, \overline{S}_0 \, I_0 + \overline{S}_1 \, S_0 \, I_1 + S_1 \, \overline{S}_0 \, I_2 + S_1 \, S_0 \, I_3$$

If this equation is implemented directly, two inverters, four 3-input AND gates, and a 4-input OR gate are required, giving a gate input count of 18. A different implementation can be obtained by factoring the AND terms, to give

$$Y = (\overline{S}_1 \, \overline{S}_0) \, I_0 + (\overline{S}_1 \, S_0) \, I_1 + (S_1 \, \overline{S}_0) \, I_2 + (S_1 \, S_0) \, I_3$$

   This implementation can be constructed by combining a 2-to-4-line decoder, four AND gates used as enabling circuits and a 4-input OR gate as shown in Figure 4-14. We will refer to the combination of AND gates and OR gates as an

☐ **TABLE 4-7**
Condensed Truth Table for 4-to-1-Line Multiplexer

S_1	S_0	Y
0	0	I_0
0	1	I_1
1	0	I_2
1	1	I_3

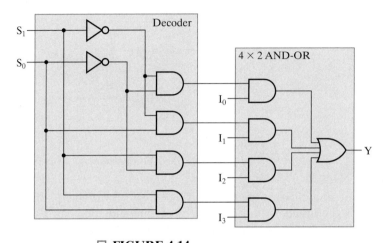

□ **FIGURE 4-14**
A Single Bit 4–to–1-Line Multiplexer

$m \times 2$ AND-OR, where m is the number of AND gates and 2 is the number of inputs to the AND gates. This resulting circuit has a gate input count of 22, which is the more costly. Nevertheless, it provides a structural basis for constructing larger n-to-2^n line multiplexers by expansion.

A multiplexer is also called a *data selector*, since it selects one of many information inputs and steers the binary information to the output line. The term "multiplexer" is often abbreviated as "MUX."

Multiplexer Expansion

Multiplexers can be expanded by considering larger values of n vectors of input bits. Expansion is based upon the circuit structure given in Figure 4-14 consisting of a decoder, enabling circuits, and an OR gate. Multiplexer design is illustrated in Examples 4-4 and 4-5.

EXAMPLE 4-4   64-to-1-Line Multiplexer

A multiplexer is to be designed for $n = 6$. This will require a 6–to–64-line decoder as given in Figure 4-9, and a 64×2 AND-OR gate. The resulting structure is shown in Figure 4-15. This structure has a gate input count of $182 + 128 + 64 = 374$. In contrast, if the decoder and the enabling circuit are replaced by inverters plus 7-input AND gates, the gate input count would be $6 + 448 + 64 = 518$. For single-bit multiplexers such as this one, combining the AND gate generating D_i with the AND gate driven by D_i into a single 3-input AND gate for every $i = 0$ through 63 reduces the gate input count to 310. For multiple-bit multiplexers, this reduction to 3-input ANDs cannot be performed without replicating the output ANDs of the decoders. As a result, in almost all cases, the original structure has a

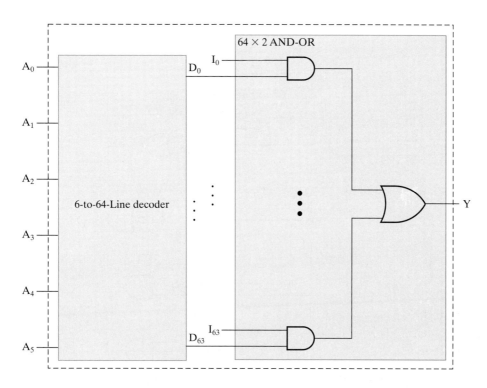

□ **FIGURE 4-15**
A 64-to-1-Line Multiplexer

lower gate input cost. The next example illustrates the expansion to a multiple-bit multiplexer.   ■

EXAMPLE 4-5   4-to-1-Line Quad Multiplexer

A quad 4-to-1-line multiplexer, which has two selection inputs and each information input replaced by a vector of four inputs, is to be designed. Since the information inputs are a vector, the output Y also becomes a four-element vector. The implementation for this multiplier requires a 2-to-4-line decoder, as given in Figure 4-7, and four 4×2 AND-OR gates. The resulting structure is shown in Figure 4-16. This structure has a gate input count of $10 + 32 + 16 = 58$. In contrast, if four 4-input multiplexers implemented with 3-input gates were placed side by side, the gate input count would be 76. So, by sharing the decoder, we reduced the gate input count.   ■

Alternative Selection Implementations

By using three-state drivers and transmission gates, it is possible to implement data selectors and multiplexers with even lower cost than is achievable with gates.

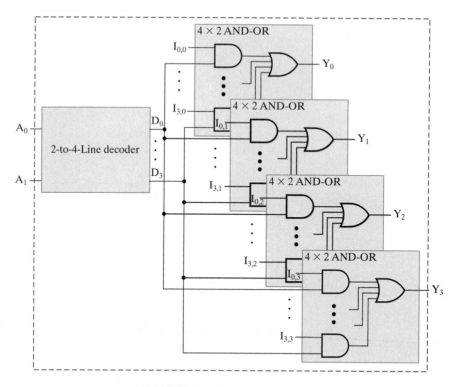

☐ **FIGURE 4-16**
A Quad 4-to-1-Line Multiplexer

THREE-STATE IMPLEMENTATIONS Three-state drivers, introduced in Chapter 2, provide an alternative implementation of multiplexers. In the implementation given in Figure 4-17(a), four 3-state drivers with their outputs connected to Y replace enabling circuits and the output OR gate to give an implementation with a gate input count of 18. The logic can be reduced further by distributing the decoding across the three-state drivers as shown in Figure 4-17(b). In this case, three pairs of enabling circuits, all with two-output decoders consisting of simply a wire and an inverter, drive the enable inputs. The gate input count for this circuit is reduced to just 14.

TRANSMISSION GATE IMPLEMENTATION As a modification of the 3-state approach in Figure 4-17(b), selection circuits can be constructed with transmission gates. This implementation, shown for 4-to-1 selection in Figure 4-18, uses transmission gates as switches. The TG circuit provides a transmission path between each *I* input and the *Y* output when the two select inputs on the transmission gates on the path have the value of 1 on the terminal without a bubble and 0 on the terminal with a bubble. With the opposite value on a select input, one of the transmission gates on the

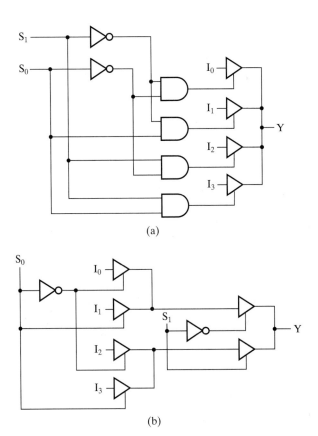

(a)

(b)

□ **FIGURE 4-17**
Selection Circuits Using Three-state Drivers

path behaves like an open switch, and the path is disconnected. The two selection inputs S_1 and S_0 control the transmission paths in the TG circuits. For example, if $S_0 = 0$ and $S_1 = 0$, there is a closed path from input I_0 to output Y, and the other three inputs are disconnected by the other TG circuits. The cost of a transmission gate is equivalent to a gate input count of one. Thus, the gate input count of this transmission gate-based multiplexer is eight.

4-6 COMBINATIONAL FUNCTION IMPLEMENTATION

Decoders and multiplexers can be used to implement Boolean functions. In addition, the programmable logic devices introduced in Chapter 3 can be viewed as containing functional blocks suitable for implementing combinational logic functions. In this section, we cover the use of decoders, multiplexers, read-only memories (ROMs), programmable logic arrays (PLAs), programmable array logic devices (PALs), and lookup tables (LUTs) for implementing combinational logic functions.

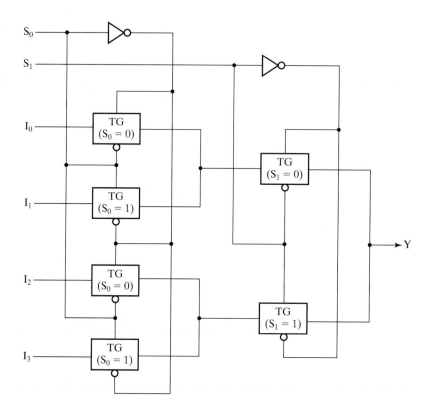

☐ **FIGURE 4-18**
4-to-1-Line Multiplexer using Transmission Gates

Using Decoders

A decoder provides the 2^n minterms of n input variables. Since any Boolean function can be expressed as a sum of minterms, one can use a decoder to generate the minterms and combine them with an external OR gate to form a sum-of-minterms implementation. In this way, any combinational circuit with n inputs and m outputs can be implemented with an n-to-2^n-line decoder and m OR gates.

The procedure for implementing a combinational circuit by means of a decoder and OR gates requires that the Boolean functions be expressed as a sum of minterms. This form can be obtained from the truth table or by plotting each function on a K-map. A decoder is chosen or designed that generates all the minterms of the input variables. The inputs to each OR gate are selected as the appropriate minterm outputs according to the list of minterms of each function. This process is shown in the next example.

EXAMPLE 4-6 Decoder and OR Gate Implementation of a Binary Adder Bit

In Chapter 1, we considered binary addition. The sum bit output S and the carry bit output C for a bit position in the addition are given in terms of the two bits being added, X and Y, and the incoming carry from the right, Z, in Table 4-8.

□ **TABLE 4-8**
Truth Table for 1-bit Binary Adder

X	Y	Z	C	S
0	0	0	0	0
0	0	1	0	1
0	1	0	0	1
0	1	1	1	0
1	0	0	0	1
1	0	1	1	0
1	1	0	1	0
1	1	1	1	1

From this truth table, we obtain the functions for the combinational circuit in sum-of-minterms form:

$$S(X,Y,Z) = \Sigma m(1,2,4,7)$$

$$C(X,Y,Z) = \Sigma m(3,5,6,7)$$

Since there are three inputs and a total of eight minterms, we need a 3-to-8-line decoder. The implementation is shown in Figure 4-19. The decoder generates all eight minterms for inputs X, Y, and Z. The OR gate for output S forms the logical sum of minterms 1, 2, 4, and 7. The OR gate for output C forms the logical sum of minterms 3, 5, 6, and 7. Minterm 0 is not used. ■

   A function with a long list of minterms requires an OR gate with a large number of inputs. A function having a list of k minterms can be expressed in its complement form with $2^n - k$ minterms. If the number of minterms in a function F is greater than $2^n/2$, then the complement of F, $\overline{F}$, can be expressed with fewer minterms. In such a case, it is advantageous to use a NOR gate instead of an OR gate. The OR portion of the NOR gate produces the logical sum of the minterms of $\overline{F}$. The output bubble of the NOR gate complements this sum and generates the normal output F.

   The decoder method can be used to implement any combinational circuit. However, this implementation must be compared with other possible implementations to determine the best solution. The decoder method may provide the best solution, particularly if the combinational circuit has many outputs based on the same inputs and each output function is expressed with a small number of minterms.

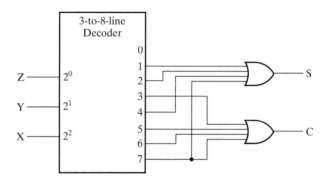

☐ **FIGURE 4-19**
Implementing a Binary Adder Using a Decoder

Using Multiplexers

In Section 4-5, we learned that a decoder combined with an $m \times 2$ AND-OR gate implements a multiplexer. The decoder in the multiplexer generates the minterms of the selection inputs. The AND-OR gate provides enabling circuits that determine whether the minterms are "attached" to the OR gate with the information inputs (I_i) used as the enabling signals. If I_i input is 1, then minterm m_i is attached to the OR gate, and, if the I_i input is a 0, then minterm m_i is replaced by a 0. Value-fixing applied to the I inputs provides a method for implementing a Boolean function of n variables with a multiplexer having n selection inputs and 2^n data inputs, one for each minterm. Further, an m-output function can be implemented by using value-fixing on a multiplexer with m-bit information vectors instead of the individual I bits, as illustrated by the next example.

EXAMPLE 4-7   Multiplexer Implementation of a Binary Adder Bit

The values for S and C from the 1-bit binary adder truth table given in Table 4-8 can be generated by using value-fixing on the information inputs of a multiplexer. Since there are three selection inputs and a total of eight minterms, we need a dual 8-to-1-line multiplexer for implementing the two outputs, S and C. The implementation based on the truth table is shown in Figure 4-20. Each pair of values, such as (0, 1) on ($I_{1,1}$, $I_{1,0}$), is taken directly from the corresponding row of the last two truth table columns.   ▪

   A more efficient method implements a Boolean function of n variables with a multiplexer that has only $n - 1$ selection inputs. The first $n - 1$ variables of the function are connected to the selection inputs of the multiplexer. The remaining variable of the function is used for the information inputs. If the final variable is Z, each data input of the multiplexer will be either Z, $\overline{Z}$, 1, or 0. The function can be implemented by attaching implementations of the four rudimentary functions from

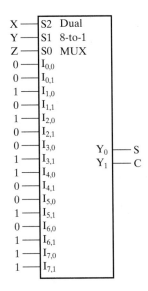

Table 4-1 to the information inputs to the multiplexer. The next example demonstrates this procedure.

EXAMPLE 4-8   Alternative Multiplexer Implementation of a Binary Adder Bit

This function can be implemented with a dual 4-to-1-line multiplexer, as shown in Figure 4-21. The design procedure can be illustrated by considering the sum S. The two variables X and Y are applied to the selection lines in that order; X is connected to the S_1 input, and Y is connected to the S_0 input. The values for the data input lines are determined from the truth table of the function. When $(X, Y) = 00$, the output S is equal to Z because $S = 0$ when $Z = 0$ and $S = 1$ when $Z = 1$. This

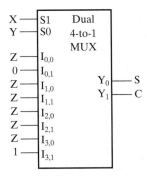

□ **FIGURE 4-21**
Implementing a 1-bit Binary Adder with a Dual 8-to-1-Line Multiplexer

requires that the variable Z be applied to information input I_{00}. The operation of the multiplexer is such that, when $(X,Y) = 00$, information input I_{00} has a path to the output that makes S equal to Z. In a similar fashion, we can determine the required input to lines I_{10}, I_{20}, and I_{30} from the value of S when $(X,Y) = 01, 10$, and 11, respectively. A similar approach can be used to determine the values for I_{01}, I_{11}, I_{21}, and I_{31}. ∎

The general procedure for implementing any Boolean function of n variables with a multiplexer with $n - 1$ selection inputs and 2^{n-1} data inputs follows from the preceding example. The Boolean function is first listed in a truth table. The first $n - 1$ variables in the table are applied to the selection inputs of the multiplexer. For each combination of the selection variables, we evaluate the output as a function of the last variable. This function can be 0, 1, the variable, or the complement of the variable. These values are then applied to the appropriate data inputs. This process is illustrated in the next example.

EXAMPLE 4-9   Multiplexer Implementation of 4-Variable Function.

As a second example, consider the implementation of the following Boolean function:

$$F(A,B,C,D) = \Sigma m(1,3,4,11,12,13,14,15)$$

This function is implemented with an 8×1 multiplexer as shown in Figure 4-22. To obtain a correct result, the variables in the truth table are connected to selection inputs S_2, S_1, and S_0 in the order in which they appear in the table (i.e., such that A is connected to S_2, B is connected to S_1, and C is connected to S_0, respectively). The values for the data inputs are determined from the truth table. The information line number is determined from the binary combination of A, B, and C. For example, when $(A, B, C) = 101$, the truth table shows that $F = D$; so the input variable D is applied to information input I_5. The binary constants 0 and 1 correspond to two fixed signal values. Recall from Section 4-2 that, in a logic schematic, these constant values are replaced by the ground and power symbols as shown in Figure 4-2. ∎

Using Read-Only Memories

On the basis of the decoders and multiplexers covered thus far, there are two approaches to the implementation of read-only memories. One approach is based on a decoder and OR gates. By inserting parallel OR gates, one for each ROM output, to sum the minterms of Boolean functions, we were able to generate any desired combinational circuit. The ROM can be viewed as a device that includes both the decoder and the OR gates within a single unit. By closing connections to the inputs of an OR gate for those minterms that are included in the function, the ROM outputs can be programmed to represent the Boolean functions of the output variables in a combinational circuit. The alternative approach is based on value-fixing on a multiple-bit multiplexer. The I_i values are used as enabling

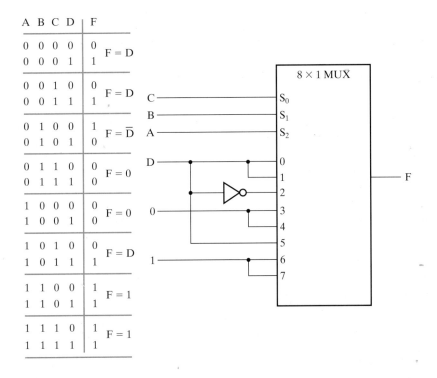

A	B	C	D	F	
0	0	0	0	0	F = D
0	0	0	1	1	
0	0	1	0	0	F = D
0	0	1	1	1	
0	1	0	0	1	F = $\overline{D}$
0	1	0	1	0	
0	1	1	0	0	F = 0
0	1	1	1	0	
1	0	0	0	0	F = 0
1	0	0	1	0	
1	0	1	0	0	F = D
1	0	1	1	1	
1	1	0	0	1	F = 1
1	1	0	1	1	
1	1	1	0	1	F = 1
1	1	1	1	1	

□ **FIGURE 4-22**

Implementing a Four-Input Function with a Multiplexer

signals to determine which minterms are connected to the OR gates within the multiplexer bits. This is illustrated by Example 4-8 which is equivalent to a ROM with three inputs and two outputs. The "programming" of this ROM approach is done by applying the truth table to the information inputs to the multiplexer. Because the decoder and OR gates approach is just a different model, it too can be "programmed" by using a truth table to determine the connections between the decoder and the OR gates. Thus, in practice, when a combinational circuit is designed by means of a ROM, it is not necessary to design the logic or to show the internal connections inside the unit. All that the designer has to do is specify the particular ROM by its name and provide the truth table for the ROM. The truth table gives all the information for programming the ROM. No internal logic diagram is needed to accompany the table. Example 4-10 shows this use of a ROM.

EXAMPLE 4-10 Implementing a Combinational Circuit Using a ROM

Design a combinational circuit using a ROM. The circuit accepts a 3-bit number and generates an output binary number equal to the square of the input number.

The first step in designing the circuit is to derive the truth table of the combinational circuit. In most cases, this is all that is needed. In other cases, we can use a partial truth table for the ROM by utilizing certain properties in the output variables. Table 4-9 is the truth table for the combinational circuit. Three inputs and six outputs are needed to accommodate all of the possible binary numbers. We note that output B_0 is always equal to input A_0, so there is no need to generate B_0 with a ROM. Moreover, output B_1 is always 0, so this output is a known constant. Thus, we actually need to generate only four outputs with the ROM; the other two are readily obtained. The minimum size of ROM needed must have three inputs and four outputs. Three inputs specify eight words; so the ROM must be of size 8×4. The implementation of the ROM is shown in Figure 4-23. The three inputs specify eight words of four bits each. The block diagram of Figure 4-23(a) shows the required connections of the combinational circuit. The truth table in Figure 4-23(b) specifies the information needed for programming the ROM. ■

☐ **TABLE 4-9**
Truth Table for Circuit of Example 4-10

Inputs			Outputs						
A_2	A_1	A_0	B_5	B_4	B_3	B_2	B_1	B_0	Decimal
0	0	0	0	0	0	0	0	0	0
0	0	1	0	0	0	0	0	1	1
0	1	0	0	0	0	1	0	0	4
0	1	1	0	0	1	0	0	1	9
1	0	0	0	1	0	0	0	0	16
1	0	1	0	1	1	0	0	1	25
1	1	0	1	0	0	1	0	0	36
1	1	1	1	1	0	0	0	1	49

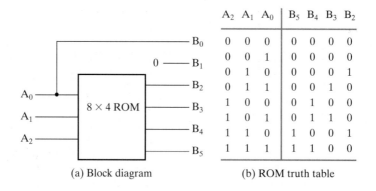

(a) Block diagram

(b) ROM truth table

☐ **FIGURE 4-23**
ROM Implementation of Example 4-10

ROM devices are widely used to implement complex combinational circuits directly from their truth tables. They are useful for converting from one code, such as Gray code, to another, such as BCD. They can generate complex arithmetic operations, such as multiplication or division, and in general, they are used in applications requiring a moderate number of inputs and a large number of outputs.

Using Programmable Logic Arrays

The PLA is similar in concept to the ROM, except that the PLA does not provide full decoding of the variables, so it does not generate all the minterms. The decoder is replaced by an array of AND gates, each of which can be programmed to generate any product term of the input variables. The product terms are then selectively connected to OR gates as in a ROM to provide the sum of products for the required functions.

The fuse map of a PLA can be specified in tabular form. For example, the programming table that specifies the PLA of Figure 4-24 is listed in Table 4-10. The table consists of three sections. The first section lists the product term numbers.

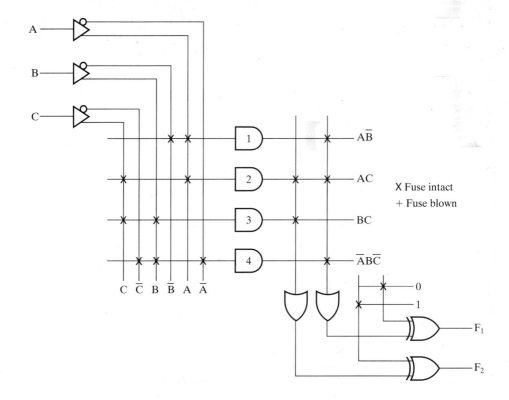

☐ **FIGURE 4-24**
PLA with Three Inputs, Four Product Terms, and Two Outputs

☐ **TABLE 4-10**
Programming Table for the PLA in Figure 4-24

Product term		Inputs			Outputs	
		A	B	C	(T) F_1	(C) F_2
$A\overline{B}$	1	1	0	—	1	—
AC	2	1	—	1	1	1
BC	3	—	1	1	—	1
$\overline{A}B\overline{C}$	4	0	1	0	1	—

The second section specifies the required paths between inputs and AND gates. The third section specifies the paths between the AND and OR gates. For each output variable, we may have a T (for true) or C (for complement) for controlling the output exclusive-OR gate. The product terms listed on the left are not part of the table; they are included for reference only. For each product term, the inputs are marked with 1, 0, or — (dash). If a variable in the product term appears in its true form, the corresponding input variable is marked with a 1. If the variable in the product term appears complemented, the corresponding input variable is marked with a 0. If the variable is absent in the product term, it is marked with a dash.

The paths between the inputs and the AND gates are specified under the column heading "Inputs" in the table. A 1 in the input column specifies a CLOSED circuit from the input variable to the AND gate. A 0 in the input column specifies a CLOSED circuit from the complement of the variable to the input of the AND gate. A dash specifies OPEN circuits for both the input variable and its complement. It is assumed that an OPEN terminal on the input of an AND gate behaves like a 1.

The paths between the AND and OR gates are specified under the column heading "Outputs." The output variables are marked with 1's for those product terms that are included in the function. Each product term that has a 1 in the output column requires a CLOSED path from the output of the AND gate to the input of the OR gate. Those product terms marked with a dash specify an OPEN circuit. It is assumed that an open terminal on the input of an OR gate behaves like a 0. Finally, a T (true) output dictates that the other input of the corresponding XOR gate be connected to 0, and a C (complement) specifies a connection to 1.

The size of a PLA is specified by the number of inputs, the number of product terms, and the number of outputs. A typical PLA has 16 inputs, 48 product terms, and eight outputs. For n inputs, k product terms, and m outputs, the internal logic of the PLA consists of n buffer-inverter gates, k AND gates, m OR gates, and m XOR gates. There are $2n \times k$ programmable connections between the inputs and the AND array, $k \times m$ programmable connections between the AND and OR arrays, and m programmable connections associated with the XOR gates.

In designing a digital system with a PLA, there is no need to show the internal connections of the unit, as was done in Figure 4-24. All that is needed is a PLA programming table from which the PLA can be programmed to supply the required logic. As with a ROM, the PLA may be mask programmable or field programmable.

In implementing a combinational circuit with a PLA, a careful investigation must be undertaken in order to reduce the number of distinct product terms, so that the complexity of the circuit may be reduced. Fewer product terms can be achieved by simplifying the Boolean function to a minimum number of terms. The number of literals in a term is less important, since all the input variables are available anyway. It is wise, however, to avoid extra literals, as these may cause problems in testing the circuit and may reduce the speed of the circuit. Both the true and complement forms of each function should be simplified to see which one can be expressed with fewer product terms and which one provides product terms that are common to other functions. This process is shown in Example 4-11.

EXAMPLE 4-11 Implementing a Combinational Circuit Using a PLA

Implement the following two Boolean functions with a PLA:

$$F_1(A, B, C) = \Sigma m(0, 1, 2, 4)$$

$$F_2(A, B, C) = \Sigma m(0, 5, 6, 7)$$

The two functions are simplified in the maps of Figure 4-25. Both the true and complement outputs of the functions are simplified in sum-of-products form. The combination that gives a minimum number of product terms is

$$F_1 = \overline{AB + AC + BC}$$

$$F_2 = AB + AC + \overline{A}\,\overline{B}\,\overline{C}$$

The simplification gives four distinct product terms: AB, AC, BC, and $\overline{A}\,\overline{B}\,\overline{C}$. The PLA programming table for this combination is shown in the figure. Note that output F_1 is the true output, even though a C is marked over it in the table. This is because $\overline{F}_1$ is generated with an AND-OR circuit and is available at the output of the OR gate. The XOR gate complements the function $\overline{F}_1$ to produce the true F_1 output. ▪

Using Programmable Array Logic Devices

In designing with a PAL device, the Boolean functions must be simplified to fit into each section as illustrated by the example PAL device in Figure 4-26. Unlike the arrangement in the PLA, a product term cannot be shared among two or more OR gates. Therefore, each function can be simplified by itself, without regard to

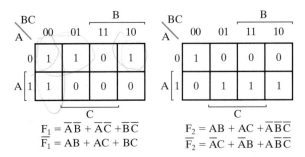

$$F_1 = \overline{A}B + \overline{A}\,\overline{C} + \overline{B}\,\overline{C}$$
$$\overline{F_1} = AB + AC + BC$$

$$F_2 = AB + AC + \overline{A}\,\overline{B}\,\overline{C}$$
$$\overline{F_2} = \overline{A}C + \overline{A}B + A\overline{B}\,\overline{C}$$

PLA programming table

Product term		Inputs		(C)	(T)
		A B C		F_1	F_2
AB	1	1 1 –		1	1
AC	2	1 – 1		1	1
BC	3	– 1 1		1	–
$\overline{A}\,\overline{B}\,\overline{C}$	4	0 0 0		–	1

☐ **FIGURE 4-25**
Solution to Example 4-11

common product terms. The number of product terms in each section is fixed, and if the number of terms in the function is too large, it may be necessary to use two or more sections to implement one Boolean function. In such a case, common terms may be useful. This process is illustrated in Example 4-12.

EXAMPLE 4-12   Implementing a Combinational Circuit Using a PAL

As an example of a PAL device incorporated into the design of a combinational circuit, consider the following Boolean functions, given in sum-of-minterms form:

$$W(A, B, C, D) = \Sigma\, m(2, 12, 13)$$

$$X(A, B, C, D) = \Sigma\, m(7, 8, 9, 10, 11, 12, 13, 14, 15)$$

$$Y(A, B, C, D) = \Sigma\, m(0, 2, 3, 4, 5, 6, 7, 8, 10, 11, 15)$$

$$Z(A, B, C, D) = \Sigma\, m(1, 2, 8, 12, 13)$$

Simplifying the four functions to a minimum number of terms results in the following Boolean functions:

$$W = AB\overline{C} + \overline{A}\,\overline{B}C\overline{D}$$

$$X = A + BCD$$

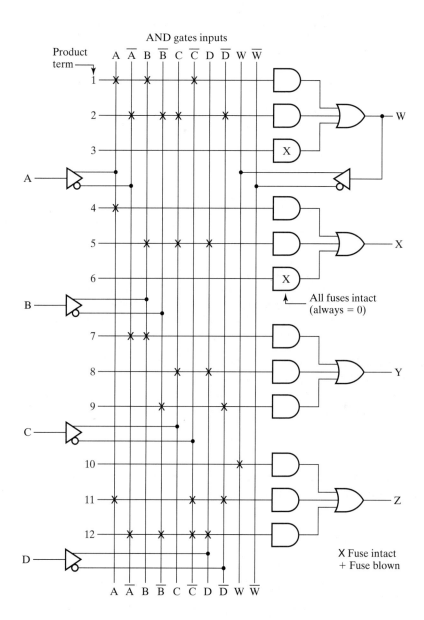

□ **FIGURE 4-26**
Connection Map for PAL® Device for Example 4-12

$$Y = \overline{A}B + CD + \overline{B}\overline{D}$$

$$Z = AB\overline{C} + \overline{A}\,\overline{B}C\overline{D} + A\overline{C}\overline{D} + \overline{A}\,\overline{B}\,\overline{C}D$$

$$= W + A\overline{C}\overline{D} + \overline{A}\,\overline{B}\,\overline{C}D$$

Note that the function for Z has four product terms. The logical sum of two of these terms is equal to W. Thus, by using W, it is possible to reduce the number of terms for Z from four to three, so that the functions can fit into the PAL device in Figure 4-26.

The PAL programming table is similar to the table used for the PLA, except that only the inputs of the AND gates need to be programmed. Table 4-11 lists the PAL programming table for the preceding four Boolean functions. The table is divided into four sections with three product terms in each, to conform with the PAL device of Figure 4-26. The first two sections need only two product terms to implement the Boolean function. By placing W in the first section of the device, the feedback connection from F1 into the array is available to reduce the function Z to three terms.

The connection map for the PAL device, as specified in the programming table, is shown in Figure 4-26. For each 1 or 0 in the table, we mark the corresponding intersection in the diagram with the symbol for a CLOSED circuit connection. For each dash, we mark the diagram with OPEN circuit connections for both the true and complement inputs. If the AND gate is not used, we leave all of its inputs as CLOSED circuits. Since the corresponding input receives both the true and the complement of each input variable, we have $A \cdot \overline{A} = 0$, and the output of the AND gate is always 0. ■

□ **TABLE 4-11**
PAL® Programming Table for Example 4-12

Product term	AND Inputs					Outputs
	A	B	C	D	W	
1	1	1	0	—	—	$W = AB\overline{C}$
2	0	0	1	0	—	$+\overline{A}\,\overline{B}C\overline{D}$
3	—	—	—	—	—	
4	1	—	—	—	—	$X = A$
5	—	1	1	1	—	$+BCD$
6	—	—	—	—	—	
7	0	1	—	—	—	$Y = \overline{A}B$
8	—	—	1	1	—	$+CD$
9	—	0	—	0	—	$+\overline{B}\,\overline{D}$
10	—	—	—	—	1	$Z = W$
11	1	—	0	0	—	$+A\overline{C}\overline{D}$
12	0	0	0	1	—	$+\overline{A}\,\overline{B}\,\overline{C}D$

Using Lookup Tables

Field-programmable gate arrays (FPGAs) and complex programmable logic devices (CPLDs), often use lookup tables (LUTs) for implementing logic. Programming a single function with m inputs is the same as programming a single output ROM (i.e., the lookup table simply stores the truth table of the function). An m-input lookup table can implement any function of m or fewer variables. Typically, $m = 4$. The key problem in programming lookup tables is dealing with functions with $> m$ input variables. Also, sharing of lookup tables among multiple functions is important. These problems can be dealt with using multiple-level logic transformations, primarily decomposition and extraction. The optimization goal is to implement the function or functions by using a minimum number of LUTs with the constraint that a single LUT can implement a subfunction of at most m variables. This can be accomplished by finding a minimum number of equations, each with at most m variables, that implement the desired function or functions. This process is illustrated for single-output and multiple-output functions with $m = 4$ in the following examples.

EXAMPLE 4-13   **Implementing a Single-Output Function with Lookup Tables**

Implement the following Boolean function with lookup tables:

$$F_1(A, B, C, D, E, F, G, H, I) = ABCDE + \overline{F}GHI\overline{DE}$$

The number of input variables for a function is called the *support s* of the function. The support for F_1 is $s = 9$. It is apparent that the minimum number of lookup tables k needed is at least 9/4 (i.e., $k = 3$). Further, for an m-output function, the minimum number of lookup tables for a single-output function must obey an even more stringent relationship, $mk \geq s + k - 1$, so k must satisfy $4k \geq 9 + k - 1$. Solving, $k = 3$, so we will look for a decomposition of F_1 into three equations, each with at most $s = 4$. Factoring F_1, we obtain

$$F_1 = (ABC)DE + (\overline{F}GHI)\overline{DE}$$

Based on this equation, F_1 can be decomposed into three equations with $s \leq 4$:

$$F_1(D, E, X_1, X_2) = X_1DE + X_2\overline{DE}$$

$$X_1(A, B, C) = ABC$$

$$X_2(F, G, H, I) = \overline{F}GHI$$

Each of these three equations can be implemented by one LUT, giving an optimum implementation for F_1.   ▪

EXAMPLE 4-14 Implementing a Multiple-Output Function with Lookup Tables

Implement the following pair of Boolean function with lookup tables:

$$F_1(A,B,C,D,E,F,G,H,I) = ABCDE + \overline{FGHI}D\overline{E}$$

$$F_2(A,B,C,D,E,F,G,H,I) = ABCEF + \overline{F}GHI$$

Each of these functions requires a support $s = 9$. Thus, at least three LUTs are required for each function. But two of the LUTs can be shared, so the minimum number of LUTs needed is $k = 6 - 2 = 4$. Factoring F_2 with sharing of equations with the decomposition of F1 from the previous example gives

$$F_2 = (ABC)EF + (\overline{F}GHI)$$

This produces an extraction for F_1 and F_2:

$$F_1(D,E,X_1,X_2) = X_1DE + X_2\overline{D}\overline{E}$$

$$F_2(E,F,X_1,X_2) = X_1EF + X_2$$

$$X_1(A,B,C) = ABC$$

$$X_2(F,G,H,I) = \overline{F}GHI$$

In this case, the extraction requires four LUTs, the minimum number. Note that in general, there is no guarantee that a decomposition or extraction can be found that requires the minimum number of LUTs calculated. ∎

4–7 HDL REPRESENTATION FOR COMBINATIONAL CIRCUITS—VHDL

Since an HDL is used for describing and designing hardware, it is very important to keep the underlying hardware in mind as you write in the language. This is particularly critical if your language description is to be synthesized. For example, if you ignore the hardware that will be generated, it is very easy to specify a large complex gate structure by using × (multiply) when a much simpler structure using only a few gates is all that is needed. For this reason, we initially emphasize description of detailed hardware with VHDL, and proceed to more abstract, higher-level descriptions later.

Selected examples in this chapter are useful for introducing VHDL as an alternative means for representing detailed digital circuits. Initially, we show structural VHDL descriptions that replace the schematic for the 2-to-4-line decoder with enable given in Figure 4-10 on page 152. This example and one using the 4-to-1-line multiplexer in Figure 4-14 on page 158 illustrate many of the fundamental concepts of VHDL. We then present higher level functional and behavioral VHDL descriptions for these circuits that further illustrate fundamental VHDL concepts.

EXAMPLE 4-15   Structural VHDL for a 2-to-4-Line Decoder

Figure 4-27 shows a VHDL description for the 2-to-4-line decoder circuit from Figure 4-10 on page 152. This example will be used to demonstrate a number of general VHDL features as well as structural description of circuits.

The text between two dashes -- and the end of the line is interpreted as a *comment*. So the description in Figure 4-27 begins with a two-line comment identifying the description and its relationship to Figure 4-10. To assist in discussion of

```
-- 2-to-4 Line Decoder with Enable: Structural VHDL Description    -- 1
-- (See Figure 4-10 for logic diagram)                             -- 2
library ieee, lcdf_vhdl;                                           -- 3
use ieee.std_logic_1164.all, lcdf_vhdl.func_prims.all;            -- 4
entity decoder_2_to_4_w_enable is                                 -- 5
  port (EN, A0, A1: in std_logic;                                 -- 6
        D0, D1, D2, D3: out std_logic);                           -- 7
end decoder_2_to_4_w_enable;                                      -- 8
                                                                  -- 9
architecture structural_1 of decoder_2_to_4_w_enable is           -- 10
  component NOT1                                                  -- 11
    port(in1: in std_logic;                                      -- 12
         out1: out std_logic);                                   -- 13
  end component;                                                  -- 14
  component AND2                                                 -- 15
    port(in1, in2: in std_logic;                                -- 16
         out1: out std_logic);                                   -- 17
  end component;                                                  -- 18
  signal A0_n, A1_n, N0, N1, N2, N3: std_logic;                 -- 19
  begin                                                          -- 20
    g0: NOT1 port map (in1 => A0, out1 => A0_n);                -- 21
    g1: NOT1 port map (in1 => A1, out1 => A1_n);                -- 22
    g2: AND2 port map (in1 => A0_n, in2 => A1_n, out1 => N0);   -- 23
    g3: AND2 port map (in1 => A0, in2 => A1_n, out1 => N1);     -- 24
    g4: AND2 port map (in1 => A0_n, in2 => A1,out1 => N2);      -- 25
    g5: AND2 port map (in1 => A0, in2 => A1, out1 => N3);       -- 26
    g6: AND2 port map (in1 => EN, in2 => N0, out1 => D0);       -- 27
    g7: AND2 port map (in1 => EN, in2 => N1, out1 => D1);       -- 28
    g8: AND2 port map (in1 => EN, in2 => N2, out1 => D2);       -- 29
    g9: AND2 port map (in1 => EN, in2 => N3, out1 => D3);       -- 30
end structural_1;                                                 -- 31
```

□ **FIGURE 4-27**

Structural VHDL Description of 2-to-4-Line Decoder

this description, comments providing line numbers have been added on the right. As a language, VHDL has a syntax that describes precisely the valid constructs that can be used in the language. This example will illustrate many aspects of the syntax. In particular, note the use of semicolons, commas and colons in the description.

Initially, we skip lines 3 and 4 of the description to focus on the overall structure. Line 5 begins the declaration of an *entity*, which is the fundamental unit of a VHDL design. In VHDL, just as for a symbol in a schematic, we need to give the design a name and to define its inputs and outputs. This is the function of the *entity declaration*. **Entity** and **is** are keywords in VHDL. Keywords, which we show in bold type, have a special meaning and cannot be used to name objects such as entities, inputs, outputs or signals. Statement **entity** decoder_2_to_4_w_enable **is** declares that a design exists with the name decoder_2_to_4_w_enable. VHDL is case insensitive (i.e., names and keywords are not distinguished by the use of uppercase or lowercase letters). DECODER_2_4_W_ENABLE is the same as Decoder_2_4_w_Enable and decoder_2_4_w_enable.

Next, a *port declaration* in lines 6 and 7 is used to define the inputs and outputs just as we would do for a symbol in a schematic. For the example design, there are three input signals: EN, A0, and A1. The fact that these are inputs is denoted by the mode **in**. Likewise, D0, D1, D2 and D3 are denoted as outputs by the mode **out**. VHDL is a strongly-typed language, so the type of the inputs and output must be declared. In this case, the type is std_logic, which represents *standard logic*. This type declaration specifies the values that may appear on the inputs and the outputs, as well as the operations that may be applied to the signals. Standard logic, among its nine values, includes the usual binary values 0 and 1 and two additional values X and U. X represents an unknown value, U an uninitalized value. We have chosen to use standard logic, which includes these values, since these values are used by typical simulation tools.

In order to use the type std_logic, it is necessary to define the values and the operations. For convenience, a *package* consisting of precompiled VHDL code is employed. Packages are usually stored in a directory referred to as a *library,* which is shared by some or all of the tool users. For std_logic, the basic package is ieee.std_logic_1164. This package defines the values and basic logic operators for types std_ulogic and std_logic. In order to use std_logic, we include line 3 to call up the **library** of packages called ieee and include line 4 containing ieee.std_logic_1164.**all** to indicate we want to use **all** of the package std_logic_1164 from the ieee library. An additional library, lcdf_vhdl, contains a package called func_prims made up of basic logic gates, latches and flip-flops described using VHDL, of which we use **all**. Library lcdf_vhdl is available in ASCII for copying from the Prentice Hall Companion Website for the text. Note that the statements in lines 3 and 4 are tied to the entity that follows. If another entity is included that uses type std_logic and the elements from func_prims, these statements must be repeated prior to that entity declaration.

The entity declaration ends with keyword **end** followed by the entity name. Thus far, we have discussed the equivalent of a schematic symbol in VHDL for the circuit.

STRUCTURAL DESCRIPTION   Next, we want to specify the function of the circuit. A particular representation of the function of an entity is called the *architecture* of the entity. Thus, the contents of line 10 declare a VHDL architecture named structural_1 for the entity decoder_2_to_4_w_enable to exist. The details of the architecture follow. In this case, we use a *structural description* that is equivalent to the schematic for the circuit given in Figure 4-10.

First, we declare the gate types we are going to use as components of our description in lines 11 through 18. Since we are building this architecture from gates, we declare an inverter called NOT1 and a 2-input AND gate called AND2 as *components*. These gate types are VHDL descriptions in package func_prims that contain the entity and architecture for each of the gates. The name and the port declaration for a component must be identical to those for the underlying entity. For NOT1, port gives the input name in1 and the output name out1. The component declaration for AND2 gives input names in1 and in2, and output name out1.

Next, before specifying the interconnection of the gates, which is equivalent to a circuit netlist, we need to name all of the nets in the circuit. The inputs and outputs already have names. The internal nets are the outputs of the two inverters and of the leftmost four AND gates in Figure 4-10. These output nets are declared as *signals* of type std_logic. A0_n and A1_n are the signals for the two inverter outputs and N0, N1, N2, and N3 are the signals for the four AND gate outputs. Likewise, all of the inputs and outputs declared as ports are signals. In VHDL, there are both signals and variables. Variables are evaluated instantaneously. In contrast, signals are evaluated at some future point in time. This time may be physical time, such as 2 ns from the current time, or may be what is called *delta time,* in which a signal is evaluated one delta time from the current time. Delta time is viewed as an infinitesimal amount of time. Some time delay in evaluation of signals is essential to the internal operation of the typical digital simulator and, of course, based on the delay of gates, is realistic in performing simulations of circuits. For simplicity, we will typically be simulating circuits for correct function, not for performance or delay problems. For such functional simulation, it is easiest to let the delays default to delta times. Thus, no delay will be explicit in our VHDL descriptions of circuits, although delays may appear in test benches.

Following the declaration of the internal signals, the main body of the architecture starts with the keyword **begin**. The circuit described consists of two inverters and eight 2-input AND gates. Line 21 gives the label g0 to the first inverter and indicates that the inverter is component NOT1. Next is a **port map**, which maps the input and output of the inverter to the signals to which they are connected. This particular form of port map uses => with the port of the gate on the left and the signal to which it is connected on the right. For example, the input of inverter g0 is A0 and the output is A0_n. Lines 22 through 30 give the remaining nine gates and the signals connected to their inputs and outputs. For example, in line 24, A0 and A1_n are inputs and N1 is the output. The architecture is completed with the keyword **end** followed by its name structural_1.

EXAMPLE 4-16   Structural VHDL for a 4–to–1 Multiplexer

In Figure 4-28, the structural description of the 4-to-1-line multiplexer from
Figure 4-14 on page 158 illustrates two additional VHDL concepts: std_logic_vector
and an alternative approach to mapping ports

In lines 6 and 7, instead of specifying S and I as individual std_logic inputs,
they are specified as *std_logic_vectors*. In specifying vectors, we use an index. Since

```
-- 4-to-1 Line Multiplexer: Structural VHDL Description        -- 1
-- (See Figure 4-14 for logic diagram)                         -- 2
library ieee, lcdf_vhdl;                                       -- 3
use ieee.std_logic_1164.all, lcdf_vhdl.func_prims.all;        -- 4
entity multiplexer_4_to_1_st is                               -- 5
  port (S: in std_logic_vector(0 to 1);                       -- 6
        I: in std_logic_vector(0 to 3);                       -- 7
        Y: out std_logic);                                    -- 8
end multiplexer_4_to_1_st;                                    -- 9
                                                              --10
architecture structural_2 of multiplexer_4_to_1_st is         --11
  component NOT1                                              --12
    port(in1: in std_logic;                                  --13
         out1: out std_logic);                               --14
  end component;                                             --15
  component AND2                                             --16
    port(in1, in2: in std_logic;                            --17
         out1: out std_logic);                               --18
  end component;                                             --19
  component OR4                                              --20
    port(in1, in2, in3, in4: in std_logic;                  --21
         out1: out std_logic);                               --22
  end component;                                             --23
  signal S_n: std_logic_vector(0 to 1);                      --24
  signal D, N: std_logic_vector(0 to 3);                     --25
  begin                                                      --26
    g0: NOT1 port map (S(0), S_n(0));                        --27
    g1: NOT1 port map (S(1), S_n(1));                        --28
    g2: AND2 port map (S_n(1), S_n(0), D(0));                --29
    g3: AND2 port map (S_n(1), S(0), D(1));                  --30
    g4: AND2 port map (S(1), S_n(0), D(2));                  --31
    g5: AND2 port map (S(1), S(0), D(3));                    --32
    g6: AND2 port map (D(0), I(0), N(0));                    --33
    g7: AND2 port map (D(1), I(1), N(1));                    --34
    g8: AND2 port map (D(2), I(2), N(2));                    --35
    g9: AND2 port map (D(3), I(3), N(3));                    --36
    g10: OR4 port map (N(0), N(1), N(2), N(3), Y);           --37
  end structural_2;                                          --38
```

☐ **FIGURE 4-28**
Structural VHDL Description of 4-to-1-Line Multiplexer

S consists of two input signals numbered 0 and 1, the index for S is 0 to 1. The components of this vector are S(0) and S(1). I consists of four input signals numbered 0 through 3, so the index for I is 0 to 3. Likewise, in lines 24 and 25, we specify signals S_n, D, and N as std_logic_vectors. D represents the decode outputs, and N represents the four internal signals between the AND gates and the OR gate.

Beginning at line 27, note how the signals within std_logic_vectors are referred to by giving the signal name and the index in parentheses. It is also possible to refer to subvectors (e.g., N(1 to 2), which refers to N(1) and N(2), the center two signals in N). Also, if one wishes to have the larger index for a vector appear first, VHDL uses a somewhat different notational approach. For example, signal N: std_logic_vector(3 downto 0) defines the first bit in signal N as N(3) and the last signal in N as N(0).

In lines 27 through 37, an alternative method is used to specify the port maps for the logic gates. Instead of explicitly giving the component input and output names, we assume that these names are in the port map in the same order as given for the component. We can then implicitly specify the signals attached to these names by listing the signals in same order as the names. For example, in line 29, S_n(1) appears first, and so is connected to in1. S_n(0) appears second, and so is connected to in2. Finally, D(0) is connected to out1.

Otherwise, this VHDL description is similar in structure to that for the 2-to-4-line decoder, except that the schematic represented is that in Figure 4-14 on page 158.

■

DATAFLOW DESCRIPTION A dataflow description describes a circuit in terms of function rather than structure and is made up of concurrent assignment statements or their equivalent. Concurrent assignment statements are executed concurrently (i.e., in parallel) whenever one of the values on the right-hand side of the statement changes. For example, whenever a change occurs in a value on the right-hand side of a Boolean equation, the left-hand side is evaluated. The use of dataflow descriptions made up of Boolean equations is illustrated in Example 4-17.

EXAMPLE 4-17   Dataflow VHDL for a 2-to-4-Line Decoder

Figure 4-29 shows a VHDL description for the 2-to-4 line decoder circuit from Figure 4-10 on page 152. This example will be used to demonstrate a dataflow description made up of Boolean equations. The library, use, and entity statements are identical to those in Figure 4-27, so they are not repeated here. The dataflow description begins in line 9. The signals A0_n and A1_n are defined by signal assignments that apply the **not** operation to the input signal A0 and A1, respectively. In line 11, A0_n, A1_n and EN are combined with an **and** operator to form D0. D1, D2, and D3 are similarly defined in lines 12 through 14. Note that this dataflow description is much simpler than the structural description in Figure 4-27.

■

```
-- 2-to-4 Line Decoder: Dataflow VHDL Description          --  1
-- (See Figure 4-10 for logic diagram)                    --  2
Use library, use, and entity entries from 2_to_4_decoder_st;   --  3
                                                          --  4
architecture dataflow_1 of decoder_2_to_4_w_enable is     --  5
                                                          --  6
signal A0_n, A1_n: std_logic;                             --  7
begin                                                     --  8
  A0_n <= not A0;                                         --  9
  A1_n <= not A1;                                         -- 10
  D0 <= A0_n and A1_n and EN;                             -- 11
  D1 <= A0 and A1_n and EN;                               -- 12
  D2 <= A0_n and A1 and EN;                               -- 13
  D3 <= A0 and A1 and EN;                                 -- 14
end dataflow_1;                                           -- 15
```

☐ **FIGURE 4-29**
Dataflow VHDL Description of 2-to-4-Line Decoder

In the next two examples, we describe the 4-to-1-line multiplexer to illustrate two alternative forms of data flow description: when-else and with-select.

EXAMPLE 4-18   VHDL for a 4-to-1-Line Multiplexer Using When-Else

In Figure 4-30, instead of using Boolean equation-like statements in the architecture to describe the multiplexer, we use a *when-else* statement. This statement is a

```
-- 4-to-1 Line Mux: Conditional Dataflow VHDL Description    --  1
-- Using When-Else (See Table 4-7 for function table)        --  2
library ieee;                                                --  3
use ieee.std_logic_1164.all;                                 --  4
entity multiplexer_4_to_1_we is                              --  5
  port (S : in std_logic_vector(1 downto 0);                 --  6
        I : in std_logic_vector(3 downto 0);                 --  7
        Y : out std_logic);                                  --  8
end multiplexer_4_to_1_we;                                   --  9
                                                             -- 10
architecture function_table of multiplexer_4_to_1_we is      -- 11
begin                                                        -- 12
  Y <= I(0) when S = "00" else                               -- 13
       I(1) when S = "01" else                               -- 14
       I(2) when S = "10" else                               -- 15
       I(3) when S = "11" else                               -- 16
       'X';                                                  -- 17
end function_table;                                          -- 18
```

☐ **FIGURE 4-30**
Conditional Dataflow VHDL Description of 4-to-1-Line Multiplexer Using When-Else

representation of the function table given as Table 4-7 on page 157. When S takes on a particular binary value, then a particular input I(i) is assigned to output Y. When the value on S is 00, then Y is assigned I(0). Otherwise, the **else** is invoked so that when the value on S is 01, then Y is assigned I(1), and so on. In standard logic, each of the bits can take on 9 different values. So the pair of bits for S can take on 81 possible values, only 4 of which have been specified so far. In order to define Y for the remaining 77 values, the final **else** followed by X (unknown) is given. This assigns the value X to Y if any of these 77 values occurs on S. This output value occurs only in simulation, however, since Y will always take on a 0 or 1 value in an actual circuit. ▪

EXAMPLE 4-19 VHDL for a 4-to-1-Line Multiplexer Using With-Select

With-select is a variation on when-else as illustrated for the 4-to-1-line multiplexer in Figure 4-31. The expression, the value of which is to be used for the decision, follows **with** and precedes **select**. The values for the expression that causes the alternative assignments then follow **when** with each of the assignment-value pairs separated by commas. In the example, S is the signal, the value of which determines the value selected for Y. When S = "00", I(0) is assigned to Y. When S = "01", I(1) is assigned to Y and so on. 'X' is assigned to Y **when others**, where **others** represents the 77 standard logic combinations not already specified. ▪

```
--4-to-1 Line Mux: Conditional Dataflow VHDL Description        -- 1
Using with Select (See Table 4-7 for function table)           -- 2
library ieee;                                                  -- 3
use ieee.std_logic_1164.all;                                  -- 4
entity multiplexer_4_to_1_ws is                               -- 5
   port (S : in std_logic_vector(1 downto 0);                 -- 6
         I : in std_logic_vector(3 downto 0);                 -- 7
         Y : out std_logic);                                  -- 8
end multiplexer_4_to_1_ws;                                    -- 9
                                                              -- 10
architecture function_table_ws of multiplexer_4_to_1_ws is    -- 11
begin                                                         -- 12
   with S select                                              -- 13
     Y <= I(0) when "00",                                     -- 14
          I(1) when "01",                                     -- 15
          I(2) when "10",                                     -- 16
          I(3) when "11",                                     -- 17
          'X' when others;                                    -- 18
end function_table_ws;                                         -- 19
```

□ **FIGURE 4-31**
Conditional Dataflow VHDL Description of 4-to-1-Line Multiplexer Using With-Select

Note that when-else permits decisions on multiple distinct signals. For example, for the demultiplexer in Figure 4-10, the first **when** can be conditioned on input EN with the subsequent **when**'s conditioned on input S. In contrast, the with-select can depend on only a single Boolean condition (e.g., either EN or S, but not both). Also, for typical synthesis tools, when-else typically results in a more complex logical structure, since each of the decisions depends not only on the condition currently being evaluated, but also on all prior decisions as well. As a consequence, the structure that is synthesized takes into account this priority order, replacing the 4×2 AND-OR by a chain of four 2-to-1 multiplexers. In contrast, there is no direct dependency between the decisions made in with-select. With-select produces a decoder and the 4×2 AND-OR gate.

We have now covered many of the VHDL fundamentals needed for describing combinational circuits. We will continue with more on VHDL by presenting means for describing arithmetic circuits in Chapter 5 and sequential circuits in Chapter 6.

4-8 HDL REPRESENTATIONS FOR COMBINATIONAL CIRCUITS—VERILOG

Since an HDL is used for describing and designing hardware, it is very important to keep the underlying hardware in mind as you write in the language. This is particularly critical if your language description is to be synthesized. For example, if you ignore the hardware that will be generated, it is very easy to specify a large complex gate structure by using × (multiply), when a much simpler structure using only a few gates is all that is needed. For this reason, initially, we emphasize describing detailed hardware with Verilog, and finishing with more abstract, higher-level descriptions.

Selected examples in this chapter are useful for introducing Verilog as an alternative means for representing detailed digital circuits. First, we show a structural Verilog description that replaces the schematic for the 2-to-4-line decoder with enable given in Figure 4-10 on page 152. This example, and one using the 4-to-1-line multiplexer in Figure 4-14 on page 158, illustrate many of the fundamental concepts of Verilog. We then present higher level functional and behavioral Verilog descriptions for these circuits that further illustrate Verilog concepts.

> **EXAMPLE 4-20   Structural Verilog for a 2-to-4-Line Decoder**
>
> The Verilog description for the 2-to-4-line decoder circuit from Figure 4-10 on page 152 is given in Figure 4-32. This description will be used to introduce a number of general Verilog features, as well as to illustrate structural circuit description.
>
> The text between two slashes // and the end of a line as shown in lines 1 and 2 of Figure 4-32 is interpreted as a comment. For multiline comments, there is an alternative notation using a / and *:
>
> ```
> /* 2-to-4 Line Decoder with Enable: Structural Verilog Desc.
> (See Figure 4-10 for logic diagram)*/
> ```

```
// 2-to-4 Line Decoder with Enable: Structural Verilog Desc.    // 1
// (See Figure 4-10 for logic diagram)                          // 2
module decoder_2_to_4_st_v(EN, A0, A1, D0, D1, D2, D3);         // 3
  input EN, A0, A1;                                             // 4
  output D0, D1, D2, D3;                                        // 5
                                                                // 6
  wire A0_n, A1_n, N0, N1, N2, N3;                              // 7
  not                                                           // 8
    go(A0_n, A0),                                               // 9
    g1(A1_n, A1);                                               // 10
  and                                                           // 11
    g3(N0, A0_n, A1_n),                                         // 12
    g4(N1, A0, A1_n),                                           // 13
    g5(N2, A0_n, A1),                                           // 14
    g6(N3, A0,A1),                                              // 15
    g7(D0, N0, EN),                                             // 16
    g8(D1, N1, EN),                                             // 17
    g9(D2, N2, EN),                                             // 18
    g10(D3, N3, EN);                                            // 19
endmodule                                                       // 20
```

□ **FIGURE 4-32**
Structural Verilog Description of 2-to-4-Line Decoder

To assist in discussion of the Verilog description, comments providing line numbers have been added on the right. As a language, Verilog has a syntax that describes precisely the valid constructs that can be used in the language. This example will illustrate many aspects of the syntax. In particular, note the use of commas and colons in the description. Commas (,) are typically used to separate elements of a list and semicolons (;) are used to terminate Verilog statements.

Line 3 begins the declaration of a **module**, which is the fundamental building block of a Verilog design. The remainder of the description defines the module, ending in line 20 with **endmodule**. Note that there is no ; after **endmodule**. Just as for a symbol in a schematic, we need to give the design a name and to define its inputs and outputs. This is the function of the *module statement* in line 3 and the *input* and *output declarations* that follow. The words **module**, **input** and **output** are keywords in Verilog. Keywords, which we show in bold type, have a special meaning and cannot be used as names of objects such as modules, inputs, outputs, or wires. The statement **module** decoder_2_to_4_st_v declares that a design or design part exists with the name decoder_2_to_4_st_v. Further, Verilog names are case sensitive (i.e., names are distinguished by the use of uppercase or lowercase letters). DECODER_2_4_st_v, Decoder_2_4_st_v, and decoder_2_4_st_V are all distinct names.

Just as we would do for a symbol in a schematic, we give the names of the decoder inputs and outputs in the module statement. Next, an *input declaration* is used to define which of the names in the module statement are inputs. For the example design, there are three input signals EN, A0, and A1. The fact that these are inputs is denoted by the keyword **input**. Similarly, an *output declaration* is used to define the outputs. D0, D1, D2, and D3 are denoted as outputs by the keyword **output**.

Inputs and outputs as well as other binary signal types in Verilog can take on one of four values. The two obvious values are 0 and 1. Added are x to represent unknown values and z to represent high impedance values. on the outputs of 3-state logic. Verilog also has strength values that, when combined with the four values given, provide 120 possible signal states. Strength values are used in electronic circuit modeling, however, so will not be considered here.

STRUCTURAL DESCRIPTION   Next, we want to specify the function of the decoder. In this case, we use a *structural description* that is equivalent to the circuit schematic given in Figure 4-10 on page 152. Note that the schematic is made up of gates. Verilog provides 14 primitive gates as keywords. Of these, we are interested in eight for now: **buf**, **not**, **and**, **or**, **nand**, **nor**, **xor** and **xnor**. **buf** and **not** have single inputs, and all other gate types may have from two to any integer number of inputs. **buf** is a buffer, which has the function $z = x$, with x as the input and z as the output. It is as an amplifier of electronic signals that can be used to provide greater fanout or smaller delays. **xor** is the exclusive-OR gate and **xnor** is the exclusive-NOR gate, the complement of the exclusive-OR. In our example, we will use just two gate types, **not** and **and** as shown in lines 8 and 11 of Figure 4-32.

Before specifying the interconnection of the gates, which is the same as a circuit netlist, we need to name all of the nets in the circuit. The inputs and outputs already have names. The internal nets are the outputs of the two inverters and of the four leftmost AND gates in Figure 4-10. In line 7, these nets are declared as *wires* by use of the keyword **wire.** Names A0_n and A1_n are used for the inverter outputs and N0, N1, N2, and N3 for the outputs of the AND gates. In Verilog, **wire** is the default net type. Notably, **input** and **output** ports have the default type **wire**.

Following the declaration of the internal signals, the circuit described contains two inverters and eight 2-input AND gates. A statement consists of a gate type followed by a list of instances of that gate type separated by commas. Each instance consists of a gate name and, enclosed in parentheses, the gate output and inputs separated by commas, with the output given first. The first statement begins on line 8 with the **not** gate type. Following is inverter g0 with A0_n as the output and A0 as the input. To complete the statement, g1 is similarly described. Lines 11 through 19 give the remaining eight gates and the signals connected to their outputs and inputs respectively. For example, in line 14, an instance of a 2-input AND gate named g5 is defined. It has output N2 and inputs A0_n and A1. The module is completed with the keyword **endmodule**.   ■

EXAMPLE 4-21   Structural Verilog for a 4-to-1-Line Multiplexer

In Figure 4-33, the structural description of the 4-to-1-line multiplexer from Figure 4-14 on page 158 illustrates the Verilog concept of a vector. In lines 4 and 5, instead of specifying S and I as single bit wires, they are specified as multiple bit

```
// 4-to-1 Line Multiplexer: Structural Verilog Description       // 1
// (See Figure 4-14 for logic diagram)                          // 2
module multiplexer_4_to_1_st_v(S, I, Y);                        // 3
  input [1:0] S;                                                 // 4
  input [3:0] I;                                                 // 5
  output Y;                                                      // 6
                                                                 // 7
  wire [1:0] not_S;                                              // 8
  wire [0:3] D, N;                                               // 9
                                                                 // 10
not                                                              // 11
   gn0(not_S[0], S[0]),                                          // 12
   gn1(not_S[1], S[1]);                                          // 13
                                                                 // 14
and                                                              // 15
   g0(D[0], not_S[1], not_S[0]),                                 // 16
   g1(D[1], not_S[1], S[0]),                                     // 17
   g2(D[2], S[1], not_S[0]),                                     // 18
   g3(D[3], S[1], S[0]);                                         // 19
   g0(N[0], D[0], I[0]),                                         // 20
   g1(N[1], D[1], I[1]),                                         // 21
   g2(N[2], D[2], I[2]),                                         // 22
   g3(N[3], D[3], I[3]);                                         // 23
                                                                 // 24
or go(Y, N[0], N[1], N[2], N[3]);                               // 25
                                                                 // 26
endmodule                                                        // 27
```

□ **FIGURE 4-33**

Structural Verilog Description of 4-to-1-Line Multiplexer

wires called *vectors*. The bits of a vector are named by a range of integers. This range is given by maximum and minimum values. By specifying these two values, we specify the width of the vector and the names of each of its bits. Vector ranges are illustrated in lines 4, 5, 8 and 9 of Figure 4-33. **input** [1:0] S indicates that S is a vector with a width of two, with the most significant bit numbered 1 and least significant bit numbered 0. The components of S are S[1] and S[0]. **input** [3:0] I declares I as a 4-bit input, with the most significant bit numbered 3 and least significant bit numbered 0. **wire** [0:3] D is also a 4-bit vector representing the four internal wires between the leftmost and rightmost AND gates, but in this case, the most significant bit is numbered 0 and the least significant bit is numbered 3. Once a vector has been declared, then the entire vector or its subcomponents can be referenced. For example, S refers to the two bits of S, and S[1] refers to the most significant bit of S. N refers to all four bits of N and N[1:2] refers to the middle two bits of N. These types of references are used in specifying the output and inputs in instances of the gates in lines 11 through 25. Otherwise, this Verilog description is similar in structure to that for the 2-to-4-line decoder, except that the schematic represented is that in Figure 4-14.

DATAFLOW DESCRIPTION A dataflow description is a form of Verilog description that is not based on structure, but rather on function. A dataflow description is made up of dataflow statements. For the first dataflow description, Boolean equations are used rather than the equivalent of a logic schematic. The Boolean equations given are executed in parallel whenever one of the values on the right-hand side of the equation changes.

EXAMPLE 4-22   Dataflow Verilog for a 2-to-4-Line Decoder

In Figure 4-34, a dataflow description is given for the 2-to-4-line decoder. This particular dataflow description uses the assignment statement consisting of the keyword **assign** followed, in this case, by a Boolean equation. In such equations, we use the bitwise Boolean operators given in Table 4-12. In line 7 of Figure 4-34, EN, ~A0 and ~A1 are combined with an & operator. This & combination is assigned to the output D0. D1, D2, and D3 are similarly defined in lines 8 through 10.   ■

In the next three examples, we describe the 4-to-1-line multiplexer to illustrate three alternative forms of data flow description: Boolean equations, binary combinations as conditions, and binary decisions as conditions.

EXAMPLE 4-23   Dataflow Verilog for a 4-to-1-Line Multiplexer

In Figure 4-35, a single Boolean equation for Y describes the multiplexer. This equation is in sum-of-products form with & for AND and | for OR. Components of the S and I vectors are used as its variables.   ■

EXAMPLE 4-24   Verilog for a 4-to-1-Line Multiplexer Using Combinations

The description in Figure 4-36 resembles the function table given as Table 4-7 on page 157 by using a conditional operator on binary combinations. If the logical value within the parentheses is true, then the value before the : is assigned to the independent variable, in this case, Y. If the logical value is false, then the value after the : is assigned. The logical equality operator is denoted by ==. Suppose we consider condition S == 2'b00. 2'b00 represents a constant. The 2 specifies that the constant contains two digits, b that the constant is given in binary, and 00 gives the constant value. Thus, the expression has value true if vector S is equal to 00;

☐ **TABLE 4-12**
Bitwise Verilog Operators

Operation	Operator
~	Bitwise NOT
&	Bitwise AND
		Bitwise OR
^	Bitwise XOR
^~ or ~^	Bitwise XNOR

```
// 2-to-4 Line Decoder with Enable: Dataflow Verilog Desc.      //  1
// (See Figure 4-10 for logic diagram)                          //  2
module decoder_2_to_4_df_v(EN, A0, A1, D0, D1, D2, D3);         //  3
  input EN, A0, A1;                                             //  4
  output D0, D1, D2, D3;                                        //  5
                                                                //  6
  assign D0 = EN & ~A1 & ~A0;                                   //  7
  assign D1 = EN & ~A1 & A0;                                    //  8
  assign D2 = EN & A1 & ~A0;                                    //  9
  assign D3 = EN & A1 & A0;                                     // 10
                                                                // 11
endmodule                                                       // 12
```

□ **FIGURE 4-34**

Dataflow Verilog Description of 2-to-4-Line Decoder

```
// 4-to-1 Line Multiplexer: Dataflow Verilog Description
// (See Figure 4-14 for logic diagram)
module multiplexer_4_to_1_df_v(S, I, Y);
  input [1:0] S;
  input [3:0] I;
  output Y;

  assign Y = (~ S[1] & ~ S[0] & I[0])|  (~ S[1] & S[0] & I[1])
           | (S[1] & ~ S[0] & I[2]) | (S[1] & S[0] & I[3]);
endmodule
```

□ **FIGURE 4-35**

Dataflow Verilog Description of 4-to-1-Line Multiplexer Using a Boolean Equation

otherwise, it is false. If the expression is true, then I[0] is assigned to Y. If the expression is false, then the next expression containing a ? is evaluated, and so on. In this case, for a condition to be evaluated, all conditions preceding it must evaluate to false. If none of the decisions evaluate to true, then the default value 1'bx is assigned to Y. Recall that default value x represents unknown. ■

EXAMPLE 4-25   Verilog for a 4-to-1-Line Multiplexer Using Binary Decisions

The final form of dataflow description is shown in Figure 4-37. It is based on conditional operators used to form a decision tree, which corresponds to a factored Boolean expression. In this case, if S[1] is 1, then S[0] is evaluated to determine whether Y is assigned I[3] or assigned I[2]. If S[1] is 0, then S[0] is evaluated to determine whether Y is assigned I[1] or I[0]. For a regular structure such as a multiplexer, this approach, based on two-way (binary) decisions, gives a simple dataflow expression. ■

```
// 4-to-1 Line Multiplexer: Dataflow Verilog Description
// (See Table 4-7 for function table)
module multiplexer_4_to_1_cf_v(S, I, Y);
   input [1:0] S;
   input [3:0] I;
   output Y;

   assign Y = (S == 2'b00) ? I[0] :
              (S == 2'b01) ? I[1] :
              (S == 2'b10) ? I[2] :
              (S == 2'b11) ? I[3] : 1'bx ;
endmodule
```

☐ **FIGURE 4-36**

Conditional Dataflow Verilog Description of 4-to-1-Line Multiplexer Using Combinations

```
// 4-to-1 Line Multiplexer: Dataflow Verilog Description
// (See Table 4-7 for function table)
module multiplexer_4_to_1_tf_v(S, I, Y);
   input [1:0] S;
   input [3:0] I;
   output Y;

   assign Y = S[1] ? (S[0] ? I[3] : I[2]) :
              (S[0] ? I[1] : I[0]) ;
endmodule
```

☐ **FIGURE 4-37**

Conditional Dataflow Verilog Description of 4-to-1-Line Multiplexer Using Binary Decisions

This completes our initial introduction to Verilog. We will continue with more on Verilog by presenting means for describing arithmetic circuits in Chapter 5 and sequential circuits in Chapter 6.

4-9 CHAPTER SUMMARY

This chapter has dealt with a number of combinational circuit types, called functional blocks, that are frequently used to design larger circuits. Rudimentary circuits that implement functions of a single variable were introduced. The design of decoders that activate one of a number of output lines in response to an input code was covered. Encoders, the inverse of decoders, generate a code associated with the active line from a set of lines. The design of multiplexers, which take data applied at the input selected and present it at the output, was illustrated.

The design of combinational logic circuits using decoders, multiplexers, and programmable logic was covered. In combination with OR gates, decoders provide a simple minterm-based approach to implementing combinational circuits. Procedures were given for using an n-to-1-line multiplexer or a single inverter and an $(n - 1)$-to-1-line multiplexer to implement any n-input Boolean function. ROMs can be programmed with truth tables. PLAs and PALs can be programmed with their own specialized programming tables. Multiple-level logic decomposition and extraction from Chapter 2 map combinational equations for lookup table implementations.

The last two sections of the chapter introduced VHDL and Verilog descriptions for combinational circuits. Each of the HDLs was illustrated by descriptions at structural, functional, and behavioral levels for various functional blocks presented earlier in the chapter.

REFERENCES

1. MANO, M. M. *Digital Design,* 3rd ed. Englewood Cliffs, NJ: Prentice Hall, 2002.
2. WAKERLY, J. F. *Digital Design: Principles and Practices,* 3rd ed. Englewood Cliffs, NJ: Prentice Hall, 2000.
3. *High-Speed CMOS Logic Data Book.* Dallas: Texas Instruments, 1989.
4. *IEEE Standard VHDL Language Reference Manual.* (ANSI/IEEE Std 1076-1993; revision of IEEE Std 1076-1987). New York: The Institute of Electrical and Electronics Engineers, 1994.
5. SMITH, D. J. *HDL Chip Design.* Madison, AL: Doone Publications, 1996.
6. PELLERIN, D. AND D. TAYLOR. *VHDL Made Easy!* Upper Saddle River, NJ: Prentice Hall PTR, 1997.
7. STEFAN, S. AND L. LINDH. *VHDL for Designers.* London: Prentice Hall Europe, 1997.
8. YALAMANCHILI, S. *VHDL Starter's Guide.* Upper Saddle River, NJ: Prentice Hall, 1998.
9. *IEEE Standard Description Language Based on the Verilog(TM) Hardware Description Language* (IEEE Std 1364-1995). New York: The Institute of Electrical and Electronics Engineers, 1995.
10. PALNITKAR, S. *Verilog HDL: A Guide to Digital Design and Synthesis.* Upper Saddle River, NJ: SunSoft Press (A Prentice Hall Title), 1996.
11. BHASKER, J. *A Verilog HDL Primer.* Allentown, PA: Star Galaxy Press, 1997.
12. THOMAS, D., AND P. MOORBY. *The Verilog Hardware Description Language* 4th ed. Boston: Kluwer Academic Publishers, 1998.
13. CILETTI, M. *Modeling, Synthesis, and Rapid Prototyping with the Verilog HDL,* Upper Saddle River, NJ: Prentice Hall, 1999.

PROBLEMS

The plus (+) indicates a more advanced problem and the asterisk (*) indicates a solution is available on the text website.

4–1. *(a) Draw an implementation diagram for a constant vector function $F = (F_7, F_6, F_5, F_4, F_3, F_2, F_1, F_0) = (1, 0, 0, 1, 0, 1, 1, 0)$ using the ground and power symbols from Figure 4-2(b).

(b) Draw an implementation diagram for a rudimentary vector function $G = (G_7, G_6, G_5, G_4, G_3, G_2, G_1, G_0) = (A, \overline{A}, 0, 1, \overline{A}, A, 1, 1)$ using inputs 1, 0, A, and $\overline{A}$.

4–2. **(a)** Draw an implementation diagram for rudimentary vector function $F = (F_7, F_6, F_5, F_4, F_3, F_2, F_1, F_0) = (1, 0, A, \overline{A}, \overline{A}, A, 0, 1)$, using the ground and power symbols in Figure 4-2(b) and the wire and inverter in Figure 4-2(c) and 4-2(d).

(b) Draw an implementation diagram for rudimentary vector function $G = (G_7, G_6, G_5, G_4, G_3, G_2, G_1, G_0) = (F_3, \overline{F}_2, 0, 1, \overline{F}_1, F_0, 1, 1)$, using the ground and power symbols and components of vector F.

4–3. **(a)** Draw an implementation diagram for the vector $G = (G_7, G_6, G_5, G_4, G_3, G_2, G_1, G_0) = (F_{11}, F_{10}, F_9, F_8, F_3, F_2, F_1, F_0)$.

(b) Draw a simple implementation for the rudimentary vector $H = (H_7, H_6, H_5, H_4, H_3, H_2, H_1, H_0) = (F_3, F_2, F_1, F_0, G_3, G_2, G_1, G_0)$.

4–4. A home security system has a master switch that is used to enable an alarm, lights, video cameras, and a call to local police in the event one or more of six sets of sensors detects an intrusion. The inputs, outputs, and operation of the enabling logic are specified as follows:

Inputs:
   S_i, i = 0, 1, 2, 3, 4, 5 - signals from six sensor sets (0 - intrusion detected, 1 - no intrusion detected)
   M - master switch (0 - security system on, 1 - security system off)

Outputs:
   A - alarm (0 - alarm on, 1 - alarm off)
   L - lights (0 - lights on, 1 - lights off)
   V - video cameras (0 - video cameras off, 1 - video cameras on)
   C - call to police (0 - call off, 1 - call on)

Operation:
   If one or more of the sets of sensors detects an intrusion and the security system is on, then all outputs are on. Otherwise, all outputs are off.

Find a minimum gate input count realization of the enabling logic using AND and OR gates and inverters.

4–5. Design a 4-to-16-line decoder using two 3-to-8-line decoders and 16 2-input AND gates.

4–6. Design a 4-to-16-line decoder with enable using five 2-to-4-line decoders with enables as shown in Figure 4-10.

4–7. *Design a 5-to-32-line decoder using four 3-to-8-line decoders and 48 2-input AND gates.

4–8. A special 4-to-6-line decoder is to be designed. The input codes used are 000 through 101. For a given code applied, the output D_i, with i equal to the decimal equivalent of the code, is 1 and all other outputs are 0. Design the decoder with a 2-to-4-line decoder, a 1-to-2-line decoder, and six 2-input AND gates, such that all decoder outputs are used at least once.

4–9. Draw the detailed logic diagram of a 3-to-8-line decoder using only NOR and NOT gates. Include an enable input.

4–10. *Design a 4-input priority encoder with inputs and outputs as in Table 4-5, but with the truth table representing the case in which input D_0 has the highest priority and input D_3 has the lowest priority.

4–11. Derive the truth table of a BCD-to-binary priority encoder.

4–12. **(a)** Design an 8-to-1-line multiplexer using a 3-to-8-line decoder and an 8×2 AND-OR.

   (b) Repeat part (a), using two 4-to-1-line multiplexers and one 2-to-1-line multiplexer.

4–13. Design a 16-to-1-line multiplexer using a 4-to-16-line decoder and a 16×2 AND-OR.

4–14. Design a dual 8-to-1-line decoder using a 3-to-8-line decoder and two 8×2 AND-ORs.

4–15. Design a dual 4-to-1-line multiplexer using a 2-to-4-line decoder and eight-3-state buffers.

4–16. Design an 8-to-1-line multiplexer using transmission gates.

4–17. Construct a 10-to-1-line multiplexer with a 3-to-8-line decoder, a 1-to-2-line decoder, and a 10×3 AND-OR. The selection codes 0000 through 1001 must be directly applied to the decoder inputs without added logic.

4–18. Construct a quad 9-to-1-line multiplexer with four single 8-to-1-line multiplexers and one quadruple 2-to-1-line multiplexer. The multiplexers should be interconnected and inputs labeled so that the selection codes 0000 through 1000 can be directly applied to the multiplexer selection inputs without added logic.

4–19. *Construct a 15-to-1-line multiplexer with two 8-to-1-line multiplexers. Interconnect the two multiplexers and label the inputs such that any added logic required to have selection codes 0000 through 1110 is minimized.

4–20. Rearrange the condensed truth table for the circuit of Figure 4-10, and verify that the circuit can function as a demultiplexer.

4–21. A combinational circuit is defined by the following three Boolean functions:

$$F_1 = \overline{X + Z} + XYZ$$

$$F_2 = \overline{X + Z} + \overline{X}YZ$$

$$F_3 = X\overline{Y}Z + \overline{X + Z}$$

Design the circuit with a decoder and external OR gates.

4–22. A combinational circuit is specified by the following three Boolean functions:

$$F_1(A, B, C) = \Sigma m(0, 3, 4)$$

$$F_2(A, B, C) = \Sigma m(1, 2, 7)$$

$$F_3(A, B, C) = \Pi M(0, 1, 2, 4)$$

Implement the circuit with a decoder and external OR gates.

4–23. Implement a binary full adder with a dual 4-to-1-line multiplexer and a single inverter.

4–24. Implement the following Boolean function with an 8-to-1-line multiplexer and a single inverter with variable D as its input:

$$F(A, B, C, D) = \Sigma m(2, 4, 6, 9, 10, 11, 15)$$

4–25. *Implement the Boolean function

$$F(A, B, C, D) = \Sigma m(1, 3, 4, 11, 12, 13, 14, 15)$$

with a 4-to-1-line multiplexer and external gates. Connect inputs A and B to the selection lines. The input requirements for the four data lines will be a

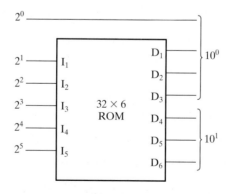

☐ **FIGURE 4-38**
Binary–to–Decimal ROM Converter

function of the variables C and D. The values of these variables are obtained by expressing F as a function of C and D for each of the four cases when AB = 00, 01, 10, and 11. These functions must be implemented with external gates.

4–26. Repeat problem 4-25 using two 3-to-8-line decoders with enables, an inverter, and OR gates with a maximum fan-in of 4.

4–27. Given a 256 × 8 ROM chip with an enable input, show the external connections necessary to construct a 1K × 16 ROM with eight chips and a decoder.

4–28. *The 32 × 6 ROM, together with the 2^0 line, as shown in Figure 4-38, converts a 6-bit binary number to its corresponding two-digit BCD number. For example, binary 100001 converts to BCD 011 0011 (decimal 33). Specify the truth table for the ROM.

4–29. Specify the size of a ROM (number of words and number of bits per word) that will accommodate the truth table for the following combinational circuit components:
(a) An 8-bit adder–subtractor with C_{in} and C_{out}.
(b) A binary multiplier that multiplies two 8-bit numbers.
(c) A code converter from a 4-digit BCD number to a binary number.

4–30. Tabulate the truth table for an 8 × 4 ROM that implements the following four Boolean functions:

$$A(X, Y, Z) = \Sigma\, m(0, 1, 2, 6, 7)$$

$$B(X, Y, Z) = \Sigma\, m(2, 3, 4, 5, 6)$$

$$C(X, Y, Z) = \Sigma\, m(2, 6)$$

$$D(X, Y, Z) = \Sigma\, m(1, 2, 3, 5, 6, 7)$$

4–31. Obtain the PLA programming table for the four Boolean functions listed in Problem 4–30. Minimize the number of product terms. Be sure to attempt to share product terms between functions that are not prime implicants of individual functions and to consider the use of complemented (**C**) outputs.

4–32. Derive the PLA programming table for the combinational circuit that squares a 3-bit number. Minimize the number of product terms.

4–33. List the PLA programming table for a BCD–to–Excess-3 code converter.

4–34. *Repeat Problem 4–33, using a PAL device.

4–35. The following is the truth table of a three-input, four-output combinational circuit. Obtain the PAL programming table for the circuit, and mark the fuses to be blown in a PAL diagram similar to the one shown in Figure 4-26.

Inputs			Outputs			
X	Y	Z	A	B	C	D
0	0	0	0	1	0	0
0	0	1	1	1	1	1
0	1	0	1	0	1	1
0	1	1	0	1	0	1
1	0	0	1	0	1	0
1	0	1	0	0	0	1
1	1	0	1	1	1	0
1	1	1	0	1	1	1

 All HDL files for circuits referred to in the remaining problems are available in ASCII form for simulation and editing on the Companion Website for the text. A VHDL or Verilog compiler/simulator is necessary for the problems or portions of problems requesting simulation. Descriptions can still be written, however, for many problems without using compilation or simulation.

4–36. Compile and simulate the 2-to-4-line decoder with enable in Figure 4-27 for sequence 000, 001, 010, 011, 100, 101, 110, 111 on E_n, A0, A1. Verify that the circuit functions as a decoder. You will need to compile library lcdf_vhdl.func_prims first since it is used in the simulation.

4–37. Rewrite the VHDL given in Figure 4-27 for the 2-to-4-line decoder using (1) std_logic_vector notation instead of std_logic notation for A and D_n and (2) implicit specification of the component input and output names by their order in package func_prims in library lcdf_vhdl given in the Companion Website Gallery. See Figure 4-28 and accompanying text for these concepts. Compile and simulate the resulting file as in problem 4-36.

4–38. Compile and simulate the 4-to-1-line multiplexer in Figure 4-28 for the sequence of all 16 combinations of 00, 10, 01, 11 on S and 1000, 0100, 0010, 0001 on D. You will need to compile library lcdf_vhdl.func_prims first since it is used in the simulation. Verify that the circuit functions as a multiplexer.

4–39. *Find a logic diagram that corresponds to the VHDL structural description in Figure 4-39. Note that complemented inputs are not available.

4–40. Using Figure 4-28 as a framework, write a structural VHDL description of the circuit in Figure 4-40. Replace X, Y, and Z with X(0:2). Consult package func_prims in library lcdf_vhdl for information on the various gate components. Compile func_prims and your VHDL, and simulate your VHDL for all eight possible input combinations to verify your description's correctness.

4–41. Using Figure 4-27 as a framework, write a structural VHDL description of the circuit in Figure 4-41. Consult package func_prims in library

```
-- Combinational Circuit 1: Structural VHDL Description
library ieee, lcdf_vhdl;
use ieee.std_logic_1164.all, lcdf_vhdl.func_prims.all;
  entity comb_ckt_1 is
  port(x1, x2, x3, x4 : in std_logic;
       f : out std_logic);
end comb_ckt_1;

architecture structural_1 of comb_ckt_1 is
  component NOT1
    port(in1: in std_logic;
         out1: out std_logic);
  end component;
  component AND2
    port(in1, in2 : in std_logic;
         out1: out std_logic);
  end component;
  component OR3
    port(in1, in2, in3 : in std_logic;
         out1: out std_logic);
  end component;
  signal n1, n2, n3, n4, n5, n6 : std_logic;
  begin
    g0: NOT1 port map (in1 => x1, out1 => n1);
    g1: NOT1 port map (in1 => n3, out1 => n4);
    g2: AND2 port map (in1 => x2, in2 => n1,
                       out1 => n2);
    g3: AND2 port map (in1 => x2, in2 => x3,
                       out1 => n3);
    g4: AND2 port map (in1 => x3, in2 => x4,
                       out1 => n5);
    g5: AND2 port map (in1 => x1, in2 => n4,
                       out1 => n6);
    g6: OR3 port map (in1 => n2, in2 => n5,
                      in3 => n6, out1 => f);
end structural_1;
```

□ **FIGURE 4-39**
VHDL for Problem 4–39

lcdf_vhdl for information on the various gate components. Compile func_prims and your VHDL, and simulate your VHDL for all 16 possible input combinations to verify your description's correctness.

4–42. Find a logic diagram representing minimum two-level logic needed to implement the VHDL dataflow description in Figure 4-42. Note that complemented inputs are available.

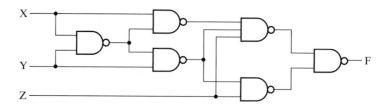

☐ **FIGURE 4-40**
Circuit for Problems 4–40, 4–43, 4-51, and 4–53

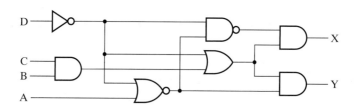

☐ **FIGURE 4-41**
Circuit for Problem 4–41 and 4-50

```
-- Combinational Circuit 2: Dataflow VHDL Description
-library ieee;
use ieee.std_logic_1164.all;
entity comb_ckt_2 is
  port(a, b, c, d, a_n, b_n, c_n, d_n: in std_logic;
       f, g : out std_logic);
-- a_n, b_n, ... are complements of a, b, ... , respectively.
end comb_ckt_2;

architecture dataflow_1 of comb_ckt_2 is
begin
   f <= b and (a or (a_n and c)) or (b_n and c and d_n);
   g <= b and (c or (a_n and c_n) or (c_n and d_n));
end dataflow_1;
```

☐ **FIGURE 4-42**
VHDL for Problem 4–42

4–43. *Write a dataflow VHDL description for the circuit in Figure 4-40 by using the Boolean equation for the output *F*.

4–44. +Write a dataflow VHDL description for the priority encoder using the "when else" dataflow concept from Figure 4-30. Compile and simulate your description with a set of input vectors that are a good test for the priority function it performs.

4-45. Write a dataflow VHDL description for an 8-to-1-line multiplexer using the "with select" dataflow concept from Figure 4-31. Compile and simulate your description with a set of input vectors that are a good test for the selection function it performs.

4-46. *Compile and simulate the 2-to-4-line decoder in Figure 4-32 for sequence 000, 001, 010, 011, 100, 101, 110, 111 on E, A0, A1. Verify that the circuit functions as a decoder.

4-47. Rewrite the Verilog description given in Figure 4-32 for the 2-to-4-line decoder using vector notation for inputs, outputs, and wires. See Figure 4-33 and accompanying text for these concepts. Compile and simulate the resulting file as in problem 4-46.

4-48. Compile and simulate the 4-to-1-line multiplexer in Figure 4-33 for the sequence of all 16 combinations of 00, 10, 01, 11 on S and 1000, 0100, 0010, 0001 on D. Verify that the circuit functions as a multiplexer.

4-49. *Find a logic diagram that corresponds to the Verilog structural description in Figure 4-43. Note that complemented inputs are not available.

4-50. Using Figure 4-32 as a framework, write a structural Verilog description of the circuit in Figure 4-41. Compile and simulate your Verilog for all 16 possible input combinations to verify your description's correctness.

4-51. Using Figure 4-33 as a framework, write a structural Verilog description of the circuit in Figure 4-40. Replace X, Y, and Z with **input** [2:0] X. Compile

```
// Combinational Circuit 1: Structural Verilog Description
module comb_ckt_1(x1, x2, x3, x4,f);
   input x1, x2, x3, x4;
   output f;

   wire n1, n2, n3, n4, n5, n6;
   not
     go(n1, x1),
     g1(n4, n3);
   and
     g2(n2, x2, n1),
     g3(n3, x2, x3),
     g4(n5, x3, x4),);
     g5(n6, x1, n4),);
   or
     g6(f, n2, n5, n6),
endmodule
```

☐ **FIGURE 4-43**
Verilog for Problem 4–49

```
// Combinational Circuit 2: Dataflow Verilog Description
module comb_ckt_1 (a, b, c, d, a_n, b_n, c_n, d_n, f, g);
// a_n, b_n, ... are complements of a, b, ... , respectively.
   input a, b, c, d, a_n, b_n, c_n, d_n;
   output f, g;

   assign f = b & (a |(a_n & c)) | (b_n & c & d_n);
   assign g = b & (c | (a_n & c_n) | (c_n & d_n));
endmodule
```

☐ **FIGURE 4-44**
Verilog for Problem 4–52

and simulate your Verilog for all eight possible input combinations to verify your description's correctness.

4–52. Find a logic diagram representing minimum 2-level logic needed to implement the Verilog dataflow description in Figure 4-44. Note that complemented inputs are available.

4–53. *Write a dataflow Verilog description for the circuit in Figure 4-40 by using the Boolean equation for the output F and using Figure 4-35 as a model.

4–54. By using the conditional dataflow concept from Figure 4-36, write a Verilog dataflow description for an 8-to–1-line multiplexer. Compile and simulate your description with a set of input vectors that are a good test for the selection function it performs.

4–55. +Write a dataflow description for the priority encoder in Figure 4-12 using the binary decision dataflow concept from Figure 4-37. Compile and simulate your description with a set of input vectors that are a good test for the priority function it performs.

5

ARITHMETIC FUNCTIONS AND CIRCUITS

I n this chapter, the focus continues to be on functional blocks, specifically, a special class of functional blocks that perform arithmetic operations. The concept of iterative circuits made up of arrays of combinational cells is introduced. Blocks designed as iterative arrays for performing addition, addition and subtraction, and multiplication are covered. The simplicity of these arithmetic circuits comes from using complement representations for numbers and complement-based arithmetic. In addition, we introduce circuit contraction that permits us to design new functional blocks. Contraction involves application of value-fixing to the inputs of existing blocks and simplification of the resulting circuits. These circuits perform operations such as incrementing a number, decrementing a number, or multiplying a number by a constant. Many of these new functional blocks are used to construct sequential functional blocks in Chapter 7.

In the generic computer diagram at the beginning of Chapter 1, adders, adder-subtractors, and multipliers are used in the processor. Incrementers and decrementers are used widely in other components as well, so concepts from this chapter apply across most components of the generic computer.

5-1 ITERATIVE COMBINATIONAL CIRCUITS

In this chapter, the arithmetic blocks are typically designed to operate on binary input vectors and produce binary output vectors. Further, the function implemented often requires that the same subfunction be applied to each bit position. Thus, a functional block can be designed for the subfunction and then used repetitively for each bit positions of the overall arithmetic block being designed. There

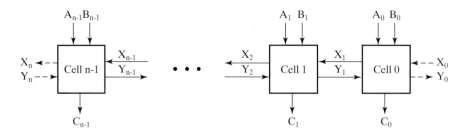

☐ **FIGURE 5-1**

Block Diagram of an Iterative Circuit

will often be one or more connections to pass values between adjacent bit positions. These internal variables are inputs or outputs of the subfunctions, but not accessible outside of the overall arithmetic block. The subfunction blocks are referred to as *cells* and the overall implementation is an *array of cells*. The cells in the array are often, but not always, identical. Due to the repetitive nature of the circuit and the association of a vector index with each of the circuit cells, the overall functional block is referred to as an *iterative array*. The use of iterative arrays, a special case of hierarchical circuits, is useful in handling vectors of bits, for example, a circuit that adds two 32-bit binary integers. At a minimum, such a circuit has 64 inputs and 32 outputs. As a consequence, beginning with truth tables and writing equations for the entire circuit is out of the question. Since iterative circuits are based on repetitive cells, the design process is considerably simplified by a basic structure that guides the design.

A block diagram for an iterative circuit that operates on two *n*-input vectors and produces an *n*-output vector is shown in Figure 5-1. In this case, there are two lateral connections between each pair of cells in the array, one from left to right and the other from right to left. Also, optional connections, indicated by dashed lines, exist at the right and left ends of the array. An arbitrary array employs as many lateral connections as needed for a particular design. The definition of the functions associated with such connections is very important in the design of the array and its cell. In particular, the number of connections used and their functions can affect both the cost and speed of an iterative circuit.

In the next section, we will define cells for performing addition in individual bit positions and then define a binary adder as an array of cells.

5-2 BINARY ADDERS

An arithmetic circuit is a combinational circuit that performs arithmetic operations such as addition, subtraction, multiplication, and division with binary numbers or with decimal numbers in a binary code. We will develop arithmetic circuits by means of hierarchical, iterative design. We begin at the lowest level by finding a circuit that performs the addition of two binary digits. This simple addition consists of four possible elementary operations: $0 + 0 = 0, 0 + 1 = 1, 1 + 0 = 1$, and $1 + 1 = 10$. The first three operations produce a sum requiring only one bit to represent it, but

□ **TABLE 5-1**
Truth Table of Half Adder

Inputs		Outputs	
X	Y	C	S
0	0	0	0
0	1	0	1
1	0	0	1
1	1	1	0

when both the augend and addend are equal to 1, the binary sum requires two bits. Because of this case, the result is always represented by two bits, the carry and the sum. The carry obtained from the addition of two bits is added to the next higher order pair of significant bits. A combinational circuit that performs the addition of two bits is called a *half adder*. One that performs the addition of three bits (two significant bits and a previous carry) is called a *full adder*. The names of the circuits stem from the fact that two half adders can be employed to implement a full adder. The half adder and the full adder are basic arithmetic blocks with which other arithmetic circuits are designed.

Half Adder

A half adder is an arithmetic circuit that generates the sum of two binary digits. The circuit has two inputs and two outputs. The input variables are the augend and addend bits to be added, and the output variables produce the sum and carry. We assign the symbols X and Y to the two inputs and S (for "sum") and C (for "carry") to the outputs. The truth table for the half adder is listed in Table 5-1. The C output is 1 only when both inputs are 1. The S output represents the least significant bit of the sum. The Boolean functions for the two outputs, easily obtained from the truth table, are

$$S = \overline{X}Y + X\overline{Y} = X \oplus Y$$

$$C = XY$$

The half adder can be implemented with one exclusive-OR gate and one AND gate, as shown in Figure 5-2.

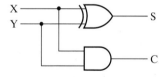

□ **FIGURE 5-2**
Logic Diagram of Half Adder

☐ **TABLE 5-2**
Truth Table of Full Adder

Inputs			Outputs	
X	Y	Z	C	S
0	0	0	0	0
0	0	1	0	1
0	1	0	0	1
0	1	1	1	0
1	0	0	0	1
1	0	1	1	0
1	1	0	1	0
1	1	1	1	1

Full Adder

A full adder is a combinational circuit that forms the arithmetic sum of three input bits. Besides the three inputs, it has two outputs. Two of the input variables, denoted by X and Y, represent the two significant bits to be added. The third input, Z, represents the carry from the previous lower significant position. Two outputs are necessary because the arithmetic sum of three bits ranges in value from 0 to 3, and binary 2 and 3 need two digits for their representation. Again, the two outputs are designated by the symbols S for "sum" and C for "carry"; the binary variable S gives the value of the bit of the sum, and the binary variable C gives the output carry. The truth table of the full adder is listed in Table 5-2. The values for the outputs are determined from the arithmetic sum of the three input bits. When all the input bits are 0, the outputs are 0. The S output is equal to 1 when only one input is equal to 1 or when all three inputs are equal to 1. The C output has a carry of 1 if two or three inputs are equal to 1. The maps for the two outputs of the full adder are shown in Figure 5-3. The simplified sum-of-product functions for the two outputs are

$$S = \overline{X}\,\overline{Y}Z + \overline{X}Y\overline{Z} + X\overline{Y}\,\overline{Z} + XYZ$$

$$C = XY + XZ + YZ$$

The two-level implementation requires seven AND gates and two OR gates. However, the map for output S is recognized as an odd function, as discussed in Section 2-7. Furthermore, the C output function can be manipulated to include the exclusive-OR of X and Y. The Boolean functions for the full adder in terms of exclusive-OR operations can then be expressed as

$$S = (X \oplus Y) \oplus Z$$

$$C = XY + Z(X \oplus Y)$$

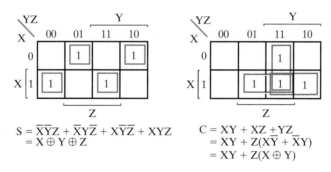

$$S = \overline{X}\overline{Y}Z + \overline{X}Y\overline{Z} + X\overline{Y}\overline{Z} + XYZ$$
$$= X \oplus Y \oplus Z$$

$$C = XY + XZ + YZ$$
$$= XY + Z(X\overline{Y} + \overline{X}Y)$$
$$= XY + Z(X \oplus Y)$$

□ **FIGURE 5-3**
Maps for Full Adder

The logic diagram for this multiple-level implementation is shown in Figure 5-4. It consists of two half adders and an OR gate.

Binary Ripple Carry Adder

A parallel binary adder is a digital circuit that produces the arithmetic sum of two binary numbers using only combinational logic. The parallel adder uses n full adders in parallel, with all input bits applied simultaneously to produce the sum. The full adders are connected in cascade, with the carry output from one full adder connected to the carry input of the next full adder. Since a 1 carry may appear near the least significant bit of the adder and yet propagate through many full adders to the most significant bit, just as a wave ripples outward from a pebble dropped in a pond, the parallel adder is referred to as a *ripple carry adder*. Figure 5-5 shows the interconnection of four full-adder blocks to form a 4-bit ripple carry adder. The augend bits of A and the addend bits of B are designated by subscripts in increasing order from right to left, with subscript 0 denoting the least significant bit. The carries are connected in a chain through the full adders. The input carry to the parallel adder is C_0, and the output carry is C_4. An n-bit ripple carry adder requires n full adders, with each output carry connected to the input carry of the next-higher-order

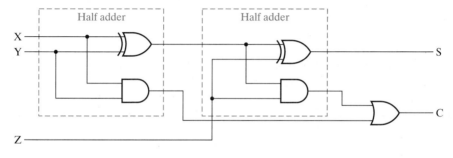

□ **FIGURE 5-4**
Logic Diagram of Full Adder

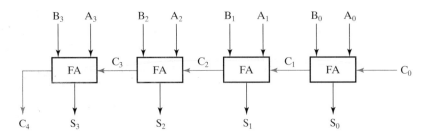

☐ **FIGURE 5-5**
4-Bit Ripple Carry Adder

full adder. For example, consider the two binary numbers $A = 1011$ and $B = 0011$. Their sum, $S = 1110$, is formed with a 4-bit ripple carry adder as follows:

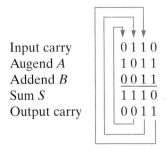

Input carry	0 1 1 0
Augend A	1 0 1 1
Addend B	0 0 1 1
Sum S	1 1 1 0
Output carry	0 0 1 1

The input carry in the least significant position is 0. Each full adder receives the corresponding bits of A and B and the input carry and generates the sum bit for S and the output carry. The output carry in each position is the input carry of the next-higher-order position, as indicated by the blue lines.

The 4-bit adder is a typical example of a digital component that can be used as a building block. It can be used in many applications involving arithmetic operations. Observe that the design of this circuit by the usual method would require a truth table with 512 entries, since there are nine inputs to the circuit. By cascading the four instances of the known full adders, it is possible to obtain a simple and straightforward implementation without directly solving this larger problem. This is an example of the power of iterative circuits and circuit reuse in design.

Carry Lookahead Adder

The ripple carry adder, although simple in concept, has a long circuit delay due to the many gates in the carry path from the least significant bit to the most significant bit. For a typical design, the longest delay path through an n-bit ripple carry adder is $2n + 2$ gate delays. Thus, for a 16-bit ripple carry adder, the delay is 34 gate delays. This delay tends to be one of the largest in a typical computer design. Accordingly, we find an alternative design, the *carry lookahead adder*, attractive. This adder is a practical design with reduced delay at the price of more complex hardware. The carry lookahead design can be obtained by a transformation of the ripple carry design in which the carry logic over fixed groups of bits of the adder is reduced to two-level logic. The transformation is shown for a 4-bit adder group in Figure 5-6.

First, we construct a new logic hierarchy, separating the parts of the full adders not involving the carry propagation path from those containing the path. We call the first part of each full adder a *partial full adder* (PFA). This separation is shown in Figure 5-6(a), which presents a diagram of a PFA and a diagram of four PFAs connected to the carry path. We have removed the OR gate and one of the AND gates from each of the full adders to form the ripple carry path.

There are two outputs, P_i and G_i, from each PFA to the ripple carry path and one input C_i, the carry input, from the carry path to each PFA. The function $P_i = A_i \oplus B_i$ is called the *propagate* function. Whenever P_i is equal to 1, an incoming carry is propagated through the bit position from C_i to C_{i+1}. For P_i equal to 0, carry propagation through the bit position is blocked. The function $G_i = A_i \cdot B_i$ and is called the *generate* function. Whenever G_i is equal to 1, the carry output from the position is 1, regardless of the value of P_i, so a carry has been generated in the position. When G_i is 0, a carry is not generated, so that C_{i+1} is 0 if the carry propagated through the position from C_i is also 0. The generate and propagate functions correspond exactly to the half adder and are essential in controlling the values in the ripple carry path. Also, as in the full adder, the PFA generates the sum function by the exclusive-OR of the incoming carry C_i and the propagate function P_i.

The carry path remaining in the 4-bit ripple carry adder has a total of eight gates in cascade, so the circuit has a delay of eight gate delays. Since only AND and OR gates are involved in the carry path, ideally, the delay for each of the four carry signals produced, C_1 through C_4, would be just two gate delays. The basic carry lookahead circuit is simply a circuit in which functions C_1 through C_3 have a delay of only two gate delays. The implementation of C_4 is more complicated in order to allow the 4-bit carry lookahead adder to be extended to multiples of 4 bits, such as 16 bits. The 4-bit carry lookahead circuit is shown in Figure 5-6(b). It is designed to directly replace the ripple carry path in Figure 5-6(a). Since the logic generating C_1 is already two-level, it remains unchanged. The logic for C_2, however, has four levels. So to find the carry lookahead logic for C_2, we must reduce the logic to two levels. The equation for C_2 is found from Figure 5-6(a), and the distributive law is applied to obtain

$$C_2 = G_1 + P_1(G_0 + P_0 C_0)$$

$$= G_1 + P_1 G_0 + P_1 P_0 C_0$$

This equation is implemented by the logic with output C_2 in Figure 5-6(b). We obtain the two-level logic for C_3 by finding its equation from the carry path in Figure 5-6(a) and applying the distributive law:

$$C_3 = G_2 + P_2(G_1 + P_1(G_0 + P_0 C_0))$$

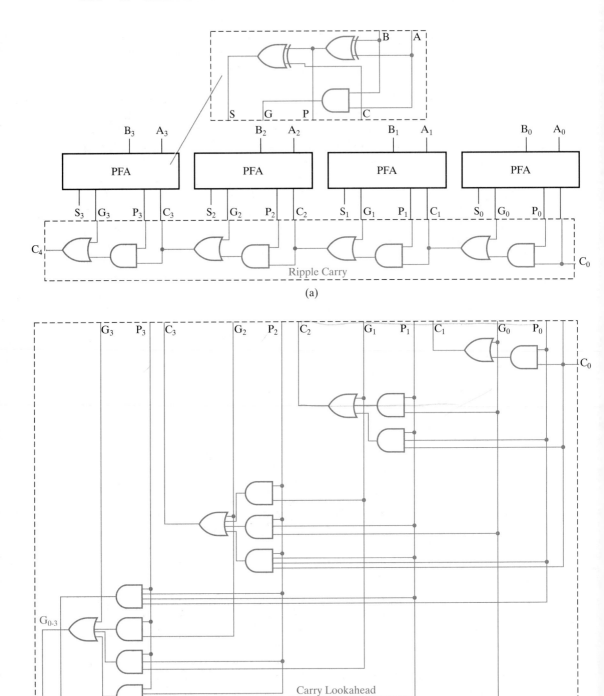

☐ **FIGURE 5-6**
Development of a Carry Lookahead Adder

$$= G_2 + P_2(G_1 + P_1G_0 + P_0C_0)$$

$$= G_2 + P_2G_1 + P_2P_1G_0 + P_2P_1P_0C_0$$

The two-level logic with output C_3 in Figure 5-6(b) implements this function.

We could implement C_4 using the same method. But some of the gates would have a fan-in of five, which may increase the delay. Also, we are interested in reusing this same circuit for higher numbered bits (e.g., 4 through 7, 8 through 11, and 12 through 15 of a 16-bit adder). For this adder, in positions 4, 8, and 12 we would like the carry to be produced as fast as possible without using excessive fan-in. Accordingly, we want to repeat the same carry lookahead trick for *4-bit groups* that we used to handle the 4 bits. This will allow us to reuse the carry lookahead circuit for each group of 4 bits, and also to use the same circuit for four 4-bit groups as if they were individual bits. So instead of generating C_4, we produce generate and propagate functions that apply to 4-bit groups instead of a single bit to act as the inputs for the group carry lookahead circuit. To propagate a carry from C_0 to C_4, we need to have all four of the propagate functions equal to 1, giving the *group propagate* function

$$P_{0-3} = P_3P_2P_1P_0$$

To represent the generation of a carry in positions 0, 1, 2, and 3, and its propagation to C_4, we need to consider the generation of a carry in each of the positions, as represented by G_0 through G_3, and the propagation of each of these four generated carries to position 4. This gives the *group generate* function

$$G_{0-3} = G_3 + P_3G_2 + P_3P_2G_1 + P_3P_2P_1G_0$$

The group propagate and group generate equations are implemented by the logic in the lower part of Figure 5-6(b). If there are only 4 bits in the adder, then the logic circuit used for C_1 can be used to generate C_4 from these two outputs. In a longer adder, a carry lookahead circuit identical to that in the figure, except for labeling, is placed at the second level to generate C_4, C_8, and C_{12}. This concept can be extended with more carry lookahead circuits in the second level and with one carry lookahead circuit in the third level to generate carries for positions 16, 32, and 48 in a 64-bit adder.

Assuming that an exclusive OR contributes 2 gate delays, the longest delay in the 4-bit carry lookahead adder is 6 gate delays, compared with 10 gate delays in the ripple carry adder. The improvement is very modest and perhaps not worth all the extra logic. But applying the carry lookahead circuit to a 16-bit adder using five copies in two levels of lookahead reduces the delay from 34 to just 10 gate delays, improving the performance of the adder by a factor of close to three. In a 64-bit adder, with the use of 21 carry lookahead circuits in three levels of lookahead, the delay is reduced from 130 gate delays to 14 gate delays, giving more than a factor of 8 in improved performance. In general, for the implementation we have shown, the delay of a carry lookahead adder designed for the best performance is $4L + 2$ gate delays, where L is the number of lookahead levels in the design.

5-3 BINARY SUBTRACTION

In Chapter 1, we briefly examined the subtraction of unsigned binary numbers. Although beginning texts cover only signed number addition and subtraction, to the complete exclusion of the unsigned alternative, unsigned number arithmetic plays an important role in computation and computer hardware design. It is used in floating-point units, in signed-magnitude addition and subtraction algorithms, and in extending the precision of fixed-point numbers. For these reasons, we will treat unsigned number addition and subtraction here. We also, however, choose to treat it first so that we can clearly justify, in terms of hardware cost, that which otherwise appears bizarre and often is accepted on faith, namely, the use of complement representations in arithmetic.

In Section 1-3, subtraction is performed by comparing the subtrahend with the minuend and subtracting the smaller from the larger. The use of a method containing this comparison operation results in inefficient and costly circuitry. As an alternative, we can simply subtract the subtrahend from the minuend. Using the same numbers as in a subtraction example from Section 1-3, we have

Borrows into:	11100
Minuend:	10011
Subtrahend:	-11110
Difference:	10101
Correct Difference:	-01011

If no borrow occurs into the most significant position, then we know that the subtrahend is not larger than the minuend and that the result is positive and correct. If a borrow does occur into the most significant position, as indicated in blue, then we know that the subtrahend is larger than the minuend. The result must then be negative, and so we need to correct its magnitude. We can do this by examining the result of the calculation when a borrow occurs:

$$M - N + 2^n$$

Note that the added 2^n represents the value of the borrow into the most significant position. Instead of this result, the desired magnitude is $N - M$. This can be obtained by subtracting the preceding formula from 2^n:

$$2^n - (M - N + 2^n) = N - M$$

In the previous example, $100000 - 10101 = 01011$, which is the correct magnitude.

In general, the subtraction of two n-digit numbers, $M - N$, in base 2 can be done as follows:

1. Subtract the subtrahend N from the minuend M.
2. If no end borrow occurs, then $M \geq N$, and the result is nonnegative and correct.

3. If an end borrow occurs, then $N > M$, and the difference, $M - N + 2^n$, is subtracted from 2^n, and a minus sign is appended to the result.

Subtraction of a binary number from 2^n to obtain an n-digit result is called taking the *2's complement* of the number. So in step 3, we are taking the *2's complement* of the difference $M - N + 2^n$. Use of the 2's complement in subtraction is illustrated by the following example.

EXAMPLE 5-1   Unsigned Binary Subtraction by 2's Complement Subtract

Perform the binary subtraction $01100100 - 10010110$. We have

Borrows into:	10011110
Minuend:	01100100
Subtrahend:	$-$ 10010110
Initial Result	11001110

The end borrow of 1 implies correction:

2^8	100000000
$-$ Initial Result	$-$ 11001110
Final Result	$-$ 00110010

To perform subtraction using this method requires a subtractor for the initial subtraction. In addition, when necessary, either the subtractor must be used a second time to perform the correction, or a separate 2's complementer circuit must be provided. So, thus far, we require a subtractor, an adder, and possibly a 2's complementer to perform both addition and subtraction. The block diagram for a 4-bit adder–subtractor using these functional blocks is shown in Figure 5-7. The inputs are applied to both the adder and the subtractor, so both operations are performed in parallel. If an end borrow value of 1 occurs in the subtraction, then the selective 2's complementer receives a value of 1 on its Complement input. This circuit then takes the 2's complement of the output of the subtractor. If the end borrow has value of 0, the selective 2's complementer passes the output of the subtractor through unchanged. If subtraction is the operation, then a 1 is applied to S of the multiplexer that selects the output of the complementer. If addition is the operation, then a 0 is applied to S, thereby selecting the output of the adder.

As we will see, this circuit is more complex than necessary. To reduce the amount of hardware, we would like to share logic between the adder and the subtractor. This can also be done using the notion of the complement. So before considering the combined adder–subtractor further, we will take a more careful look at complements.

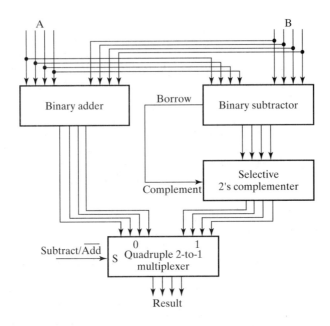

☐ **FIGURE 5-7**
Block Diagram of Binary Adder–Subtractor

Complements

There are two types of complements for each base-r system: the *radix complement*, which we saw earlier for base 2, and the *diminished radix complement*. The first is referred to as the *r's complement* and the second as the *(r − 1)'s complement*. When the value of the base r is substituted in the names, the two types are referred to as the 2's and 1's complements for binary numbers and the 10's and 9's complements for decimal numbers, respectively. Since our interest for the present is in binary numbers and operations, we will deal with only 1's and 2's complements.

Given a number N in binary having n digits, the *1's complement of N* is defined as $(2^n − 1) − N$. 2^n is represented by a binary number that consists of a 1 followed by n 0's. $2^n − 1$ is a binary number represented by n 1's. For example, if $n = 4$, we have $2^4 = (10000)_2$ and $2^4 − 1 = (1111)_2$. Thus, the 1's complement of a binary number is obtained by subtracting each digit from 1. When subtracting binary digits from 1, we can have either $1 − 0 = 1$ or $1 − 1 = 0$, which causes the original bit to change from 0 to 1 or from 1 to 0, respectively. Therefore, the 1's complement of a binary number is formed by changing all 1's to 0's and all 0's to 1's—that is, applying the NOT or complement operation to each of the bits. Following are two numerical examples:

The 1's complement of 1011001 is 0100110.

The 1's complement of 0001111 is 1110000.

In similar fashion, the 9's complement of a decimal number, the 7's complement of an octal number, and the 15's complement of a hexadecimal number are obtained by subtracting each digit from 9, 7, and F (decimal 15), respectively.

Given an n-digit number N in binary, the *2's complement of N* is defined as $2^n - N$ for $N \neq 0$ and 0 for $N = 0$. The reason for the special case of $N = 0$ is that the result must have n bits, and subtraction of 0 from 2^n gives an $(n + 1)$-bit result, $100...0$. This special case is achieved by using only an n-bit subtractor or otherwise dropping the 1 in the extra position. Comparing with the 1's complement, we note that the 2's complement can be obtained by adding 1 to the 1's complement, since $2^n - N = \{[(2^n - 1) - N] + 1\}$. For example, the 2's complement of binary 101100 is $010011 + 1 = 010100$ and is obtained by adding 1 to the 1's complement value. Again, for $N = 0$, the result of this addition is 0, achieved by ignoring the carry out of the most significant position of the addition. These concepts hold for other bases as well. As we will see later, they are very useful in simplifying 2's complement and subtraction hardware.

Also, the 2's complement can be formed by leaving all least significant 0's and the first 1 unchanged and then replacing 1's with 0's and 0's with 1's in all other higher significant bits. Thus, the 2's complement of 1101100 is 0010100 and is obtained by leaving the two low-order 0's and the first 1 unchanged and then replacing 1's with 0's and 0's with 1's in the other four most significant bits. In other bases, the first nonzero digit is subtracted from the base r, and the remaining digits to the left are replaced with $r - 1$ minus their values.

It is also worth mentioning that the complement of the complement restores the number to its original value. To see this, note that the 2's complement of N is $2^n - N$, and the complement of the complement is $2^n - (2^n - N) = N$, giving back the original number.

Subtraction with Complements

Earlier, we expressed a desire to simplify hardware by sharing adder and subtractor logic. Armed with complements, we are prepared to define a binary subtraction procedure that uses addition and the corresponding complement logic. The subtraction of two n-digit unsigned numbers, $M - N$, in binary can be done as follows:

1. Add the 2's complement of the subtrahend N to the minuend M. This performs $M + (2^n - N) = M - N + 2^n$.

2. If $M \geq N$, the sum produces an end carry, 2^n. Discard the end carry, leaving result $M - N$.

3. If $M < N$, the sum does not produce an end carry since it is equal to $2^n - (N - M)$, the 2's complement of $N - M$. Perform a correction, taking the 2's complement of the sum and placing a minus sign in front to obtain the result $- (N - M)$.

The examples that follow further illustrate the foregoing procedure. Note that, although we are dealing with unsigned numbers, there is no way to get an unsigned result for the case in step 3. When working with paper and pencil, we recognize, by the absence of the end carry, that the answer must be changed to a negative number. If the minus sign for the result is to be preserved, it must be stored separately from the corrected n-bit result.

EXAMPLE 5-2   Unsigned Binary Subtraction by 2's Complement Addition

Given the two binary numbers $X = 1010100$ and $Y = 1000011$, perform the subtraction $X - Y$ and $Y - X$ using 2's complement operations. We have

$$
\begin{array}{rl}
X = & 1010100 \\
\text{2's complement of } Y = & 0111101 \\
\text{Sum} = & 10010001 \\
\text{Discard end carry } 2^7 = & -\underline{10000000} \\
\textit{Answer: } X - Y = & 0010001 \\
Y = & 1000011 \\
\text{2's complement of } X = & \underline{0101100} \\
\text{Sum} = & 1101111
\end{array}
$$

There is no end carry.

Answer: $Y - X = -(2$'s complement of $1101111) = -0010001$. ■

Subtraction of unsigned numbers also can be done by means of the 1's complement. Remember that the 1's complement is one less than the 2's complement. Because of this, the result of adding the minuend to the complement of the subtrahend produces a sum that is one less than the correct difference when an end carry occurs. Discarding the end carry and adding one to the sum is referred to as an *end-around carry*.

EXAMPLE 5-3   Unsigned Binary Subtraction by 1's Complement Addition

Repeat Example 5-2 using 1's complement operations. Here, we have

$$X - Y = 1010100 - 1000011$$

$$
\begin{array}{rl}
X = & 1010100 \\
\text{1's complement of } Y = & +\underline{0111100} \\
\text{Sum} = & 10010000 \\
\text{End-around carry} & \quad\llcorner\!\!\longrightarrow + 1 \\
\textit{Answer: } X - Y = & 0010001
\end{array}
$$

$$Y - X = 1000011 - 1010100$$

$$Y = \quad 1000011$$

$$\text{1's complement of } X = \; +\; \underline{0101011}$$

$$\text{Sum} = \quad 1101110$$

There is no end carry.

Answer: $Y - X = -(1\text{'s complement of } 1101110) = -0010001.$ ■

Note that the negative result is obtained by taking the 1's complement of the sum, since this is the type of complement being used.

5-4 BINARY ADDER–SUBTRACTORS

Using either the 2's or 1's complement, we have eliminated the subtraction operation and need only the appropriate complementer and an adder. When performing a subtraction we complement the subtrahend N, and when performing an addition we do not complement N. These operations can be accomplished by using a selective complementer and adder interconnected to form an adder–subtractor. We have used 2's complement, since it is most prevalent in modern systems. The 2's complement can be obtained by taking the 1's complement and adding 1 to the least significant bit. The 1's complement can be implemented easily with inverter circuits, and we can add 1 to the sum by making the input carry of the parallel adder equal to 1. Thus, by using 1's complement and an unused adder input, the 2's complement is obtained inexpensively. In 2's complement subtraction, as the correction step after adding, we complement the result and append a minus sign if an end carry does not occur. The correction operation is performed by using either the adder–subtractor a second time with $M = 0$ or a selective complementer as in Figure 5-7.

The circuit for subtracting $A - B$ consists of a parallel adder as shown in Figure 5-5, with inverters placed between each B terminal and the corresponding full-adder input. The input carry C_0 must be equal to 1. The operation that is performed becomes A plus the 1's complement of B plus 1. This is equal to A plus the 2's complement of B. For unsigned numbers, it gives $A - B$ if $A \geq B$ or the 2's complement of $B - A$ if $A < B$.

The addition and subtraction operations can be combined into one circuit with one common binary adder. This is done by including an exclusive-OR gate with each full adder. A 4-bit adder–subtractor circuit is shown in Figure 5-8. Input S controls the operation. When $S = 0$ the circuit is an adder, and when $S = 1$ the circuit becomes a subtractor. Each exclusive-OR gate receives input S and one of the inputs of B, B_i. When $S = 0$, we have $B_i \oplus 0$. If the full adders receive the value of B, and the input carry is 0, the circuit performs A plus B. When $S = 1$, we have $B_i \oplus 1 = \overline{B_i}$ and $C_0 = 1$. In this case, the circuit performs the operation A plus the 2's complement of B.

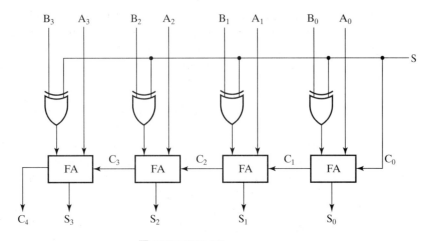

☐ **FIGURE 5-8**

Adder–Subtractor Circuit

Signed Binary Numbers

In the previous section, we dealt with the addition and subtraction of unsigned numbers. We will now extend this approach to signed numbers, including a further use of complements that eliminates the correction step.

Positive integers and the number zero can be represented as unsigned numbers. To represent negative integers, we need a notation for negative values. In ordinary arithmetic, a negative number is indicated by a minus sign and a positive number by a plus sign. Because of hardware limitations, computers must represent everything with 1's and 0's, including the sign of a number. As a consequence, it is customary to represent the sign with a bit placed in the most significant position of an n-bit number. The convention is to make the sign bit 0 for positive numbers and 1 for negative numbers.

It is important to realize that both signed and unsigned binary numbers consist of a string of bits when represented in a computer. The user determines whether the number is signed or unsigned. If the binary number is signed, then the leftmost bit represents the sign and the rest of the bits represent the number. If the binary number is assumed to be unsigned, then the leftmost bit is the most significant bit of the number. For example, the string of bits 01001 can be considered as 9 (unsigned binary) or +9 (signed binary), because the leftmost bit is 0. Similarly, the string of bits 11001 represents the binary equivalent of 25 when considered as an unsigned number or −9 when considered as a signed number. The latter is because the 1 in the leftmost position designates a minus sign and the remaining four bits represent binary 9. Usually, there is no confusion in identifying the bits because the type of number representation is known in advance. The representation of signed numbers just discussed is referred to as the *signed-magnitude* system. In this system, the number consists of a magnitude and a symbol (+ or −) or a bit (0 or 1) indicating the sign. This is the representation of signed numbers used in ordinary arithmetic.

In implementing signed-magnitude addition and subtraction for n-bit numbers, the single sign bit in the leftmost position and the $n - 1$ magnitude bits are processed separately. The magnitude bits are processed as unsigned binary numbers. Thus, subtraction involves the correction step. To avoid this step, we use a different system for representing negative numbers, referred to as a *signed-complement* system. In this system, a negative number is represented by its complement. While the signed-magnitude system negates a number by changing its sign, the signed-complement system negates a number by taking its complement. Since positive numbers always start with 0 (representing a plus sign) in the leftmost position, their complements will always start with a 1, indicating a negative number. The signed-complement system can use either the 1's or the 2's complement, but the latter is the most common. As an example, consider the number 9, represented in binary with eight bits. $+9$ is represented with a sign bit of 0 in the leftmost position, followed by the binary equivalent of 9, to give 00001001. Note that all eight bits must have a value, and therefore, 0's are inserted between the sign bit and the first 1. Although there is only one way to represent $+9$, we have three different ways to represent -9 using eight bits:

In signed-magnitude representation:	10001001
In signed-1's complement representation:	11110110
In signed-2's complement representation:	11110111

In signed magnitude, -9 is obtained from $+9$ by changing the sign bit in the leftmost position from 0 to 1. In signed 1's complement, -9 is obtained by complementing all the bits of $+9$, including the sign bit. The signed 2's complement representation of -9 is obtained by taking the 2's complement of the positive number, including the 0 sign bit.

Table 5-3 lists all possible 4-bit signed binary numbers in the three representations. The equivalent decimal number is also shown. Note that the positive numbers in all three representations are identical and have 0 in the leftmost position. The signed 2's complement system has only one representation for 0, which is always positive. The other two systems have a positive 0 and a negative 0, which is something not encountered in ordinary arithmetic. Note that all negative numbers have a 1 in the leftmost bit position; this is the way we distinguish them from positive numbers. With 4 bits, we can represent 16 binary numbers. In the signed-magnitude and the 1's complement representations, there are seven positive numbers and seven negative numbers, and two signed zeros. In the 2's complement representation, there are seven positive numbers, one zero, and eight negative numbers.

The signed-magnitude system is used in ordinary arithmetic, but is awkward when employed in computer arithmetic due to the separate handling of the sign and the correction step required for subtraction. Therefore, the signed complement is normally used. The 1's complement imposes difficulties because of its two representations of 0 and is seldom used for arithmetic operations. It is useful as a logical operation, since the change from 1 to 0 or 0 to 1 is equivalent to a logical complement operation. The following discussion of signed binary arithmetic deals exclusively with the signed-2's complement representation of negative numbers because it prevails in actual use. By using 1's complementation and the end-around carry,

□ **TABLE 5-3**
Signed Binary Numbers

Decimal	Signed 2's Complement	Signed 1's Complement	Signed Magnitude
+7	0111	0111	0111
+6	0110	0110	0110
+5	0101	0101	0101
+4	0100	0100	0100
+3	0011	0011	0011
+2	0010	0010	0010
+1	0001	0001	0001
+0	0000	0000	0000
−0	—	1111	1000
−1	1111	1110	1001
−2	1110	1101	1010
−3	1101	1100	1011
−4	1100	1011	1100
−5	1011	1010	1101
−6	1010	1001	1110
−7	1001	1000	1111
−8	1000	—	—

the same procedures as those for signed 2's complement can be applied to signed 1's complement.

Signed Binary Addition and Subtraction

The addition of two numbers, $M + N$, in the signed-magnitude system follows the rules of ordinary arithmetic: If the signs are the same, we add the two magnitudes and give the sum the sign of M. If the signs are different, we subtract the magnitude of N from the magnitude of M. The absence or presence of an end borrow then determines the sign of the result, based on the sign of M, and determines whether or not a 2's complement correction is performed. For example, since the signs are different, (0 0011001) + (1 0100101) causes 0100101 to be subtracted from 0011001. The result is 1110100, and an end borrow of 1 occurs. The end borrow indicates that the magnitude of M is smaller than the magnitude of N. So the sign of the result is opposite that of M and is therefore a minus. The end borrow indicates that the magnitude of the result, 1110100, must be corrected by taking its 2's complement. Combining the sign and the corrected magnitude of the result, we obtain 1 0001100.

In contrast to this signed-magnitude case, the rule for adding numbers in the signed-complement system does not require comparison or subtraction, but only addition. The procedure is simple and can be stated as follows for binary numbers:

The addition of two signed binary numbers with negative numbers represented in signed-2's complement form is obtained from the addition of the two numbers, including their sign bits. A carry out of the sign bit position is discarded.

Numerical examples of signed binary addition are given in Example 5-4. Note that negative numbers will already be in 2's complement form and that the sum obtained after the addition, if negative, is left in that same form.

EXAMPLE 5-4   Signed Binary Addition Using 2's Complement

+ 6	00000110	− 6	11111010	+ 6	00000110	− 6	11111010
+ 13	00001101	+13	00001101	− 13	11110011	− 13	11110011
+ 19	00010011	+ 7	00000111	− 7	11111001	− 19	11101101

In each of the four cases, the operation performed is addition, including the sign bits. Any carry out of the sign bit position is discarded, and negative results are automatically in 2's complement form. ■

The complement form for representing negative numbers is unfamiliar to people accustomed to the signed-magnitude system. To determine the value of a negative number in signed-2's complement, it is necessary to convert the number to a positive number in order to put it in a more familiar form. For example, the signed binary number 11111001 is negative because the leftmost bit is 1. Its 2's complement is 00000111, which is the binary equivalent of +7. We therefore recognize the original number to be equal to −7.

The subtraction of two signed binary numbers when negative numbers are in 2's complement form is very simple and can be stated as follows:

Take the 2's complement of the subtrahend (including the sign bit) and add it to the minuend (including the sign bit). A carry out of the sign bit position is discarded.

This procedure stems from the fact that a subtraction operation can be changed to an addition operation if the sign of the subtrahend is changed. That is,

$$(\pm A) - (+B) = (\pm A) + (-B)$$
$$(\pm A) - (-B) = (\pm A) + (+B)$$

But changing a positive number to a negative number is easily done by taking its 2's complement. The reverse is also true, because the complement of a negative number that is already in complement form produces the corresponding positive number. Numerical examples are shown in Example 5-5.

EXAMPLE 5-5   Signed Binary Subtraction Using 2's Complement

− 6	11111010		11111010	+ 6	00000110		00000110
− (−13)	− 11110011		+ 00001101	− (−13)	− 11110011		+ 00001101
+ 7			00000111	+ 19			00010011

The end carry is discarded. ■

It is worth noting that binary numbers in the signed-complement system are added and subtracted by the same basic addition and subtraction rules as are unsigned numbers. Therefore, computers need only one common hardware circuit to handle both types of arithmetic. The user or programmer must interpret the results of such addition or subtraction differently, depending on whether it is assumed that the numbers are signed or unsigned. Thus, the same adder–subtractor designed for unsigned numbers can be used for signed numbers.

If the signed numbers are in 2's complement representation, then the circuit in Figure 5-8 can be used with no correction step required. For 1's complement, the input from S to C_0 of the adder must be replaced by an input from C_n to C_0.

Overflow

To obtain a correct answer when adding and subtracting, we must ensure that the result has a sufficient number of bits to accommodate the sum. If we start with two n-bit numbers, and the sum occupies $n + 1$ bits, we say that an *overflow* occurs. This is true for binary or decimal numbers, whether signed or unsigned. When one performs addition with paper and pencil, an overflow is not a problem, since we are not limited by the width of the page. We just add another 0 to a positive number and another 1 to a negative number, in the most significant position, to extend them to $n + 1$ bits and then perform the addition. Overflow is a problem in computers because the number of bits that hold a number is fixed, and a result that exceeds the number of bits cannot be accommodated. For this reason, computers detect and can signal the occurrence of an overflow. The overflow condition may be handled automatically by interrupting the execution of the program and taking special action. An alternative is to monitor for overflow conditions using software.

The detection of an overflow after the addition of two binary numbers depends on whether the numbers are considered to be signed or unsigned. When two unsigned numbers are added, an overflow is detected from the end carry out of the most significant position. In unsigned subtraction, the magnitude of the result is always equal to or smaller than the larger of the original numbers, making overflow impossible. In the case of signed 2's complement numbers, the most significant bit always represents the sign. When two signed numbers are added, the sign bit is treated as a part of the number, and an end carry of 1 does not necessarily indicate an overflow.

With signed numbers, an overflow cannot occur for an addition if one number is positive and the other is negative: Adding a positive number to a negative number produces a result whose magnitude is equal to or smaller than the larger of the original numbers. An overflow may occur if the two numbers added are both positive or both negative. To see how this can happen, consider the following 2's complement example: Two signed numbers, $+70$ and $+80$, are stored in two 8-bit registers. The range of binary numbers, expressed in decimal, that each register can accommodate is from $+127$ to -128. Since the sum of the two stored numbers is $+150$, it exceeds the capacity of an 8-bit register. This is also true for -70 and -80. These two additions, together with the two most significant carry bit values, are as follows:

Carries: 0 1		Carries: 1 0	
+ 70	0 1000110	− 70	1 0111010
+ 80	0 1010000	− 80	1 0110000
+ 150	1 0010110	−150	0 1101010

Note that the 8-bit result that should have been positive has a negative sign bit and that the 8-bit result that should have been negative has a positive sign bit. If, however, the carry out of the sign bit position is taken as the sign bit of the result, then the 9-bit answer so obtained will be correct. But since there is no position in the result for the 9th bit, we say that an overflow has occurred.

An overflow condition can be detected by observing the carry into the sign bit position and the carry out of the sign bit position. If these two carries are not equal, an overflow has occurred. This is indicated in the 2's complement example just completed, where the two carries are explicitly shown. If the two carries are applied to an exclusive-OR gate, an overflow is detected when the output of the gate is equal to 1. For this method to work correctly for 2's complement, it is necessary either to apply the 1's complement of the subtrahend to the adder and add 1 or to have overflow detection on the circuit that forms the 2's complement. The latter condition is due to overflow when complementing the maximum negative number.

Simple logic that provides overflow detection is shown in Figure 5-9. If the numbers are considered unsigned, then the C output being equal to 1 detects a carry (an overflow) for an addition and indicates that no correction step is required for a subtraction. C being equal to 0 detects no carry (no overflow) for an addition and indicates that a correction step is required for a subtraction.

If the numbers are considered signed, then the output V is used to detect an overflow. If $V = 0$ after a signed addition or subtraction, it indicates that no overflow has occurred and the result is correct. If $V = 1$, then the result of the operation contains $n + 1$ bits, but only the rightmost n of those bits fit in the n-bit result, so an overflow has occurred. The $(n+1)$th bit is the actual sign, but it cannot occupy the sign bit position in the result.

5-5 BINARY MULTIPLICATION

Multiplication of binary numbers is performed in the same way as with decimal numbers. The multiplicand is multiplied by each bit of the multiplier, starting from the least significant bit. Each such multiplication forms a partial product. Successive partial products are shifted one bit to the left. The final product is obtained from the sum of the partial products.

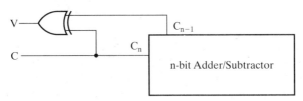

□ **FIGURE 5-9**
Overflow Detection Logic for Addition and Subtraction

To see how a binary multiplier can be implemented with a combinational circuit, consider the multiplication of two 2-bit numbers, as shown in Figure 5-10. The multiplicand bits are B_1 and B_0, the multiplier bits are A_1 and A_0, and the product is $C_3 C_2 C_1 C_0$. The first partial product is formed by multiplying $B_1 B_0$ by A_0. The multiplication of two bits such as A_0 and B_0 produces a 1 if both bits are 1; otherwise it produces a 0. This is identical to an AND operation. Therefore, the partial product can be implemented with AND gates as shown in the diagram. The second partial product is formed by multiplying $B_1 B_0$ by A_1 and is shifted one position to the left. The two partial products are added with two half-adder (HA) circuits. Usually there are more bits in the partial products, and it will be necessary to use full adders to produce the sum of the partial products. Note that the least significant bit of the product does not have to go through an adder, since it is formed by the output of the first AND gate.

A combinational circuit binary multiplier with more bits can be constructed in a similar fashion. A bit of the multiplier is ANDed with each bit of the multiplicand in as many levels as there are bits in the multiplier. The binary output in each level of AND gates is added in parallel with the partial product of the previous level to form a new partial product. The last level produces the product. For J multiplier bits and K multiplicand bits, we need $J \times K$ AND gates and $(J - 1)$ K-bit adders to produce a product of $J + K$ bits. As an example of a combinational circuit binary multiplier, consider a circuit that multiplies a binary number of four bits by a number of three bits. Let the multiplicand be represented by $B_3 B_2 B_1 B_0$ and the multiplier by $A_2 A_1 A_0$. Since $K = 4$ and $J = 3$, we need 12 AND gates and two 4-bit adders to produce a product of 7 bits. The logic diagram of this kind of

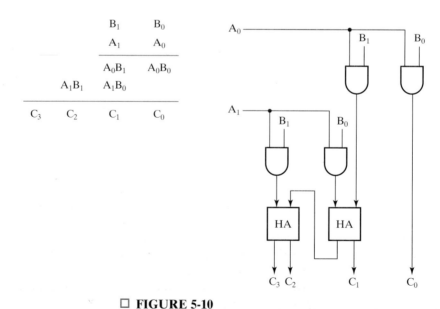

□ **FIGURE 5-10**
A 2-Bit by 2-Bit Binary Multiplier

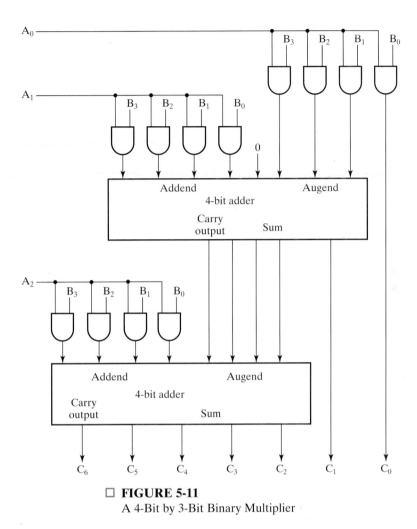

□ **FIGURE 5-11**
A 4-Bit by 3-Bit Binary Multiplier

multiplier circuit is shown in Figure 5-11. Note that the Carry Output bit enters the adder at the next level down in the multiplier.

5-6 OTHER ARITHMETIC FUNCTIONS

There are other arithmetic functions beyond +, −, and ×, that are quite important. Among these functions are incrementing, decrementing, multiplication and division by a constant, greater than comparison, and less than comparison. Each of these functions can be implemented for multiple-bit operands by using an iterative array of 1-bit cells. Instead of using these basic approaches, a combination of rudimentary functions and a new technique called contraction is used. Contraction begins with a circuit such as a binary adder, a carry-lookahead adder, or a binary multiplier. This approach simplifies design by converting existing circuits into useful, less-complicated circuits instead of designing the latter circuits directly.

Contraction

Value-fixing, transferring, and inverting on inputs can be combined with function blocks as done in Chapter 4 to implement new functions. We can implement new functions by using similar techniques on a given circuit or on its equations and then contracting it for a specific application to a simpler circuit. We will call the procedure *contraction*. The goal of contraction is to accomplish the design of a logic circuit or functional block by using results from past designs. It can be applied by the designer in designing a target circuit or can be applied by logic synthesis tools to simplify an initial circuit with value-fixing, transferring, and inverting on its inputs in order to obtain a target circuit. In both cases, contraction can also be applied to circuit outputs that are unused, to simplify a source circuit to a target circuit. First, we illustrate contraction by using Boolean equations.

EXAMPLE 5-6   Contraction of Full Adder Equations

The circuit Add1 to be designed is to form the sum S_i and carry C_{i+1} for the single bit addition $A_i + 1 + C_i$. This addition is a special case with $B_i = 1$ of the addition performed by a full adder, $A_i + B_i + C_i$. Thus, equations for the new circuit can be obtained by taking the full adder equations,

$$S_i = A_i \oplus B_i \oplus C_i$$
$$C_{i+1} = A_i B_i + A_i C_i + B_i C_i$$

setting $B_i = 1$, and simplifying the results, to obtain

$$S_i = A_i \oplus 1 \oplus C_i = \overline{A_i \oplus C_i}$$
$$C_{i+1} = A_i \cdot 1 + A_i C_i + 1 \cdot C_i = A_i + C_i$$

Suppose that this Add1 circuit is used in place of each of the four full adders in a 4-bit ripple carry adder. Instead of $S = A + B + C_0$, the computation being performed is $S = A + 1111 + C_0$. In 2's complement, this computation is $S = A - 1 + C_0$. If $C_0 = 0$, this implements the *decrement* operation $S = A - 1$, using considerably less logic than that used for a 4-bit addition or subtraction. ∎

Contraction can be applied to equations, as done here, or directly on circuit diagrams with rudimentary functions applied to function block inputs. In order to successfully apply contraction, the desired function must be able to be obtained from the initial circuit by application of rudimentary functions on its inputs. Next we consider contraction based on unused outputs.

Placing an unknown value, X, on the output of a circuit means that output will not be used. Thus, the output gate and any other gates that drive only that output gate can be removed. The rules for contracting equations with X's on one or more outputs are as follows:

1. Delete all equations with X's on the circuit outputs.
2. If an intermediate variable does not appear in any remaining equation, delete its equation.

3. If an input variable does not appear in any remaining equation, delete it.

4. Repeat 2 and 3 until no new deletions are possible.

The rules for contracting a logic diagram with X's on one or more outputs are as follows:

1. Beginning at the outputs, delete all gates with X's on their outputs and place X's on their input wires.

2. If all input wires driven by a gate are labeled with X's, delete the gate and place X's on its inputs.

3. If all input wires driven by an external input are labeled with X's, delete the input.

4. Repeat 2 and 3 until no new deletions are possible.

In the next subsection, contraction of a logic diagram is illustrated for the increment operation.

Incrementing

Incrementing means adding a fixed value to an arithmetic variable, most often a fixed value of 1. An n-bit *incrementer* that performs the operation $A + 1$ can be obtained by using a binary adder that performs the operation $A + B$ with $B = 0...01$. The use of $n = 3$ is large enough to determine the incrementer logic to construct the circuit needed for an n-bit incrementer.

Figure 5-12(a) shows a 3-bit adder with the inputs fixed to represent the computation $A + 1$ and with the output from the most significant carry bit C_3 fixed at value X. Operand $B = 001$ and the incoming carry $C_0 = 0$, so that $A + 001 + 0$ is computed. Alternatively, $B = 000$ and incoming carry $C_0 = 1$ could have been used.

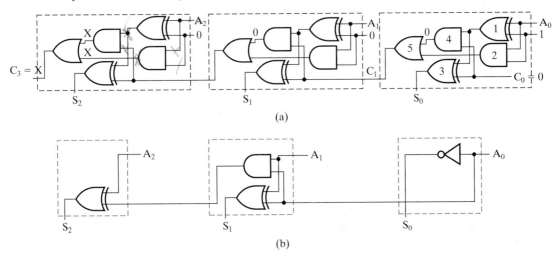

(a)

(b)

□ **FIGURE 5-12**
Contraction of Adder to Incrementer

Based on value-fixing, there are three distinct contraction cases for the cells in the adder:

1. The least significant cell on the right with $B_0 = 1$ and $C_0 = 0$,
2. The typical cell in the middle with $B_1 = 0$, and
3. The most significant cell on the left with $B_2 = 0$ and $C_3 = $ X.

For the right cell, the output of gate 1 becomes $\overline{A_0}$ so it can be replaced by an inverter. The output of gate 2 becomes A_0, so it can be replaced by a wire connected to A_0. Applying $\overline{A_0}$ and 0 to gate 3, it can be replaced by a wire, connecting A_0 to the output S_0. The output of gate 4 is 0, so it can be replaced with a 0 value. Applying this 0 and A_0 from gate 2 to gate 5, gate 5 can be replaced by a wire connecting A_0 to C_1. The resulting circuit is shown as the right cell in Figure 5-12(b).

Applying the same technique to the typical cell with $B_1 = 0$ yields

$$S_1 = A_1 \oplus C_1$$
$$C_2 = A_1 C_1$$

giving the circuit shown as the middle cell in Figure 5-12(b). For the left cell with $B_2 = 0$ and $C_3 = $ X, the effects of X are propagated first to save effort. Since gate A has X on its output, it is removed and X's are placed on its two inputs. Since all gates driven by gates B and C have X's on their inputs, they can be removed and X's can be placed on their inputs. Gates D and E cannot be removed, since they are each driving a gate without an X on its input. The resulting circuit is shown as the left cell in Figure 5-12(b).

For an incrementer with $n > 3$ bits, the least significant incrementer cell is used in position 0, the typical cell in positions 1 through $n - 2$, and the most significant cell in position $n - 1$. In this example, the rightmost cell in position 1 is contracted, but, if desired, it could be replaced with the cell in position 2 with $B_0 = 0$ and $C_0 = 1$. Likewise, the output C_3 could be generated, but not used. In both cases, logic cost and power efficiency are sacrificed to make all of the cells identical.

Decrementing

Decrementing is the addition of a fixed negative value to an arithmetic variable, most often, a fixed value of -1. A decrementer has already been designed in Example 5-6. Alternatively, a decrementer could be designed by using an adder–subtractor as a starting circuit and applying $B = 0...01$ and $C_0 = 0$, and selecting the subtraction operation by setting S to 1. Beginning with an adder–subtractor, we can also use contraction to design a circuit that increments for $S = 0$ and decrements for $S = 1$ by applying $B = 0...01$, $C_0 = 0$, and letting S remain a variable. In this case, the result is a cell of the complexity of a full adder in the typical bit positions. In fact, by going back to basics and redefining the carry function and designing the cell using this redefinition, the cost can be lowered somewhat. This illustrates that contraction, while it yields an implementation, may not produce a result with the least cost or best performance.

Multiplication by Constants

Assuming the circuit in Figure 5-11 is used as a basis for multiplication, multiplication by a constant can be achieved by simply applying the constant as the multiplier A. If the value for a particular bit position is 1, than the multiplicand will be applied to an adder. If the value for a particular bit position is 0, then 0 will be applied to an adder and the adder will be removed by contraction. In both cases, the AND gates will be removed. The process is illustrated in Figure 5-13(a). For this case, the multiplier has been set to 101. In the contraction process, since $0 + B = B$, the carryout value is always 0. The end result of the contraction is a circuit that conveys the two least significant bits of B to the outputs C_1 and C_0. The circuit adds the two most significant bits of B to B shifted two positions to the left and applies the result to product outputs C_6 through C_2.

An important special case occurs when the constant equal 2^i (i.e., for multiplication $2^i \times B$). In this case, only one 1 appears in the multiplier and all logic is eliminated from the circuit resulting in only wires. In this case, for the 1 in position i, the result is B followed by i 0's. The functional block that results is simply a combination of skewed transfers and value fixing to 0. The function of this block is called a *left shift by i bit positions with zero fill*. *Zero fill* refers to the addition of 0's to the right of (or to the left of) an operand such as B. Shifting is a very important operation applied to both numerical and nonnumerical data. The contraction resulting from a multiplication by 2^2 (i.e., a left shift of 2 bit positions) is shown in Figure 5-13(b).

Division by Constants

Since we have not covered the division operation, our discussion of division by constants will be restricted to division by powers of 2 (i.e., by 2^i in binary). Since multiplication by 2^i results in addition of i 0's to the right of the multiplicand, by analogy, division by 2^i results in removal of the i least significant bits of the dividend. The remaining bits are the quotient and the bits discarded are the remainder. The function of this block is called a *right shift by i bit positions*. Just as for left shifting, right shifting is likewise a very important operation. The function block for division by 2^2 (i.e., right shifting by two bit positions) is shown in Figure 5-13 (c).

Zero Fill and Extension

Zero fill, as defined previously for multiplication by a constant, can also be used to increase the number of bits in an operand. For example, suppose that a byte 01101011 is to be used as an input to a circuit that requires an input of 16 bits. One possible way of producing the 16-bit input is to zero-fill with eight 0's on the left to produce 0000000001101011. Another is to zero-fill on the right to produce

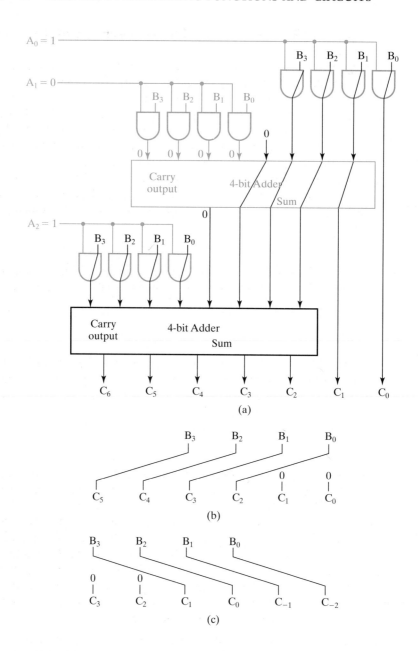

☐ **FIGURE 5-13**

Contractions of Multiplier: (a) for $101 \times B$, (b) for $100 \times B$, and (c) for $B \div 100$

0110101100000000. The former approach would be appropriate for operations such as addition or subtraction. The latter approach could be used to produce a low-precision 16-bit multiplication result in which the byte represents the most significant eight bits of the actual product with the lower byte of the product discarded.

In contrast to zero fill, *sign extension* is used to increase the number of bits in an operand represented by using a complement representation for signed numbers. If the operand is positive, then bits can be added on the left by extending the sign of the number (0 for positive and 1 for negative). Byte 01101011, which represents 107 in decimal, extended to 16 bits becomes 0000000001101011. Byte 10010101, which in 2's complement represents −107, extended to 16 bits becomes 1111111110010101. The reason for using sign extension is to preserve the complement representation for signed numbers. For example, if 10010101 were extended with 0's, the magnitude represented would be very large, and further, the leftmost bit, which should be a 1 for a minus sign, would be incorrect in the 2's-complement representation.

DECIMAL ARITHMETIC   The supplement that discusses decimal arithmetic functions and circuit implementations is available on the Companion Website for the text.

5-7 HDL REPRESENTATIONS—VHDL

Thus far, all of the VHDL descriptions used have contained only a single entity. Descriptions that represent circuits using hierarchies have multiple entities, one for each distinct element of the hierarchy, as shown in the next example.

EXAMPLE 5-7   Hierarchical VHDL for a 4-Bit Ripple Carry Adder

The example in Figures 5-14 and 5-15 uses three entities to build a hierarchical description of a 4-bit ripple carry adder. The style used for the architectures will be a mix of structural and dataflow description. The three entities are a half adder, a full adder that uses half adders, and the 4-bit adder itself. The architecture of half_adder consists of two dataflow assignments, one for s and one for c. The architecture of full_adder uses half_adder as a component. In addition, three internal signals, hs, hc, and tc, are declared. These signals are applied to two half adders and are also used in one dataflow assignment to construct the full adder in Figure 5-4. In the adder_4 entity, four full adder components are simply connected together using the signals given in Figure 5-5.

Note that C0 and C4 are an input and an output, respectively, but C(0) through C(4) are internal signals (i.e., neither inputs nor outputs). C(0) is assigned C0 and C4 is assigned C(4). The use of C(0) and C(4) separately from C0 and C4 is not essential here, but is useful to illustrate a VHDL constraint. Suppose we wanted to add overflow detection to the adder as shown in Figure 5-9. If C(4) is not defined separately, then one might attempt to write

V <= C(3) **xor** C4

```vhdl
-- 4-bit Adder: Hierarchical Dataflow/Structural
-- (See Figures 5-4 and 5-5 for logic diagrams)
library ieee;
use ieee.std_logic_1164.all;
entity half_adder is
  port (x, y : in std_logic;
        s, c : out std_logic);
end half_adder;

architecture dataflow_3 of half_adder is
  begin
    s <= x xor y;
    c <= x and y;
end dataflow_3;

library ieee;
use ieee.std_logic_1164.all;
entity full_adder is
  port (x, y, z : in std_logic;
        s, c : out std_logic);
end full_adder;

architecture struc_dataflow_3 of full_adder is
  component half_adder
    port(x, y : in std_logic;
         s, c : out std_logic);
  end component;
  signal hs, hc, tc: std_logic;
  begin
    HA1: half_adder
      port map (x, y, hs, hc);
    HA2: half_adder
      port map (hs, z, s, tc);
    c <= tc or hc;
end struc_dataflow_3;

library ieee;
use ieee.std_logic_1164.all;
entity adder_4 is
  port(B, A : in std_logic_vector(3 downto 0);
       C0 : in std_logic;
       S : out std_logic_vector(3 downto 0);
       C4: out std_logic);
end adder_4;
```

☐ **FIGURE 5-14**
Hierarchical Structural/Dataflow Description of 4-Bit Full Adder

```
architecture structural_4 of adder_4 is
  component full_adder
    port(x, y, z : in std_logic;
         s, c : out std_logic);
  end component;
  signal C: std_logic_vector(4 downto 0);
  begin
    Bit0: full_adder
      port map (B(0), A(0), C(0), S(0), C(1));
    Bit1: full_adder
      port map (B(1), A(1), C(1), S(1), C(2));
    Bit2: full_adder
      port map (B(2), A(2), C(2), S(2), C(3));
    Bit3: full_adder
      port map (B(3), A(3), C(3), S(3), C(4));
    C(0) <= C0;
    C4 <= C(4);
end structural_4;
```

□ **FIGURE 5-15**

Hierarchical Structural/Dataflow Description of 4-Bit Full Adder (Continued)

In VHDL, this is incorrect. An output cannot be used as an internal signal. Thus, it is necessary to define an internal signal to use in place of C4 (e.g., C(4)) giving

$$V <= C(3) \textbf{ xor } C(4)$$

Behavioral Description

The 4-bit adder provides an opportunity to illustrate description of circuits at a levels higher than the logic level. Such levels of description are referred to as the behavioral level or the register transfer level. We will specifically study register transfers in Chapter 7. Without studying register transfers, however, we can still show a behavioral level description.

EXAMPLE 5-8   Behavioral VHDL for a 4-Bit Ripple Carry Adder

A behavioral description for the 4-bit adder is given in Figure 5-16. In the architecture of the entity adder_4_b, the addition logic is described by a single statement using + and &. The + represents addition and the & represents an operation called *concatenation*. A concatenation operator combines two signals into a single signal having its number of bits equal to the sum of the number of bits in the original signals. In the example, '0' & A represents the signal vector

$$'0' A(3) A(2) A(1) A(0)$$

```
-- 4-bit Adder: Behavioral Description
library ieee;
use ieee.std_logic_1164.all;
use ieee.std_logic_unsigned.all;

entity adder_4_b is
  port(B, A : in std_logic_vector(3 downto 0);
       C0 : in std_logic;
       S : out std_logic_vector(3 downto 0);
       C4: out std_logic);
end adder_4_b;

architecture behavioral of adder_4_b is
signal sum : std_logic_vector(4 downto 0);
begin
  sum <= ('0' & A) + ('0' & B) + ("0000" & C0);
  C4 <= sum(4);
  S <= sum(3 downto 0);
end behavioral;
```

☐ **FIGURE 5-16**

Behavioral Description of 4-Bit Full Adder

with $1 + 4 = 5$ signals. Note that `'0'`, which appears on the left in the concatenation expression appears on the left in the signal listing. The inputs to the addition are all converted to 5-bit quantities for consistency, since the output including C4 is 5 bits. This conversion is not essential, but is a safe approach.

Since + cannot be performed on the `std_logic` type, we need an additional package to define addition for the `std_logic` type. In this case, we are using `std_logic_arith`, a package present in the `ieee` library. Further, we wish to specifically define the addition to be unsigned, so we use the `unsigned` extension. Also, concatenation in VHDL cannot be used on the left side of an assignment statement. To obtain C4 and S as the result of the addition, a 5-bit signal sum is declared. The signal sum is assigned the result of the addition including the carry out. Following are two additional assignment statements which split sum into outputs C4 and S. ■

This completes our introduction to VHDL for arithmetic circuits. We will continue with more on VHDL by presenting means for describing sequential circuits in Chapter 6.

5-8 HDL REPRESENTATIONS—VERILOG

Thus far, all of the descriptions used have contained only a single module. Descriptions that represent circuits using hierarchy have multiple modules, one for each distinct element of the hierarchy, as shown in the next example.

```
// 4-bit Adder: Hierarchical Dataflow/Structural
// (See Figures 5-4 and 5-5 for logic diagrams)

module half_adder_v(x, y, s, c);
   input x, y;
   output s, c;

   assign s = x ^ y;
   assign c = x & y;

endmodule

module full_adder_v(x, y, z, s, c);
   input x, y, z;
   output s, c;

   wire hs, hc, tc;

   half_adder_v     HA1(x, y, hs, hc),
                    HA2(hs, z, s, tc);
   assign c = tc | hc;

endmodule

module adder_4_v(B, A, C0, S, C4);
   input[3:0] B, A;
   input C0;
   output[3:0] S;
   output C4;

   wire[3:1] C;

   full_adder_v     Bit0(B[0], A[0], C0, S[0], C[1]),
                    Bit1(B[1], A[1], C[1], S[1], C[2]),
                    Bit2(B[2], A[2], C[2], S[2], C[3]),
                    Bit3(B[3], A[3], C[3], S[3], C4);
endmodule
```

☐ **FIGURE 5-17**

Hierarchical Dataflow/Structural Verilog Description of 4-bit Adder

EXAMPLE 5-9   Hierarchical Verilog for a 4-Bit Ripple Carry Adder

The description in Figure 5-17 uses three modules to represent a hierarchical design for a 4-bit ripple carry adder. The style used for the modules will be a mix of structural and dataflow description. The three modules are a half adder, a full adder built around half adders, and the 4-bit adder itself.

The half_adder module consists of two dataflow assignments, one for s and one for c. The full_adder module uses the half_adder as a component as in Figure 5-4. In the full_adder, three internal wires, hs, hc, and tc are declared. Inputs, outputs and these wire names are applied to the two half adders and `tc` and `hc` are ORed to form carry `c`. Note that the same names can be used on different modules (e.g., x, y, s, and c are used in both the `half_adder` and `full_adder`).

In the `adder_4` module, four full adders are simply connected together using the signals given in Figure 5-5. Note that C0 and C4 are an input and an output, respectively, but C(3) through C(1) are internal signals (i.e., neither inputs nor outputs). ▪

Behavioral Description

The 4-bit adder provides an opportunity to illustrate description of circuits at a levels higher than the logic level. Such levels of description are referred to as the behavioral level or the register transfer level. We will specifically study register transfers in Chapter 7. Without studying register transfers, however, we can still show the behavioral level description for the 4-bit adder.

EXAMPLE 5-10   Behavioral Verilog for a 4-Bit Ripple Carry Adder

Figure 5-18 shows the Verilog description for the adder. In module `adder_4_b_v`, the addition logic is described by a single statement using + and {}. The + represent addition and the {} represent an operation called *concatenation*. The operation + performed on wire data types is unsigned. Concatenation combines two signals into a single signal having its number of bits equal to the sum of the number of bits in the original signals. In the example, {C4, S} represents the signal vector

$$C4 \; S[3] \; S[2] \; S[1] \; S[0]$$

with 1 + 4 = 5 signals. Note that C4, which appears on the left in the concatenation expression, appears on the left in the signal listing. ▪

This completes our introduction to Verilog for arithmetic circuits. We will continue with more on Verilog by presenting means for describing sequential circuits in Chapter 6.

```
// 4-bit Adder: Behavioral Verilog Description

module adder_4_b_v(A, B, C0, S, C4);
   input[3:0] A, B;
   input C0;
   output[3:0] S;
   output C4;

   assign {C4, S} = A + B + C0;
endmodule
```

☐ **FIGURE 5-18**
Behavioral Description of 4-Bit Full Adder Using Verilog

5-9 CHAPTER SUMMARY

This chapter introduced circuits for performing arithmetic. The implementation of binary adders, including the carry lookahead adder for improved performance, was treated in detail. The subtraction of unsigned binary numbers using 2's and 1's complements was presented, as was the representation of signed binary numbers and their addition and subtraction. The adder–subtractor, developed for unsigned binary, was found to apply directly to the addition and subtraction of signed 2's complement numbers as well. A very brief introduction to binary multiplication, using combinational circuits made up of AND gates and binary adders, was given.

Additional arithmetic operations introduced included incrementing, decrementing, multiplication and division by a constant, and shifting. The implementations for these operations were obtained by a design technique we called contraction. Zero fill and sign extension of operands was also introduced.

The last two sections of the chapter provided an introduction to VHDL and Verilog descriptions for arithmetic circuits. Both HDLs were illustrated by studying descriptions at the functional and behavioral levels for various functional blocks in the chapter.

REFERENCES

1. MANO, M. M. *Digital Design,* 3rd ed. Englewood Cliffs, NJ: Prentice Hall, 2002.

2. WAKERLY, J. F. *Digital Design: Principles and Practices,* 3rd ed. Englewood Cliffs, NJ: Prentice Hall, 2000.

3. *High-Speed CMOS Logic Data Book.* Dallas: Texas Instruments, 1989.

4. *IEEE Standard VHDL Language Reference Manual.* (ANSI/IEEE Std 1076-1993; revision of IEEE Std 1076-1987). New York: The Institute of Electrical and Electronics Engineers, 1994.

5. SMITH, D. J. *HDL Chip Design.* Madison, AL: Doone Publications, 1996.

6. PELLERIN, D. AND D. TAYLOR. *VHDL Made Easy!* Upper Saddle River, NJ: Prentice Hall PTR, 1997.

7. STEFAN, S. AND L. LINDH. *VHDL for Designers.* London: Prentice Hall Europe, 1997.

8. *IEEE Standard Description Language Based on the Verilog® Hardware Description Language* (IEEE Std 1364-1995). New York: The Institute of Electrical and Electronics Engineers, 1995.

9. PALNITKAR, S. *Verilog HDL: A Guide to Digital Design and Synthesis.* Upper Saddle River, NJ: SunSoft Press (A Prentice Hall Title), 1996.

10. BHASKER, J. *A Verilog HDL Primer.* Allentown, PA: Star Galaxy Press, 1997.

11. THOMAS, D., AND P. MOORBY. *The Verilog Hardware Description Language* 4th ed. Boston: Kluwer Academic Publishers, 1998.

12. CILETTI, M. *Modeling, Synthesis, and Rapid Prototyping with the Verilog HDL,* Upper Saddle River, NJ: Prentice Hall, 1999.

PROBLEMS

The plus (+) indicates a more advanced problem and the asterisk (*) indicates a solution is available on the Companion Website for the text.

5-1. Design a combinational circuit that forms the 2-bit binary sum $S_1 S_0$ of two 2-bit numbers $A_1 A_0$ and $B_1 B_0$ and has both a carry input C_0 and carry output C_2. Design the entire circuit implementing each of the three outputs with a two-level circuit plus inverters for the input variables. Begin the design with the following equations for each of the two bits of the adder:

$$S_i = \overline{A_i}\,\overline{B_i}\,C_i + \overline{A_i}\,B_i\,\overline{C_i} + A_i\,\overline{B_i}\,\overline{C_i} + A_i\,B_i\,C_i$$

$$C_{i+1} = A_i B_i + A_i C_i + B_i C_i$$

5-2. *The logic diagram of the first stage of a 4-bit adder, as implemented in integrated circuit type 74283, is shown in Figure 5-19. Verify that the circuit implements a full adder.

5-3. *Obtain the 1's and 2's complements of the following unsigned binary numbers: 10011100, 10011101, 10101000, 00000000, and 10000000.

5-4. Perform the indicated subtraction with the following unsigned binary numbers by taking the 2's complement of the subtrahend:

(a) $11111 - 10000$ **(c)** $1011110 - 1011110$

(b) $10110 - 1111$ **(d)** $101 - 101000$

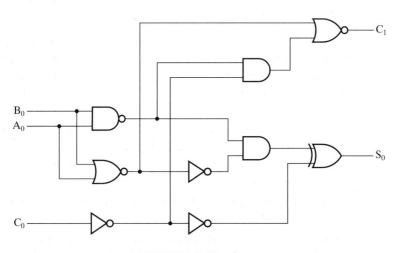

☐ **FIGURE 5-19**
Circuit for Problem 5–2

5–5. Repeat problem 5-4, assuming the numbers are 2's complement signed numbers. Use extension to equalize the length of the operands. Does overflow occur during the complement operations for any of the given numbers? Does overflow occur for the overall subtraction for any of the given numbers?

5–6. *Perform the arithmetic operations $(+36) + (-24)$ and $(-35) - (-24)$ in binary using signed-2's complement representation for negative numbers.

5–7. The following binary numbers have a sign in the leftmost position and, if negative, are in 2's complement form. Perform the indicated arithmetic operations and verify the answers.

(a) $100111 + 111001$ **(c)** $110001 - 010010$

(b) $001011 + 100110$ **(d)** $101110 - 110111$

Indicate if overflow occurs for each computation.

5–8. +Design three versions of the combinational circuit whose input is a 4-bit number and whose output is the 2's complement of the input number, for each of the following cases:

(a) The circuit is a simplified two-level circuit, plus inverters as needed for the input variables.

(b) The circuit is made up of four identical two-input, two-output cells, one for each bit. The cells are connected in cascade, with lines similar to a carry between them. The value applied to the rightmost carry bit is 0.

(c) The circuit is redesigned with carry lookahead-like logic in order to speed up the circuit in part (b) for use in larger circuits with $4n$ input bits.

5–9. Use contraction beginning with a 4-bit adder with carry out to design a 4-bit increment-by-2 circuit with carry out that adds the binary value 0010 to its 4-bit input. The function to be implemented is $S = A + 0010$.

5–10. Use contraction beginning with an 8-bit adder–subtractor without carry out to design an 8-bit circuit without carry out that increments its input by 00000010 for input $S = 0$ and decrements its input by 00000010 for input $S = 1$. Perform the design by designing the distinct 1-bit cells needed and indicating the type of cell use in each of the eight bit positions.

5–11. +**(a)** Use contraction beginning with a 4-bit carry lookahead adder with input carry and group carry and group propagate functions to design a 4-bit carry-lookahead-based circuit that increments its 4-bit input by the binary equivalent of 2.

(b) Repeat part (a), designing a 4-bit carry-lookahead–based circuit that adds 0000 to its 4-bit input.

(c) Construct a 16-bit carry-lookahead circuit that increments its 16-bit input by 2, giving the 16-bit output plus an output carry, by using the circuits designed in (a) and (b) an additional contraction of a 4-bit carry-lookahead circuit that produces C_4, C_8, and C_{12} and G_{0-15} and P_{0-15} with inputs P_{0-3}, G_{0-3}, P_{4-7}, P_{8-11}, and P_{12-15}, and the additional logic needed to produce C_{16}. (Note that, due to part (b), G_{4-7}, G_{8-11}, and G_{12-15} are equal to 0.)

5–12. Design a combinational circuit that compares two 4-bit unsigned numbers A and B to see whether B is greater than A. The circuit has one output X, so that $X = 1$ if $A < B$ and $X = 0$ if $A \geqslant B$.

5–13. +Repeat Problem 5-12 by using three-input, one-output circuits, one for each of the four bits. The four circuits are connected together in cascade by carry-like signals. One of the inputs to each cell is a carry input, and the single output is a carry output.

5–14. Repeat Problem 5-12 by applying contraction to a 4-bit subtractor and using the borrow out as X.

5–15. Design a combinational circuit that compares 4-bit unsigned numbers A and B to see whether $A = B$ or $A < B$. Use an iterative circuit as in Problem 5-14.

5–16. +Design a 5-bit signed-magnitude adder–subtractor. Divide the circuit for design into (1) sign generation and add–subtract control logic, (2) an unsigned number adder–subtractor using 2's complement of the minuend for subtraction, and (3) selective 2's complement result correction logic.

5–17. *The adder–subtractor circuit of Figure 5-8 has the following values for input select S and data inputs A and B:

	S	A	B
(a)	0	0111	0111
(b)	1	0100	0111
(c)	1	1101	1010
(d)	0	0111	1010
(e)	1	0001	1000

Determine, in each case, the values of the outputs S_3, S_2, S_1, S_0, and C_4.

5–18. *Design a binary multiplier that multiplies two 4-bit unsigned numbers. Use AND gates and binary adders.

5–19. Design a circuit that multiplies a 4-bit multiplicand by the constant 1010 by applying contraction to the solution to Problem 5-18.

5–20. **(a)** Design a circuit that multiplies a 4-bit multiplicand by the constant 1000.

(b) Design a circuit that divides an 8-bit dividend by the constant 1000 giving both an 8-bit quotient and an 8-bit remainder.

All files referred to in the remaining problems are available in ASCII form for simulation and editing on the Companion Website for the text. A VHDL or Verilog compiler/simulator is necessary for the problems or portions of problems requesting simulation. Descriptions can still be written, however, for many problems without using compilation or simulation.

5–21. Compile and simulate the 4-bit adder in Figures 5-14 and 5-15. Apply combinations that check out the rightmost full adder for all eight input combinations; this also serves as a check for the other full adders. Also, apply combinations that check the carry chain connections between all full adders by demonstrating that a 0 and a 1 can be propagated from C0 to C4.

5–22. *Compile and simulate the behavioral description of the 4-bit Adder in Figure 5-16. Assuming a ripple carry implementation, apply combinations that check out the rightmost full adder for all eight input combinations. Also apply combinations that check the carry chain connections between all full adders by demonstrating that a 0 and a 1 can be propagated from C0 to C4.

5–23. +Using Figure 5-16 as a guide and a "when else" on S, write a high-level behavior VHDL description for the adder–subtractor in Figure 5-8. Compile and simulate your description. Assuming a ripple carry implementation, apply combinations that check out one of the full adder–subtractor stages for all 16 possible input combinations. Also, apply combinations to check the carry chain connections in between the full adders by demonstrating that a 0 and a 1 can be propagated from C0 to C4.

5–24. +Write a hierarchical dataflow VHDL description similar to that in Figures 5-14 and 5-15 for the logic of the 4-bit carry lookahead adder in Figure 5-6. Compile and simulate your description. Devise a set of tests that does a good job of exercising the logic in the adder.

5–25. Compile and simulate the 4-bit Adder in Figure 5-17. Apply combinations that check out the rightmost full adder for all eight input combinations; this also serves as a check for the other full adders. Also, apply combinations that check the carry chain connections between all full adders by demonstrating that a 0 and a 1 can be propagated from C0 to C4.

5–26. *Compile and simulate the behavioral description of the 4-bit adder in Figure 5-18. Assuming a ripple carry implementation, apply all eight input combinations to check out the rightmost full adder. Also, apply combinations to check the carry chain connections between all full adders by demonstrating that a 0 and a 1 can be propagated from C0 to C4.

5–27. Using Figure 5-18 as a guide and a "binary decision" on S from Figure 4-37, write a high-level behavior Verilog description for the adder–subtractor in Figure 5-8. Compile and simulate your description. Assuming a ripple carry implementation, apply input combinations to your design that will (1) cause all 16 possible input combinations to be applied to the full adder–subtractor stage for bit 2, and (2) simultaneously cause the carry output of bit 2 to appear at one of your design's outputs. Also, apply combinations that check the carry chain connections between all full adders by demonstrating that a 0 and a 1 can be propagated from C0 to C4.

5–28. +Write a hierarchical Verilog dataflow description similar to that in Figure 5-17 for the logic of the 4-bit carry lookahead adder in Figure 5-6. Compile and simulate your description. Devise a set of tests that does a good job of exercising the logic in the adder.

CHAPTER

6

SEQUENTIAL CIRCUITS

To this point, we have studied only combinational logic. Although such logic is capable of interesting operations, such as addition and subtraction, the performance of useful sequences of operations using combinational logic alone requires cascading many structures together. The hardware to do this, however, is very costly and inflexible. In order to perform useful or flexible sequences of operations, we need to be able to construct circuits that can store information between the operations. Such circuits are called sequential circuits. This chapter begins with an introduction to sequential circuits, which is followed by a study of the basic elements for storing binary information, called latches and flip-flops. We distinguish flip-flops from latches and study various types of each. We then analyze sequential circuits consisting of both flip-flops and combinational logic. State tables and state diagrams provide a means for describing the behavior of sequential circuits. Subsequent sections of the chapter develop the techniques for designing sequential circuits and verifying their correctness. In the last two sections, we provide VHDL and Verilog hardware description language representations for storage elements and for the type of sequential circuits in this chapter.

Latches, flip-flops, and sequential circuits are fundamental components in the design of almost all digital logic. In the generic computer given at the beginning of Chapter 1, latches and flip-flops are widespread in the design. The exception is memory circuits, since large portions of memory are designed as electronic circuits rather than as logic circuits. Nevertheless, due to the wide use of logic-based storage, this chapter contains fundamental material for any in-depth understanding of computers and digital systems and how they are designed.

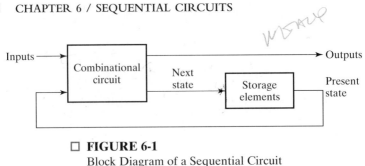

□ **FIGURE 6-1**
Block Diagram of a Sequential Circuit

6-1 SEQUENTIAL CIRCUIT DEFINITIONS

The digital circuits considered thus far have been combinational. Although every digital system is likely to include a combinational circuit, most systems encountered in practice also include storage elements, requiring that the systems be described as sequential circuits.

A block diagram of a sequential circuit is shown in Figure 6-1. A combinational circuit and storage elements are interconnected to form the sequential circuit. The storage elements are circuits that are capable of storing binary information. The binary information stored in these elements at any given time defines the *state* of the sequential circuit at that time. The sequential circuit receives binary information from its environment via the inputs. These inputs, together with the present state of the storage elements, determine the binary value of the outputs. They also determine the values used to specify the next state of the storage elements. The block diagram demonstrates that the outputs in a sequential circuit are a function not only of the inputs, but also of the present state of the storage elements. The next state of the storage elements is also a function of the inputs and the present state. Thus, a sequential circuit is specified by a time sequence of inputs, internal states, and outputs.

There are two main types of sequential circuits, and their classification depends on the times at which their inputs are observed and their internal state changes. The behavior of a *synchronous sequential circuit* can be defined from the knowledge of its signals at discrete instants of time. The behavior of an *asynchronous sequential circuit* depends upon the inputs at any instant of time and the order in continuous time in which the inputs change.

Information is stored in digital systems in many ways, including the use of logic circuits. Figure 6-2 (a) shows a buffer. This buffer has a propagation delay t_{pd}. Since information present at the buffer input at time t appears at the buffer output at time $t + t_{pd}$, the information has effectively been stored for time t_{pd}. But, in general, we wish to store information for an indefinite time that is typically much longer than the time delay of one or even many gates. This stored value is to be changed at arbitrary times based on the inputs applied to the circuit and should not depend on the specific time delay of a gate.

Suppose that the output of the buffer in Figure 6-2(a) is connected to its input as shown in Figures 6-2(b) and (c). Suppose further that the value on the input to the buffer in part (b) has been 0 for at least time t_{pd}. Then the output produced by the buffer will be 0 at time $t + t_{pd}$. This output is applied to the input so

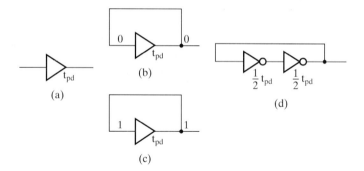

□ **FIGURE 6-2**
Logic Structures for Storing Information

that the output will also be 0 at time $t + 2\,t_{pd}$. This relationship between input and output holds for all t, so the 0 will be stored indefinitely. The same argument can be made for storing a 1 in the circuit in Figure 6-2(c).

The example of the buffer illustrates that storage can be constructed from logic with delay connected in a closed loop. Any loop that produces such storage must also have a property possessed by the buffer, namely, that there must be no inversion of the signal around the loop. A buffer is usually implemented by using two inverters, as shown in Figure 6-2(d). The signal is inverted twice, that is,

$$\overline{\overline{X}} = X$$

giving no net inversion of the signal around the loop. In fact, this example is an illustration of one of the most popular methods of implementing storage in computer memories. (See Chapter 9.) However, although the circuits in Figures 6-2(b) through (d) are able to store information, there is no way for the information to be changed. By replacing the inverters with NOR or NAND gates, the information can be changed. Asynchronous storage circuits called latches are made in this manner and are discussed in the next section.

In general, more complex asynchronous circuits are difficult to design, since their behavior is highly dependent on the propagation delays of the gates and on the timing of the input changes. Thus, circuits that fit the synchronous model are the choice of most designers. Nevertheless, some asynchronous design is necessary. A very important case is the use of asynchronous latches as blocks to build storage elements, called flip-flops, that store information in synchronous circuits.

A synchronous sequential circuit employs signals that affect the storage elements only at discrete instants of time. Synchronization is achieved by a timing device called a *clock generator* which produces a periodic train of *clock pulses*. The pulses are distributed throughout the system in such a way that synchronous storage elements are affected only in some specified relationship to every pulse. In practice, the clock pulses are applied with other signals that specify the required change in the storage elements. The outputs of storage elements can change their value only in the presence of clock pulses. Synchronous sequential circuits that use clock pulses as inputs to storage elements are called *clocked sequential circuits*. These are the type of

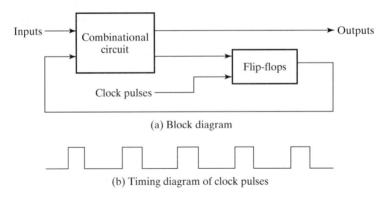

(a) Block diagram

(b) Timing diagram of clock pulses

☐ **FIGURE 6-3**
Synchronous Clocked Sequential Circuit

circuit most frequently encountered in practice, since they operate correctly in spite of wide differences in circuit delays and are relatively easy to design.

The storage elements used in clocked sequential circuits are called flip-flops. For simplicity, assume circuits with a single clock signal. A *flip-flop* is a binary storage device capable of storing one bit of information and having timing characteristics to be defined in Section 6-3. The block diagram of a synchronous clocked sequential circuit is shown in Figure 6-3. The flip-flops receive their inputs from the combinational circuit and also from a clock signal with pulses that occur at fixed intervals of time, as shown in the timing diagram. The flip-flops can change state only in response to a clock pulse. For synchronous operation, when a clock pulse is absent, the flip-flop outputs cannot change even if the outputs of the combinational circuit driving their inputs change in value. Thus, the feedback loops shown in the figure between the combinational logic and the flip-flops are broken. As a result, a transition from one state to the other occurs only at fixed time intervals dictated by the clock pulses, giving synchronous operation. The sequential circuit outputs are shown as outputs of the combinational circuit. This is valid even when some sequential circuit outputs are actually the flip-flop outputs. In this case, the combinational circuit part between the flip-flop outputs and the sequential circuit outputs consists of connections only.

A flip-flop has one or two outputs, one for the normal value of the bit stored and an optional one for the complemented value of the bit stored. Binary information can enter a flip-flop in a variety of ways, a fact that gives rise to different types of flip-flops. Our focus will be on the most prevalent flip-flop used today, the D flip-flop. In section 6-6, other flip-flop types will be considered. In preparation for studying flip-flops and their operation, necessary groundwork is presented in the next section on latches from which the flip-flops are constructed.

6-2 LATCHES

A storage element can maintain a binary state indefinitely (as long as power is delivered to the circuit), until directed by an input signal to switch states. The major differences among the various types of latches and flip-flops are the number of inputs they possess and the manner in which the inputs affect the binary state. The most basic storage elements are latches, from which flip-flops are usually constructed. Although latches are most often used within flip-flops, they can also be used with more complex clocking methods to implement sequential circuits directly. The design of such circuits is, however, beyond the scope of the basic treatment given here. In this section, the focus is on latches as basic primitives for constructing storage elements.

SR and *S̄R̄* Latches

The *SR* latch is a circuit constructed from two cross-coupled NOR gates. It is derived from the single-loop storage element in Figure 6-2(d) by simply replacing the inverters with NOR gates, as shown in Figure 6-4(a). This replacement allows the stored value in the latch to be changed. The latch has two inputs, labeled S for set and R for reset, and two useful states. When output $Q = 1$ and $\overline{Q} = 0$, the latch is said to be in the *set state*. When $Q = 0$ and $\overline{Q} = 1$, it is in the *reset state*. Outputs Q and $\overline{Q}$ are normally the complements of each other. When both inputs are equal to 1 at the same time, an undefined state with both outputs equal to 0 occurs.

Under normal conditions, both inputs of the latch remain at 0 unless the state is to be changed. The application of a 1 to the S input causes the latch to go to the set (1) state. The S input must go back to 0 before R is changed to 1 to avoid occurrence of the undefined state. As shown in the function table in Figure 6-4(b), two input conditions cause the circuit to be in the set state. The initial condition is $S = 1, R = 0$, to bring the circuit to the set state. Applying a 0 to S with $R = 0$ leaves the circuit in the same state. After both inputs return to 0, it is possible to enter the reset state by applying a 1 to the R input. The 1 can then be removed from R, and the circuit remains in the reset state. Thus, when both inputs are equal to 0, the latch can be in either the set or the reset state, depending on which input was most recently a 1.

If a 1 is applied to both the inputs of the latch, both outputs go to 0. This produces an undefined state because it violates the requirement that the outputs be the complement of each other. It also results in an indeterminate or unpredictable next state when both inputs return to 0 simultaneously. In normal operation, these problems are avoided by making sure that 1's are not applied to both inputs simultaneously.

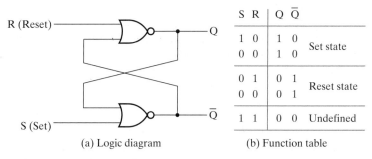

S R	Q Q̄	
1 0	1 0	Set state
0 0	1 0	
0 1	0 1	Reset state
0 0	0 1	
1 1	0 0	Undefined

(a) Logic diagram   (b) Function table

☐ **FIGURE 6-4**
SR Latch with NOR Gates

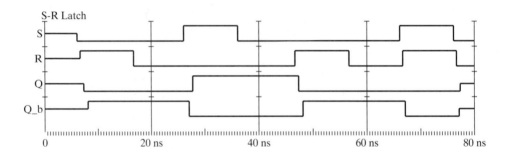

☐ **FIGURE 6-5**
Logic Simulation of *SR* Latch Behavior

The behavior of the *SR* latch described in the preceding paragraph is illustrated by the ModelSim® logic simulator waveforms shown in Figure 6-5. Initially, the inputs and the state of the latch are unknown, as indicated by a logic level halfway between 0 and 1. When *R* becomes 1 with *S* at 0, the latch is reset with *Q* first becoming 0 and, in response, *Q_b* (which represents $\overline{Q}$) becomes 1. Next, when *R* becomes 0, the latch remains reset, storing the 0 value present on *Q*. When S becomes 1 with *R* at 0, the latch is set, with *Q_b* going to 0 first and, in response, *Q* going to 1 next. The delays in the changes of *Q* and *Q_b* after an input changes are directly related to the delays of the two NOR gates used in the latch implementation. When *S* returns to 0, the latch remains set, storing the 1 value present on *Q*. When *R* becomes 1 with S equal to 0, the latch is reset with *Q* changing to 0 and *Q_b* responding by changing to 1. The latch remains reset when *R* returns to 0. When *S* and *R* both become 1, both *Q* and *Q_b* become 0. When *S* and *R* simultaneously return to 0, both *Q* and *Q_b* take on unknown values. This form of indeterminate state behavior for the (S, R) sequence of inputs (1, 1), (0, 0) results from assuming simultaneous input changes and equal gate delays. The actual indeterminate behavior that occurs depends on circuit delays and slight differences in the times at which *S* and *R* change in the actual circuit. Regardless of the simulation results, these indeterminate behaviors are viewed as undesirable, and the input combination (1,1) is avoided. In general, the latch state changes only in response to input changes and remains unchanged otherwise.

The $\overline{S}\,\overline{R}$ latch with two cross-coupled NAND gates is shown in Figure 6-6. It operates with both inputs normally at 1, unless the state of the latch has to be changed. The application of a 0 to the *S* input causes output *Q* to go to 1, putting the latch in the set state. When the *S* input goes back to 1, the circuit remains in the set state. With both inputs at 1, the state of the latch is changed by placing a 0 on the *R* input. This causes the circuit to go to the reset state and stay there even after both inputs return to 1. The condition that is undefined for this NAND latch is when both inputs are equal to 0 at the same time, an input combination that should be avoided.

Comparing the NAND latch with the NOR latch, note that the input signals for the NAND require the complement of those values used for the NOR. Because the NAND latch requires a 0 signal to change its state, it is referred to as an $\overline{S}\,\overline{R}$

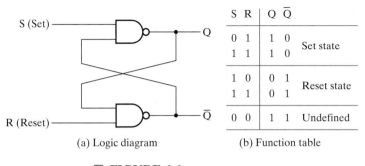

S	R	Q	Q̄	
0	1	1	0	Set state
1	1	1	0	
1	0	0	1	Reset state
1	1	0	1	
0	0	1	1	Undefined

(a) Logic diagram   (b) Function table

□ **FIGURE 6-6**
$\overline{S}\,\overline{R}$ Latch with NAND Gates

latch. The bar above the letters designates the fact that the inputs must be in their complement form in order to act upon the circuit state.

The operation of the basic NOR and NAND latches can be modified by providing an additional control input that determines when the state of the latch can be changed. An *SR* latch with a control input is shown in Figure 6-7. It consists of the basic NAND latch and two additional NAND gates. The control input *C* acts as an enable signal for the other two inputs. The output of the NAND gates stays at the logic-1 level as long as the control input remains at 0. This is the quiescent condition for the $\overline{SR}$ latch composed of two NAND gates. When the control input goes to 1, information from the *S* and *R* inputs is allowed to affect the $\overline{SR}$ latch. The set state is reached with $S = 1$, $R = 0$, and $C = 1$. To change to the reset state, the inputs must be $S = 0$, $R = 1$, and $C = 1$. In either case, when *C* returns to 0, the circuit remains in its current state. Control input $C = 0$ disables the circuit so that the state of the output does not change, regardless of the values of *S* and *R*. Moreover, when $C = 1$ and both the *S* and *R* inputs are equal to 0, the state of the circuit does not change. These conditions are listed in the function table accompanying the diagram.

An undefined state occurs when all three inputs are equal to 1. This condition places 0's on both inputs of the basic $\overline{SR}$ latch, giving an undefined state. When the control input goes back to 0, one cannot conclusively determine the next state, since

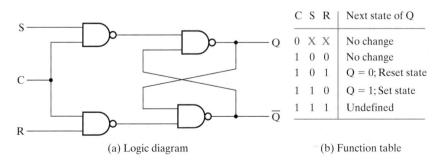

C	S	R	Next state of Q
0	X	X	No change
1	0	0	No change
1	0	1	Q = 0; Reset state
1	1	0	Q = 1; Set state
1	1	1	Undefined

(a) Logic diagram   (b) Function table

□ **FIGURE 6-7**
SR Latch with Control Input

the $\overline{SR}$ latch sees inputs $(0, 0)$ followed by $(1, 1)$. The SR latch with control input is an important circuit because other latches and flip-flops are constructed from it. Sometimes the SR latch with control input is referred to as an SR (or RS) flip-flop: however, according to our terminology, it does not qualify as a flip-flop, since the circuit does not fulfill the flip-flop requirements presented in the next section.

D Latch

One way to eliminate the undesirable undefined state in the SR latch is to ensure that inputs S and R are never equal to 1 at the same time. This is done in the D latch, shown in Figure 6-8. This latch has only two inputs: D (data) and C (control). The complement of D input goes directly to the $\overline{S}$ input, and D is applied to the $\overline{R}$ input. As long as the control input is 0, the $\overline{SR}$ latch has both inputs at the 1 level, and the circuit cannot change state regardless of the value of D. The D input is sampled when $C = 1$. If D is 1, the Q output goes to 1, placing the circuit in the set state. If D is 0, output Q goes to 0, placing the circuit in the reset state.

The D latch receives its designation from its ability to hold *data* in its internal storage. The binary information present at the data input of the D latch is transferred to the Q output when the control input is enabled (1). The output follows changes in the data input, as long as the control input is enabled. When the control input is disabled (0), the binary information that was present at the data input at the time the transition occurred is retained at the Q output until the control input is enabled again.

The D latch in VLSI circuits is often constructed with transmission gates (TGs), as shown in Figure 6-9. The TG was defined in Figure 2-35. The C input controls two TGs. When $C = 1$, the TG connected to input D conducts, and the TG connected to output Q disconnects. This produces a path from input D through two

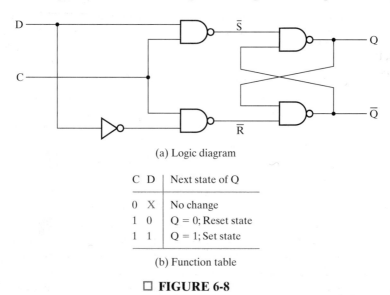

(a) Logic diagram

C	D	Next state of Q
0	X	No change
1	0	Q = 0; Reset state
1	1	Q = 1; Set state

(b) Function table

☐ **FIGURE 6-8**
D Latch

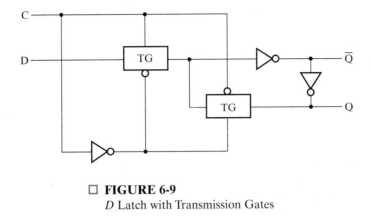

□ **FIGURE 6-9**
D Latch with Transmission Gates

inverters to output *Q*. Thus, the output follows the data input as long as *C* remains active (1). When *C* changes to 0, the first TG disconnects input *D* from the circuit, and the second TG connects the two inverters at the output into a loop. Hence, the value that was present at input *D* at the time that *C* went from 1 to 0 is retained at the *Q* output by the loop.

6-3 FLIP-FLOPS

The state of a latch in a flip-flop is allowed to switch by a momentary change in value on the control input. This change is called a *trigger*, and it enables, or triggers, the flip-flop. The *D* latch with clock pulses on its control input is triggered every time a pulse to the logic-l level occurs. As long as the pulse remains at the active (1) level, any changes in the data input will change the state of the latch. In this sense, the latch is *transparent*, since its input value can be seen from the outputs.

As the block diagram of Figure 6-3 shows, a sequential circuit has a feedback path from the outputs of the flip-flops to the combination circuit. As a consequence, the data inputs of the flip-flops are derived in part from the outputs of the same and other flip-flops. When latches are used for the storage elements, a serious difficulty arises. The state transitions of the latches start as soon as the clock pulse changes to the logic-1 level. The new state of a latch may appear at its output while the pulse is still active. This output is connected to the inputs of some of the latches through a combinational circuit. If the inputs applied to the latches change while the clock pulse is still in the logic-1 level, the latches will respond to *new state values* of other latches instead of the *original state values*, and a succession of changes of state instead of a single one may occur. The result is an unpredictable situation, since the state may keep changing and continue to change until the clock returns to 0. The final state depends on how long the clock pulse stays at level logic 1. Because of this unreliable operation, the output of a latch cannot be applied directly or through combinational logic to the input of the same or another latch when all the latches are triggered by a single clock signal.

Flip-flop circuits are constructed in such a way as to make them operate properly when they are part of a sequential circuit that employs a single clock.

Note that the problem with the latch is that it is transparent: As soon as an input changes, shortly thereafter the corresponding output changes to match it. This transparency is what allows a change on a latch output to produce additional changes at other latch outputs while the clock pulse is at logic 1. The key to the proper operation of flip-flops is to prevent them from being transparent. In a flip-flop, before an output can change, the path from its inputs to its outputs is broken. So a flip-flop cannot "see" the change of its output or of the outputs of other, like flip-flops at its input during the same clock pulse. Thus, the new state of a flip-flop depends only on the immediately preceding state, and the flip-flops do not go through multiple changes of state.

There are two ways that latches are combined to form a flip-flop. One way is to combine two latches such that (1) the inputs presented to the flip-flop when a clock pulse is present control its state and (2) the state of the flip-flop changes only when a clock pulse is not present. Such a circuit is called a *master-slave* flip-flop. Another way is to produce a flip-flop that triggers only during a signal *transition* from 0 to 1 (or from 1 to 0) on the clock and that is disabled at all other times, including for the duration of the clock pulse. Such a circuit is said to be an *edge-triggered* flip-flop. Next, the implementations of these two flip-flop triggering approaches are presented. It is necessary to consider the SR flip-flop for the master-slave triggering approach since a properly-constructed D flip-flop has the same behavior for both triggering types.

Master-Slave Flip-Flops

The master-slave *SR* flip-flop, consisting of two latches and an inverter, is shown in Figure 6-10. The symbol with *S*, *C*, and *R* on it is that for the *SR* latch with control input (Figure 6-7), which is referred to here as a clocked *SR* latch. The left clocked *SR* latch in Figure 6-10 is called the master, the right the slave. When the clock input *C* is 0, the output of the inverter is 1. The slave latch is then enabled, and its output *Q* is equal to the master output *Y*. The master latch is disabled, because *C* is 0. When a logic-1 clock pulse is applied, the values on *S* and *R* control the value stored in the master latch *Y*. The slave, however, is disabled as long as the pulse remains at the 1 level, because its *C* input is equal to 0. Any changes in the external

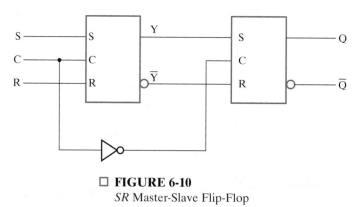

☐ **FIGURE 6-10**
SR Master-Slave Flip-Flop

S and R inputs change the master output Y, but cannot affect the slave output Q. When the pulse returns to 0, the master is disabled and is isolated from the S and R inputs. At the same time, the slave is enabled, and the current value of Y is transferred to the output of the flip-flop at Q.

A ModelSim logic simulation illustrating master-slave flip-flop SR behavior is shown in Figure 6-11. Initially, all values are unknown including the clock C. When S and R both go to 0, and the clock goes from 1 to 0, the output of the master, Y and the output of the slave, Q, both remain unknown, since the prior value is effectively being stored. S is at 1 with R at 0 to set the flip-flop in response to the next clock pulse. When C becomes 1, Y sets to 1. When C becomes 0, the slave copies the value of Y setting Q to 1. After S returns to 0, Y and Q remain unchanged, storing the 1 value through the next clock period. Next, R becomes 1. After the clock pulse transition from 0 to 1, the master latch is reset with Y changing to 0. The slave latch is not affected, because its C input is 0. Since the master is an internal circuit, its change of state is not presented at output Q. Even if the inputs S and R change during this interval and the state of the master latch responds by changing, the output of the flip-flop remains in its previous state. When the pulse returns to 0, the information from the master is allowed to pass through to the slave. For the simulation example, the value $Y = 0$ is copied to the slave latch making the external output $Q = 0$. Note that these changes are delayed from the pulse changes by gate delays. Also, the external inputs S and R can change anytime after the clock pulse goes through its negative transition. This is because, as the C input reaches 0, the master is disabled, and S and R have no effect until the next clock pulse. The next sequence of signal changes illustrates the "one's catching" behavior of the SR master-slave flip-flop. A narrow pulse to 1 occurs on S at the beginning of a clock pulse. The master latch responds to the 1 on S by changing Y to 1. Then S goes to 0 and a narrow 1 pulse occurs on R. The master latch responds to the 1 on R by changing Y back to 0. Since there are no further 1 values on S or R, the master continues to store 0 which is copied to the slave latch, changing Q to 0, in response to the clock changing to 0. Thus, the master latch "caught" both the 1 on S and the 1 on R. Since the 1 on R was caught

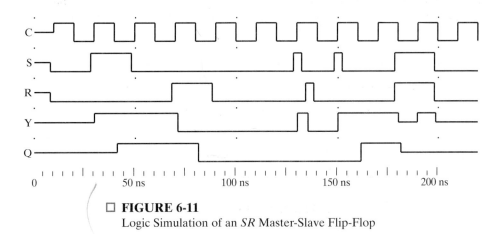

☐ **FIGURE 6-11**
Logic Simulation of an SR Master-Slave Flip-Flop

last, the output Q remained at 0. In general, the "correct" response is assumed to be the response to the input values when the clock goes to 0. So, in this case, the response happens to be correct, although more by accident with the changing values in the master. For the next clock pulse, a narrow 1 pulse occurs on S setting the master output Y to 1. The clock then goes to 0 and the value 1 is transferred to the slave latch and appears on Q. In this case, the correct value on Q should be 0 since Q was 0 before the clock pulse and both S and R are 0 just before the clock goes to 0. Since Q equals 1, due to "1's catching" on S, the flip-flop is in the wrong state. For the final clock pulse of interest, both S and R become 1 before the clock goes to 0. This applies the invalid combination to the master latch making both Y and $\overline{Y}$ equal to 1. When the clock changes to 0, the $\overline{S}\,\overline{R}$ latch within the master sees its inputs change from $(0, 0)$ to $(1, 1)$, causing the master latch to enter an unknown state which is immediately transferred to the inputs of the slave which also enters an unknown state. This demonstrates that $S = 1, R = 1$ is an invalid input combination for the SR master-slave flip-flop.

Now consider a sequential system containing many master-slave flip-flops, with the outputs of some flip-flops going to inputs of other flip-flops. Assume that the clock pulses to all of the flip-flops are synchronized and occur at the same time. At the beginning of each clock pulse, some of the masters change state, but all the slaves remain in their previous states. This means that the flip-flop slaves are still in their original states, while the flip-flop masters have changed to the new states. After the clock pulse returns to 0, some of the flip-flop slaves change state, but none of the new states have an effect on any of the masters until the next pulse. Thus, the states of flip-flops in a synchronous system can change simultaneously for the same clock pulse, even though outputs of flip-flops are connected to inputs of the same or other flip-flops. This is possible because the inputs affect the state of the flip-flop only while the clock pulse is 1 and the new state appears at the outputs only after the clock pulse has returned to 0, ensuring that the flip-flops are *not* transparent.

For reliable sequential circuit operation, all signals must propagate from the outputs of flip-flops, through the combinational circuit, and back to inputs of master-slave flip-flops, while the clock pulse remains at the logic-0 level. Any changes that occur at the inputs of flip-flops after the clock pulse goes to the logic-1 level, whether intentional or not, affect the flip-flop state and may result in the storage of incorrect values. Suppose that the delay in the combinational circuit is such that S is still changing after the clock pulse has gone to the logic-1 level. Suppose also that, as a consequence, the master is set to 1 by the presence of $S = 1$. When S finally stops changing, it is at 0, indicating that the state of the flip-flop was *not* to be changed from 0. Thus, the 1 value in the master, which will be transferred to the slave, is in error. There are two consequences of this behavior. First, the master-slave flip-flop is also referred to as a *pulse-triggered* flip-flop, since it can respond to input values that cause a change in state and occur anytime during its clock pulse. Second, the circuit must be designed so that combinational circuit delays are short enough to prevent S and R from changing during the clock pulse.

A master-slave D flip-flop can be constructed from the SR master-slave flip-flop by simply replacing the master SR latch with a master D latch. The resulting

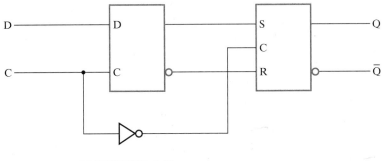

circuit is shown in Figure 6-12. The resulting circuit changes its value on the negative edge of the clock pulse just as the master-slave SR flip-flop does. However, the D type of flip-flop does not demonstrate the usual pulse-triggered behavior. Instead it demonstrates edge-triggered behavior, in this case, negative edge-triggered behavior. Thus, a master-slave D flip-flop constructed as shown, is also an edge-triggered flip-flop.

Edge-Triggered Flip-Flop

An *edge-triggered* flip-flop ignores the pulse while it is at a constant level and triggers only during a *transition* of the clock signal. Some edge-triggered flip-flops trigger on the positive edge (0-to-1 transition), whereas others trigger on the negative edge (1-to-0 transition) as illustrated in the previous subsection. The logic diagram of a D-type positive-edge-triggered flip-flop to be analyzed in detail here appears in Figure 6-13. This flip-flop takes exactly the form of a master-slave flip-flop, with the master a D latch and the slave an SR latch or a D latch. Also, an inverter is added to the clock input. Because the master latch is a D latch, the flip-flop exhibits edge-triggered rather than master-slave or pulse-triggered behavior. For the clock input equal to 0, the master latch is enabled and transparent and follows the

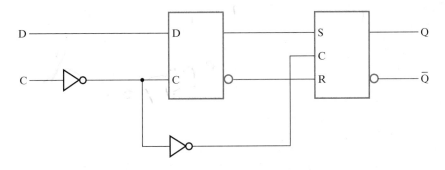

□ **FIGURE 6-13**
Positive Edge-Triggered D Flip-Flop

D input value. The slave latch is disabled and holds the state of the flip-flop fixed. When the positive edge occurs, the clock input changes to 1. This disables the master latch so that its value is fixed and enables the slave latch so that it copies the state of the master latch. The state of the master latch to be copied is the state that is present at the positive edge of the clock. Thus, the behavior appears to be edge-triggered. With the clock input equal to 1, the master latch is disabled and cannot change, so the state of both the master and the slave remain unchanged. Finally, when the clock input changes from 1 to 0, the master is enabled and begins following the *D* value. But during the 1–to–0 transition, the slave is disabled before any change in the master can reach it. Thus, the value stored in the slave remains unchanged during this transition. An alternative implementation is given in Problem 6–3 at the end of the chapter.

Standard Graphics Symbols

The standard graphics symbols for the different types of latches and flip-flops are shown in Figure 6-14. A flip-flop or latch is designated by a rectangular block with

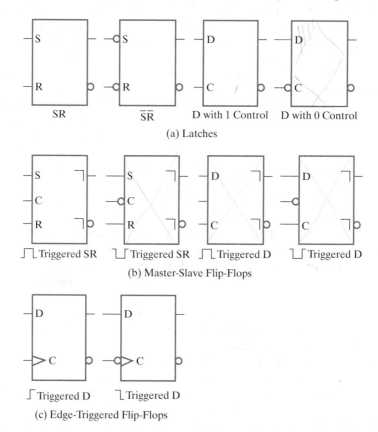

☐ **FIGURE 6-14**
Standard Graphics Symbols for Latches and Flip-Flops

inputs on the left and outputs on the right. One output designates the normal state of the flip-flop, and the other, with a bubble, designates the complement output. The graphics symbol for the SR latch or SR flip-flop has inputs S and R indicated inside the block. In the case of the $\overline{S}\,\overline{R}$ latch, bubbles are added to the inputs to indicate that setting and resetting occur for 0-level inputs. The graphics symbol for the D latch or D flip-flop has inputs D and C indicated inside the block.

Below each symbol, a descriptive title, which is not part of the symbol, is given. In the titles, ⎍⎍ denotes a positive pulse, ⎍⎍ a negative pulse, ⎍ a positive edge, and ⎍ a negative edge.

Triggering by the 0 level rather than the 1 level is denoted on the latch symbols by adding a bubble at the triggering input. The master-slave is a pulse-triggered flip-flop and is indicated as such with a right-angle symbol called a *postponed output indicator* in front of the outputs. This symbol shows that the output signal changes at the end of the pulse. To denote that the master-slave flip-flop will respond to a negative pulse (i.e., a pulse to 0 with the inactive clock value at 1), a bubble is placed on the C input. To denote that the edge-triggered flip-flop responds to an edge, an arrowhead-like symbol in front of the letter C designates a *dynamic input*. This *dynamic indicator* symbol denotes the fact that the flip-flop responds to edge transitions of the input clock pulses. A bubble outside the block adjacent to the dynamic indicator designates a negative-edge transition for triggering the circuit. The absence of a bubble designates a positive-edge transition for triggering.

Often, all of the flip-flops used in a circuit are of the same triggering type, such as positive-edge triggered. All of the flip-flops will then change in relation to the same clocking event. When using flip-flops having different triggering in the same sequential circuit, one may still wish to have all of the flip-flop outputs change relative to the same clocking event. Those flip-flops that behave in a manner opposite from the adopted polarity transition can be changed by the addition of inverters to their clock inputs. A preferred procedure is to provide both positive and negative pulses from the master clock generator that are carefully aligned. We apply positive pulses to positive-pulse-triggered (master-slave) and negative-edge-triggered flip-flops and negative pulses to negative-pulse-triggered (master-slave) and positive-edge-triggered flip-flops. In this way, all flip-flop outputs will change at the same time. Finally, to prevent specific timing problems, some designers use flip-flops having different triggering (i.e., both positive and negative edge-triggered flip-flops) with a single clock. In these cases, flip-flop outputs are purposely made to change at different times.

In this text, it is assumed that all flip-flops are of the positive-edge-triggered type, unless otherwise indicated. This provides a uniform graphics symbol for the flip-flops and consistent timing diagrams.

Note that there is no input to the D flip-flop that produces a "no change" condition. This condition can be accomplished either by disabling the clock pulses on the C input or by leaving the clock pulses undisturbed and connecting the output back into the D input using a multiplexer when the state of the flip-flop must remain the same. The technique that disables clock pulses is referred to as *clock gating*. This technique typically uses fewer gates and saves power, but is often avoided because the gated clock pulses into the flip-flops are delayed. The delay, called *clock skew,* causes gated clock and nongated clock flip-flops to change at different times. This can make the circuit unreliable, since the outputs of some flip-flops may reach others while their inputs are still affecting their state.

Direct Inputs

Flip-flops often provide special inputs for setting and resetting them asynchronously (i.e., independently of the clock input *C*). The inputs that asynchronously set the flip-flop are called *direct set,* or *preset.* The inputs that asynchronously reset the flip-flop are called *direct reset,* or *clear.* Application of a logic 1 (or a logic 0 if a bubble is present) to these inputs affects the flip-flop output without the use of the clock. When power is turned on in a digital system, the states of its flip-flops can be anything. The direct inputs are useful for bringing flip-flops in a digital system to an initial state prior to the normal clocked operation.

The IEEE standard graphics symbol for a positive-edge-triggered *D* flip-flop with direct set and direct reset is shown in Figure 6-15(a). The notations, *C*1 and 1*D,* illustrate control dependency. An input labeled *Cn,* where *n* is any number, controls all the other inputs starting with the number *n.* In the figure, *C*1 controls input 1*D. S* and *R* have no 1 in front of them, and therefore, they are not controlled by the clock at *C*1. The *S* and *R* inputs have circles on the input lines to indicate that they are active at the logic-0 level (i.e., a 0 applied will result in the set or reset action).

The function table in Figure 6-15(b) specifies the operation of the circuit. The first three rows in the table specify the operation of the direct inputs *S* and *R.* These inputs behave like NAND $\overline{SR}$ latch inputs (see Figure 6-6), operating independently of the clock, and are therefore asynchronous inputs. The last two rows in the function table specify the clocked operation for values of *D.* The clock at *C* is shown with an upward arrow to indicate that the flip-flop is a positive-edge-triggered type. The *D* input effects are controlled by the clock in the usual manner.

Figure 6-15(c) shows a less formal symbol for the positive-edge-triggered flip-flop with direct set and reset. The positioning of *S* and *R* at the top and bottom of the symbol rather than on the left edge implies that resulting output changes are not controlled by the clock *C.*

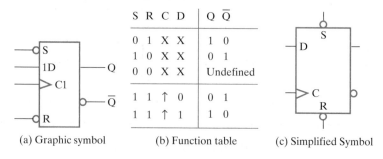

S	R	C	D	Q	$\overline{Q}$
0	1	X	X	1	0
1	0	X	X	0	1
0	0	X	X	Undefined	
1	1	↑	0	0	1
1	1	↑	1	1	0

(a) Graphic symbol   (b) Function table   (c) Simplified Symbol

☐ **FIGURE 6-15**
D Flip-Flop with Direct Set and Reset

Flip-Flop Timing

There are timing parameters associated with the operation of both pulse-triggered and edge-triggered flip-flops. These parameters are illustrated for a master-slave *SR* flip-flop and for a negative-edge-triggered *D* flip-flop in Figure 6-16. The parameters for the positive-edge-triggered *D* flip-flop are the same except that they are referenced to the positive clock edge rather than the negative clock edge.

The timing of the response of a flip-flop to its inputs and clock *C* must be taken into account when using the flip-flops. For both flip-flops, there is a minimum time called the *setup time*, t_s, for which the *S* and *R* or *D* inputs must be maintained at a constant value prior to the occurrence of the clock transition that causes the output to change. Otherwise, the master could be changed erroneously in the case of the master-slave flip-flop or be at an intermediate value at the time the slave copies it in the case of the edge-triggered flip-flop. Similarly, there is a minimum time called the *hold time*, t_h, for which the *S* and *R* or *D* inputs must not change after the application of the clock transition that causes the output to change. Otherwise, the master might respond to the input change and be changing at the time the slave latch copies it. In addition, there is a minimum *clock pulse width* t_w, to

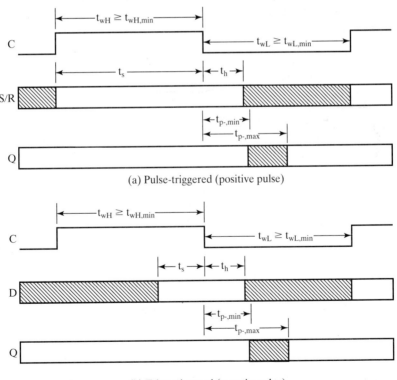

(a) Pulse-triggered (positive pulse)

(b) Edge-triggered (negative edge)

□ **FIGURE 6-16**
Flip-flop Timing Parameters

insure that the master has time enough to capture the input values correctly. Among these parameters, the one that differs most between the pulse-triggered and edge-triggered flip-flops is the setup time as shown in Figure 6-16. The pulse-triggered flip-flop has its setup time equal to the clock pulse width, whereas the setup time for the edge-triggered flip-flop can be much smaller than the clock pulse width. As a consequence, edge-triggering tends to provide faster designs since the flip-flop inputs can change later with respect to the upcoming triggering clock edge.

The *propagation delay times,* t_{PHL}, t_{PLH}, *or* t_{pd} of the flip-flops are defined as the interval between the triggering clock edge and the stabilization of the output to a new value. These times are defined in the same fashion as those for an inverter except that the values are measured from the triggering clock edge rather than the inverter input. In Figure 6-16, all of these parameters are denoted by $t_{P.}$ and are given minimum and maximum values. Since the changes of the flip-flop outputs are to be separated from the control by the flip-flop inputs, the minimum propagation delay time should be longer than the hold time for correct operation. These and other parameters are specified in manufacturers' data books for specific integrated circuit products.

Similar timing parameters can be defined for latches and direct inputs, with additional propagation delays needed to model the transparent behavior of latches.

6-4 SEQUENTIAL CIRCUIT ANALYSIS

The behavior of a sequential circuit is determined from the inputs, outputs, and present state of the circuit. The outputs and the next state are a function of the inputs and the present state. The analysis of a sequential circuit consists of obtaining a suitable description that demonstrates the time sequence of inputs, outputs, and states.

A logic diagram is recognized as a synchronous sequential circuit if it includes flip-flops with the clock inputs driven directly or indirectly by a clock signal and if the direct sets and resets are unused during the normal functioning of the circuit. The flip-flops may be of any type, and the logic diagram may or may not include combinational gates. In this section, an algebraic representation for specifying the logic diagram of a sequential circuit is given. A state table and state diagram are presented that describe the behavior of the circuit. Specific examples will be used throughout the discussion to illustrate the various procedures.

Input Equations

The logic diagram of a sequential circuit consists of flip-flops and, usually, combinational gates. The knowledge of the type of flip-flops used and a list of Boolean functions for the combinational circuit provide all the information needed to draw the logic diagram of the sequential circuit. The part of the combinational circuit that generates the signals for the inputs of flip-flops can be described by a set of Boolean functions called *flip-flop input equations*. We will adopt the convention of using the flip-flop input symbol to denote the flip-flop input equation variable and using the name of the flip-flop output as the subscript for the variable. From this example, it becomes apparent that a flip-flop input equation is a Boolean expression for a combinational circuit. The subscripted symbol is an output variable of the combinational circuit. This output is always connected to the input of a flip-flop—thus the name "flip-flop input equation."

The flip-flop input equations constitute a convenient algebraic expression for specifying the logic diagram of a sequential circuit. They imply the type of flip-flop from the letter symbol, and they fully specify the combinational circuit that drives the flip-flops. Time is not included explicitly in these equations, but is implied from the clock at the C input of the flip-flops. An example of a sequential circuit is given in Figure 6-17. The circuit has two D-type flip-flops, an input X, and an output Y. It can be specified by the following equations:

$$D_A = AX + BX$$

$$D_B = \overline{A}X$$

$$Y = (A + B)\overline{X}$$

The first two equations are for flip-flop inputs, and the third equation specifies the output Y. Note that the input equations use the symbol D, which is the same as the input symbol of the flip-flops. The subscripts A and B designate the outputs of the respective flip-flops.

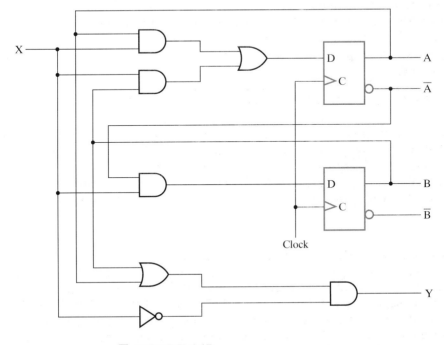

□ **FIGURE 6-17**
Example of a Sequential Circuit

State Table

The functional relationships among the inputs, outputs, and flip-flop states of a sequential circuit can be enumerated in a *state table*. The state table for the circuit of Figure 6-17 is shown in Table 6-1. The table consists of four sections, labeled *present state*, *input*, *next state*, and *output*. The present-state section shows the states of flip-flops A and B at any given time t. The input section gives each value of X for each possible present state. Note that for each possible input combination, each of the present states is repeated. The next-state section shows the states of the flip-flops one clock period later, at time $t + 1$. The output section gives the value of Y at time t for each combination of present state and input.

The derivation of a state table consists of first listing all possible binary combinations of present state and inputs. In Table 6-1, there are eight binary combinations, from 000 to 111. The next-state values are then determined from the logic diagram or from the flip-flop input equations. For a D flip-flop, the relationship $A(t + 1) = D_A(t)$ holds. This means that the next state of flip-flop A is equal to the present value of its input D. The value of the D input is specified in the flip-flop input equation as a function of the present state of A and B and input X. Therefore, the next state of flip-flop A must satisfy the equation

$$A(t+1) = D_A = AX + BX$$

The next-state section in the state table under column A has three 1's where the present state and input value satisfy the conditions $(A,X) = 11$ or $(B,X) = 11$. Similarly, the next state of flip-flop B is derived from the input equation

$$B(t+1) = D_B = \overline{A}X$$

and is equal to 1 when the present state of A is 0 and input X is equal to 1. The output column is derived from the output equation

$$Y = A\overline{X} + B\overline{X}$$

☐ **TABLE 6-1**
State Table for Circuit of Figure 6-17

Present State		Input	Next State		Output
A	B	X	A	B	Y
0	0	0	0	0	0
0	0	1	0	1	0
0	1	0	0	0	1
0	1	1	1	1	0
1	0	0	0	0	1
1	0	1	1	0	0
1	1	0	0	0	1
1	1	1	1	0	0

□ **TABLE 6-2**
Two-Dimensional State Table for the Circuit in Figure 6-17

Present state		Next state				Output	
		X = 0		X = 1		X = 0	X = 1
A	B	A	B	A	B	Y	Y
0	0	0	0	0	1	0	0
0	1	0	0	1	1	1	0
1	0	0	0	1	0	1	0
1	1	0	0	1	0	1	0

The state table of any sequential circuit with D-type flip-flops is obtained in this way. In general, a sequential circuit with m flip-flops and n inputs needs 2^{m+n} rows in the state table. The binary numbers from 0 through $2^{m+n} - 1$ are listed in the combined present-state and input columns. The next-state section has m columns, one for each flip-flop. The binary values for the next state are derived directly from the D flip-flop input equations. The output section has as many columns as there are output variables. Its binary values are derived from the circuit or from the Boolean functions in the same manner as in a truth table.

Table 6-1 is one-dimensional in the sense that the present state and input combinations are combined into a single column of combinations. A two-dimensional state table having the present state tabulated in the left column and the inputs tabulated across the top row is also frequently used. The next-state entries are made in each cell of the table for the present-state and input combination corresponding to the location of the cell. A similar two-dimensional table is used for the outputs if they depend upon the inputs. Such a state table is shown in Table 6-2. Sequential circuits in which the outputs depend on the inputs, as well as on the states, are referred to as *Mealy model* circuits. Otherwise, if the outputs depend only on the states, then a one-dimensional column suffices. In this case, the circuits are referred to as *Moore model* circuits. Each model is named after its originator.

As an example of a Moore model circuit, suppose we want to obtain the logic diagram and state table of a sequential circuit that is specified by the flip-flop input equation

$$D_A = A \oplus X \oplus Y$$

and output equation

$$Z = A$$

The D_A symbol implies a D-type flip-flop with output designated by the letter A. The X and Y variables are taken as inputs and Z as the output. The logic diagram and state table for this circuit are shown in Figure 6-18. The state table has one column for the present state and one column for the inputs. The next state and output are also in single columns. The next state is derived from the flip-flop input

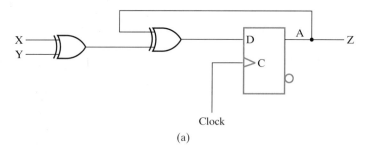

Clock

(a)

Present state	Inputs	Next state	Output
A	X Y	A	Z
0	0 0	0	0
0	0 1	1	0
0	1 0	1	0
0	1 1	0	0
1	0 0	1	1
1	0 1	0	1
1	1 0	0	1
1	1 1	1	1

(b) State table

☐ **FIGURE 6-18**
Logic Diagram and State Table for $D_A = A \oplus X \oplus Y$

equation which specifies an odd function. (See Section 2-8.) The output column is simply a copy of the column for the present-state variable A.

State Diagram

The information available in a state table may be represented graphically in the form of a state diagram. In this type of diagram, a state is represented by a circle, and transitions between states are indicated by directed lines connecting the circles. Examples of state diagrams are given in Figure 6-19. Figure 6-19(a) shows the state diagram for the sequential circuit in Figure 6-17 and its state table in Table 6-1. The state diagram provides the same information as the state table and is obtained directly from it. The binary number inside each circle identifies the state of the flip-flops. For Mealy model circuits, the directed lines are labeled with two binary numbers separated by a slash. The input value during the present state precedes the slash, and the value following the slash gives the output value during the present state with the given input applied. For example, the directed line from state 00 to state 01 is labeled 1/0, meaning that when the sequential circuit is in the present state 00 and the input is 1, the output is 0. After the next clock transition, the circuit goes to the next state, 01. If the input changes to 0, then the output becomes 1, but if the input remains at 1, the output stays at 0. This information is obtained from the state diagram along the two directed

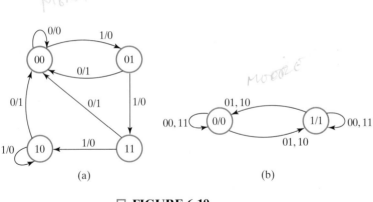

□ **FIGURE 6-19**
State Diagrams

lines emanating from the circle with state 01. A directed line connecting a circle with itself indicates that no change of state occurs.

The state diagram of Figure 6-19(b) is for the sequential circuit of Figure 6-18. Here, only one flip-flop with two states is needed. There are two binary inputs, and the output depends only on the state of the flip-flop. For such a Moore model circuit, the slash on the directed lines is not included, since the outputs depend only on the state and not on the input values. Instead, the output is included under a slash below the state in a circle. There are two input conditions for each state transition in the diagram, and they are separated by a comma. When there are two input variables, each state may have up to four directed lines coming out of the corresponding circle, depending upon the number of states and the next state for each binary combination of the input values.

There is no difference between a state table and a state diagram, except for their manner of representation. The state table is easier to derive from a given logic diagram and input equations. The state diagram follows directly from the state table. The state diagram gives a pictorial view of state transitions and is the form more suitable for human interpretation of the operation of the circuit. For example, the state diagram of Figure 6-19(a) clearly shows that, starting at state 00, the output is 0 as long as the input stays at 1. The first 0 input after a string of 1's gives an output of 1 and sends the circuit back to the initial state of 00. The state diagram of Figure 6-19(b) shows that the circuit stays at a given state as long as the two inputs have the same value (00 or 11). There is a state transition between the two states only when the two inputs are different (01 or 10).

Sequential Circuit Timing

In addition to analyzing the function of a circuit, it is also important to analyze its performance in terms of the *maximum input-to-output delay* and the *maximum clock frequency*, f_{max} at which it can operate. First of all, the clock frequency is just the inverse of clock period t_p shown in Figure 6-20. So, the maximum allowable clock frequency corresponds to the minimum allowable clock period t_p. To determine how small we can make the clock period, we need to determine the longest

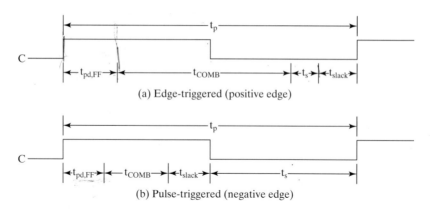

(a) Edge-triggered (positive edge)

(b) Pulse-triggered (negative edge)

☐ **FIGURE 6-20**
Sequential Circuit Timing Parameters

delay from the triggering edge of the clock to the next triggering edge of the clock. These delays are measured on all such paths in the circuit down which changing signals propagate. Each of these path delays has three components: (1) a flip-flop propagation delay, $t_{pd,FF}$, (2) a combinational logic delay through the chain of gates along the path, $t_{pd,COMB}$, and (3) a flip-flop setup time, t_s. As a signal change propagates down the path, it is delayed successively by an amount equal to each of these delays. Note that we have used t_{pd}, instead of the more detailed values, t_{PLH} and t_{PHL}, for both the flip-flops and combinational logic gates to simplify the delay calculations. Figure 6-20 summarizes the delay picture for both the edge-triggered and pulse-triggered flip-flops.

After a positive edge on a clock, if a flip-flop is to change, its output changes at time $t_{pd,FF}$ after the clock edge. This change enters the combinational logic path and must propagate down the path to a flip-flop input. This requires an additional time, $t_{pd,COMB}$, for the signal change to reach the second flip-flop. Finally, before the next positive clock edge, this change must be held on the flip-flop input for setup time t_s. This path, $P_{FF,FF}$ as well as other possible paths are illustrated in Figure 6-21. For paths $P_{IN,FF}$ driven by primary inputs, $t_{pd,FF}$ is replaced by t_i, which is the latest time that the input changes after the positive clock edge. For a path $P_{FF,OUT}$ driving primary outputs, t_s is replaced by t_o, which is the latest time that the output is permitted to change prior to the next clock edge. Finally, in a Mealy model circuit, combinational paths from input to output, $P_{IN,OUT}$, that use both t_i and t_o can appear. Each path has a slack time, t_{slack}, the extra time allowed in the clock period beyond that required by the path. From Figure 6-21, the following equation for a path of type $P_{FF,FF}$ results:

$$t_p = t_{slack} + (t_{pd,FF} + t_{pd,COMB} + t_s)$$

In order to guarantee that a changing value is captured by the receiving flip-flop, t_{slack} must be greater than or equal to zero for all of the paths. This requires that

$$t_p \geq \max \left(t_{pd,FF} + t_{COMB} + t_s \right) = t_{p,min}$$

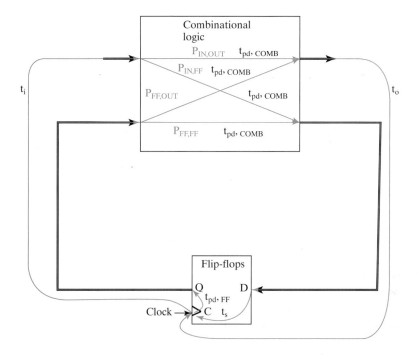

□ **FIGURE 6-21**
Sequential Circuit Timing Paths

where the maximum is taken over all paths down which signals propagate from flip-flop to flip-flop. The next example presents representative calculations for paths $P_{FF,FF}$.

EXAMPLE 6-1   Clock Period and Frequency Calculations

Suppose that all flip-flops used are the same and have $t_{pd} = 0.2$ ns (nanosecond = 10^{-9} seconds) and $t_s = 0.1$ ns. Then the longest path beginning and ending with a flip-flop will be the path with the largest $t_{pd,COMB}$. Further, suppose that the largest $t_{pd,COMB}$ is 1.3 ns and that t_p has been set to 1.5 ns. From the previous equation for t_p, we can write

$$1.5 \text{ ns} = t_{slack} + 0.2 + 1.3 + 0.1 = t_{slack} + 1.6 \text{ns}$$

Solving, we have $t_{slack} = -0.1$ ns, so this value of t_p is too small. In order for t_{slack} to be greater than or equal to zero for the longest path, $t_p \geq t_{p,min} = 1.6$ ns. The maximum frequency $f_{max} = 1/1.6$ ns $= 625$ MHz (megahertz $= 10^6$ cycles per second). We note that, if t_p is too large to meet the circuit specifications, we must either employ faster logic cells or change the circuit design to reduce the problematic path delays through the circuit while still performing the desired function. ■

It is interesting to note that the hold time for a flip-flop, t_h, does not appear in the clock period equation. It relates to another timing constraint equation dealing with one or both of two specific situations. In one case, output changes arrive at the inputs of one or more flip-flops too soon. In the other case, the clock signals reaching one or more flip-flops are somehow delayed, a condition referred to as *clock skew*. Clock skew also can affect the maximum clock frequency.

Simulation

Sequential circuit simulation involves issues not present in combinational circuits. First of all, rather than a set of input patterns for which the order of application is immaterial, the patterns must be applied in a sequence. This sequence includes timely application of input patterns as well as clock pulses. Second, there must be some means to place the circuit in a known state. Realistically, initialization to a known state is accomplished by application of an initialization subsequence at the beginning of the simulation. In the simplest case, this subsequence is a reset signal. For flip-flops lacking a circuit reset (or set), a longer sequence typically consisting of an initial reset followed by a sequence of normal input patterns is required. A simulator may also have a means of setting the initial state which is useful to avoid long sequences that may be needed to get to an initial state. Aside from getting to an initial state, a third issue is observing the state to verify correctness. In some circuits, application of an additional sequence of inputs is required to determine the state of the circuit at a given point. The simplest alternative is to set up the simulation so that the state of the circuit can be observed directly; the approach to doing this varies depending on the simulator and whether or not the circuit contains hierarchy. A crude approach that works with all simulators is to add a circuit output with a path from each state variable signal.

A final issue to be dealt with in more detail is the timing of application of inputs and observation of outputs relative to the active clock edge. Initially, we discuss the timing for *functional simulation* having as its objective determination or verification of the function of the circuit. In functional simulation, components of the circuit have no delay or a very small delay. Much more complex is *timing simulation* in which the circuit elements have realistic delays and verification of the proper operation of the circuit in terms of timing is the simulation objective.

Some simulators, by default, use a very small component delay for functional simulation so that the order of changes in signals can be observed provided that the time scale used for display is small enough. Suppose that the component delays and the setup and hold times for flip-flops are all 0.1 ns for such a simulation and that the longest delay from positive clock edge to positive clock edge is 1.2 ns in your circuit. If you happen to use a clock period of 1.0 ns for your simulation, when the result depends on the longest delay, the simulation results will be in error! So for functional simulation with such a simulator, either a longer clock period should be chosen for the simulation or the default delay needs to be changed by the user to a smaller value.

In addition to the clock period, the time of application of inputs relative to the positive clock edge is important. For functional simulation, to allow for any small, default component delays, the inputs for a given clock cycle should be

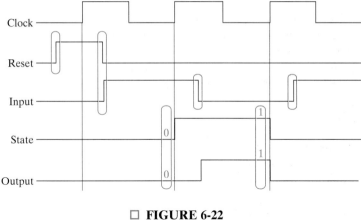

□ **FIGURE 6-22**
Simulation Timing

changed well before the positive clock edge, preferably early in the clock cycle while the clock is still at a 1 value. This is also an appropriate time to change the reset signal values to insure that the reset signal is controlling the state rather than the clock edge or a meaningless combination of clock and reset.

A final issue is the time at which to examine a simulation result in functional simulation. At the very latest, the state variable values and outputs should be at their final values just before the positive clock edge. Although it may be possible to observe the values at other locations, this location provides a foolproof observation time for functional simulation.

The ideas just presented are summarized in Figure 6-22. Input changes in Reset and Input, encircled in blue, occur at about the 25% point in the clock cycle. Signal values on State and Output, as well as on Input and Reset, all encircled in blue and listed, are observed just before the 100% point in the clock cycle.

6-5 SEQUENTIAL CIRCUIT DESIGN

The design of clocked sequential circuits starts from a set of specifications and culminates in a logic diagram or a list of Boolean functions from which the logic diagram can be obtained. In contrast to a combinational circuit, which is fully specified by a truth table, a sequential circuit requires a state table for its specification. Thus, the first step in the design of a sequential circuit is to obtain a state table or an equivalent representation such as a state diagram.

A synchronous sequential circuit is made up of flip-flops and combinational gates. The design of the circuit consists of choosing the flip-flops and finding a combinational circuit structure which, together with the flip-flops, produces a circuit that fulfills the stated specifications. The minimum number of flip-flops is determined by the number of states in the circuit; n flip-flops can represent up to 2^n binary states. The combinational circuit is derived from the state table by evaluating the flip-flop input equations and output equations. In fact, once the type and number of flip-flops are determined, the design process transforms a sequential circuit problem into a combinational circuit problem. In this way, the techniques of combinational circuit design can be applied.

Design Procedure

The following procedure for the design of sequential circuits in similar to that for combinational circuits but has some additional steps:

1. **Specification:** Write a specification for the circuit, if not already available.
2. **Formulation:** Obtain either a state diagram or a state table from the statement of the problem.
3. **State Assignment:** If only a state diagram is available from step 1, obtain the state table. Assign binary codes to the states in the table.
4. **Flip-Flop Input Equation Determination:** Select the flip-flop type or types. Derive the flip-flop input equations from the next-state entries in the encoded state table.
5. **Output Equation Determination:** Derive output equations from the output entries in the state table.
6. **Optimization:** Optimize the flip-flop input equations and output equations.
7. **Technology Mapping:** Draw a logic diagram of the circuit using flip-flops, ANDs, ORs, and inverters. Transform the logic diagram to a new diagram using the available flip-flop and gate technology.
8. **Verification:** Verify the correctness of the final design.

For convenience, we usually omit the technology mapping in step 7 and use only flip-flops, AND gates, OR gates, and inverters in the schematic.

Finding State Diagrams and State Tables

The specification for a circuit is often in the form of a verbal description of the behavior of the circuit. This description needs to be interpreted in order to find a state diagram or state table in the formulation step of the design procedure. This is the often the most creative part of the design procedure, with many of the subsequent steps performed automatically by computer-based tools.

Fundamental to the formulation of state diagrams and tables is an intuitive understanding of the concept of a state. A state is used to "remember" something about the history of input combinations applied to the circuit at either triggering clock edges or during triggering pulses. In some cases, the states may literally store input values retaining a complete history of the sequence appearing on the inputs. In most cases, however, a state is an *abstraction* of the sequence of input combinations at the triggering points. For example, a given state S_1 may represent the fact that among the sequence of values applied to a single bit input X, "the value 1 has appeared on X for the last three consecutive clock edges." Thus, the circuit would be in state S_1 after sequences ... 00111 or ... 0101111, but would not be in state S_1

after sequences ... 00011 or ... 011100. A state S_2 might represent the fact that the sequence of 2-bit input combinations applied are "in order 00, 01, 11, 10 with any number of consecutive repetitions of each combination permitted and 10 as the most recently applied combination." The circuit would be in state S_2 for the following example sequences: 00, 00, 01, 01, 01, 11, 10, 10 or 00, 01, 11, 11, 11, 10. The circuit would not be in state S_2 for sequences: 00, 11, 10, 10 or 00, 00, 01, 01, 11, 11. In formulating a state diagram or state table it is useful to write down the abstraction represented by each state. In some cases, it may be easier to describe the abstraction by referring to values that have occurred on the outputs as well as on the inputs. For example, state S_3 might represent the abstraction that "the output bit Z_2 is 1, and the input combination has bit X_2 at 0." In this case, Z_2 equal to 1 might uniquely represent a complex set of past sequences of input combinations that would be more difficult to describe in detail.

As one formulates a state table or state diagram, new states are added. There is potential for the set of states to become unnecessarily large or potentially even infinite in size! Instead of adding a new state for every current state and possible applied input combination, it is essential that states be reused as next states to prevent uncontrolled state growth as outlined above. The mechanism for doing this is a knowledge of the abstraction that each state represents. To illustrate, consider state S_1 defined previously as an abstraction "the value 1 has appeared at the last three consecutive clock edges." If S_1 has been entered due to the sequence ... 00111 and the next input is a 1, giving sequence ... 001111, is a new state needed or can the next state be S_1? By examining the new sequence, we see that the last three input values are 1's, which matches the abstraction defined for state S_1. So, state S_1 can be used as the next state for current state S_1 and input value 1, avoiding the definition of a new state.

Since one does not know the state of the flip-flops when the power is first turned on, it is customary to provide a *master reset* signal to initialize the state of the flip-flops in at least some component circuits. This avoids starting these circuits in an unknown state. Typically, the master reset signal is applied to the flip-flops at their asynchronous (direct) inputs (see Figure 6-23) before clocked operation starts. In most cases flip-flops are reset to 0, but some may be set to 1, depending upon the initial state desired. If there is a large number of flip-flops, based upon the circuit operation, only a portion of them may be initialized.

When the power in a digital system is first turned on, the state of the flip-flops is unknown. It is possible to apply an input sequence with the circuit in an unknown state, but that sequence must be able to bring a portion of the circuit to a known state before meaningful outputs can be expected. In fact, many of the larger sequential circuits we design in subsequent chapters will be of this type. In this chapter, however, the circuits that we design must have a known *initial state*, and further, a hardware mechanism must be provided to get the circuit from any unknown state into this state. This mechanism is a *reset* or *master reset* signal. Regardless of all other inputs applied to the circuit, the reset places the circuit in its initial state. In fact, the initial state is often called the *reset state*. The reset signal is usually activated automatically when the circuit is powered up. In addition, it may be activated electronically or by pushing a reset button.

(a) Asynchronous Reset   (b) Synchronous Reset

☐ **FIGURE 6-23**
Asynchronous and Synchronous Reset for D Flip-flops

The reset may be asynchronous, taking place without clock triggering. In this case, the reset is applied to the direct inputs on the circuit flip-flops. as shown in Figure 6-23 (a). This design assigns 00...0 to the initial state of the flip-flops to be reset. If an initial state with a different code is desired, then the *Reset* signal can be selectively connected to direct set inputs instead of direct reset inputs. It is important to note that these inputs should not be used in the normal synchronous circuit design process. Instead, they are reserved only for an asynchronous reset that returns the system, of which the circuit is a component, to an initial state. Using these direct inputs as a part of the synchronous circuit design violates the fundamental synchronous circuit definition, since it permits a flip-flop state to change asynchronously within direct clock triggering.

Alternatively, the reset may be synchronous and require a clock triggering event to occur. The reset must be incorporated into the synchronous design of the circuit. A simple approach to synchronous reset for D flip-flops, without formally including the reset bit in the input combinations, is to add the AND gate shown in Figure 6-23 (b) after doing the normal circuit design. This design also assigns 00 ... 0 to the initial state. If a different initial state code is desired, then OR gates with *Reset* as an input can selectively replace the AND gates with inverted *Reset*.

To illustrate the formulation process, two examples follow, each resulting in a different style of state diagram.

EXAMPLE 6-2   Finding a State Diagram for a Sequence Recognizer

The first example is a circuit that recognizes the occurrence of a particular sequence of bits, regardless of where it occurs in a longer sequence. This "sequence-recognizer" has one input X and one output Z. It has Reset applied to the direct reset inputs on its flip-flops to initialize the state of the circuit to all zeros. The circuit is to recognize the occurrence of the sequence of bits 1101 on X by making Z equal to 1 when the previous three inputs to the circuit were 110 and current input is a 1. Otherwise, Z equals 0.

The first step in the formulation process is to determine whether the state diagram or table must be a Mealy model or Moore model circuit. The portion of the preceding specification that says "... making Z equal to 1 when the previous three inputs to the circuit are 110 and the current input is a 1" implies that the output is determined from not only the current state, but also the current input. As a

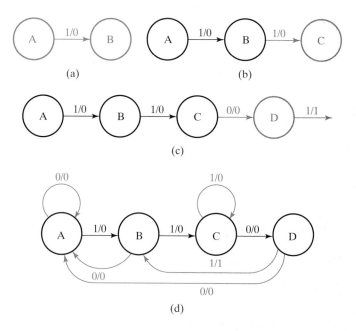

□ **FIGURE 6-24**
Construction of a State Diagram for Example 6.2

consequence, a Mealy model circuit with the output dependent on both state and inputs is required.

Recall that a key factor in the formulation of any state diagram is to recognize that states are used to "remember" something about the history of the inputs. For example, for the sequence 1101, to be able to produce the output value 1 coincident with the final 1 in the sequence, the circuit must be in a state that "remembers" that the previous three inputs were 110. With this concept in mind, we begin to formulate the state diagram by defining an arbitrary initial state A as the reset state and the state in which "none of the sequence to be recognized has occurred." If a 1 occurs on the input, since 1 is the first bit in the sequence, this event must be "remembered," and the state after the clock pulse cannot be A. So a second state, B, is established to represent the occurrence of the first 1 in the sequence. Further, to represent the occurrence of the first 1 in the sequence, a transition is placed from A to B and labeled with a 1. Since this is not the final 1 in the sequence 1101, its output is a 0. This initial portion of the state diagram is given in Figure 6-24 (a).

The next bit of the sequence is a 1. When this 1 occurs in state B, a new state is needed to represent the occurrence of two 1's in a row on the input—that is, the occurrence of an additional 1 while in state B. So a state C and the associated transition are added, as shown in Figure 6-24 (b). The next bit of the sequence is a 0. When this 0 occurs in state C, a state is needed to represent the occurrence of the two 1's in a row followed by a 0. So the additional state D with a transition having a 0 input and 0 output is added. Since state D represents the occurrence of 110 as the previous three input bit values on X, the occurrence of a 1 in state D completes the sequence to be recognized, so the transition for the input value 1 from state D has an output value of 1. The resulting partial state diagram, which completely represents the occurrence of the sequence to be recognized, is shown in Figure 6-24(c).

Note in Figure 6-24(c) that, for each state, a transition is specified for only one of the two possible input values. Also, the state that is the destination of the transition from D for input 1 is not yet defined. The remaining transitions must be based on the idea that the recognizer is to identify the sequence 1101, regardless of where it occurs in a longer sequence. Suppose that an initial part of the sequence 1101 is represented by a state in the diagram. Then, the transition from that state for an input value that represents the next input value in the sequence must enter a state such that the 1 output occurs if the remaining bits of the sequence are applied. For example, state C represents the first two bits, 11, of sequence 1101. If the next input values is 0, then the state that is entered, in this case, D, gives a 1 output if the remaining bit of the sequence, 1, is applied.

Next, evaluate where the transition for the 1 input from the D state is to go. Since the transition input is a 1, it could be the first or second bit in the sequence to be recognized. But because the circuit is in state D, it is evident that the prior input was a 0. So this 1 input is the first 1 in the sequence, since it cannot be preceded by a 1. The state that represents the occurrence of a first 1 in the sequence is B, so the transition with input 1 from state D is to state B. This transition is shown in the diagram in Figure 6-24(d). Examining state C, we can trace back through states B and A to see that the occurrence of a 1 input in C is at least the second 1 in the sequence. The state representing the occurrence of two 1's in sequence is C, so the new transition is to state C. Since the combination of two 1's is not the sequence to be recognized, the output for the transition is 0. Repeating this same analysis for missing transitions from states B and A, the final state diagram in Figure 6-24(d) is obtained. The resulting state table is given in two-dimensional form in Table 6-3.

One issue that arises in the formulation of any state diagram is whether, in spite of best designer efforts, excess states have been used. This is not the case in the preceding example, since each state represents input history that is essential for

☐ **TABLE 6-3**
State Table for State Diagram in Figure 6-21

Present State	Next State		Output Z	
	X = 0	X = 1	X = 0	X = 1
A	A	B	0	0
B	A	C	0	0
C	D	C	0	0
D	A	B	0	1

recognition of the stated sequence. If, however, excess states are present, then it may be desirable to combine states into the fewest needed. This can be done using ad hoc methods and formal state minimization procedures. Due to the complexity of the latter, particularly in the case in which don't-care entries appear in the state table, formal procedures are not covered here. For the interested student, state minimization procedures are found in the references listed at the end of the chapter. The next example illustrates an additional method for avoiding extra states.

EXAMPLE 6-3   Finding a State Diagram for a BCD–to–Excess-3 Decoder

In Chapter 3, a BCD–to–excess-3 decoder was designed. In this example, the function of the circuit is similar except that the inputs, rather than being presented to the circuit simultaneously, are presented serially in successive clock cycles, least significant bit first. In Table 6-4(a), the input sequences and corresponding output sequences are listed with the least significant bit first. For example, during four successive clock cycles, if 1010 is applied to the input, the output will be 0001. In order to produce each output bit in the same clock cycle as the corresponding input bit, the output depends on the present input value as well as the state. The specifications also state that the circuit must be ready to receive a new 4-bit sequence as soon as the prior sequence has completed. The input to this circuit is labeled X and the output is labeled Z. In order to focus on the patterns for past inputs, the rows of Table 6-4(a) are sorted according to the first bit value, the second bit value, and the third bit value of the input sequences. Table 6-4(b) results.

The state diagram begins with an initial state as shown in Figure 6-25(a). Examining the first column of bits in Table 6-4(b) indicates that a 0 produces a 1 output and a 1 produces a 0 output. Next, we ask the question, "Do we need to

□ **TABLE 6-4**
Sequence Tables for Code Converter Example

(a) Sequences in Order of Digits Represented								(b) Sequences in Order of Common Prefixes							
BCD Input				Excess-3 Output				BCD Input				Excess-3 Output			
1	2	3	4	1	2	3	4	1	2	3	4	1	2	3	4
0	0	0	0	1	1	0	0	0	0	0	0	1	1	0	0
1	0	0	0	0	0	1	0	0	0	0	1	1	1	0	1
0	1	0	0	1	0	1	0	0	0	1	0	1	1	1	0
1	1	0	0	0	1	1	0	0	1	0	0	1	0	1	0
0	0	1	0	1	1	1	0	0	1	1	0	1	0	0	1
1	0	1	0	0	0	0	1	1	0	0	0	0	0	1	0
0	1	1	0	1	0	0	1	1	0	0	1	0	0	1	1
1	1	1	0	0	1	0	1	1	0	1	0	0	0	0	1
0	0	0	1	1	1	0	1	1	1	0	0	0	1	1	0
1	0	0	1	0	0	1	1	1	1	1	0	0	1	0	1

remember the value of the first bit?" In Table 6-4(b), when the first bit is a 0, a 0 in the second bit results in an output of 1 and a 1 in the second bit gives an output of 0. In contrast, if the first bit is a 1, a 0 in the second bit causes an output of 0, and a 1 in the second bit gives output 1. It is clear that the output for the second bit cannot be determined without "remembering" the value of the first bit. Thus, the first input equal to 0 and the first input equal to 1 must give different states as shown in Figure 6-25(a), which also shows the input/output values for the arcs to the new states.

Next, it must be determined whether the inputs following the two new states need to have two states to remember the second bit value. In the first two columns of inputs in Table 6-4(b), sequence 00 produces outputs for the third bit that are 0 for input 0 and 1 for input 1. On the other hand, for sequence 01, the outputs for the third bit are 1 for input 0 and 0 for input 1. Since these are different for the same input values in the third bit, separate states are necessary, as shown in Figure 6-25(b). A similar analysis for input sequences 10 and 11, which examines the outputs for both the third and fourth bits, shows that the value of the second bit has no effect on the output values. Thus, in Figure 6-25(b), there is only a single next state for state B1 = 1.

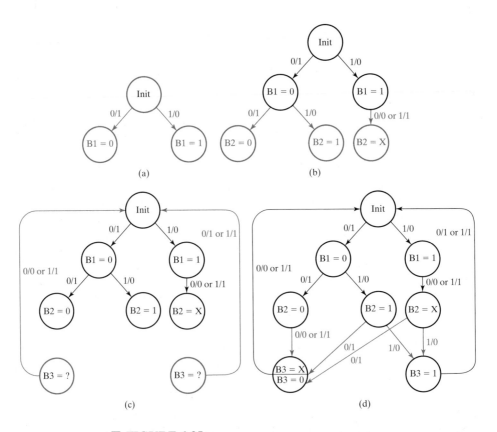

☐ **FIGURE 6-25**
Construction of a State Diagram for Example 6.3

At this point, six potential new states might result from the three states just added. Note, however, that these states are needed only to define the outputs for the fourth input bit since it is known that the next state thereafter will be *Init* in preparation for applying the next input sequence of four bits. How many states does one need to specify the different possibilities for the output value in the last bit? Looking at the final column, a 1 input always produces a 1 output and a 0 may produce either a 0 or a 1 output. Thus, at most two states are necessary, one that has a 0 output to a 0 and one that has a 1 output to a 0. The output for a 1 input is the same for both states. In Figure 6-25(c), we have added these two states. For the circuit to be ready to receive the next sequence, the next state for these new states is *Init*.

Remaining is the determination of the blue arcs shown in Figure 6-25(d). The arcs from each of the bit B2 states can be defined based on the third bit in the input/output sequences. The next state can be chosen based on the response to input 0 in the fourth bit of the sequence. The B2 state reaches the B3 state on the left with B3 = 0 or B3 = 1 as indicated by B3 = X on the upper halve of the B3 state. The other two B2 states reach this same state with B3 = 1 as indicated on the lower half of the state. These same two B2 states reach the B3 state on the right with B3 = 0 as indicated by the label on the state. ▨

State Assignment

In contrast to the states in the analysis examples, the states in the diagrams constructed have been assigned symbolic names rather than binary codes. It is necessary to replace these symbolic names with binary codes in order to proceed with the design. In general, if there are m states, then the codes must contain n bits, where $2^n \geq m$, and each state must be assigned a unique code. So, for the circuit in Table 6-3 with four states, the codes assigned to the states require two bits.

We begin by assigning a code to the initial reset state. If 1101 occurs as the first four inputs to the circuit after *Reset* = 1, it should be recognized. But if 101, 01, or 1 occurs as the first input sequence, they should not be recognized. The only state that can provide this property is state A. So, with direct resets used on the flip-flops, the code 00 must be assigned to state A. As a basis for encoding the remaining states, extensive work on the assignment of codes to states exists, but it is too complex for our treatment here. These methods have focused primarily on attempting to select codes in such a way that the logic required to implement the flip-flop input equations and output equations is minimized. In our example, we simply assign the state codes in Gray code order, beginning with state A. The Gray code is selected in this case simply because it makes it easier for the next-state and output functions to be placed on a Karnaugh map. The state table with the codes assigned is shown in Table 6-5.

Designing with D Flip-Flops

The remainder of the sequential circuit design procedure will be illustrated by the next example. We wish to design a clocked sequential circuit that operates according to the state table for Example 6-2, the sequence recognizer, shown in Table 6-5.

□ **TABLE 6-5**
Table 6-3 with Names Replaced by Binary Codes

Present State	Next State		Output Z	
AB	X = 0	X = 1	X = 0	X = 1
00	00	01	0	0
01	00	11	0	0
11	10	11	0	0
10	00	01	0	1

This state table, with the binary codes assigned to the states, specifies four states, two input values, and two output values. Two flip-flops are needed to represent the four states. We label the flip-flop outputs with the letters A and B, the input with X, and the output with Z.

Steps 1 through 3 of the design procedure have been completed for this circuit. Beginning step 4, D flip-flops are chosen. To complete step 4, the flip-flop input equations are obtained from the next-state values listed in the table. For step 5, the output equation is obtained from the values of Z in the table. The flip-flop input equations and output equation can be expressed as a sum of minterms of the present-state variables A and B and the input variable X:

$$A(t+1) = D_A(A,B,X) = \Sigma\, m(3,6,7)$$

$$B(t+1) = D_B(A,B,C) = \Sigma\, m(1,3,5,7)$$

$$Z(A,B,X) = \Sigma\, m(5)$$

The Boolean functions are simplified by using the maps plotted in Figure 6-26. The simplified functions are

$$D_A = AB + BX$$

$$D_B = X$$

$$Z = A\overline{B}X$$

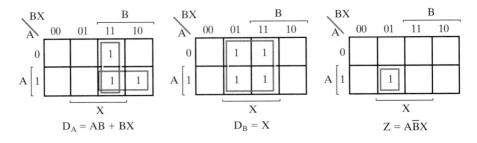

$$D_A = AB + BX \qquad D_B = X \qquad Z = A\overline{B}X$$

□ **FIGURE 6-26**
Maps for Input Equations and Output Z

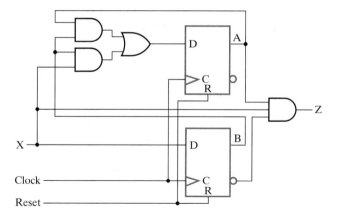

□ **FIGURE 6-27**
Logic Diagram for Sequential Circuit with D Flip-Flops

The logic diagram of the sequential circuit is shown in Figure 6-27.

Designing with Unused States

A circuit with n flip-flops has 2^n binary states. The state table from which the circuit was originally derived, however, may have any number of states $m \le 2^n$. States that are not used in specifying the sequential circuit are not listed in the state table. In simplifying the input equations, the unused states can be treated as don't-care conditions. The state table in Table 6-6 defines three flip-flops, A, B, and C, and one input, X. There is no output column, which means that the flip-flops serve as outputs of the circuit. With three flip-flops, it is possible to specify eight states, but the state table lists only five. Thus, there are three unused states that are not included

□ **TABLE 6-6**
State Table for Designing with Unused States

Present State			Input	Next State		
A	B	C	X	A	B	C
0	0	1	0	0	0	1
0	0	1	1	0	1	0
0	1	0	0	0	1	1
0	1	0	1	1	0	0
0	1	1	0	0	0	1
0	1	1	1	1	0	0
1	0	0	0	1	0	1
1	0	0	1	1	0	0
1	0	1	0	0	0	1
1	0	1	1	1	0	0

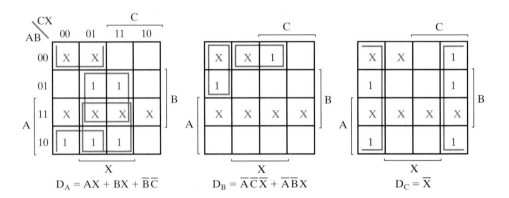

☐ **FIGURE 6-28**
Maps for Optimizing Input Equations

in the table: 000, 110, and 111. When an input of 0 or 1 is included with the unused present-state values, six unused combinations are obtained for the present-state and input columns: 0000, 0001, 1100, 1101, 1110, and 1111. These six combinations are not listed in the state table and hence may be treated as don't-care minterms.

The three input equations for the D flip-flops are derived from the next-state values and are simplified in the maps of Figure 6-28. Each map has six don't-care minterms in the squares corresponding to binary 0, 1, 12, 13, 14, and 15. The optimized equations are

$$D_A = AX + BX + \overline{B}\,\overline{C}$$

$$D_B = \overline{A}\,\overline{C}\,\overline{X} + \overline{A}\,\overline{B}X$$

$$D_C = \overline{X}$$

The logic diagram can be obtained directly from the input equations and will not be drawn here.

It is possible that outside interference or a malfunction will cause the circuit to enter one of the unused states. Thus, it is sometimes desirable to specify, fully or at least partially, the next state values or the output values for the unused states. Depending on the function and application of the circuit, a number of ideas may be applied. First, the outputs for the unused states may be specified so that any actions that result from entry into and transitions between the unused states are not harmful. Second, an additional output may be provided or an unused output code employed which indicates that the circuit has entered an incorrect state. Third, to ensure that a return to normal operation is possible without resetting the entire system, the next-state behavior for the unused states may be specified. Typically, next states are selected such that one of the normally occurring states is reached within a few clock cycles, regardless of the input values. The

decision as to which of the three options to apply, either individually or in combination, is based on the application of the circuit or the policies of a particular
design group.

Verification

Sequential circuits can be verified by showing that the circuit produces the original
state diagram or state table. In the simplest cases, all possible input combinations
are applied with the circuit in each of the states and the state variables and outputs
are observed. For small circuits, the actual verification can be performed manually.
More generally, simulation is used. In manual simulation, it is straightforward to
apply each of the state–input combinations and verify that the output and the next
state are correct.

Verification with simulation is less tedious, but typically requires a sequence
of input combinations and applied clocks. In order to check out a state–input
combination, it is first necessary to apply a sequence of input combinations to
place the circuit in the desired state. It is most efficient to find a single sequence
to test all the state–input combinations. The state diagram is ideal for generating
and optimizing such a sequence. A sequence must be generated to apply each
input combination in each state while observing the output and next state that
appears after the positive clock edge. The sequence length can be optimized by
using the state diagram. The reset signal can be used as an input during this
sequence. In particular, it is used at the beginning to reset the circuit to its initial
state.

In Example 6-4, both manual and simulation-based verification are illustrated.

EXAMPLE 6-4   Verifying the Sequence Recognizer

The state diagram for the sequence recognizer appears in Figure 6-24(d) and the
logic diagram appears in Figure 6-27. There are four states and two input combinations, giving a total of eight state-input combinations to verify. The next state can
be observed as the state on the flip-flop outputs after the positive clock edge. For D
flip-flops, the next state is the same as the D input just before the clock edge. For
other types of flip-flops, the flip-flop inputs just before the clock edge are used to
determine the next state of the flip-flop. Initially, beginning with the circuit in an
unknown state, we apply a 1 to the Reset input. This input goes to the direct reset
input on the two flip-flops in Figure 6-27. Since there is no bubble on these inputs,
the 1 value resets both flip-flops to 0, giving state A (0,0). Next, we apply input 0,
and manually simulate the circuit in Figure 6-27 to find that the output is 0 and the
next state is A (0,0), which agrees with the transition for input 0 while in state A.
Next, simulating state A with input 1, next state B (0,1) and output 0 result. For
state B, input 0 gives output 0 and next state A (0,0), and input 1 gives output 0 and
next state C(1,1). This same process can be continued for each of the two input
combinations for states C and D.

For verification by simulation, an input sequence that applies all state–input
combination pairs is to be generated accompanied by the output sequence and

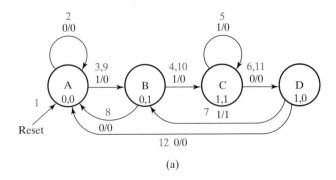

(a)

Clock Edge:	0	1	2	3	4	5	6	7	8	9	10	11	12	13
Input R:	X	1	0	0	0	0	0	0	0	0	0	0	0	
Input X:	X	0	0	1	1	1	0	1	0	1	1	0	0	
State (A,B):	X,X	0,0*	0,0	0,0	0,1	1,1	1,1	1,0	0,1	0,0	0,1	1,1	1,0	0,0
Output Z:	X	0	0	0	0	0	0	1	0	0	0	0	0	

(b)

☐ **FIGURE 6-29**
Test Sequence Generation for Simulation in Example 6.5

state sequence for checking output and next-state values. Optimization requires that the number of clock periods used exceed the number of state–input combination pairs by as few periods as possible (i.e., the repetition of state–input combination pairs should be minimized). This can be interpreted as drawing the shortest path through the state diagram that passes through each state–input combination pair at least once.

In Figure 6-29(a), for convenience, the codes for the states are shown and the path through the diagram is denoted by a sequence of blue integers beginning with 1. These integers correspond to the positive clock edge numbers in Figure 6-29(b), where the verification sequence is to be developed. The values shown for the clock edge numbers are those present just before the positive edge of the clock (i.e., during the setup time interval). Clock edge 0 is at $t = 0$ in the simulation and gives unknown values for all signals. We begin with value 1 applied to Reset (1) to place the circuit in state A. Input value 0 is applied first (2) so that the state remains A, followed by 1 (3) checking the second input combination for state A. Now in state B, we can either move forward to state C or go back to state A. It is not apparent which choice is best, so we arbitrarily apply 1 (4) and go to state C. In state C, 1 is applied (5) so the state remains C. Next, a 0 is applied to check the final input for state C. Now in state D, we have an

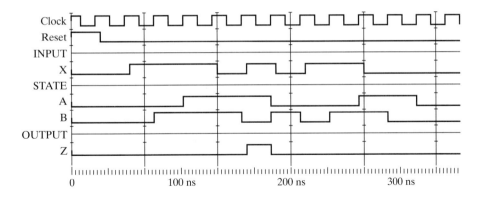

□ **FIGURE 6-30**
Simulation for Example 6-5

arbitrary choice to return to state A or to state B. If we return to state B by applying 1 (7), then we can check the transition from B to A for input 0 (8). Then, the only remaining transition to check is state D for input 0. To reach state D from state A, we must apply the sequence 1, 1, 0 (9) (10) (11) and then apply 0 (12) to check the transition from D to A. We have checked eight transitions with a sequence consisting of reset plus 11 inputs. Although this test sequence is of optimum length, optimality is not guaranteed by the procedure used. However, it usually produces an efficient sequence.

In order to simulate the circuit, we enter the schematic in Figure 6-27 using the Xilinx ISE 4.2 Schematic Editor and enter the sequence from Figure 6-29(b) as a waveform using the Xilinx ISE 4.2 HDL Bencher. While entering the waveform, it is important that the input X changes well before the clock edge. This insures that there is time available to display the current output and to permit input changes to propagate to the flip-flop inputs before the setup time begins. This is illustrated by the INPUT waveforms in Figure 6-30 in which X changes shortly after the positive clock edge providing a good portion of the clock period for the change to propagate to the flip-flops. The circuit is simulated with the MTI Model-Sim simulator. We can then compare the values just before the positive clock edge on the STATE and OUTPUT waveforms in Figure 6-30 with the values shown on the state diagram for each clock period in Figure 6-29. In this case, the comparison verifies that the circuit operation is correct. ■

6-6 OTHER FLIP-FLOP TYPES

This section introduces *JK* and *T* flip-flops and the representations of their behavior used in analysis and design.

Because of their lesser importance in contemporary design relative to *D* flip-flops, the analysis and design examples illustrating their use are given on the Companion Website for the text.

JK and T Flip-Flops

Four types of flip-flops are characterized in Table 6-7, including the *SR* and *D* from section 6-3 given for reference, and the *JK* and *T* introduced here. With the exception of the *SR* flip-flop which is master-slave, the symbol for a positive-edge-triggered version of each flip-flop type is given. A logic diagram for an implementation of each flip-flop type is either referenced or given. A new concept, the *characteristic table,* defines the logical properties of flip-flop operation in tabular form. Specifically, the table defines the next state as a function of the present state and inputs. $Q(t)$ refers to the present state prior to the application of a clock pulse. $Q(t + 1)$ represents the state one clock period later (i.e., the *next state*). Note that the triggering edge (or pulse) at input C is not listed in the characteristic table, but is assumed to occur between time t and $t + 1$. Next to the characteristic table, the *characteristic equation* for each flip-flop type is given. These equations define the next state after the clock pulse for each of the flip-flops as a function of the preset inputs and the present state before the clock pulse. The final column of the table consists of *excitation tables* for each of the flip-flop types. These tables define the input value or values required to obtain each possible next state value after the clock pulse, given the present state value before the clock pulse. Excitation tables can be used to determine the flip-flop input equations from state table information.

Historically, the *JK* flip-flop was a modified version of the master-slave *SR* flip-flop. While the SR flip-flop produces undefined outputs and indeterminate behavior for $S = R = 1$, the *JK* flip-flop causes the output to complement its current value. The master-slave version of the *JK* flip-flop has pulse-triggered behavior and, in addition, exhibits a property called "1's catching." Once $J = 1$ or $K = 1$ occurs, such that the master changes to the opposite state, the master cannot be changed back to its original state before the clock pulse ends, regardless of the values on J and K. This worsens the setup time problem that already exists for the pulse-triggered flip-flop. The same solution applies as for the *SR* flip-flop (i.e., making the setup time, t_S, the entire duration of the triggering pulse). To avoid this additional contribution to the length of the clock cycle, we use only edge-triggered *JK* flip-flops built upon an edge-triggered *D* flip-flop.

In Table 6-7, the symbol for a positive-edge-triggered *JK* flip-flop is shown as well as its logic diagram using a positive-edge-triggered *D* flip-flop. The characteristic table given describes the behavior of the *JK* flip-flop. The *J* input behaves like the *S* input to set the flip-flop. The *K* input is similar to the *R* input for resetting the flip-flop. The only difference between the *SR* and *JK* flip-flops is their response to the condition when both inputs are equal to 1. As can be verified from the logic diagram, this condition complements the state of the *JK* flip-flop. When $J = 1$ and $Q = 0$, then $D = 1$, complementing the *JK* flip-flop outputs. When $K = 1$ and $Q = 1$, then $D = 0$, complementing the *JK* flip-flop outputs. This demonstrates that, regardless of the value of Q, the condition $J = 1$ and $K = 1$ causes the outputs of the flip-flop to be complemented in response to a clock pulse. The next state behavior is summarized in the characteristic table column of Table 6-7. The clock input is not explicitly shown, but a clock pulse is assumed to have occurred between the present state and the next state of Q.

□ TABLE 6-7
Flip-Flop Logic, Characteristic Tables and Equations, and Excitation Tables

Type D

Symbol: D flip-flop with D input and C clock input

Logic Diagrams: See Figure 6-13

Characteristic Table:

D	Q(t+1)	Operation
0	0	Reset
1	1	Set

Characteristic Equation: $Q(t+1) = D(t)$

Excitation Table:

Q(t+1)	D	Operation
0	0	Reset
1	1	Set

Type SR

Symbol: SR flip-flop with S, C, R inputs

Logic Diagrams: See Figure 6-10

Characteristic Table:

S	R	Q(t+1)	Operation
0	0	$Q(t)$	No change
0	1	0	Reset
1	0	1	Set
1	1	?	Undefined

Characteristic Equation: $Q(t+1) = S(t) + \overline{R}(t) \cdot Q(t)$

Excitation Table:

Q(t)	Q(t+1)	S	R	Operation
0	0	0	X	No change
0	1	1	0	Set
1	0	0	1	Reset
1	1	X	0	No change

Type JK

Symbol: JK flip-flop with J, C, K inputs

Logic Diagrams: JK flip-flop logic diagram

Characteristic Table:

J	K	Q(t+1)	Operation
0	0	$Q(t)$	No change
0	1	0	Reset
1	0	1	Set
1	1	$\overline{Q}(t)$	Complement

Characteristic Equation: $Q(t+1) = J(t) \cdot \overline{Q}(t) + \overline{K}(t) \cdot Q(t)$

Excitation Table:

Q(t)	Q(t+1)	J	K	Operation
0	0	0	X	No change
0	1	1	X	Set
1	0	X	1	Reset
1	1	X	0	No Change

Type T

Symbol: T flip-flop with T input and C clock input

Logic Diagrams: T flip-flop logic diagram

Characteristic Table:

T	Q(t+1)	Operation
0	$Q(t)$	No change
1	$\overline{Q}(t)$	Complement

Characteristic Equation: $Q(t+1) = T(t) \oplus Q(t)$

Excitation Table:

Q(t+1)	T	Operation
$Q(t)$	0	No change
$\overline{Q}(t)$	1	Complement

The T (toggle) flip-flop is equivalent to the JK flip-flop with J and K tied together so that $J = K = T$. With this connection, only the combinations $J = 0$, $K = 0$ and $J = 1$, $K = 1$ are applied. If we take the characteristic equation for the JK flip-flop and make this connection, the equation becomes

$$Q(t+1) = T\,\overline{Q} + \overline{T}\,Q = T \oplus Q$$

The symbol for the T flip-flop and its logic diagram based on the preceding equation are given in Table 6-7. The characteristic equation for the T flip-flop is that just given, and the characteristic table in Table 6-7 shows that for $T = 0$, the T flip-flop outputs remain unchanged, and for $T = 1$, the outputs are complemented. Since the T flip-flop can only hold its state unchanged or complement its state, there is no way to establish an initial state using only the T input without adding external sampling of the current output in the next state logic outside of the flip-flop. Thus, the T flip-flop is usually initialized to a known state by using a direct set or direct reset.

6-7 HDL REPRESENTATION FOR SEQUENTIAL CIRCUITS— VHDL

In Chapter 4, VHDL was used to describe combinational circuits. Likewise, VHDL can describe storage elements and sequential circuits. In this section, descriptions of a positive-edge-triggered D flip-flop and a sequence recognizer circuit illustrate such uses of VHDL. These descriptions involve new VHDL concepts, the most important of which is the *process*. Thus far, concurrent statements have described combinations of conditions and actions in VHDL. A concurrent statement, however, is limited in the complexity that can be represented. Typically, the sequential circuits to be described are complex enough that description within a concurrent statement is very difficult. A process can be viewed as a replacement for a concurrent statement that permits considerably greater descriptive power. Multiple processes may execute concurrently, and a process may execute concurrently with concurrent statements.

The body of a process typically implements a sequential program. Signal values, which are assigned during the process, however, change only when the process is completed. If the portion of a process executed is

```
B <= A;
C <= B;
```

then, at the completion of the process, B will contain the original contents of A and C will contain the original contents of B. In contrast, after execution of these two statements in a program, C would contain the original contents of A. To achieve program-like behavior, VHDL uses another construct called a *variable*. In contrast to a signal which evaluates after some delay, a variable evaluates immediately. Thus, if B is a variable in the execution of

```
B := A;
C := B;
```

B will instantaneously evaluate to the contents of A and C will evaluate to the new contents of B, so that C finally contains the original contents of A. Variables appear only within processes. Note the use of := instead of <= for variable assignment.

EXAMPLE 6-5   VHDL for Positive-Edge-Triggered D Flip-Flop with Reset

Basic process structure is illustrated by an example process describing the architecture of a positive-edge-triggered *D* flip-flop in Figure 6-31. The process begins with the keyword **process**. Optionally, **process** can be preceded by a process name followed by a colon. Following in parentheses are two signals, CLK and RESET. This is the *sensitivity list* for the process. If either CLK or RESET change, then the process is executed. In general, a process is executed whenever a signal or variable in its sensitivity list changes. It is important to note that the sensitivity list is not a parameter list containing all inputs and outputs. For example, D does not appear, since a change in its value cannot initiate a possible change in the value of Q. Following the sensitivity list at the beginning of the process is the keyword **begin,** and at the end of the process the keyword **end** appears. The word **process** following **end** is optional.

Within the body of the process, there are additional VHDL conditional structures that can appear. Notable in the Figure 6-31 example is **if-then-else**. The general structure of an **if-then-else** in VHDL is

```
-- Positive-Edge-Triggered D Flip-Flop with Reset:
-- VHDL Process Description
library ieee;
use ieee.std_logic_1164.all;
entity dff is
    port(CLK, RESET, D : in std_logic;
         Q : out std_logic);
end dff;


architecture pet_pr of dff is
-- Implements positive edge-triggered bit state storage
-- with asynchronous reset.

begin
process (CLK, RESET)
   begin
     if (RESET = '1') then
       Q <= '0';
     elsif (CLK'event and CLK = '1') then
          Q <= D;
        end if;
      end if;
   end process;
end;
```

□ **FIGURE 6-31**
VHDL Process Description of Positive-Edge-Triggered Flip-Flop with Reset

```
if condition then
    sequence of statements
{elsif condition then
    sequence of statements}
else
    sequence of statements
end if;
```

The statements within braces { } can appear from zero to any number of times. The **if-then-else** within a process is similar in effect to the **when else** concurrent assignment statement. Illustrating, we have

```
if A = '1' then
    Q <= X;
elsif B = '0' then
    Q <= Y;
else
    Q <= Z;
end if;
```

If A is 1, then flip-flop Q is loaded with the contents of X. If A is 0 and B is 0, then flip-flop Q is loaded with the contents of Y. Otherwise, Q is loaded with the contents of Z. The end result for the four combination of values on A and B is

```
A = 0, B = 0    Q <= Y
A = 0, B = 1    Q <= Z
A = 1, B = 0    Q <= X
A = 1, B = 1    Q <= X
```

More complex conditional execution of statements can be achieved by nesting if-then-else structures, as in the following code:

```
if A = '1' then
    if C = '0' then
        Q <= W;
    else
        Q <= X;
    end if;
elsif B = '0' then
    Q <= Y;
else
    Q <= Z;
end if;
```

The end result for the eight combinations of values on A, B, and C is

```
A = 0, B = 0, C = 0    Q <= Y
A = 0, B = 0, C = 1    Q <= Y
A = 0, B = 1, C = 0    Q <= Z
A = 0, B = 1, C = 1    Q <= Z
A = 1, B = 0, C = 0    Q <= W
A = 1, B = 0, C = 1    Q <= X
```

```
A = 1, B = 1, C = 0     Q <= W
A = 1, B = 1, C = 1     Q <= X
```

With the information introduced thus far, the positive-edge-triggered *D* flip-flop in Figure 6-31 can now be studied. The sensitivity list for the process includes CLK and RESET, so the process is executed if either CLK or RESET or both change value. If D changes value, the value of Q is not to change for an edge-triggered flip-flop, so D does not appear on the sensitivity list. Based on the **if-then-else**, if RESET is 1, the flip-flop output Q is reset to 0. Otherwise, if the clock value changes, which is represented by appending 'event to CLK, and the new clock value is 1, which is represented by CLK = '1', a positive edge has occurred on CLK. The result of the positive edge occurrence is the loading of the value on D into the flip-flop so that it appears on output Q. Note that, due to the structure of the **if-then-else**, RESET equal to 1 dominates the clocked behavior of the *D* flip-flop causing the output Q to go to 0. Similar simple descriptions can be used to represent other flip-flop types and triggering approaches. ▪

EXAMPLE 6-6 VHDL for the Sequence Recognizer

A more complex example in Figures 6-32 and 6-33 represents the sequence recognizer state diagram in Figure 6-24(d). The architecture in this description consists of three distinct processes, which can execute simultaneously and interact via shared signal values. New concepts included are type declarations for defining new types and case statements for handling conditions.

The type declaration permits us to define new types analogous to existing types such as std_logic. A type declaration begins with the keyword **type** followed by the name of the new type, the keyword **is**, and, within parentheses, the list of values for signals of the new type. Using the example from Figure 6-31, we have

> **type** state_type **is** (A, B, C, D);

The name of the new type is state_type and the values in this case are the names of the states in Figure 6-24(d). Once a **type** has been declared, it can be used for declaring signals or variables. From the example in Figure 6-31,

> **signal** state, next_state : state_type;

indicates that state and next_state are signals that are of the type state_type. Thus, state and next state can have values A, B, C, and D.

The basic **if-then-else** (without using the **elsif**) makes a two-way decision based on whether a condition is TRUE or FALSE. In contrast, the **case** statement can make a multiway decision based on which of a number of statements is TRUE. A simplified form for the generic **case** statement is

> **case** expression **is**
>   {**when** choices =>
>   sequence of statements;}
> **end case;**

```vhdl
-- Sequence Recognizer: VHDL Process Description
-- (See Figure 6-24(d) for state diagram)
library ieee;
use ieee.std_logic_1164.all;
entity seq_rec is
   port(CLK, RESET, X: in std_logic;
        Z: out std_logic);
end seq_rec;

architecture process_3 of seq_rec is
   type state_type is (A, B, C, D);
   signal state, next_state : state_type;
begin

-- Process 1 - state_register: implements positive edge-triggered
-- state storage with asynchronous reset.
   state_register: process (CLK, RESET)
   begin
     if (RESET = '1') then
        state <= A;
     elsif (CLK'event and CLK = '1') then
           state <= next_state;
        end if;
     end if;
end process;

-- Process 2 - next_state_function: implements next state as
-- a function of input X and state.
   next_state_func: process (X, state)
   begin
      case state is
         when A =>
      if X = '1' then
         next_state <= B;
      else
         next_state <= A;
           end if;
        when B =>
      if X = '1' then
         next_state <= C;
      else
         next_state <= A;
           end if;
```

☐ **FIGURE 6-32**
VHDL Process Description of a Sequence Recognizer

```
-- Sequence Recognizer: VHDL Process Description (continued)
  when C =>
     if X = '1' then
        next_state <= C;
     else
        next_state <= D;
           end if;
        when D =>
     if X = '1' then
        next_state <= B;
     else
        next_state <= A;
           end if;
     end case;
  end process;

-- Process 3 - output_function: implements output as function
-- of input X and state.
  output_func: process (X, state)
  begin
     case state is
        when A =>
     Z <= '0';
  when B =>
     Z <= '0';
  when C =>
     Z <= '0';
        when D =>
  if X = '1' then
     Z <= '1';
  else
     Z <= '0';
        end if;
     end case;
  end process;
end;
```

□ **FIGURE 6-33**

VHDL Process Description of a Sequence Recognizer (continued)

The choices must be values that can be taken on by a signal of the type used in the expression. The **case** statement has an effect similar to the **with-select** concurrent assignment statement.

In the example in Figures 6-32 and 6-33, Process 2 uses a **case** statement to define the next state function for the sequence recognizer. The **case** statement makes a multiway decision based on the current state of the circuit, A, B, C, or D. **If-then-else** statements are used for each of the state alternatives to make a binary decision based on whether input X is 1 or 0. Concurrent assignment statements are then used to assign the next state based on the eight possible combinations of state value and input value. For example, consider the state alternative **when** B. If X equals 1, then the next state will be C; if X equals 0, then the next state will be A. This corresponds to the two transitions out of state B in Figure 6-24(d). For more complex circuits, case statements can also be used for handing the input conditions.

With this brief introduction to the **case** statement, the overall sequencer recognizer can now be studied. Each of the three processes has a distinct function, but the processes interact to provide the overall sequence recognizer. Process 1 describes the storage of the state. Note that the description is like that of the positive-edge-triggered flip-flop. There are two differences, however. The signals involved are of type state_type instead of type std_logic. Second, the state that results from applying RESET is state A rather than state 0. Also, since we are using state names such as A, B, and C, the number of state variables (i.e., the number of flip-flops) is unspecified and the state codes are unknown. Process 1 is the only one of the three processes that contains storage.

Process 2 describes the next state function, as discussed earlier. The sensitivity list in this case contains signals X and state. In general, for describing combinational logic, all inputs must appear in the sensitivity list, since, whenever an input changes, the process must be executed.

Process 3 describes the output function. The same **case** statement framework as in Process 2 with state as the expression is used. Instead of assigning state names to next state, values 0 and 1 are assigned to Z. If the value assigned is the same for both values 0 and 1 on X, no **if-then-else** is needed, so an **if-then-else** appears only for state D. If there are multiple input variables, more complex **if-then-else** combinations or a **case** statement, as illustrated earlier, can be used to represent the conditioning of the outputs on the inputs. This example is a Mealy state machine, in which the output is a function of the circuit inputs. If it were a Moore state machine, with the output dependent only on the state, input X would not appear on the sensitivity list, and there would be no **if-then-else** structures in the **case** statement. ∎

There is a common pitfall present whenever an **if-then-else** or **case** statement is employed. During synthesis, unexpected storage elements in the form of latches or flip-flops appear. For the simple **if-then-else** used in Figure 6-31, using this pitfall gives a specification that synthesizes to a flip-flop. In addition to the two input signals, RESET and CLK, the signal CLK'event is produced by applying the predefined attribute 'event to the CLK signal. CLK'event is TRUE if the value of CLK changes. All possible combinations of values are represented in Table 6-8. Whenever RESET is 0 and the CLK is fixed at 0 or 1 or has a negative edge, no action is specified. In VHDL, it is assumed that, for any combinations of conditions that have unspecified actions in **if-then-else** or **case** statements, the left-hand side of an assignment statement remains unchanged. This is equivalent to Q <= Q, causing storage to occur. Thus, all combinations of conditions must have the resulting action specified when no storage is intended. If this is not a natural situation, an **others** can be used in the **if-then else** or **case**. If there are binary values used in the **case** statement, just as in Section 4-7, an **others** must also be used to handle combinations including the seven values other than 0 and 1 permitted for std_logic.

Together, the three processes used for the sequence recognizer describe the state storage, the next state function, and the output function for a sequential

□ **TABLE 6-8**
Illustration of generation of storage in VHDL

Inputs			Action
RESET = 1	CLK = 1	CLK'event	
FALSE	FALSE	FALSE	Unspecified
FALSE	FALSE	TRUE	Unspecified
FALSE	TRUE	FALSE	Unspecified
FALSE	TRUE	TRUE	Q <= D
TRUE	—	—	Q <= '0'

circuit. Since these are all of the components of a sequential circuit at the state diagram level, the description is complete. The use of three distinct processes is only one methodology for sequential circuit description. Pairs of processes or all three processes can be combined for more elegant descriptions. Nevertheless, the three-process description is the easiest for new users of VHDL and also works well with synthesis tools.

To synthesize the circuit into actual logic, a state assignment is needed, in addition to a technology library. Many synthesis tools will make the state assignment independently or based on a directive from the user. It is also possible for the user to specify explicitly the state assignment. This can be done in VHDL by using an enumeration type. The encoding for the state machine in Figures 6-32 and 6-33 can be specified by adding the following after the **type** state_type declaration:

```
attribute enum_encoding: string;
attribute enum_encoding of state_type:
type is "00, 01, 10, 11";
```

This is not a standard VHDL construct, but it is recognized by many synthesis tools. Another option is not to use a type declaration for the states, but to declare the state variables as signals and use the actual codes for the states. In this case, if states appear in the simulation output, they will appear as the encoded state values.

6-8 HDL REPRESENTATION FOR SEQUENTIAL CIRCUITS— VERILOG

In Chapter 4, Verilog was used to describe combinational circuits. Likewise, Verilog can describe storage elements and sequential circuits. In this section, descriptions of a positive edge-triggered D flip-flop and a sequence recognizer circuit illustrate such uses of Verilog. These descriptions will involve new Verilog concepts, the most important of which are the process and the register type for nets.

Thus far, continuous assignment statements have been used to describe combinations of conditions and actions in Verilog. A continuous assignment statement is limited in what can be described, however. A *process* can be viewed as a replacement for a continuous assignment statement that permits considerably greater descriptive power. Multiple processes may execute concurrently and a process may execute concurrently with continuous assignment statements.

Within a process, procedural assignment statements, which are not continuous assignments, are used. Because of this, the assigned values need to be retained over time. This retention of information can be achieved by using the *register* type rather than the wire type for nets. The keyword for the register type is **reg**. Note that just because a net is of type **reg** does not mean that an actual register is associated with its implementation. There are additional conditions that need to be present to cause an actual register to exist.

There are two basic types of processes, the **initial** process and the **always** process. The **initial** process executes only once, beginning at t = 0. The **always** process also executes at t =0, but executes repeatedly thereafter. To prevent rampant, uncontrolled execution, some timing control is needed in the form of delay or event-based waiting. The # operator followed by an integer can be used to specify delay. The @ operator can be viewed as "wait for event." @ is followed by an expression that describes the event or events the occurrence of which will cause the process to execute.

The body of a process is like a sequential program. The process begins with the keyword **begin** and ends with the keyword **end**. There are procedural assignment statements that make up the body of the process. These assignment statements are classified as blocking or nonblocking. Blocking assignments use = as the assignment operator and nonblocking assignments use <= as the operator. *Blocking assignments* are executed sequentially, much like a program in a procedural language such as C. *Nonblocking assignments* evaluate the right-hand side, but do not make the assignment until all right-hand sides have been evaluated. Blocking assignments can be illustrated by the following process body, in which A, B, and C are of type **reg**:

```
begin
    B = A;
    C = B;
end
```

The first statement transfers the contents of A into B. The second statement then transfers the new contents of B into C. At process completion, C contains the original contents of A.

Suppose that the same process body uses nonblocking assignments:

```
begin
    B <= A;
    C <= B;
end
```

The first statement transfers the original contents of A into B and the second statement transfers the original contents of B into C. At process completion, C contains the original contents of B, not those of A. Effectively, the two statement have executed concurrently instead of in sequence. Nonblocking assignments, except in the cases in which we want registers (of type **reg**) to be evaluated sequentially, will be used.

EXAMPLE 6-7   Verilog for Positive-Edge-Triggered D Flip-Flop with Reset

These new concepts can now be applied to the Verilog description of a positive-edge-triggered *D* flip-flop given in Figure 6-34. The module and its inputs and outputs are declared. Q is declared as of type **reg** since it will store information. The process begins with the keyword **always**. Following is **@ (posedge** CLK **or posedge** RESET) This is the *event control* statement for the process that initiates process execution if an event (i.e., a specified change in a specified signal occurs). For the *D* flip-flop, if either CLK or RESET changes to 1, then the process is executed. It is important to note that the event control statement is not a parameter list containing all inputs. For example, D does not appear, since a change in its value cannot initiate a possible change in the value of Q. Following the event control statement at the beginning of the process is the keyword **begin**, and at the end of the process the keyword **end** appears.

Within the body of the process, there are additional Verilog conditional structures that can appear. Notable in the Figure 6-34 example is **if-else**. The general structure of an **if-else** in Verilog is

```
if (condition)
   begin procedural statements end
{else if (condition)
   begin procedural statements end}
{else
   begin procedural statements end}
```

If there is a single procedural statement, then the **begin** and **end** are unnecessary:

```
if (A == 1)
   Q <= X;
else if (B == 0)
   Q <= Y;
else
   Q <= Z;
```

Note that a double equals signs is used in conditions. If A is 1, then flip-flop Q is loaded with the contents of X. If A is 0 and B is 0, then flip-flop Q is loaded with the contents of Y. Otherwise, Q is loaded with the contents of Z. The end result for the four combination of values on A and B is

```
A = 0, B = 0  Q <= Y
A = 0, B = 1  Q <= Z
A = 1, B = 0  Q <= X
A = 1, B = 1  Q <= X
```

The **if-else** within a process is similar in effect to the conditional operator in a continuous assignment statement introduced earlier. The conditional operator can be used within a process, but the **if-else** cannot be used in a continuous assignment statement.

```
// Positive-Edge-Triggered D Flip-Flop with Reset:
// Verilog Process Description

module dff_v(CLK, RESET, D, Q);
    input CLK, RESET, D;
    output Q;
    reg Q;

always @(posedge CLK or posedge RESET)
begin
    if (RESET)
        Q <= 0;
    else
        Q <= D;
end
endmodule
```

□ **FIGURE 6-34**
Verilog Process Description of Positive-Edge-Triggered Flip-Flop with Reset

More complex conditional execution of statements can be achieved by nesting **if-else** structures. For example, we might have

```
if (A == 1)
    if (C == 0)
        Q <= W;
    else
        Q <= X;
else if (B == 0)
    Q <= Y;
else
    Q <= Z;
```

In this type of structure, an **else** is associated with the closest **if** preceding it that does not already have an **else**. The end result for the eight combinations of values on A, B, and C is

```
A = 0, B = 0, C = 0      Q <= Y
A = 0, B = 0, C = 1      Q <= Y
A = 0, B = 1, C = 0      Q <= Z
A = 0, B = 1, C = 1      Q <= Z
A = 1, B = 0, C = 0      Q <= W
A = 1, B = 0, C = 1      Q <= X
A = 1, B = 1, C = 0      Q <= W
A = 1, B = 1, C = 1      Q <= X
```

Returning to the **if-else** in the positive-edge-triggered D flip-flop shown in Figure 6-34, assuming that a positive edge has occurred on either CLK or RESET, if RESET is 1, the flip-flop output Q is reset to 0. Otherwise, the value on D is stored in the flip-flop so that Q equals D. Due to the structure of the **if-else**, RESET equal to 1 dominates the clocked behavior of the *D* flip-flop causing the output Q to go to 0. Similar simple descriptions can be used to represent other flip-flop types and triggering approaches. ■

EXAMPLE 6-8 Verilog for the Sequence Recognizer

A more complex example in Figure 6-35 represents the sequence recognizer state diagram in Figure 6-24(d). The architecture in this description consists of three distinct processes that can execute simultaneously and interact via shared signal values. New concepts included are state encoding and case statements for handling conditions.

In Figure 6-35, the module seq_rec_v and input and output variables CLK, RESET, X, and Z are declared. Next, registers are declared for state and next_state. Note that since next_state need not be stored, it could also be declared as a wire, but, since it is assigned within an **always**, it must be declared as a **reg**. Both registers are two bits, with the most significant bit (MSB) numbered 1 and the least significant bit (LSB) numbered 0.

Next, a name is given to each of the states taken on by state and next_state, and binary codes are assigned to them. This can be done using a parameter statement or a compiler directive **define**. We will use the parameter statement, since the compiler directive requires a somewhat inconvenient ' before each state throughout the description. From the diagram in Figure 6-24(d), the states are A, B, C, and D. In addition, the parameter statements give the state codes assigned to each of these states. The notation used to define the state codes is 2'b followed by the binary code. The 2 denotes that there are two bits in the code and the 'b denotes that the base of the code given is binary.

The **if-else** (without using the **else if**) makes a two-way decision based on whether a condition is TRUE or FALSE. In contrast, the **case** statement can make a multiway decision based on which one of a number of statements is TRUE. A simplified form for the generic **case** statement is

```
case expression
   {case expression : statements}
endcase
```

in which the braces { } represent one or more such entries.

The case expression must have values that can be taken on by a signal of the type used in expression. Typically, there are sequences of multiple statements. In the example in Figure 6-35, the **case** statement for the next state function makes a multiway decision based on the current state of the circuit, A, B, C, or D. For each of the case expressions, conditional statements of various types are used to make a binary decision based on whether input X is 1 or 0. Nonblocking assignment statements are then used to assign the next state based on the eight possible combinations of state value and input value. For example, consider the expression B. If X equals 1, then the next state will be C; if X equals 0, then the next state will be A. This corresponds to the two transitions out of state B in Figure 6-24(d).

With this brief introduction to the **case** statement, the overall sequencer recognizer can now be understood. Each of the three processes has a distinct function, but

```verilog
// Sequence Recognizer: Verilog Process Description
// (See Figure 6-24(d) for state diagram)
module seq_rec_v(CLK, RESET, X, Z);
  input CLK, RESET, X;
  output Z;
  reg [1:0] state, next_state;
  parameter A = 2'b00, B = 2'b01, C = 2'b10, D = 2'b11;
  reg Z;
// state register: implements positive edge-triggered
// state storage with asynchronous reset.
always @(posedge CLK or posedge RESET)
begin
  if (RESET == 1)
    state <= A;
  else
    state <= next_state;
end
// next state function: implements next state as function
// of X and state
always @(X or state)
begin
  case (state)
    A:  if (X == 1)
          next_state <= B;
        else
          next_state <= A;
    B:  if(X) next_state <= C;else next_state <= A;
    C:  if(X) next_state <= C;else next_state <= D;
    D:  if(X) next_state <= B;else next_state <= A;
  endcase
end
// output function: implements output as function
// of X and state
always @(X or state)
begin
  case (state)
    A: Z <= 0;
    B: Z <= 0;
    C: Z <= 0;
    D: Z <= X ? 1 : 0;
  endcase
end
endmodule
```

☐ **FIGURE 6-35**
Verilog Process Description of a Sequence Recognizer

the processes interact to provide the overall sequence recognizer. The first process describes the state register for storing the sequence recognizer state. Note that the description resembles that of the positive-edge-triggered flip-flop. There are two differences, however. First, there are two bits in the state register. Second, the state that results from applying RESET is state A rather than state 0. The first process is the only one of the three processes that has storage associated with it.

The second process describes the next state function as discussed earlier. The event control statement contains signals X and state. In general, for describing combinational logic, all inputs must appear in the event control statement, since, whenever an input changes, the process must be executed.

The final process describes the output function and uses the same **case** statement framework as in the next state function process. Instead of assigning state names, values 0 and 1 are assigned to Z. If the value assigned is the same for both values 0 and 1 on X, no conditional statement is needed, so a conditional statement appears only for state D. If there are multiple input variables, more complex **if-else** combinations, as illustrated earlier, can be used to represent the conditioning of the outputs on the inputs. This example is a Mealy state machine in which the output is a function of the circuit inputs. If it were a Moore state machine, with the output dependent only on the state, input X would not appear on the event control statement and there would be no conditional structures within the **case** statement.

■

There is a common pitfall present whenever an **if-else** or **case** statement is employed. During synthesis, unexpected storage elements in the form of latches or flip-flops appear. For the very simple **if-else** used in Figure 6-34, this pitfall is employed to give a specification that synthesizes to a flip-flop. In addition to the two input signals, RESET and CLK, events **posedge** CLK and **posedge** RESET are produced, which are TRUE if the value of the respective signal changes from 0 to 1. Selected combinations of values for RESET and the two events are shown in Table 6-9. Whenever RESET has no positive edge, or RESET is 0 and CLK is fixed at 0 or 1 or has a negative edge, no action is specified. In Verilog, the assumption is

□ **TABLE 6-9**
Illustration of generation of storage in Verilog

Inputs		Action
posedge RESET and RESET = 1	**posedge CLK**	
FALSE	FALSE	Unspecified
FALSE	TRUE	Q <= D
TRUE	FALSE	Q <= 0
TRUE	TRUE	Q <= 0

that, for any combination of conditions with unspecified actions in **if-else** or **case** statements, the left-hand side of an assignment statement will remain unchanged. This is equivalent to Q <= Q causing storage to occur. Thus, all combinations of conditions must have the resulting action specified when no storage is intended. To prevent undesirable latches and flip-flops from occurring, for **if-else** structures, care must be taken to include **else** in all cases if storage is not desired. In a **case** statement, a **default** statement which defines what happens for all choices not specified should be added. Within the **default** statement, a specific next state can be specified, which in the example is state A.

Together, the three processes used for the sequence recognizer describe the state storage, the next state function, and the output function for the sequential circuit. Since these are all of the components of a sequential circuit at the state diagram level, the description is complete. The use of three distinct processes is only one methodology for sequential circuit description. For example, the next state and output processes could be easily combined. Nevertheless, the three-process description is the easiest for new users of Verilog and also works well with synthesis tools.

6-9 CHAPTER SUMMARY

Sequential circuits are the foundation upon which most digital design is based. Flip-flops are the basic storage elements for synchronous sequential circuits. Flip-flops are constructed of more fundamental elements called latches. By themselves, latches are transparent and, as a consequence, are very difficult to use in synchronous sequential circuits using a single clock. When latches are combined to form flip-flops, nontransparent storage elements very convenient for use in such circuits are formed. There are two triggering methods used for flip-flops: master-slave and edge triggering. In addition, there are a number of flip-flop types, including *D*, *SR*, *JK*, and *T*.

Sequential circuits are formed using these flip-flops and combinational logic. Sequential circuits can be analyzed to find state tables and state diagrams that represent the behavior of the circuits. Also, analysis can be performed by using logic simulation.

These same state diagrams and state tables can be formulated from verbal specifications of digital circuits. By assigning binary codes to the states and finding flip-flop input equations, sequential circuits can be designed. The design process also includes issues such as finding logic for the circuit outputs, resetting the state at power-up, and controlling the behavior of the circuit when it enters states unused in the original specification. Finally, logic simulation plays an important role in verifying that the circuit designed meets the original specification.

As an alternative to the use of logic diagrams, state diagrams, and state tables, sequential circuits can be defined in VHDL or Verilog descriptions. These descriptions, typically at the behavioral level, provide a powerful, flexible approach to sequential circuit specification for both simulation and automatic circuit synthesis. These representations involve processes that provide added descriptive power beyond the concurrent assignment statements of VHDL and the continuous assignment statement of Verilog. The processes, which permit programlike coding and use if-then-else and case conditional statements, can also be used to efficiently describe combinational logic.

REFERENCES

1. MANO, M. M. *Digital Design,* 3rd ed. Englewood Cliffs, NJ: Prentice Hall, 2002.

2. ROTH, C. H. *Fundamentals of Logic Design*, 4th ed. St. Paul: West, 1992.

3. WAKERLY, J. F. *Digital Design: Principles and Practices,* 3rd ed. Upper Saddle River, NJ: Prentice Hall, 2000.

4. *IEEE Standard VHDL Language Reference Manual.* (ANSI/IEEE Std 1076-1993; revision of IEEE Std 1076-1987). New York: The Institute of Electrical and Electronics Engineers, 1994.

5. PELLERIN, D. AND D. TAYLOR. *VHDL Made Easy!* Upper Saddle River, NJ: Prentice Hall PTR, 1997.

6. STEFAN, S. AND L. LINDH. *VHDL for Designers.* London: Prentice Hall Europe, 1997.

7. *IEEE Standard Description Language Based on the Verilog(TM) Hardware Description Language* (IEEE Std 1364-1995). New York: The Institute of Electrical and Electronics Engineers, 1995.

8. PALNITKAR, S. *Verilog HDL: A Guide to Digital Design and Synthesis.* Upper Saddle River, NJ: SunSoft Press (A Prentice Hall Title), 1996.

9. CILETTI, M., *Modeling, Synthesis, and Rapid Prototyping with the Verilog HDL*, Upper Saddle River, NJ: Prentice Hall, 1999.

10. THOMAS, D. E., AND P. R. MOORBY. *The Verilog Hardware Description Language* 4th ed. Boston: Kluwer Academic Publishers, 1998.

PROBLEMS

 The plus (+) indicates a more advanced problem and the asterisk (*) indicates a solution is available on the Companion Website for the text.

6–1. Perform a manual or computer-based logic simulation similar to that given in Figure 6-5 for the $\overline{SR}$ latch shown in Figure 6-6. Construct the input sequence, keeping in mind that changes in state for this type of latch occur in response to 0 rather than 1.

6–2. Perform a manual or computer-based logic simulation similar to that given in Figure 6-5 for the *SR* latch with control input *C* in Figure 6-7. In particular, examine the behavior of the circuit when *S* and *R* are changed while *C* has the value 1.

6–3. A popular alternative design for a positive-edge-triggered *D* flip-flop is shown in Figure 6-36. Manually or automatically simulate the circuit to determine whether its functional behavior is identical to that of the circuit in Figure 6-13.

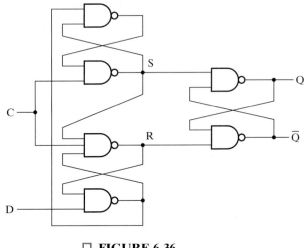

☐ **FIGURE 6-36**
Circuit for Problem 6-3

6–4. A set of waveforms applied to *SR* and *D* flip-flops is shown in Figure 6-37. These waveforms are applied to the flip-flops shown along with the values of their timing parameters.

(a) Indicate the locations on the waveforms at which there are input combination or timing parameter violations in signal S1 for flip-flop 1.

(b) Indicate the locations on the waveforms at which there are input combination or timing parameter violations in signal R1 for flip-flop 1.

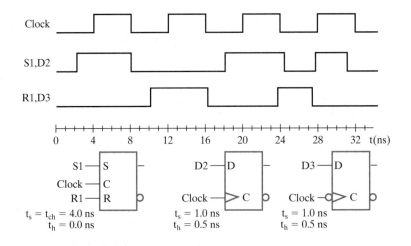

☐ **FIGURE 6-37**
Waveforms and Flip-flops for Problem 6-4

(c) List the times at which there are timing parameter violations in signal D2 for flip-flop 2.

(d) List the times at which there are timing parameter violations in signal D3 for flip-flop 3.

Violations should be indicated even if the state of the flip-flop is such that the violations will not affect the next state.

6–5. A sequential circuit with two D flip-flops A and B, two inputs X and Y, and one output Z is specified by the following input equations:

$$D_A = \overline{X}A + XY \quad D_B = \overline{X}A + XB \quad Z = XB$$

(a) Draw the logic diagram of the circuit.

(b) Derive the state table.

(c) Derive the state diagram.

6–6. *A sequential circuit has three D flip-flops A, B, and C, and one input X. The circuit is described by the following input equations:

$$D_A = (B\overline{C} + \overline{B}C)X + (BC + \overline{B}\,\overline{C})\overline{X}$$

$$D_B = A$$

$$D_C = B$$

(a) Derive the state table for the circuit.

(b) Draw two state diagrams, one for $X = 0$ and the other for $X = 1$.

6–7. A sequential circuit has one flip-flop Q, two inputs X and Y, and one output S. The circuit consists of a D flip-flop with S as its output and logic implementing the function

$$D = X \oplus Y \oplus S$$

with D as the input to the D flip-flop. Derive the state table and state diagram of the sequential circuit.

6–8. Starting from state 00 in the state diagram of Figure 6-19(a), determine the state transitions and output sequence that will be generated when an input sequence of 10011011110 is applied.

6–9. Draw the state diagram of the sequential circuit specified by the state table in Table 6-10.

6–10. *A sequential circuit has two SR flip-flops, one input X, and one output Y. The logic diagram of the circuit is shown in Figure 6-38. Derive the state table and state diagram of the circuit.

☐ **TABLE 6-10**
State Table for Circuit of Problem 6-9

Present State		Input s	Input s	Next State		Output
A	B	X	Y	A	B	Z
0	0	0	0	0	0	0
0	0	0	1	0	1	0
0	0	1	0	1	0	1
0	0	1	1	1	1	1
0	1	0	0	0	1	1
0	1	0	1	1	0	1
0	1	1	0	1	0	0
0	1	1	1	0	0	0
1	0	0	0	1	1	1
1	0	0	1	1	1	0
1	0	1	0	1	1	1
1	0	1	1	1	0	0
1	1	0	0	0	0	0
1	1	0	1	0	0	1
1	1	1	0	0	0	0
1	1	1	1	0	1	1

6–11. A sequential circuit is given in Figure 6-38. The timing parameters for the gates and flip-flops are as follows:

Inverter: $t_{pd} = 0.5$ ns
XOR Gate: $t_{pd} = 2.0$ ns
Flip-flop: $t_{pd} = 2.0$ ns, $t_s = 1.0$ ns and $t_h = 0.25$ ns

(a) Find the longest path delay from an external circuit input passing through gates only to an external circuit output.

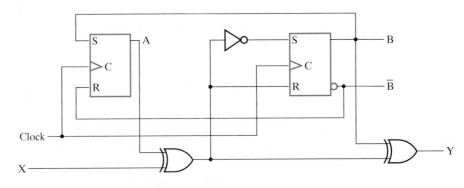

☐ **FIGURE 6-38**
Circuit for Problem 6-10, 6-11, and 6-12

(b) Find the longest path delay in the circuit from an external input to positive clock edge.

(c) Find the longest path delay from positive clock edge to output.

(d) Find the longest path delay from positive clock edge to positive clock edge.

(e) Determine the maximum frequency of operation of the circuit in megahertz (MHz).

6–12. Repeat problem 6-11 assuming that the circuit consists of two copies of the circuit in Figure 6-38 with input X of the second circuit copy driven by input Y of the first circuit copy.

6–13. A sequential circuit is given in Figure 6-17.

(a) Add the necessary logic and/or connections to the circuit to provide an asynchronous reset to state A = 0, B = 1 for signal Reset = 1.

(b) Add the necessary logic and/or connections to the circuit to provide a synchronous reset to state A = 0, B = 0 for signal Reset = 0.

6–14. *Design a sequential circuit with two D flip-flops A and B and one input X. When $X = 0$, the state of the circuit remains the same. When $X = 1$, the circuit goes through the state transitions from 00 to 10 to 11 to 01, back to 00, and then repeats.

6–15. *A serial two's complementer is to be designed. A binary integer of arbitrary length is presented to the serial two's complementer, least significant bit first, on input X. When a given bit is presented on input X, the corresponding output bit is to appear during the same clock cycle on output Z. To indicate that a sequence is complete and that the circuit is to be initialized to receive another sequence, input Y becomes 1 for one clock cycle. Otherwise, Y is 0.

(a) Find the state diagram for the serial two's complementer.

(b) Find the state table for the serial two's complementer.

6–16. A Universal Serial Bus (USB) communication link requires a circuit that produces the sequence 00000001. You are to design a synchronous sequential circuit that starts producing this sequence for input $E = 1$. Once the sequence starts, it completes. If $E = 1$, during the last output in the sequence, the sequence repeats. Otherwise, if $E = 0$, the output remains constant at 1.

(a) Draw the Moore state diagram for the circuit.

(b) Find the state table and make a state assignment.

(c) Design the circuit using D flip-flops and logic gates. A reset should be included to place the circuit in the appropriate initial state at which E is examined to determine if the sequence or constant 1's is to be produced.

6–17. Repeat Problem 6-16 for the sequence 01111110 that is used in a different communication network protocol.

6–18. +The sequence in Problem 6-17 is a flag used in a communication network that represents the beginning of a message. This flag must be unique. As a consequence, at most five 1's in sequence may appear anywhere else in the message. Since this is unrealistic for normal message content, a trick called zero-insertion is used. The normal message, which can contain strings of 1's longer than 5, enters input X of a sequential zero-insertion circuit. The circuit has two outputs Z and S. When a fifth 1 in sequence appears on X, a 0 is inserted in the stream of outputs appearing on Z and the output $S = 1$ indicating to the circuit supplying the zero-insertion circuit with inputs that it must stall and not apply a new input for one clock cycle. This is necessary since the insertion of 0's in the output sequence causes it to be longer than the input sequence without the stall. Zero-insertion is illustrated by the following example sequences:

Sequence on X without any stalls: 0111110011111100001011110101
Sequence on X with stalls: 01111110011111110000101110101
Sequence on Z: 0111110001111101100001011110101
Sequence on S: 0000001000000010000000000000000

(a) Find the state diagram for the circuit

(b) Find the state table for the circuit and make a state assignment.

(c) Find an implementation of the circuit using D flip-flops and logic gates.

6–19. In many communication and networking systems, the signal transmitted on the communication line uses a non-return-to-zero (NRZ) format. USB uses a specific version referred to as non-return-to-zero inverted (NRZI). A circuit that converts any message sequence of 0's and 1's to a sequence in the NRZI format is to be designed. The mapping for such a circuit is as follows:

(a) If the message bit is a 0, then the NRZI format message contains an immediate change from 1 to 0 or change from 0 to 1, depending on the current NRZI value.

(b) If the message bit is a 1, then the NRZI format message remains fixed at 0 or 1, depending on the current NRZI value.

This transformation is illustrated by the following example which assumes that the initial value of the NRZI message is 1:

Message: 10001110011010
NRZI Message: 10100001000101

(a) Find the Mealy model state diagram for the circuit.

(b) Find the state table for the circuit and make a state assignment.

(c) Find an implementation of the circuit using D flip-flops and logic gates.

6–20. +Repeat problem 6-19, designing a sequential circuit that transforms an NRZI message into a normal message. The mapping for such a circuit is as follows:

(a) If a change from 0 to 1 or from 1 to 0 occurs between adjacent bits in the NRZI message, then the message bit is a 0.

(b) If no change occurs between adjacent bits in the NRZI message, then the message bit is a 1.

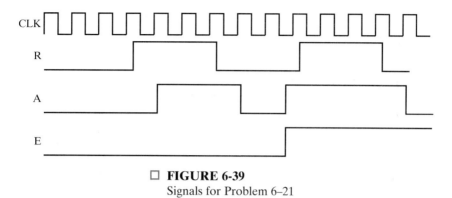

□ **FIGURE 6-39**
Signals for Problem 6–21

6–21. A pair of signals Request (R) and Acknowledge(A) is used to coordinate transactions between a CPU and its I/O system. The interaction of these signals is often referred to as a "handshake." These signals are synchronous with the clock and, for a transaction, are to always have their transitions appear in the order shown in Figure 6-39. A handshake checker is to be designed that will verify the transition order. The checker has inputs, R and A, asynchronous reset signal, RESET, and has output, Error(E). If the transitions in a handshake are in order, $E = 0$. If the transitions are out of order, then E becomes 1 and remains at 1 until the an asynchronous reset signal (RESET = 1) is applied to the CPU.

(a) Find the state diagram for the handshake checker.
(b) Find the state table for the handshake checker.

6–22. A serial leading 1's detector is to be designed. A binary integer of arbitrary length is presented to the serial leading 1's detector, most significant bit first, on input X. When a given bit is presented on input X, the corresponding output bit is to appear during the same clock cycle on output Z. As long as the bits applied to X are 0, $Z = 0$. When the first 1 is applied to X, $Z = 1$. For all bit values applied to X after the first 1 is applied, $Z = 0$. To indicate that a sequence is complete and that the circuit is to be initialized to receive another sequence, input Y becomes 1 for one clock cycle. Otherwise, Y is 0.

(a) Find the state diagram for the serial leading 1's detector.
(b) Find the state table for the serial leading 1's detector.

6–23. *A sequential circuit has two flip-flops A and B, one input X and one output Y. The state diagram is shown in Figure 6-40. Design the circuit with D flip-flops.

6–24. *A set-dominant master-slave flip-flop has set and reset inputs. It differs from a conventional master-slave SR flip-flop in that, when both S and R are equal to 1, the flip-flop is set.

(a) Obtain the characteristic table of the set-dominant flip-flop.
(b) Find the state diagram for the set-dominant flip-flop.
(c) Design the set-dominant flip-flop by using an SR flip-flop and logic gates (including inverters).

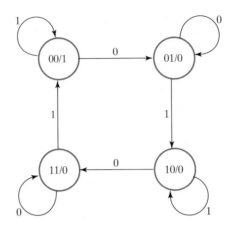

☐ **FIGURE 6-40**
State Diagram for Problem 6–23

6–25. Find the logic diagram for the circuit having the state table given in Table 6-5. Use *D* flip-flops.

6–26. +The state table for a twisted ring counter is given in Table 6-11. This circuit has no inputs, and its outputs are the uncomplemented outputs of the flip-flops. Since it has no inputs, it simply goes from state to state whenever a clock pulse occurs.
(a) Design the circuit using *D* flip-flops and assuming that the unspecified next states are don't-care conditions.
(b) Add the necessary logic to the circuit to initialize it to state 000 on power-up master reset.
(c) In the subsection "Designing with Unused States" of Section 6-5, three techniques for dealing with situations in which a circuit accidentally enters an unused state are discussed. If the circuit you designed in parts (a) and (b) is used in a child's toy, which of the three techniques given would you apply? Justify your decision.

☐ **TABLE 6-11**
State Table for Problem 6-26

Present State	Next State
ABC	ABC
000	100
100	110
110	111
111	011
011	001
001	000

(d) Based on your decision in part (c), redesign the circuit if necessary.

(e) Repeat part (c) for the case in which the circuit is used to control engines on a commercial airliner. Justify your decision.

(f) Repeat part (d) based on your decision in part (e).

6–27. Do a manual verification of the solution (either yours or the one posted on the text website) to Problem 6-24. Consider all transitions where S and R change with the clock equal to 0.

6–28. Do an automatic logic simulation-based verification of your design in Problem 6-25. The input sequence used in the simulation should include all transitions in Table 6-6. The simulation output should include the input X and the state variables A, B, and output Z.

6–29. *Generate a verification sequence for the circuit described by the state table in Table 6-10. To reduce the length of the simulation sequence, assume that the simulator can handle X inputs and use X's whenever possible. Assume that a Reset input is available to initialize the state to $A = 0$, $B = 0$ and that all transitions in the state diagram must be exercised.

6–30. Design the circuit specified by Table 6-10 and use the sequence from Problem 6-29 (either yours or the one posted on the text web site) to perform an automatic logic simulation-based verification of your design.

6–31. *Obtain a timing diagram similar to Figure 6-11 for a positive-edge-triggered JK flip-flop during four clock pulses. Show the timing signals for C, J, K, Y, and Q. Assume that initially the output Q is equal to 1, with $J = 0$ and $K = 1$ for the first pulse. Then, for successive pulses, J goes to 1, followed by K going to 0 and then J going back to 0. Assume that each input changes near the negative edge of the pulse.

 All files referred to in the remaining problems are available in ASCII form for simulation and editing on the Companion Website for the text. A VHDL or Verilog compiler/simulator is necessary for the problems or portions of problems requesting simulation. Descriptions can still be written, however, for many of the problems without using compilation or simulation.

6–32. *Write a VHDL description for the multiplexer in Figure 4-14 by using a process containing a case statement rather than the continuous assignment statements as shown in Section 4-7.

6–33. Repeat Problem 6-32 by using a VHDL process containing if-then-else statements.

6–34. +Write a VHDL description for the sequential circuit with the state diagram given by Figure 6-25(d). Include an asynchronous RESET signal to initialize the circuit to state `Init`. Compile your description, apply an input sequence to pass through every arc of the state diagram at least once, and verify the correctness of the state and output sequence by comparing them to the state diagram.

6–35. Write a VHDL description for the circuit specified in Problem 6-15.

6–36. Write a VHDL description for the circuit specified in Problem 6-19.

6–37. *Write a VHDL description for a *JK* negative-edge-triggered flip-flop with clock CLK. Compile and simulate your description. Apply a sequence that causes all eight combinations of inputs *J* and *K* and stored value *Q* to be applied in some clock cycle.

6–38. Write a Verilog description for the multiplexer in Figure 4-14 by using a process containing a case statement rather than the continuous assignment statements as shown in Section 4-8.

6–39. *Repeat Problem 6-38 by using a Verilog process containing if-else statements.

6–40. +Write a Verilog description for the sequential circuit given by the state diagram in Figure 6-25(d). Include an asynchronous RESET signal to initialize the circuit to state `Init`. Compile your description, apply an input sequence to pass through every arc of the state diagram at least once, and verify the correctness of the state and output sequence by comparing them to the state diagram.

6–41. Write a Verilog description for the circuit specified in Problem 6-15.

6–42. Write a Verilog description for the circuit specified in Problem 6-19.

6–43. *Write a Verilog description for a *JK* negative-edge-triggered flip-flop with clock CLK. Compile and simulate your description. Apply a sequence that causes all eight combinations of inputs *J* and *K* and stored value *Q* to be applied in some clock cycle.

CHAPTER

7

REGISTERS AND REGISTER TRANSFERS

I n Chapters 4 and 5, we studied combinational functional blocks. In Chapter 6, we examined sequential circuits. In this chapter, we bring the two ideas together and present sequential functional blocks, generally referred to as registers and counters. In Chapter 6, the circuits that were analyzed or designed did not have any particular structure, and the number of flip-flops was quite small. In contrast, the circuits we consider here have more structure, with multiple stages or cells that are identical or close to identical. Also, because of this structure, it is easy to add more stages to produce circuits with many more flip-flops than the circuits in Chapter 6. Registers are particularly useful for storing information during the processing of data and counters assist in sequencing the processing.

In a digital system, a datapath and a control unit are frequently present at the top levels of the design hierarchy. A *datapath* consists of processing logic and a collection of registers that performs data processing. A *control unit* is made up of logic that determines the sequence of data-processing operations performed by the datapath. Register transfer notation describes elementary data-processing actions referred to as *microoperations*. Register transfers move information between registers, between registers and memory, and through processing logic. Dedicated transfer hardware using multiplexers and shared transfer hardware called *buses* implement these movements of data.

In the generic computer at the beginning of Chapter 1, registers are used extensively for temporary storage of data in areas aside from memory. Registers of this kind are often large, with at least 32 bits. Special registers called shift registers are used less frequently, appearing primarily in the input–output parts of the system. Counters are used in the various parts of the computer to control or keep track of the sequence of activities. Overall, sequential functional blocks are used widely in the generic computer. In particular, the CPU and FPU parts of the processor each contain large numbers of registers that are involved in register transfers and execution of microoperations. It is in the CPU and the FPU that data transfers, additions, subtractions, and other microoperations take place. Finally, the connections shown between various electronic parts of the computer are buses, which we discuss for the first time in this chapter.

7-1 REGISTERS AND LOAD ENABLE

A register includes a set of flip-flops. Since each flip-flop is capable of storing one bit of information, an n-bit register, composed of n flip-flops, is capable of storing n bits of binary information. By the broadest definition, a *register* consists of a set of flip-flops, together with gates that implement their state transitions. This broad definition includes the various sequential circuits considered in Chapter 6. More commonly, the term *register* is applied to a set of flip-flops, possibly with added combinational gates, that perform data-processing tasks. The flip-flops hold data, and the gates determine the new or transformed data to be transferred into the flip-flops.

A *counter* is a register that goes through a predetermined sequence of states upon the application of clock pulses. The gates in the counter are connected in a way that produces the prescribed sequence of binary states. Although counters are a special type of registers, it is common to differentiate them from registers.

Registers and counters are sequential functional blocks that are used extensively in the design of digital systems in general and in digital computers in particular. Registers are useful for storing and manipulating information; counters are employed in circuits that sequence and control operations in a digital system.

The simplest register is a register that consists of only flip-flops without external gates. Figure 7-1(a) shows such a register constructed from four D-type flip-flops. The common *Clock* input triggers all flip-flops on the rising edge of each pulse, and the binary information available at the four D inputs is transferred into the 4-bit register. The four Q outputs can be sampled to obtain the binary information stored in the register. The $\overline{Clear}$ input goes to the $\overline{R}$ inputs of all four flip-flops and is used to clear the register to all 0's prior to its clocked operation. This input is labeled $\overline{Clear}$ rather than *Clear*, since a 0 must be applied to reset the flip-flops asynchronously. Activation of the asynchronous $\overline{R}$ inputs to flip-flops during normal clocked operation can lead to circuit designs that are highly delay dependent and that can, therefore, easily malfunction. Thus, we maintain $\overline{Clear}$ at logic 1 during normal clocked operation, allowing it to be logic 0 only when a system reset is desired. We note that the ability to clear a register to all 0's is optional; whether a clear operation is provided depends upon the use of the register in the system.

The transfer of new information into a register is referred to as *loading* the register. If all the bits of the register are loaded simultaneously with a common clock pulse, we say that the loading is done in parallel. A positive clock transition applied to the *Clock* input of the register of Figure 7-1(a) loads all four D inputs into the flip-flops in parallel.

Figure 7-1(b) shows a symbol for the register in Figure 7-1(a). This symbol permits the use of the register in a design hierarchy. The symbol has all inputs to the logic circuit on its left and all outputs from the circuit on the right. The inputs include the clock input with the dynamic indicator to represent positive-edge triggering of the flip-flops. We note that the name *Clear* appears inside the symbol, with a bubble in the signal line on the outside of the symbol. This notation indicates that application of a logic 0 to the signal line activates the clear operation on

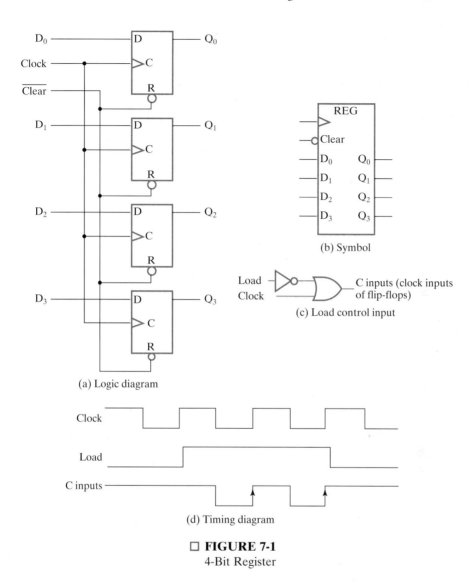

(a) Logic diagram

(b) Symbol

(c) Load control input

(d) Timing diagram

□ **FIGURE 7-1**
4-Bit Register

the flip-flops in the register. If the signal line were labeled outside the symbol, the label would be $\overline{Clear}$.

Register with Parallel Load

Most digital systems have a master clock generator that supplies a continuous train of clock pulses. The pulses are applied to all flip-flops and registers in the system. In effect, the master clock acts like a heart that supplies a constant beat to all parts of the system. For the design in Figure 7-1(a), the clock must be prevented from reaching the clock input to the circuit if the contents of the register are to be left unchanged. Thus, a separate control signal is used to control the clock cycles during which clock pulses are to have an effect on the register. The clock pulses are prevented from reaching the register when its content is not to be changed. This approach can be implemented with a load control input *Load* combined with the clock, as shown in Figure 7-1(c). The output of the OR gate is applied to the *C* inputs of the register flip-flops. The equation for the logic shown is

$$C\ inputs\ =\ \overline{Load} + Clock$$

When the *Load* signal is 1, *C inputs* = *Clock*, so the register is clocked normally, and new information is transferred into the register on the positive transitions of the clock. When the *Load* signal is 0, *C inputs* = 1. With this constant input applied, there are no positive transitions on *C inputs*, so the contents of the register remain unchanged. The effect of the *Load* signal on the signal *C inputs* is shown in Figure 7-1(d). Note that the clock pulses that appear on *C inputs* are pulses to 0, which end with the positive edge that triggers the flip-flops. These pulses and edges appear when *Load* is 1 and are replaced by a constant 1 when *Load* is 0. In order for this circuit to work correctly, *Load* must be constant at the correct value, either 0 or 1, throughout the interval when *Clock* is 0. One situation in which this occurs is if *Load* comes from a flip-flop that is triggered on a positive edge of *Clock*, a normal circumstance if all flip-flops in the system are positive-edge triggered. Since the clock is turned on and off at the register *C* inputs by the use of a logic gate, the technique is referred to as *clock gating*.

Inserting gates in the clock pulse path produces different propagation delays between *Clock* and the inputs of flip-flops with and without clock gating. If the clock signals arrive at different flip-flops or registers at different times, *clock skew* is said to exist. But to have a truly synchronous system, we must ensure that all clock pulses arrive simultaneously throughout the system so that all flip-flops trigger at the same time. For this reason, in routine designs, control of the operation of the register without using clock gating is advisable. Otherwise, delays must be controlled to drive the clock skew as close to zero as possible. This is applicable in aggressive low power or high speed designs.

A 4-bit register with a control input *Load* that is directed through gates into the *D* inputs of the flip-flops, instead of through the *C* inputs, is shown in Figure 7-2(c). This register is based on a bit cell shown in Figure 7-2(a) consisting of a 2-to-1 multiplexer and a *D* flip-flop. The signal *EN* selects between the data bit *D* entering the cell and the value *Q* at the output of the cell. For *EN* = 1, *D* is selected and the cell is loaded. For *EN* = 0, *Q* is selected and the output is loaded back into the flip-flop, preserving its current state. The feedback connection from output to input of the flip-flop is necessary because the *D* flip-flop, unlike other flip-flop types does not have a "no change" input condition: With each clock pulse, the *D* input determines the next state of the output. To leave the output unchanged, it is necessary to make the *D* input equal to the present value of the output. The logic in Figure 7-2(a) can be viewed as a new type of *D* flip-flop, a *D flip-flop with enable*, having the symbol shown in Figure 7-2(b).

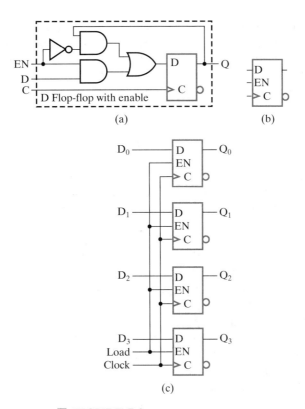

□ **FIGURE 7-2**
4-Bit Register with Parallel Load

The register is implemented by placing four D flip-flops with enables in parallel and connecting the *Load* input to the *EN* inputs. When *Load* is 1, the data on the four inputs is transferred into the register with the next positive clock edge. When *Load* is 0, the current value remains in the register at the next positive clock edge. Note that the clock pulses are applied continuously to the *C* inputs. *Load* determines whether the next pulse accepts new information or leaves the information in the register intact. The transfer of information from inputs to register is done simultaneously for all four bits during a single positive pulse transition. This method of transfer is traditionally preferred over clock gating, since it avoids clock skew and the potential for malfunctions of the circuit.

7-2 REGISTER TRANSFERS

A digital system is a sequential circuit made up of interconnected flip-flops and gates. In Chapter 6, we learned that sequential circuits can be specified by means of state tables. To specify a large digital system with state tables is very difficult, if not impossible, because the number of states is prohibitively large. To overcome this difficulty, digital systems are designed using a modular, hierarchical approach. The system is partitioned into subsystems or modules, each of which performs some functional task. The modules are constructed hierarchically from functional blocks such as registers, counters, decoders, multiplexers, buses, arithmetic elements, flip-flops, and primitive gates. The various subsystems communicate with data and control signals to form a digital system.

In most digital system designs, we partition the system into two types of modules: a *datapath*, which performs data-processing operations, and a *control unit*, which determines the sequence of those operations. Figure 7-3 shows the general relationship between a datapath and a control unit. *Control signals* are binary signals that activate the various data-processing operations. To activate a sequence of such operations, the control unit sends the proper sequence of control signals to the datapath. The control unit, in turn, receives status bits from the datapath. These status bits describe aspects of the state of the datapath. The status bits are used by the control unit in defining the specific sequence of the operations to be performed. Note that the datapath and control unit may also interact with other parts of a digital system, such as memory and input-output logic, through the paths labeled data inputs, data outputs, control inputs, and control outputs.

Datapaths are defined by their registers and the operations performed on binary data stored in the registers. Examples of register operations are load, clear, shift, and count. The registers are assumed to be basic components of the digital system. The movement of the data stored in registers and the processing performed on the data are referred to as *register transfer operations*. The register transfer operations of digital systems are specified by the following three basic components:

1. the set of registers in the system,

2. the operations that are performed on the data stored in the registers, and

3. the control that supervises the sequence of operations in the system.

A register has the capability to perform one or more *elementary operations* such as load, count, add, subtract, and shift. For example, a right-shift register is a register that can shift data to the right. A counter is a register that increments a number by one. A single flip-flop is a 1-bit register that can be set or cleared. In fact, by this definition, the flip-flops and closely associated gates of any sequential circuit can be called registers.

An elementary operation performed on data stored in registers is called a *microoperation*. Examples of microoperations are loading the contents of one

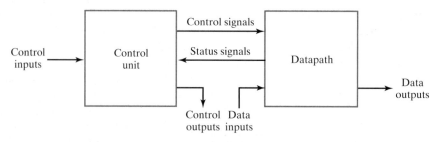

☐ **FIGURE 7-3**
Interaction between Datapath and Control Unit

register into another, adding the contents of two registers, and incrementing the contents of a register. A microoperation is usually, but not always, performed in parallel on a vector of bits during one clock cycle. The result of the microoperation may replace the previous binary data in the register. Alternatively, the result may be transferred to another register, leaving the previous data unchanged. The sequential functional blocks introduced in this chapter are registers that implement one or more microoperations.

The control unit provides signals that sequence the microoperations in a prescribed manner. The results of a current microoperation may determine both the sequence of control signals and the sequence of future microoperations to be executed. Note that the term "microoperation," as used here, does not refer to any particular way of producing the control signals: specifically, it does not imply that the control signals are generated by a control unit based on a technique called microprogramming.

This chapter introduces registers, their implementations and register transfers using a simple register transfer language (RTL) to represent registers and specify the operations on their contents. The register transfer language uses a set of expressions and statements that resemble statements used in HDLs and programming languages. This notation can concisely specify part or all of a complex digital system such as a computer. The specification then serves as a basis for a more detailed design of the system.

7-3 REGISTER TRANSFER OPERATIONS

We denote the registers in a digital system by uppercase letters (sometimes followed by numerals) that indicate the function of the register. For example, a register that holds an address for the memory unit is usually called an address register and can be designated by the name *AR*. Other designations for registers are *PC* for program counter, *IR* for instruction register, and *R2* for register 2. The individual flip-flops in an *n*-bit register are typically numbered in sequence from 0 to *n*–1, starting with 0 in the least significant (often the rightmost) position and increasing toward the most significant position. Since the 0 bit is on the right, this order can be referred to as *little-endian*, in the same manner as for bytes in Chapter 1. The reverse order, with bit 0 on the left, is referred to as *big-endian*. Figure 7-4 shows representations of registers in block diagram form. The most common way to represent a register is by a rectangular box with the name of the register inside, as in part (a) of the figure. The individual bits can be identified as in part (b). The numbering of bits represented by just the leftmost and rightmost values at the top of a register box is illustrated by a 16-bit register *R2* in part (c). A 16-bit program counter, *PC*, is partitioned into two sections in part (d) of the figure. In this case, bits 0 through 7 are assigned the symbol *L* (for low-order byte), and bits 8 through 15 are assigned the symbol *H* (for high-order byte). The label *PC(L)*, which may also be written *PC(7:0)*, refers to the low-order byte of the register, and *PC(H)* or *PC(15:8)* refers to the high-order byte.

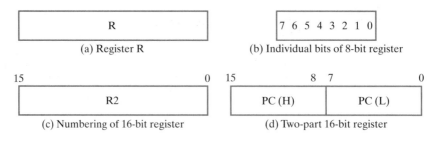

☐ **FIGURE 7-4**
Block Diagrams of Registers

Data transfer from one register to another is designated in symbolic form by means of the replacement operator ($\leftarrow$). Thus, the statement

$$R2 \leftarrow R1$$

denotes a transfer of the contents of register $R1$ into register $R2$. In other words, the statement designates the copying of the contents of $R1$ into $R2$. The register $R1$ is referred to as the *source* of the transfer and the register $R2$ as the *destination*. By definition, the contents of the source register do not change as a result of the transfer; only the contents of the destination register, $R2$, change.

A statement that specifies a register transfer implies that datapath circuits are available from the outputs of the source register to the inputs of the destination register and that the destination register has a parallel load capability. Normally, we want a given transfer to occur not for every clock pulse, but only for specific values of the control signals. This can be specified by a *conditional statement*, symbolized by the *if-then* form

$$\text{if}(K_1 = 1)\text{then}(R2 \leftarrow R1)$$

where K_1 is a control signal generated in the control unit. In fact, K_1 can be any Boolean function that evaluates to 0 or 1. A more concise way of writing the if-then form is

$$K_1: R2 \leftarrow R1$$

This control condition, terminated with a colon, symbolizes the requirement that the transfer operation be executed by the hardware only if $K_1 = 1$.

Every statement written in register transfer notation presupposes a hardware construct for implementing the transfer. Figure 7-5 shows a block diagram that depicts the transfer from $R1$ to $R2$. The n outputs of register $R1$ are connected to the n inputs of register $R2$. The letter n is used to indicate the number of bits in the register transfer path from $R1$ to $R2$. When the width of the path is known, n is replaced by an actual number. Register $R2$ has a load control input that is activated by the control signal K_1. It is assumed that the signal is synchronized with the same clock as the one applied to the register. The flip-flops are assumed to be positive-edge triggered by this clock. As shown in the timing diagram, K_1 is set to 1 on

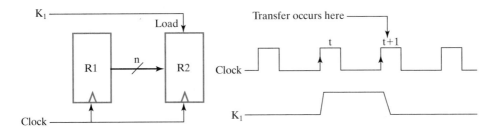

□ **FIGURE 7-5**
Transfer from $R1$ to $R2$ when $K_1 = 1$

the rising edge of a clock pulse at time t. The next positive transition of the clock at time $t + 1$ finds $K_1 = 1$, and the inputs of $R2$ are loaded into the register in parallel. In this case, K_1 returns to 0 on the positive clock transition at time $t + 1$, so that only a single transfer from $R1$ to $R2$ occurs.

Note that the clock is not included as a variable in the register transfer statements. It is assumed that all transfers occur in response to a clock transition. Even though the control condition K_1 becomes active at time t, the actual transfer does not occur until the register is triggered by the next positive transition of the clock, at time $t + 1$.

The basic symbols we use in register transfer notation are listed in Table 7-1. Registers are denoted by an uppercase letter, possibly followed by one or more uppercase letters and numerals. Parentheses are used to denote a part of a register by specifying the range of bits in the register or by giving a symbolic name to a portion of the register. The left-pointing arrow denotes a transfer of data and the direction of transfer. A comma is used to separate two or more register transfers that are executed at the same time. For example, the statement

$$K_3: R2 \leftarrow R1, R1 \leftarrow R2$$

denotes an operation that exchanges the contents of two registers simultaneously for a positive clock edge at which $K_3 = 1$. Such an exchange is possible with registers made of flip-flops, but presents a difficult timing problem with registers made

□ **TABLE 7-1**
Basic Symbols for Register Transfers

Symbol	Description	Examples
Letters (and numerals)	Denotes a register	$AR, R2, DR, IR$
Parentheses	Denotes a part of a register	$R2(1), R2(7{:}0), AR(L)$
Arrow	Denotes transfer of data	$R1 \leftarrow R2$
Comma	Separates simultaneous transfers	$R1 \leftarrow R2, R2 \leftarrow R1$
Square brackets	Specifies an address for memory	$DR \leftarrow M[AR]$

☐ **TABLE 7-2**
Textbook RTL, VHDL, and Verilog Symbols for Register Transfers

Operation	Text RTL	VHDL	Verilog
Combinational Assignment	=	<= (concurrent)	assign = (nonblocking)
Register Transfer	←	<= (concurrent)	<= (nonblocking)
Addition	+	+	+
Subtraction	–	–	–
Bitwise AND	∧	and	&
Bitwise OR	∨	or	|
Bitwise XOR	⊕	xor	∧
Bitwise NOT	¯	not	~
Shift left (logical)	sl	sll	<<
Shift right (logical)	sr	srl	>>
Vectors/Registers	A(3:0)	A(3 downto 0)	A[3:0]
Concatenation	||	&	{ , }

of latches. Square brackets are used in conjunction with a memory transfer. The letter *M* designates a memory word, and the register enclosed inside the square brackets provides the address of the word in memory. This is explained in more detail in Chapter 10.

7-4 A NOTE FOR VHDL AND VERILOG USERS ONLY

Although there are some similarities, the register transfer language used here differs from both VHDL and Verilog. In particular, there is different notation used in each of the three languages. Table 7-2 compares the notation for many identical or similar register transfer operations in the three languages. As you study this chapter and others to follow, this table will assist you in relating descriptions in the text RTL to the corresponding descriptions in VHDL or Verilog.

7-5 MICROOPERATIONS

A microoperation is an elementary operation performed on data stored in registers or in memory. The microoperations most often encountered in digital systems are of four types:

1. *Transfer* microoperations, which transfer binary data from one register to another.
2. *Arithmetic* microoperations, which perform arithmetic on data in registers.
3. *Logic* microoperations, which perform bit manipulation on data in registers.
4. *Shift* microoperations, which shift data in registers.

A given microoperation may be of more than one type. For example, a 1's complement operation is both an arithmetic microoperation and a logic microoperation.

Transfer microoperations were introduced in the previous section. This type of microoperation does not change the binary data bits as they move from the source register to the destination register. The other three types of microoperations can produce new binary data and, hence, new information. In digital systems, basic sets of operations are used to form sequences that implement more complicated operations. In this section, we define a basic set of microoperations, symbolic notation for these microoperations, and descriptions of the digital hardware that implements them.

Arithmetic Microoperations

We define the basic arithmetic microoperations as add, subtract, increment, decrement, and complement. The statement

$$R0 \leftarrow R1 + R2$$

specifies an add operation. It states that the contents of register $R2$ are to be added to the contents of register $R1$ and the sum transferred to register $R0$. To implement this statement with hardware, we need three registers and a combinational component that performs the addition, such as a parallel adder. The other basic arithmetic operations are listed in Table 7-3. Subtraction is most often implemented through complementation and addition. Instead of using the minus operator, we can specify 2's complement subtraction by the statement

$$R0 \leftarrow R1 + \overline{R2} + 1$$

where $\overline{R2}$ specifies the 1's complement of $R2$. Adding 1 to $\overline{R2}$ gives the 2's complement of $R2$. Finally, adding the 2's complement of $R2$ to the contents of $R1$ is equivalent to $R1 - R2$.

The increment and decrement microoperations are symbolized by a plus-one and minus-one operation, respectively. These operations are implemented by using

□ **TABLE 7-3**
  Arithmetic Microoperations

Symbolic designation	Description
$R0 \leftarrow R1 + R2$	Contents of $R1$ plus $R2$ transferred to $R0$
$R2 \leftarrow \overline{R2}$	Complement of the contents of $R2$ (1's complement)
$R2 \leftarrow \overline{R2} + 1$	2's complement of the contents of $R2$
$R0 \leftarrow R1 + \overline{R2} + 1$	$R1$ plus 2's complement of $R2$ transferred to $R0$ (subtraction)
$R1 \leftarrow R1 + 1$	Increment the contents of $R1$ (count up)
$R1 \leftarrow R1 - 1$	Decrement the contents of $R1$ (count down)

a special combinational circuit, an adder–subtractor, or a binary up–down counter with parallel load.

Multiplication and division are not listed in Table 7-3. Multiplication can be represented by the symbol $*$ and division by $/$. These two operations are not included in the basic set of arithmetic microoperations because they are assumed to be implemented by sequences of basic microoperations. In contrast, multiplication can be considered as a microoperation if implemented by a combinational circuit as illustrated in Section 5-4. In such a case, the result is transferred into a destination register at the clock edge after all signals have propagated through the entire combinational circuit.

There is a direct relationship between the statements written in register transfer notation and the registers and digital functions required for their implementation. To illustrate, consider the following two statements:

$$\overline{X}K_1 : R1 \leftarrow R1 + R2$$

$$XK_1 : R1 \leftarrow R1 + \overline{R2} + 1$$

Control variable K_1 activates an operation to add or subtract. If, at the same time, control variable X is equal to 0, then $\overline{X}K_1 = 1$, and the contents of $R2$ are added to the contents of $R1$. If X is equal to 1, then $XK_1 = 1$, and the contents of $R2$ are subtracted from the contents of $R1$. Note that the two control conditions are Boolean functions and reduce to 0 when $K_1 = 0$, a condition that inhibits the execution of both operations simultaneously.

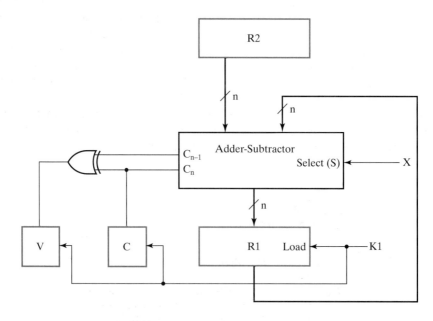

☐ **FIGURE 7-6**
Implementation of Add and Subtract Microoperations

A block diagram showing the implementation of the preceding two statements is given in Figure 7-6. An n-bit adder–subtractor, similar to the one shown in Figure 5-8, receives its input data from registers $R1$ and $R2$. The sum or difference is applied to the inputs of $R1$. The Select input S of the adder–subtractor selects the operation in the circuit. When $S = 0$, the two inputs are added, and when $S = 1$, $R2$ is subtracted from $R1$. Applying the control variable X to the S input activates the required operation. The output of the adder–subtractor is loaded into $R1$ on any positive clock edge at which $\overline{X} K_1 = 1$ or $XK_1 = 1$. We can simplify this to just K_1, since

$$\overline{X}K_1 + XK_1 = (\overline{X} + X)K_1 = K_1$$

Thus, the control variable X selects the operation, and the control variable K_1 loads the result into $R1$.

Based on the discussion of overflow in Section 5-3, the overflow output is transferred to flip-flop V, and the output carry from the most significant bit of the adder–subtractor is transferred to flip-flop C, as shown in Figure 7-6. These transfers occur when $K_1 = 1$ and are not represented in the register transfer statements; if desired, we could show them as additional simultaneous transfers.

Logic Microoperations

Logic microoperations are useful in manipulating the bits stored in a register. These operations consider each bit in the register separately and treat it as a binary variable. The symbols for the four basic logic operations are shown in Table 7-4. The NOT microoperation, represented by a bar over the source register name, complements all bits and thus is the same as the 1's complement. The symbol $\wedge$ is used to denote the AND microoperation and the symbol $\vee$ to denote the OR microoperation. By using these special symbols, it is possible to distinguish between the add microoperation represented by a $+$ and the OR microoperation. Although the $+$ symbol has two meanings, one can distinguish between them by noting where the symbol occurs. If the $+$ occurs in a microoperation, it denotes addition. If the $+$ occurs in a control or Boolean function, it denotes OR. The OR microoperation will always use the $\vee$ symbol. For example, in the statement

$$(K_1 + K_2){:}R1 \leftarrow R2 + R3, R4 \leftarrow R5 \vee R6$$

□ **TABLE 7-4**
Logic Microoperations

Symbolic designation	Description
$R0 \leftarrow \overline{R1}$	Logical bitwise NOT (1's complement)
$R0 \leftarrow R1 \wedge R2$	Logical bitwise AND (clears bits)
$R0 \leftarrow R1 \vee R2$	Logical bitwise OR (sets bits)
$R0 \leftarrow R1 \oplus R2$	Logical bitwise XOR (complements bits)

the + between K_1 and K_2 is an OR operation between two variables in a control condition. The + between $R2$ and $R3$ specifies an add microoperation. The OR microoperation is designated by the symbol $\vee$ between registers $R5$ and $R6$. The logic microoperations can be easily implemented with a group of gates, one for each bit position. The NOT of a register of n bits is obtained with n NOT gates in parallel. The AND microoperation is obtained using a group of n AND gates, each of which receives a pair of corresponding inputs from the two source registers. The outputs of the AND gates are applied to the corresponding inputs of the destination register. The OR and exclusive-OR microoperations require a similar arrangement of gates.

The logic microoperations can change bit values, clear a group of bits, or insert new bit values into a register. The following examples show how the bits stored in the 16-bit register $R1$ can be selectively changed by using a logic microoperation and a logic operand stored in the 16-bit register $R2$.

The AND microoperation can be used for clearing one or more bits in a register to 0. The Boolean equations $X \cdot 0 = 0$ and $X \cdot 1 = X$ dictate that, when ANDed with 0, a binary variable X produces a 0, but when ANDed with 1, the variable remains unchanged. A given bit or group of bits in a register can be cleared to 0 if ANDed with 0. Consider the following example:

10101101 10101011	$R1$	(data)
00000000 11111111	$R2$	(mask)
00000000 10101011	$R1 \leftarrow R1 \wedge R2$	

The 16-bit logic operand in $R2$ has 0's in the high-order byte and 1's in the low-order byte. By ANDing the contents of $R2$ with the contents of $R1$, it is possible to clear the high-order byte of $R1$ and leave the bits in the low-order byte unchanged. Thus, the AND operation can be used to selectively clear bits of a register. This operation is sometimes called *masking out* the bits, because it masks or deletes all 1's in the *data* in $R1$, based on bit positions that are 0 in the *mask* provided in $R2$.

The OR microoperation is used to set one or more bits in a register. The Boolean equations $X + 1 = 1$ and $X + 0 = X$ dictates that, when ORed with 1, the binary variable X produces a 1, but when ORed with 0, the variable remains unchanged. A given bit or group of bits in a register can be set to 1 if ORed with 1. Consider the following example:

10101101 10101011	$R1$	(data)
11111111 00000000	$R2$	(mask)
11111111 10101011	$R1 \leftarrow R1 \vee R2$	

The high-order byte of $R1$ is set to all 1's by ORing it with all 1's in the $R2$ operand. The low-order byte remains unchanged because it is ORed with 0's.

The XOR (exclusive-OR) microoperation can be used to complement one or more bits in a register. The Boolean equations $X \oplus 1 = \overline{X}$ and $X \oplus 0 = X$ dictate that, when a binary variable X is XORed with 1, it is complemented, but when XORed with 0, the variable remains unchanged. By XORing a bit or group of bits in register $R1$ with 1's in selected positions in $R2$, it is possible to complement the bits in the selected positions in $R1$. Consider the following example:

10101101 10101011	$R1$	(data)
11111111 00000000	$R2$	(mask)
01010010 10101011	$R1 \leftarrow R1 \oplus R2$	

The high-order byte in $R1$ is complemented after the XOR operation with $R2$, and the low-order byte is unchanged.

Shift Microoperations

Shift microoperations are used for lateral movement of data. The contents of a source register can be shifted either right or left. A *left shift* is toward the most significant bit, and a *right shift* is toward the least significant bit. Shift microoperations are used in the serial transfer of data. They are also used for manipulating the contents of registers in arithmetic, logical, and control operations. The destination register for a shift microoperation may be the same as or different from the source register. We use strings of letters to represent the shift microoperations defined in Table 7-5. For example,

$$R0 \leftarrow \text{sr } R0, R1 \leftarrow \text{sl } R2$$

are two microoperations that respectively specify a 1-bit shift to the right of the contents of register $R0$ and a transfer of the contents of $R2$ shifted one bit to the left into register $R1$. The contents of $R2$ are not changed by these shifts.

For a left-shift microoperation, we call the rightmost bit of the destination register the *incoming bit*. For a right-shift microoperation, we define the leftmost bit of the destination register as the incoming bit. The incoming bit may have different values, depending upon the type of shift microoperation. Here we assume that, for sr and sl, the incoming bit is 0, as shown in the examples in Table 7-5. The *outgoing bit* is the leftmost bit of the source register for the left-shift operation and the rightmost bit of the source register for the right-shift operation. For the left and right shifts shown, the outgoing bit value is simply discarded. In chapter 11, we will

□ **TABLE 7-5**
Examples of Shifts

Type	Symbolic designation	Eight-bit examples	
		Source $R2$	After shift: Destination $R1$
shift left	$R1 \leftarrow \text{sl } R2$	10011110	00111100
shift right	$R1 \leftarrow \text{sr } R2$	11100101	01110010

explore other types of shifts differing in the way incoming and outgoing bits are treated.

7-6 MICROOPERATIONS ON A SINGLE REGISTER

This section covers the implementation of one or more microoperations with a single register as the destination of all primary results. The single register may also serve as a source of an operand for binary and unary operations. Due to the close ties between a single set of storage elements and the microoperations, the combinational logic implementing the microoperations is assumed to be a part of the register and is called *dedicated logic* of the register. This is in contrast to logic which is shared by multiple destination registers. In this case, the combinational logic implementing the microoperations is called *shared logic* for the set of destination registers.

The combinational logic implementing the microoperations described in the previous section can use one or more functional blocks from chapters 4 and 5 or can be designed specifically for the register. Initially, functional blocks will be used in combination with D flip-flops or D-flip-flops with enable. A simple technique using multiplexers for selection is introduced to allow multiple microoperations on a single register. Next, single and multiple function registers that perform shifting and counting are designed.

Multiplexer-Based Transfers

There are occasions when a register receives data from two or more different sources at different times. Consider the following conditional statement having an *if-then-else* form:

$$\text{if}(K_1 = 1)\text{then}(R0 \leftarrow R1) \text{ else if } (K_2 = 1) \text{ then } (R0 \leftarrow R2)$$

The value in register $R1$ is transferred to register $R0$ when control signal K_1 equals 1. When $K_1 = 0$, the value in register $R2$ is transferred to $R0$ when K_2 equals 1. Otherwise, the contents of $R0$ remains unchanged. The conditional statement may be broken into two parts using the following control conditions:

$$K_1: R0 \leftarrow R1, \overline{K}_1 K_2: R0 \leftarrow R2$$

This specifies hardware connections from two registers, $R1$ and $R2$, to one common destination register $R0$. In addition, making a selection between two source registers must be based on values of the control variables K_1 and K_2.

The block diagram for a circuit with 4-bit registers that implements the conditional register transfer statements using a multiplexer is shown in Figure 7-7(a). The quad 2-to-1 multiplexer selects between the two source registers. For $K_1 = 1$, $R1$ is loaded into $R0$, irrespective of the value of K_2. For $K_1 = 0$ and $K_2 = 1$, $R2$ is loaded into $R0$. When both K_1 and K_2 are equal to 0, the multiplexer selects $R2$ as the input to $R0$, but, because the control function, $K_2 + K_1$, connected to the LOAD input of $R0$ equals 0, the contents of $R0$ remain unchanged.

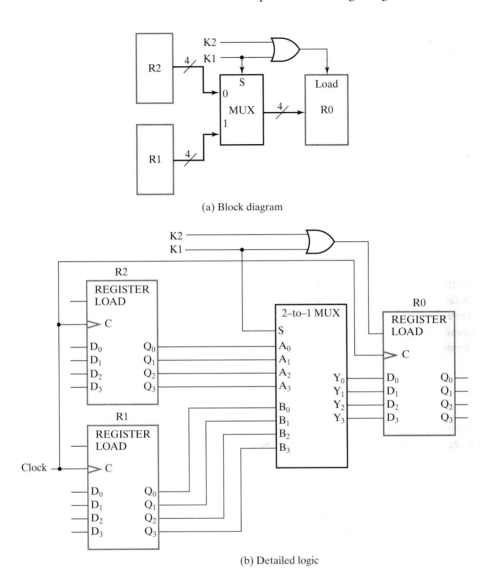

(a) Block diagram

(b) Detailed logic

□ **FIGURE 7-7**

Use of Multiplexers to Select between Two Registers

The detailed logic diagram for the hardware implementation is shown in Figure 7-7(b). The diagram uses functional block symbols based upon detailed logic for the registers in Figure 7-2 and for a quad 2-to-1 multiplexer from Chapter 4. Note that since this diagram represents just a part of a system, there are inputs and outputs that are not yet connected. Also, the clock is not shown in the block diagram, but is shown in the detailed diagram. It is important to relate the information given in a block diagram such as Figure 7-7(a) with the detailed wiring connections in the corresponding logic diagram in Figure 7-7(b). In order

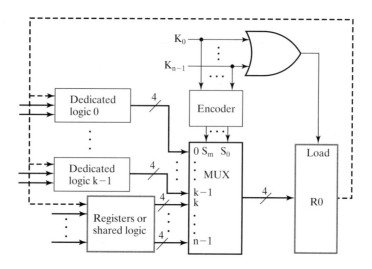

☐ **FIGURE 7-8**

Generalization of Multiplexer Selection for n Sources

to save space, we often omit the detailed logic diagrams in designs. However, it is possible to obtain a logic diagram with detailed wiring from the corresponding block diagram and a library of functional blocks. In fact, such a procedure is performed by computer programs used for automated logic synthesis.

The preceding example can be generalized by allowing the multiplexer to have n sources and these sources to be register outputs or combinational logic implementing microoperations. This generalization results in the block diagram shown in Figure 7-8. The diagram assumes that each source is either the outputs of a register or of combinational logic implementing one or more microinstructions. In those cases in which the microoperations are dedicated to the register, the corresponding dedicated logic is included as a part of the register. In Figure 7-8, the first k sources are dedicated logic and the last $n - k$ sources are either registers or shared logic. The control signals that select a given source are either a single control variable or the OR of all control signals corresponding to the microoperations associated with the source. To force R0 to load for a microoperation, these control signals are ORed together to form the *Load* signal. Since it is assumed that only one of the control signals is 1 at any time, these signals must be encoded to provide the selection codes for the multiplexer. Two modifications to the given structure are possible. The control signals could be applied directly to a $2 \times n$ AND-OR circuit (i.e., a multiplexer with the decoder deleted). Alternatively, the control signals could already be encoded, omitting the use of the all-zero code, so that the OR gate still forms the Load signal correctly.

Shift Registers

A register capable of shifting its stored bits laterally in one or both directions is called a *shift register*. The logical configuration of a shift register consists of a

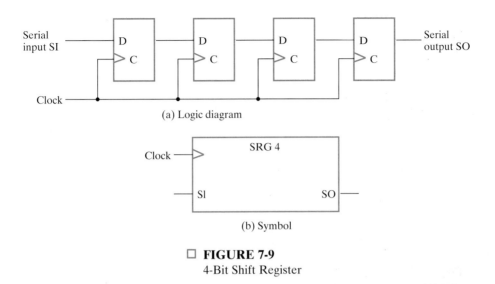

(a) Logic diagram

(b) Symbol

□ **FIGURE 7-9**
4-Bit Shift Register

chain of flip-flops, with the output of one flip-flop connected to the input of the next flip-flop. All flip-flops have a common clock pulse input that activates the shift.

The simplest possible shift register uses only flip-flops, as shown in Figure 7-9(a). The output of a given flip-flop is connected to the D input of the flip-flop at its right. The clock is common to all flip-flops. The *serial input SI* is the input to the leftmost flip-flop. The *serial output SO* is taken from the output of the rightmost flip-flop. A symbol for the shift register is given in Figure 7-9(b).

Sometimes it is necessary to control the register so that it shifts only on select positive clock edges. For the shift register in Figure 7-9, the shift can be controlled by connecting the clock through the logic shown in Figure 7-1(c), with *Shift* replacing *Load*. Again, due to clock skew, this is usually not the most desirable approach. Thus, we learn later that the shift operation can be controlled through the D inputs of the flip-flops rather than through the clock inputs C.

SHIFT REGISTER WITH PARALLEL LOAD   If all flip-flop outputs of a shift register are accessible, then information entered serially by shifting can be taken out in parallel from the flip-flop outputs. If a parallel load capability is also added to a shift register, then data entered in parallel can be shifted out serially. Thus, a shift register with accessible flip-flop outputs and parallel load can be used for converting incoming parallel data to outgoing serial data and vice versa.

The logic diagram for a 4-bit shift register with parallel load and the symbol for this register are shown in Figure 7-10. There are two control inputs, one for the shift and the other for the load. Each stage of the register consists of a D flip-flop, an OR gate, and three AND gates. The first AND gate enables the shift operation. The second AND gate enables the input data. The third AND gate restores the contents of the register when no operation is required.

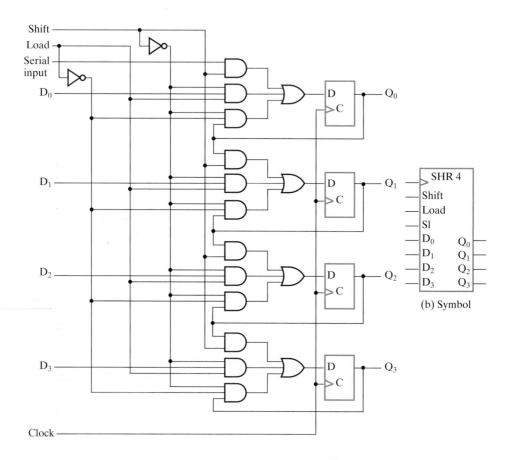

☐ **FIGURE 7-10**
Shift Register with Parallel Load

The operation of this register is specified in Table 7-6 and is also given by the register transfers:

$$\textit{Shift}: Q \leftarrow slQ$$

$$\overline{\textit{Shift}} \cdot \textit{Load}: Q \leftarrow D$$

The "No Change" operation is implicit if neither of the conditions for transfers is satisfied. When both the shift and load control inputs are 0, the third AND gate in each stage is enabled, and the output of each flip-flop is applied to its own D input. A positive transition of the clock restores the contents to the register, and the output is unchanged. When the shift input is 0 and the load input is 1, the second AND gate in each stage is enabled, and the input D_i is applied to the D input of the corresponding flip-flop. The next positive clock transition transfers the parallel input data into the register. When the shift input is equal to 1, the first AND gate in each

□ **TABLE 7-6**
Function Table for the Register of Figure 7-10

Shift	Load	Operation
0	0	No change
0	1	Load parallel data
1	$\times$	Shift down from Q_0 to Q_3

stage is enabled and the other two are disabled. Since the Load input is disabled by the $\overline{\text{Shift}}$ input on the second AND gate, we mark it with a don't-care condition in the Shift row of the table. When a positive edge occurs on the clock, the shift operation causes the data from the serial input SI to be transferred to flip-flop Q_0, the output of Q_0 to be transferred to flip-flop Q_1, and so on down the line. Note that because of the way the circuit is drawn, the shift occurs in the downward direction. If we rotate the page a quarter turn counterclockwise, the register shifts from left to right.

Shift registers are often used to interface digital systems that are distant from each other. For example, suppose it is necessary to transmit an n-bit quantity between two points. If the distance is far, it will be expensive to use n lines to transmit the n bits in parallel. It may be more economical to use a single line and transmit the information serially, one bit at a time. The transmitter loads the n-bit data in parallel into a shift register and then transmits the data serially along the common line. The receiver accepts the data serially into a shift register. When all n bits are accumulated, they can be taken in parallel from the outputs of the register. Thus, the transmitter performs a parallel-to-serial conversion of data, and the receiver does a serial-to-parallel conversion.

BIDIRECTIONAL SHIFT REGISTER   A register capable of shifting in only one direction is called a *unidirectional shift register*. A register that can shift in both directions is called a *bidirectional shift register*. It is possible to modify the circuit of Figure 7-10, by adding a fourth AND gate in each stage, for shifting the data in the upward direction. An investigation of the resultant circuit will reveal that the four AND gates, together with the OR gate in each stage, constitute a multiplexer with the selection inputs controlling the operation of the register.

One stage of a bidirectional shift register with parallel load is shown in Figure 7-11(a). Each stage consists of a D flip-flop and a 4-to-1-line multiplexer. The two selection inputs S_1 and S_0 select one of the multiplexer inputs to apply to the D flip-flop. The selection lines control the mode of operation of the register according to the function table of Table 7-7 and the register transfers:

$$\overline{S}_1 \cdot S_0 : Q \leftarrow \text{sl}Q$$

$$S_1 \cdot \overline{S}_0 : Q \leftarrow \text{sr}Q$$

$$S_1 \cdot S_0 : Q \leftarrow D$$

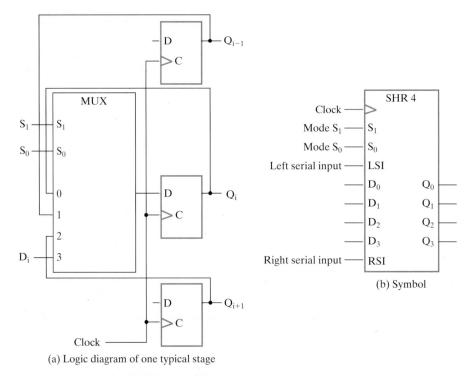

(a) Logic diagram of one typical stage

☐ **FIGURE 7-11**
Bidirectional Shift Register with Parallel Load

☐ **TABLE 7-7**
Function Table for the Register of Figure 7-7

| Mode control | | Register |
S_1	S_0	Operation
0	0	No change
0	1	Shift down
1	0	Shift up
1	1	Parallel load

The "No Change" operation is implicit if none of the conditions for transfers is satisfied. When the mode control $S_1 S_0 = 00$, input 0 of the multiplexer is selected. This forms a path from the output of each flip-flop into its own input. The next clock transition transfers the current stored value back into each flip-flop, and no change of state occurs. When $S_1 S_0 = 01$, the terminal marked 1 on the multiplexer has a path to the D input of each flip-flop. These paths cause a shift-down operation. The serial input is transferred into the first stage, and the content of each stage Q_{i-1}, is transferred into stage Q_i. When $S_1S_0 = 10$, a shift-up operation results in a second serial input that enters the last stage. In addition, the value in each stage Q_{i+1} is transferred into stage Q_i. Finally, when $S_1S_0 = 11$, the binary information on each parallel input line is transferred into the corresponding flip-flop, resulting in a parallel load.

Figure 7-11(b) shows a symbol for the bidirectional shift register from Figure 7-11(a). Note that both a left serial input (*LSI*) and a right serial input (*RSI*) are provided. If serial outputs are desired, Q_3 is used for left shift and Q_0, for right shift.

Ripple Counter

A register that goes through a prescribed sequence of distinct states upon the application of a sequence of input pulses is called a *counter*. The input pulses may be clock pulses or may originate from some other source, and they may occur at regular or irregular intervals of time. In our discussion of counters, we assume clock pulses, but other signals can be substituted for the clock. The sequence of states may follow the binary number sequence or any other prescribed sequence of states. A counter that follows the binary number sequence is called a *binary counter*. An *n*-bit binary counter consists of *n* flip-flops and can count in binary from 0 through $2^n - 1$.

Counters are available in two categories: ripple counters and synchronous counters. In a ripple counter, the flip-flop output transitions serve as the sources for triggering the changes in other flip-flops. In other words, the C inputs of some of the flip-flops are triggered not by the common clock pulse, but rather by the transitions that occur on other flip-flop outputs. In a synchronous counter, the C inputs of all flip-flops receive the common clock pulse, and the change of state is determined from the present state of the counter. Synchronous counters are discussed in the next two subsections. Here we present the binary ripple counter and explain its operation.

The logic diagram of a 4-bit binary ripple counter is shown in Figure 7-12. The counter is constructed from D flip-flops connected such that the application of a positive edge to the C input of each flip-flop causes the flip-flop to complement its state. The complemented output of each flip-flop is connected to the C input of the next most significant flip-flop. The flip-flop holding the least significant bit receives the incoming clock pulses. Positive-edge triggering makes each flip-flop complement its value when the signal on its C input goes through a positive transition. The positive transition occurs when the complemented output of the previous flip-flop, to which C is connected, goes from 0 to 1. A 1-level signal on *Reset* driving the R inputs clears the register to all zeros asynchronously.

To understand the operation of a binary ripple counter, let us examine the upward counting sequence given in the left half of Table 7-8. The count starts at binary 0 and increments by one with each count pulse. After the count of 15, the counter goes back to 0 to repeat the count. The least significant bit (Q_0) is complemented by each count pulse. Every time that Q_0 goes from 1 to 0, $\overline{Q_0}$ goes from 0 to 1, complementing Q_1. Every time that Q_1 goes from 1 to 0, it complements Q_2. Every time that Q_2 goes

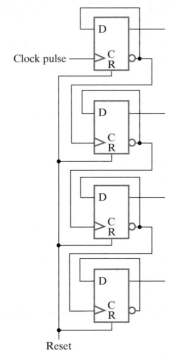

Clock pulse

Reset

☐ **FIGURE 7-12**
4-Bit Ripple Counter

☐ **TABLE 7-8**
Counting Sequence of Binary Counter

Upward Counting Sequence				Downward Counting Sequence			
Q_3	Q_2	Q_1	Q_0	Q_3	Q_2	Q_1	Q_0
0	0	0	0	1	1	1	1
0	0	0	1	1	1	1	0
0	0	1	0	1	1	0	1
0	0	1	1	1	1	0	0
0	1	0	0	1	0	1	1
0	1	0	1	1	0	1	0
0	1	1	0	1	0	0	1
0	1	1	1	1	0	0	0
1	0	0	0	0	1	1	1
1	0	0	1	0	1	1	0
1	0	1	0	0	1	0	1
1	0	1	1	0	1	0	0
1	1	0	0	0	0	1	1
1	1	0	1	0	0	1	0
1	1	1	0	0	0	0	1
1	1	1	1	0	0	0	0

from 1 to 0, it complements Q_3, and so on for any higher order bits in the ripple counter. For example, consider the transition from count 0011 to 0100. Q_0 is complemented with the count pulse positive edge. Since Q_0 goes from 1 to 0, it triggers Q_1 and complements it. As a result, Q_1, goes from 1 to 0, which complements Q_2, changing it from 0 to 1. Q_2 does not trigger Q_3, because $\overline{Q}_2$ produces a negative transition, and the flip-flops respond only to positive transitions. Thus, the count from 0011 to 0100 is achieved by changing the bits one at a time. The counter goes from 0011 to 0010 (Q_0 from 1 to 0), then to 0000 (Q_1 from 1 to 0), and finally to 0100 (Q_2 from 0 to 1). The flip-flops change one at a time in quick succession as the signal propagates through the counter in a ripple fashion from one stage to the next.

A ripple counter that counts downward gives the sequence in the right half of Table 7-8. Downward counting can be accomplished by connecting the true output of each flip-flop to the C input of the next flip-flop.

The advantage of ripple counters is their simple hardware. Unfortunately, they are asynchronous circuits and, with added logic, can become circuits with delay dependence and unreliable operation. This is particularly true for logic that provides feedback paths from counter outputs back to counter inputs. Also, due to the length of time required for the ripple to finish, large ripple counters can be slow circuits. As a consequence, synchronous binary counters are favored in all but low-power designs where ripple counters have an advantage. (See Problem 7-11.)

Synchronous Binary Counters

Synchronous counters, in contrast to ripple counters, have the clock applied to the C inputs of all flip-flops. Thus, the common clock pulse triggers all flip-flops simultaneously rather than one at a time, as in a ripple counter. A synchronous binary counter that counts up by 1 can be constructed from the incrementer in Figure 5-12 and D flip-flops as shown in Figure 7-13(a). The carry output CO is added by not placing an X value on the C_4 output before the contraction of an adder to the incrementer in Figure 5-12. Output CO is used to extend the counter to more stages.

Note that the flip-flops trigger on the positive-edge transition of the clock. The polarity of the clock is not essential here, like it was for the ripple counter. The synchronous counter can be triggered with either the positive or the negative clock transition.

SERIAL AND PARALLEL COUNTERS We will use the synchronous counter in Figure 7-13 to demonstrate two alternative designs for binary counters. In Figure 7-13(a), a chain of 2-input AND gates is used to provide information to each stage about the state of the prior stages in the counter. This is analogous to the carry logic in the ripple carry adder. A counter that uses such logic is said to have *serial gating* and is referred to as a *serial counter*. The analogy to the ripple carry adder suggests that there might be counter logic analogous to the carry lookahead adder. Such logic can be derived by contracting a carry lookahead adder, with the result shown in Figure 7-13(b). This logic can simply replace that in the blue box in Figure 7-13(a) to produce a counter with *parallel gating*, called a *parallel counter*.

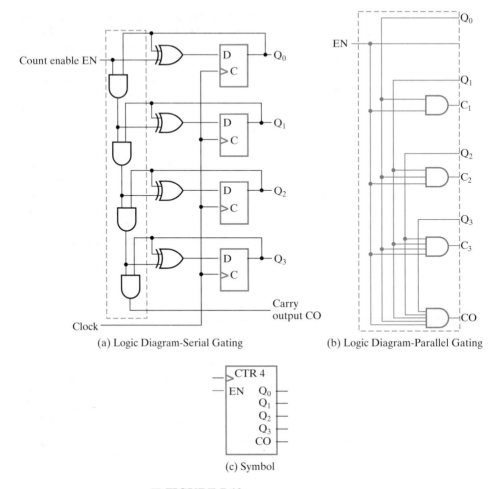

(a) Logic Diagram-Serial Gating

(b) Logic Diagram-Parallel Gating

(c) Symbol

☐ **FIGURE 7-13**
4-Bit Synchronous Binary Counter

The advantage of parallel gating logic is that, in going from state 1111 to state 0000, only one AND gate delay occurs instead of the four AND gate delays that occur for the serial counter. This reduction in delay allows the counter to operate much faster.

If we connect two 4-bit parallel counters together by connecting the CO output of one to the EN input of the other, the result is an 8-bit serial-parallel counter. This counter has two 4-bit parallel parts connected in series with each other. The idea can be extended to counters of any length. Again, employing the analogy to carry lookahead adders, additional levels of gating logic can be introduced to replace the serial connections between the 4-bit segments. The added reduction in delay that results is useful for constructing large, fast counters.

The symbol for the 4-bit counter using positive-edge triggering is shown in Figure 7-13(c).

UP-DOWN BINARY COUNTER A synchronous count-down binary counter goes through the binary states in reverse order from 1111 to 0000 and back to 1111 to repeat the count. The logic diagram of a synchronous count-down binary counter is similar to the circuit for the binary up-counter, except that a decrementer is used instead of an incrementer. The two operations can be combined to form a counter that can count both up and down, which is referred to as an up–down binary counter. Such a counter can be designed by contracting the adder–subtractor in Figure 5-8 into an incrementer–decrementer and adding the D flip-flops. The counter counts up for $S = 0$ and down for $S = 1$.

Alternatively, an up-down counter with ENABLE can be designed directly from counter behavior. It needs a mode input to select between the two operations. We designate this mode select input by S, with $S = 0$ for up-counting and $S = 1$ for down-counting. Let variable EN be a count enable input, with $EN = 1$ for normal up- or down-counting and $EN = 0$ for disabling both counts. A 4-bit up–down binary counter can be described by the following flip-flop input equations:

$$D_{A0} = Q_0 \oplus EN$$

$$D_{A1} = Q_1 \oplus ((Q_0 \cdot \overline{S} + \overline{Q}_0 \cdot S) \cdot EN)$$

$$D_{A2} = Q_2 \oplus ((Q_0 \cdot Q_1 \cdot \overline{S} + \overline{Q}_0 \cdot \overline{Q}_1 \cdot S) \cdot EN)$$

$$D_{A3} = Q_3 \oplus ((Q_0 \cdot Q_1 \cdot Q_2 \cdot \overline{S} + \overline{Q}_0 \cdot \overline{Q}_1 \cdot \overline{Q}_2 \cdot S) \cdot EN)$$

The logic diagram of the circuit can be easily obtained from the input equations, but is not included here. It should be noted that the equations, as written, provide parallel gating using distinct carry logic for up-counting and down-counting. It is also possible to use two distinct serial gating chains as well. In contrast, the counter derived using the incrementer–decrementer uses only a single carry chain. Overall, the logic cost is similar.

BINARY COUNTER WITH PARALLEL LOAD Counters employed in digital systems quite often require a parallel-load capability for transferring an initial binary number into the counter prior to the count operation. This function can be implemented by an incrementer with an ENABLE, n ENABLEs, and n 2-input OR gates as shown in Figure 7-14. The n ENABLEs are used to enable and disable the parallel load of input data, D, using the signal Load. Note that ENABLE on the incrementer is used to enable or disable counting using Count $\overline{\text{Load}}$. With the Load and Count inputs both at 0, the outputs do not change, even when pulses are applied to the C inputs. If the load input is maintained at logic 0, the Count input controls the operation of the counter, and the outputs change to the next binary count for each positive transition of the clock. The data applied to the D inputs is loaded into the flip-flops when Load equals 1, regardless of the value of Count because $\overline{\text{Load}}$ is ANDed with Count. Counters with parallel load are very useful in the design of digital computers. In subsequent chapters, we will refer to them as registers with load and increment operations.

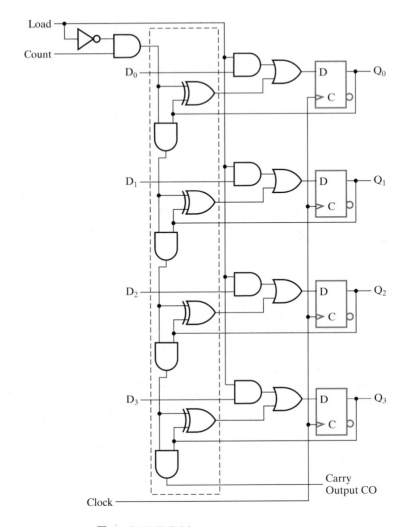

☐ **FIGURE 7-14**
4-Bit Binary Counter with Parallel Load

The binary counter with parallel load can be converted into a synchronous BCD counter (without load input) by connecting an external AND gate to it, as shown in Figure 7-15. The counter starts with an all-zero output, and the count input is always active. As long as the output of the AND gate is 0, each positive clock edge increments the counter by one. When the output reaches the count of 1001, both Q_0 and Q_3 become 1, making the output of the AND gate equal to 1. This condition makes *Load* active; so on the next clock transition, the counter does not count, but is loaded from its four inputs. Since all four inputs are connected to logic 0, 0000 is loaded into the counter following the count of 1001. Thus, the circuit counts from 0000 through 1001, followed by 0000, as required for a BCD counter.

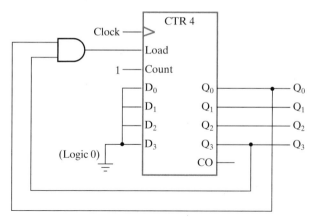

□ **FIGURE 7-15**
BCD Counter

Other Counters

Counters can be designed to generate any desired number of states in sequence. A *divide-by-N counter* (also known as a *modulo-N counter*) is a counter that goes through a repeated sequence of N states. The sequence may follow the binary count or may be any other arbitrary sequence. In either case, the design of the counter follows the procedure presented in Chapter 6 for the design of synchronous sequential circuits. To demonstrate this procedure, we will present the design of two counters: a BCD counter and a counter with an arbitrary sequence of states.

BCD COUNTER   As shown in the previous section, a BCD counter can be obtained from a binary counter with parallel load. It is also possible to design a BCD counter directly using individual flip-flops and gates. Assuming D-type flip-flops for the counter, we list the present states and corresponding next states in Table 7-9. An output Y is included in the table. This output is equal to 1 when the present state is 1001. In this way, Y can enable the count of the next decade while its own decade switches from 1001 to 0000.

The flip-flop input equations for D are obtained from the next-state values listed in the table and can be simplified by means of K-maps. The unused states for minterms 1010 through 1111 are used as don't-care conditions. The simplified input equations for the BCD counter are

$$D_1 = \overline{Q}_1$$
$$D_2 = Q_2 \oplus Q_1\overline{Q}_8$$
$$D_4 = Q_4 \oplus Q_1Q_2$$
$$D_8 = Q_8 \oplus (Q_1Q_8 + Q_1Q_2Q_4)$$
$$Y = Q_1Q_8$$

☐ **TABLE 7-9**
State Table and Flip-Flop Inputs for BCD Counter

Present State				Next State				Output
				$D_8 =$	$D_4 =$	$D_2 =$	$D_1 =$	
Q_8	Q_4	Q_2	Q_1	$Q_8(t+1)$	$Q_4(t+1)$	$Q_2(t+1)$	$Q_1(t+1)$	Y
0	0	0	0	0	0	0	1	0
0	0	0	1	0	0	1	0	0
0	0	1	0	0	0	1	1	0
0	0	1	1	0	1	0	0	0
0	1	0	0	0	1	0	1	0
0	1	0	1	0	1	1	0	0
0	1	1	0	0	1	1	1	0
0	1	1	1	1	0	0	0	0
1	0	0	0	1	0	0	1	0
1	0	0	1	0	0	0	0	1

☐ **TABLE 7-10**
State Table and Flip-Flop Inputs for Counter

Present State			Next State		
			DA =	DB =	DC=
A	B	C	A(t+1)	B(t+1)	C(t+1)
0	0	0	0	0	1
0	0	1	0	1	0
0	1	0	1	0	0
1	0	0	1	0	1
1	0	1	1	1	0
1	1	0	0	0	0

Synchronous BCD counters can be cascaded to form counters for decimal numbers of any length. The cascading is done by replacing D_1 with $D_1 = Q_1 \oplus Y$ where Y is from the next lower BCD counter. Also, Y needs to be ANDed with the product terms to the right of each of the XOR symbol in each of the equations for D_2 through D_8.

ARBITRARY COUNT SEQUENCE   Suppose we wish to design a counter that has a repeated sequence of six states, as listed in Table 7-10. In this sequence, flip-flops B and C repeat the binary count 00, 01, 10, while flip-flop A alternates between 0 and 1 every three counts. Thus, the count sequence for the counter is not straight binary, and two states, 011 and 111, are not included in the count. The D flip-flop

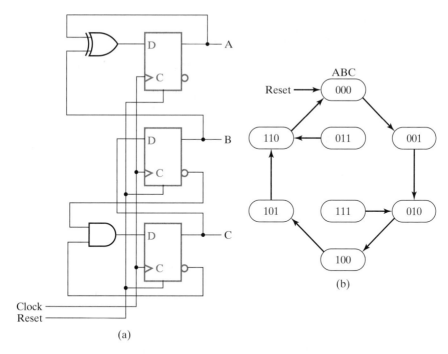

□ **FIGURE 7-16**
Counter with Arbitrary Count

input equations can be simplified using minterms 3 and 7 as don't-care conditions. The simplified functions are

$$D_A = A \oplus B$$
$$D_B = C$$
$$D_C = \overline{B}\ \overline{C}$$

The logic diagram of the counter is shown in Figure 7-16(a). Since there are two unused states, we analyze the circuit to determine their effect. The state diagram obtained is drawn in Figure 7-16(b). This diagram indicates that if the circuit ever goes to one of the unused states, the next count pulse transfers it to one of the valid states, and the circuit then continues to count correctly.

7-7 REGISTER CELL DESIGN

In Section 5-1, we discussed iterative combinational circuits. In this chapter, we connect such circuits to flip-flops to form sequential circuits. A single-bit cell of an iterative combinational circuit connected to a flip-flop that provides the output forms a two-state sequential circuit called a *register cell*. We can design an *n*-bit register with one or more associated microoperations by designing a register cell and making *n* copies of it. Depending on whether the output of the flip-flop is an input to the iterative circuit cell, the register cell may have its next state dependent on its present state and inputs or on its inputs only. If the dependency is only on inputs, then cell design for the iterative combinational circuit and attachment of the iterative circuit to flip-flops is appropriate. If, however, the state of the flip-flop is fed back to the inputs of the iterative circuit cell, sequential design methods can also be applied. The next example illustrates simple register cell design in such a case.

EXAMPLE 7-1   Register Cell Design

A register A is to implement the following register transfers:

AND: $A \leftarrow A \wedge B$
EXOR: $A \leftarrow A \oplus B$
OR: $A \leftarrow A \vee B$

Unless specified otherwise, we assume that

1. Only one of AND, EXOR, and OR is equal to 1, and
2. For all of AND, EXOR, and OR equal to 0, the content of A remains unchanged.

A simple design approach for a register cell with conditions 1 and 2 uses a register with parallel load constructed from D flip-flops with Enable (Enable = Load) from Figure 7-2. For this approach, the expression for Load is the OR of all control signals that cause a transfer to occur. The expression for D_i consists of an OR of the AND of each control signal with the operation on the right-hand side of the corresponding transition.

For this example, the resulting equations for LOAD and D_i are

$$\text{LOAD} = \text{AND} + \text{EXOR} + \text{OR}$$
$$D_i = A(t+1)_i = \text{AND} \cdot A_i B_i + \text{EXOR} \cdot (A_i \overline{B}_i + \overline{A}_i B_i) + \text{OR} \cdot (A_i + B_i)$$

The equation for D_i has an implementation similar to that used for the selection part of a multiplexer in which a set of ENABLE blocks drive an OR gate. AND, EXOR, and OR are enabling signals, and the remaining part of the respective terms in D_i consists of the function enabled.

Using D flip-flops for the register storage and no clock gating, a multiplexer must also be implemented in each cell:

$$D_{i,FF} = \text{LOAD} \cdot D_i + \overline{\text{LOAD}} \cdot A_i$$

This equation is given to show the hidden cost inside of the basic parallel load register cell.

A more complex approach is to design directly for D flip-flops using a sequential circuit design approach rather than the ad hoc approach based on parallel load flip-flops.

We can formulate a coded state table with A as the state variable and output, and AND, EXOR, OR, and B as inputs, as shown in Table 7-11.

By formulating the flip-flop input equation for $D_i = A(t + 1)_i$,

$$D_i = A(t+1)_i = \text{AND} \cdot A_i B_i + \text{EXOR} \cdot (A_i \overline{B}_i + \overline{A}_i B_i) + \text{OR} \cdot (A_i + B_i)$$
$$+ \overline{\text{AND}} \cdot \overline{\text{EXOR}} \cdot \overline{\text{OR}} \cdot A_i$$

□ **TABLE 7-11**
State Table and Flip-Flop Inputs for Counter

Present State A	Next State A(t + 1)						
	(AND = 0) ·(EXOR=0) ·(OR=0)	(OR = 1) ·(B=0)	(OR = 1) ·(B=1)	(EXOR = 1) ·(B=0)	(EXOR = 1) ·(B=1)	(AND = 1) ·(B=0)	(AND = 1) ·(B=1)
0	0	0	1	0	1	0	0
1	1	1	1	1	0	0	1

Due to the relationship between the OR operator and the AND and EXOR operators and other algebraic reductions, this can be simplified to

$$A(t+1)_i = (\text{OR} + \text{AND}) \cdot A_i \cdot B_i + (\text{OR} + \text{EXOR}) \cdot (A_i \overline{B}_i + \overline{A}_i B_i) + \overline{\text{AND} + \text{EXOR}} \cdot A_i$$

The terms OR + AND, OR + EXOR, and $\overline{\text{AND} + \text{EXOR}}$ do not depend on the values A_i and B_i associated with any of the cells. The logic for these terms can be shared by all of the register cells. Using C_1, C_2, and C_3 as intermediate variables, the following set of equations results:

$C_1 = \text{OR} + \text{AND}$
$C_2 = \overline{\text{OR} + \text{EXOR}}$
$C_3 = \overline{\text{AND} + \text{EXOR}}$
$D_i = A(t+1)_i = C_1 A_i B_i + C_2 (A_i \overline{B}_i + \overline{A}_i B_i) + C_3 \cdot A_i$

An implementation for register cell A_i and the logic shared by all of the cells is given in Figure 7-17, with the implementation for logic shared by the register cells in A. Before comparing these results with those from the simple approach, we can apply similar simplification and logic sharing to the results of the simple approach:

$C_1 = \text{OR} + \text{AND}$
$C_2 = \text{OR} + \text{EXOR}$
$D_i = A(t+1)_i = C_1 A_i B_i + C_2 (A_i \overline{B}_i + \overline{A}_i B_i)$
$\text{LOAD} = C_1 + C_2$
$D_{i,FF} = \text{LOAD} \cdot D_i + \overline{\text{LOAD}} \cdot A_i$

If these equations are used directly the cost of the simple approach is somewhat higher. However, if these equations are provided to a minimization tool instead of being used directly, the same equations as the more complex method will result. Thus, the ease of using the simplest approach does not necessarily cause any increase in hardware cost. ■

In the preceding example, there are no lateral connections between adjacent cells. Among the operations requiring lateral connections are shifts, arithmetic operations, and comparisons. One approach to the design of these structures is to combine combinational designs given in Chapter 5 with selection logic and flip-flops. A generic approach for multifunctional registers using flip-flops with parallel load is shown in Figure 7-8. This simple approach bypasses register cell design, but

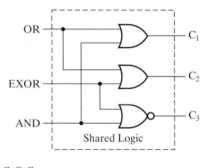

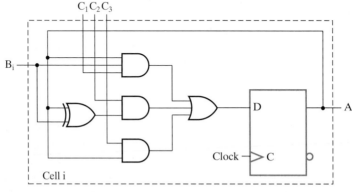

☐ **FIGURE 7-17**
Logic Diagram - Register Cell Design Example 7-1

if directly implemented, can result in excessive logic and too many lateral connections. The alternative is to do a custom register cell design. In such designs, a critical factor is the definition of the lateral connection(s) needed. Also, different operations can be defined by controlling input to the least significant cell of the cell cascade. The custom design approach is illustrated in the next example by the design of a multifunctional register cell.

EXAMPLE 7-2   Register Cell Design

A register A is to implement the following register transfers:

SHL: $A \leftarrow \text{sl } A$
EXOR: $A \leftarrow A \oplus B$
ADD: $A \leftarrow A + B$

Unless specified otherwise, we assume that

1. Only one of SHL, EXOR, and ADD is equal to 1, and
2. For all of SHL, EXOR, and ADD equal to 0, the content of A remains unchanged.

A simple approach to designing a register cell with conditions 1 and 2 is to use a register with parallel load controlled by LOAD. For this approach, the expression for LOAD is the OR of all control signals that cause a transfer to occur.

The implementation for D_i consists of an AND-OR, with each AND having a control signal and the logic for the operation on the right-hand side as its inputs.

For this example, the resulting equations for LOAD and D_i are

$$\text{LOAD} = \text{SHL} + \text{EXOR} + \text{ADD}$$
$$D_i = A(t+1)_i = \text{SHL} \cdot A_{i-1} + \text{EXOR} \cdot (A_i \oplus B_i) + \text{ADD} \cdot ((A_i \oplus B_i) \oplus C_i)$$
$$C_{i+1} = (A_i \oplus B_i)C_i + A_i B_i$$

These equations can be used without modification or can be optimized.

Now, suppose, instead, that we do a custom design assuming that all of the register cells are identical. This means that the least and most significant cells will be the same as those internal to the cell chain. Because of this, the value of C_0 must be specified and the use, if any, of C_n must be determined for each of the three operations. For the left shift, a zero fill of the vacated rightmost bit is assumed, giving $C_0 = 0$. Since C_0 is not involved in the EXOR operation, it can be assumed to be a don't-care. Finally, for the addition, C_0 either can be assumed to be 0 or can be left as a variable to permit a carry from a previous addition to be injected. We assume that C_0 equals 0 for addition, since no additional carry-in is specified by the register transfer statement.

Our first formulation goal is to minimize lateral connections between cells. Two of the three operations, left shift and addition, require a lateral connection to the left (i.e., toward the most significant end of the cell chain). Our goal is to use one signal for both operations, say, C_i. It already exists for the addition but must be redefined to handle both the addition and the left shift. Also in our custom design, the parallel load flip-flop will be replaced by a D flip-flop. We can now formulate the state table for the register cell shown in Table 7-12:

$$D_i = A(t+1)_i = \overline{\text{SHL}} \cdot \overline{\text{EXOR}} \cdot \overline{\text{ADD}} \cdot A_i + \text{SHL} \cdot C_i + \text{EXOR} \cdot (A_i \oplus B_i) + \text{ADD} \cdot (A_i \oplus B_i \oplus C_i)$$
$$C_{i+1} = \text{SHL} \cdot A_i + \text{ADD}(\cdot (A_i \oplus B_i)C_i + A_i B_i)$$

The term $A_i \oplus B_i$ appears in both the EXOR and ADD terms. In fact, if $C_i = 0$ during the EXOR operation, then the functions for the sum in ADD and for EXOR can be identical. In the C_{i+1} equation, since SHL and ADD are both 0 when EXOR is 1, C_i is 0 for all cells in the cascade except the least significant one. For the least

□ **TABLE 7-12**
State Table and Flip-Flop Inputs for Register Cell Design in Example 7-2

Present State A_i	Inputs					Next State $A_i(t+1)$/Output C_{i+1}					
	SHL = 0	SHL = 1	1	1	1	EXOR = 1	1	ADD = 1	1	1	1
	EXOR = 0	B_i = 0	0	1	1	B_i = 0	1	B_i = 0	0	1	1
	ADD = 0	C_i = 0	1	0	1			C_i = 0	1	0	1
0	0/X	0/0	1/0	0/0	1/0	0/X	1/X	0/0	1/0	1/0	0/1
1	1/X	0/1	1/1	0/1	1/1	1/X	0/X	1/0	0/1	0/1	1/1

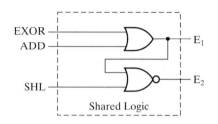

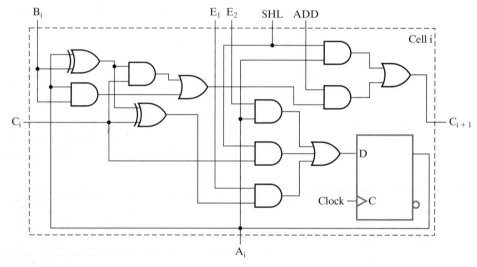

□ **FIGURE 7-18**
Logic Diagram - Register Cell Design Example 7-2

significant cell, the specification states that $C_0 = 0$. Thus, input values C_i are 0 for all cells in register A. So we can combine the ADD and EXOR operations as follows:

$$D_i = A(t+1)_i = \overline{SHL} \cdot \overline{EXOR} \cdot \overline{ADD} \cdot A_i + SHL \cdot C_i + (EXOR + ADD) \cdot ((A_i \oplus B_i) \oplus C_i)$$

The expressions $\overline{SHL} \cdot \overline{EXOR} \cdot \overline{ADD}$ and EXOR + ADD, that are independent of A_i, B_i, and C_i, can be shared by all cells. The resulting equations are

$$E_1 = EXOR + ADD$$
$$E_2 = \overline{E1 + SHL}$$
$$D_i = E_2 \cdot A_i + SHL \cdot C_i + E_1 \cdot ((A_i \oplus B_i) \oplus C_i)$$
$$C_{i+1} = SHL \cdot A_i + ADD \cdot ((A_i \oplus B_i)C_i + A_iB_i)$$

The resulting register cell appears in Figure 7-18. Comparing this result with the register cell for the simple design, we note the following two differences:

1. Only one lateral connection between cells exists instead of two.
2. Logic has been very efficiently shared by the addition and the EXOR operation.

The custom cell design has produced connection and logic savings not present in the block level design with or without optimization. ■

7-8 MULTIPLEXER AND BUS–BASED TRANSFERS FOR MULTIPLE REGISTERS

A typical digital system has many registers. Paths must be provided to transfer data from one register to another. The amount of logic and the number of interconnections may be excessive if each register has its own dedicated set of multiplexers. A more efficient scheme for transferring data between registers is a system that uses a shared transfer path called a *bus*. A bus is characterized by a set of common lines, with each line driven by selection logic. Control signals for the logic select a single source and one or more destinations on any clock cycle for which a transfer occurs.

In Section 7-4, we saw that multiplexers and parallel load registers can be used to implement dedicated transfers from multiple sources. A block diagram for such transfers between three registers is shown in Figure 7-19(a). There are three n-bit 2-to-1 multiplexers, each with its own select signal. Each register has its own load signal. The same system based on a bus can be implemented by using a single n-bit 3-to-1 multiplexer and parallel load registers. If a set of multiplexer outputs is shared as a common path, these output lines are a bus. Such a system with a single

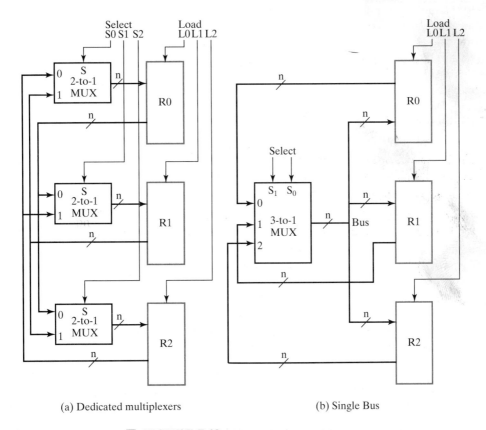

(a) Dedicated multiplexers

(b) Single Bus

□ **FIGURE 7-19**

Single Bus versus Dedicated Multiplexers

☐ **TABLE 7-13**
Examples of Register Transfers Using the Single Bus in Figure 7-19(b)

Register Transfer	Select		Load		
	S1	S0	L2	L1	L0
$R0 \leftarrow R2$	1	0	0	0	1
$R0 \leftarrow R1, R2 \leftarrow R1$	0	1	1	0	1
$R0 \leftarrow R1, R1 \leftarrow R0$			Impossible		

bus for transfers between three registers is shown in Figure 7-19(b). The control input pair, Select, determines the contents of the single source register that will appear on the multiplexer outputs (i.e., on the bus). The load inputs determine the destination register or registers to be loaded with the bus data.

In Table 7-13, transfers using the single-bus implementation of Figure 7-19(b) are illustrated. The first transfer is from $R2$ to $R0$. Select equals 10, selecting input $R2$ to the multiplexer. Load signal $L0$ for register $R0$ is 1, with all other loads at 0, causing the contents of $R2$ on the bus to be loaded into $R0$ on the next positive clock transition. The second transfer in the table illustrates the loading of the contents of $R1$ into both $R0$ and $R2$. The source $R1$ is selected because Select is equal to 01. In this case, $L2$ and $L0$ are both 1, causing the contents of $R1$ on the bus to be loaded into registers $R0$ and $R2$. The third transfer, an exchange between $R0$ and $R1$, is impossible in a single clock cycle, since it requires two simultaneous sources, $R0$ and $R1$, on the single bus. Thus, this transfer requires at least two buses or a bus combined with a dedicated path from one of the registers to the other. Note that such a transfer can be executed on the dedicated multiplexers in Figure 7-19(a). So, for a single-bus system, simultaneous transfers with different sources in a single clock cycle are impossible, whereas for the dedicated multiplexers, any combination of transfers is possible. Hence, the reduction in hardware that occurs for a single bus in place of dedicated multiplexers results in limitations on simultaneous transfers.

If we assume that only single-source transfers are needed, then we can use Figure 7-19 to compare the complexity of the hardware in dedicated versus bus-based systems. First of all, assume a multiplexer design, as in Figure 4-16. In Figure 7-19(a), there are $2n$ AND gates and n OR gates per multiplexer (not counting inverters), for a total of $9n$ gates. In contrast, in Figure 7-19(b), the bus multiplexer requires only $3n$ AND gates and n OR gates, for a total of $4n$ gates. Also, the data input connections to the multiplexers are reduced from $6n$ to $3n$. Thus, the cost of the selection hardware is reduced by about half.

Three-State Bus

A bus can be constructed with the three-state buffers introduced in Section 2-8 instead of multiplexers. This has the potential for additional reductions in the number of connections. But why use three-state buffers instead of a multiplexer, particularly for implementing buses? The reason is that many three-state buffer outputs can be connected together to form a bit line of a bus, and this bus is implemented using only one level of logic gates. On the other hand, in a multiplexer, such a large number of sources means a high fan-in OR, which requires multiple levels of OR gates, introducing more logic and increasing delay. In contrast, three-state buffers provide a practical way to construct fast buses with many sources, so they are often preferred in such cases. More important, however, is the fact that signals can travel in two directions on a three-state bus. Thus, the three-state bus can use the same interconnection to carry signals into and out of a logic circuit. This feature, which is most important when crossing chip boundaries, is illustrated in Figure 7-20(a). The figure shows a register with n lines that serve as both inputs and outputs lying across the boundary of the shaded area. If the three-state buffers are enabled, then the lines are outputs; if the three-state buffers are disabled, then the lines can be inputs. The symbol for this structure is also given in the figure. Note that the bidirectional bus lines are represented by a two-headed arrow. Also, a small inverted triangle denotes the three-state outputs of the register.

Figure 7-20(b) and Figure 7-20(c) show a multiplexer-implemented bus and a three-state bus, respectively, for comparison. The symbol from Figure 7-20(a) for a register with bidirectional input-output lines is used in Figure 7-20(c). In contrast to the situation in Figure 7-19, where dedicated multiplexers were replaced by a bus, these two implementations are identical in terms of their register transfer

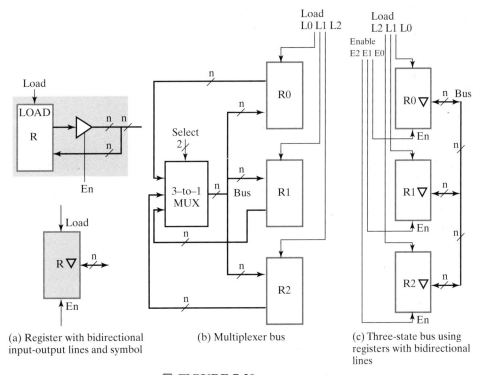

(a) Register with bidirectional input-output lines and symbol

(b) Multiplexer bus

(c) Three-state bus using registers with bidirectional lines

□ **FIGURE 7-20**
Three-State Bus versus Multiplexer Bus

capability. Note that, in the three-state bus, there are only three data connections to the set of register blocks for each bit of the bus. The multiplexer-implemented bus has six data connections per bit to the set of register blocks. This reduction in the number of data connections by half, along with the ability to easily construct a bus with many sources, makes the three-state bus an attractive alternative. The use of such bidirectional input-output lines is particularly effective between logic circuits in different physical packages.

7-9 SERIAL TRANSFER AND MICROOPERATIONS

A digital system is said to operate in a serial mode when information in the system is transferred or manipulated one bit at a time. Information is transferred one bit at a time by shifting the bits out of one register and into a second register. This transfer method is in contrast to parallel transfer, in which all the bits of the register are transferred at the same time.

The serial transfer of information from register A to register B is done with shift registers, as shown in the block diagram of Figure 7-21(a). The serial output of register A is connected to the serial input of register B. The serial input of register A receives 0's while its data are transferred to register B. It is also possible for register A to receive other binary information, or if we want to maintain the data in register A, we can connect its serial output to its serial input so that the information is circulated back into the register. The initial content of register B is shifted out through its serial output and is lost unless it is transferred back into register A, to a third shift register, or to other storage. The shift control input *Shift* determines when and how many times the registers are shifted. The registers using *Shift* are controlled by means of the logic from Figure 7-2, which allows the clock pulses to pass to the shift register clock inputs only when *Shift* has the value logic 1.

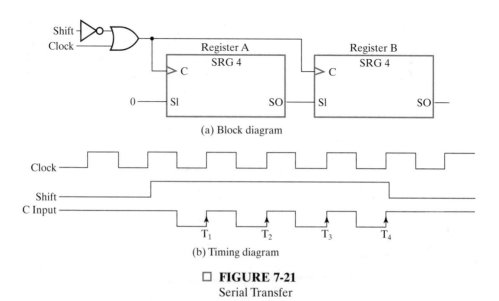

(a) Block diagram

(b) Timing diagram

☐ FIGURE 7-21
Serial Transfer

In Figure 7-21, each shift register has four stages. The logic that supervises the transfer must be designed to enable the shift registers, through the *Shift* signal, for a fixed time of four clock pulses. Shift register enabling is shown in the timing diagram for the clock gating logic in Figure 7-21(b). Four pulses find *Shift* in the active state, so that the output of the logic connected to the clock inputs of the registers produces four pulses: T_1, T_2, T_3, and T_4. Each positive transition of these pulses causes a shift in both registers. After the fourth pulse, *Shift* changes back to 0 and the shift registers are disabled. We note again that, for positive-edge triggering, the pulses on the clock inputs are 0, and the inactive level when no pulses are present is a 1 rather than a 0.

Now suppose that the binary content of register *A* before the shift is 1011, that of register *B* is 0010, and the *SI* of register *A* is logic 0. Then the serial transfer from *A* to *B* occurs in four steps, as shown in Table 7-14. With the first pulse T_1, the rightmost bit of *A* is shifted into the leftmost bit of *B*, the leftmost bit of *A* receives a 0 from the serial input, and at the same time, all other bits of *A* and *B* are shifted one position to the right. The next three pulses perform identical operations, shifting the bits of *A* into *B* one at a time while transferring 0's to *A*. After the fourth shift, the logic supervising the transfer changes the *Shift* signal to 0 and the shifts stop. Register *B* contains 1011, which is the previous value of *A*. Register *A* contains all 0's.

The difference between serial and parallel modes of operation should be apparent from this example. In the parallel mode, information is available from all bits of a register, and all bits can be transferred simultaneously during one clock pulse. In the serial mode, the registers have a single serial input and a single serial output, and information is transferred one bit at a time.

Serial Addition

Operations in digital computers are usually done in parallel because of the faster speed attainable. Serial operations are slower, but have the advantage of requiring less hardware. To demonstrate the serial mode of operation, we will show the operation of a serial adder. Also, we compare the serial adder to the parallel counterpart presented in Section 5-2 to illustrate the time–space trade-off in design.

□ **TABLE 7-14**
  Example of Serial Transfer

Timing pulse	Shift Register A				Shift Register B			
Initial value	1	0	1	1	0	0	1	0
After T_1	0	1	0	1	1	0	0	1
After T_2	0	0	1	0	1	1	0	0
After T_3	0	0	0	1	0	1	1	0
After T_4	0	0	0	0	1	0	1	1

The two binary numbers to be added serially are stored in two shift registers. Bits are added, one pair at a time, through a single full-adder (FA) circuit, as shown in Figure 7-22. The carry out of the full adder is transferred into a D flip-flop. The output of this carry flip-flop is then used as the carry input for the next pair of significant bits. The sum bit on the S output of the full adder could be transferred into a third shift register, but we have chosen to transfer the sum bits into register A as the contents of the register are shifted out. The serial input of register B can receive a new binary number as its contents are shifted out during the addition.

The operation of the serial adder is as follows: Register A holds the augend, register B holds the addend, and the carry flip-flop has been reset to 0. The serial outputs of A and B provide a pair of significant bits for the full adder at X and Y. The output of the carry flip-flop provides the carry input at Z. When *Shift* is set to 1, the OR gate enables the clock for both registers and the flip-flop. Each clock pulse shifts both registers once to the right, transfers the sum bit from S into the leftmost flip-flop of A, and transfers the carry output into the carry flip-flop. Shift control logic enables the registers for as many clock pulses as there are bits in the registers (four pulses in this example). For each pulse, a new sum bit is transferred to A, a new carry is transferred to the flip-flop, and both registers are shifted once to the right. This process continues until the shift control logic changes *Shift* to 0. Thus, the addition is accomplished by passing each pair of bits and the previous carry through a single full-adder circuit and transferring the sum, one bit at a time, back into register A.

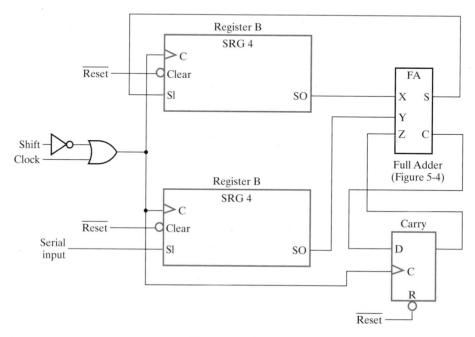

□ **FIGURE 7-22**
Serial Addition

Initially, we can reset register *A*, register *B*, and the *Carry* flip-flop to 0. Then we shift the first number into *B*. Next, the first number from *B* is added to the 0 in *A*. While *B* is being shifted through the full adder, we can transfer a second number to it through its serial input. The second number can be added to the contents of register *A* at the same time that a third number is transferred serially into register *B*. Serial addition may be repeated to form the addition of two, three, or more numbers, with their sum accumulated in register *A*.

A comparison of the serial adder with the parallel adder described in Section 5-2 provides an example of space-time trade-off. The parallel adder has *n* full adders for *n*-bit operands, whereas the serial adder requires only one full adder. Excluding the registers from both, the parallel adder is a combinational circuit, whereas the serial adder is a sequential circuit because it includes the carry flip-flop. The serial circuit also takes *n* clock cycles to complete an addition. Identical circuits, such as the *n* full adders in the parallel adder, connected together in a chain constitute an example of an *iterative logic array*. If the values on the carries between the full adders are regarded as state variables, then the states from the least significant end to the most significant end are the same as the states appearing in sequence on the flip-flop output in the serial adder. Note that in the iterative logic array the states appear in space, but in the sequential circuit the states appear in time. By converting from one of these implementations to the other, one can make a space-time trade-off. The parallel adder in space is *n* times larger than the serial adder (ignoring the area of the carry flip-flop), but it is *n* times faster. The serial adder, although it is *n* times slower, is *n* times smaller in space. This gives the designer a significant choice in emphasizing speed or area, where more area translates into more cost.

7-10 HDL REPRESENTATION FOR SHIFT REGISTERS AND COUNTERS—VHDL

Examples of shift register and a binary counter illustrate the use of VHDL in representing registers and operations on register content.

EXAMPLE 7-3 VHDL for a 4-Bit Shift Register

The VHDL code in Figure 7-23 describes a 4-bit left shift register at the behavioral level. A RESET input is present that directly resets the register contents to zero. The shift register contains flip-flops and so has a process description resembling that of a *D* flip-flop. The four flip-flops are represented by the signal shift, of type std_logic_vector of size four. Q cannot be used to represent the flip-flops since it is an output and the flip-flop outputs must be used internally. The left shift is achieved by applying the concatenation operator & to the right three bits of shift and to shift input SI. This quantity is transferred to shift moving the contents one bit to the left and loading the value of SI into the rightmost bit. Following the process that performs the shift are two statements, one which assigns the

```
-- 4-bit Shift Register with Reset

library ieee;
use ieee.std_logic_1164.all;

entity srg_4_r is
   port(CLK, RESET, SI : in std_logic;
      Q : out std_logic_vector(3 downto 0);
      SO : out std_logic);
end srg_4_r;

architecture behavioral of srg_4_r is
signal shift : std_logic_vector(3 downto 0);
begin
process (RESET, CLK)
begin
   if (RESET = '1') then
      shift <= "0000";
   elsif (CLK'event and (CLK = '1')) then
      shift <= shift(2 downto 0) & SI;
   end if;
end process;
   Q <= shift;
   SO <= shift(3);
end behavioral;
```

☐ **FIGURE 7-23**
Behavioral VHDL Description of 4-bit Left Shift Register with Direct Reset

value in shift to output Q and the other which defines the shift out signal SO as the contents of the leftmost bit of shift. ∎

EXAMPLE 7-4 VHDL for a 4-Bit Counter

The VHDL code in Figure 7-24 describes a 4-bit counter at the behavioral level. A RESET input is present that directly resets the counter contents to zero. The counter contains flip-flops and, therefore, has a process description resembling that of a D flip-flop. The four flip-flops are represented by the signal count, of type std_logic_vector and of size four. Q cannot be used to represent the flip-flops since it is an output and the flip-flop outputs must be used internally. Counting up is achieved by adding 1 in the form of "0001" to count. Since addition is not a normal operation on type std_logic_vector, it is necessary to use an additional package from the ieee library, std_logic_unsigned.all, which defines unsigned number operations on type std_logic. Following the process that performs reset and counting are two statements, one which assigns the value in count to output Q and the other which defines the count out signal CO. A **when-else** statement is used in which CO is set to 1 only for the maximum count with EN equal to 1. ∎

```
-- 4-bit Binary Counter with Reset

library ieee;
use ieee.std_logic_1164.all;
use ieee.std_logic_unsigned.all;

entity count_4_r is
   port(CLK, RESET, EN : in std_logic;
        Q : out std_logic_vector(3 downto 0);
        CO : out std_logic);
end count_4_r;

architecture behavioral of count_4_r is
signal count : std_logic_vector(3 downto 0);
begin
process (RESET, CLK)
begin
   if (RESET = '1') then
      count <= "0000";
   elsif (CLK'event and (CLK = '1') and (EN = '1')) then
       count <= count + "0001";
   end if;
end process;
Q <= count;
CO <= '1' when count = "1111" and EN = '1' else '0';
end behavioral;
```

☐ **FIGURE 7-24**
Behavioral VHDL Description of 4-bit Binary Counter with Direct Reset

7-11 HDL REPRESENTATION FOR SHIFT REGISTERS AND COUNTERS—VERILOG

Examples of shift register and binary counter illustrate the use of Verilog in representing registers and operations on register content.

EXAMPLE 7-5 Verilog Code for a Shift Register

The Verilog description in Figure 7-25 describes a left shift register at the behavioral level. A RESET input is present that directly resets the register contents to zero. The shift register contains flip-flops, so has a process description beginning with **always** resembling that of a D flip-flop. The four flip-flops are represented by the vector Q, of type **reg** with bits numbered 3 down to 0. The left shift is achieved by applying { } to concatenate the right three bits of Q and shift input SI. This quantity is transferred to Q, moving the contents one bit to the left and loading the value of SI into the rightmost bit. Just prior to the process that performs the shift

```
// 4-bit Left Shift Register with Reset

module srg_4_r_v (CLK, RESET, SI, Q,SO);
    input CLK, RESET, SI;
    output [3:0] Q;
    output SO;

reg [3:0] Q;

assign SO = Q[3];

always@(posedge CLK or posedge RESET)
begin
    if (RESET)
        Q <= 4'b0000;
    else
        Q <= {Q[2:0], SI};
end
endmodule
```

☐ **FIGURE 7-25**

Behavioral Verilog Description of 4-bit Left Shift Register with Direct Reset

is a continuous assignment statement that assigns the contents of the leftmost bit of Q to the shift output signal SO. ∎

EXAMPLE 7-6   Verilog Code for a Counter

The Verilog description in Figure 7-26 describes a 4-bit binary counter at the behavioral level. A RESET input is present that directly resets the register contents to zero. The counter contains flip-flops and, therefore, the description contains a process resembling that for a D flip-flop. The four flip-flops are represented by the signal Q of type **reg** and size four. Counting up is achieved by adding 1 to Q. Prior to the process that performs reset and counting is a conditional continuous assignment statement that defines the count out signal CO. CO is set to 1 only for the maximum count and EN equal to 1. Note that logical AND is denoted by &&. ∎

7-12   CHAPTER SUMMARY

Registers are sets of flip-flops, or interconnected sets of flip-flops, and combinational logic. The simplest registers are flip-flops that are loaded with new contents from their inputs on every clock cycle. More complex are registers in which the flip-flops can be loaded with new contents under the control of a signal on only selected clock cycles. Register transfers are a means of representing and specifying elementary processing operations. Register transfers can be related to corresponding digital system hardware, both at the block diagram

```
// 4-bit Binary Counter with Reset

module count_4_r_v (CLK, RESET, EN, Q, CO);
   input CLK, RESET, EN;
   output [3:0] Q;
   output CO;

reg [3:0] Q;

assign CO = (count == 4'b1111 && EN == 1'b1) ? 1 : 0;
always@(posedge CLK or posedge RESET)
   begin
   if (RESET)
     Q <= 4'b0000;
   else if (EN)
     Q <= Q + 4'b0001;
   end
endmodule
```

□ **FIGURE 7-26**
Behavioral Verilog Description of 4-bit Binary Counter with Direct Reset

level and at the detailed logic level. Microoperations are elementary operations performed on data stored in registers. Arithmetic microoperations include addition and subtraction, which are described as register transfers and are implemented with corresponding hardware. Logic microoperations—that is, the bitwise application of logic primitives such as AND, OR, and XOR, combined with a binary word—provide masking and selective complementing on other binary words. Left- and right-shift microoperations move data laterally one or more bit positions at a time.

Shift registers add a new dimension to data transfer, since they are designed to move information laterally one or more bit positions at a time. When combined with the ability to be loaded with data, a shift register can be used to convert data presented in parallel into data presented serially. Likewise, if the outputs of the register are accessible, a shift register can be used to convert data presented serially into data presented in parallel. This lateral movement of data can also be used in hardware structures that perform serial arithmetic operations.

Counters are used to provide a sequence of values, often in binary counting order. The simplest of counters has no inputs other than an asynchronous reset for initialization to zero. This kind of counter simply counts clock pulses. More complex versions can also be loaded with data and have input signals that enable them to count.

Multiplexers select among multiple transfer paths entering a register. Buses are shared register transfer paths for multiple registers and offer reduced hardware in trade for limitations on possible simultaneous transfers. In addition to multiplexers, three-state buffers enhance the implementation of buses by providing bidirectional transfer paths and reduced connections.

REFERENCES

1. MANO, M. M. *Digital Design,* 3rd ed. Englewood Cliffs, NJ: Prentice Hall, 2002.

2. WAKERLY, J. F. *Digital Design: Principles and Practices,* 3rd ed. Upper Saddle River, NJ: Prentice Hall, 2000.

3. *IEEE Standard VHDL Language Reference Manual.* (ANSI/IEEE Std 1076-1993; revision of IEEE Std 1076-1987). New York: The Institute of Electrical and Electronics Engineers, 1994.

4. *IEEE Standard Description Language Based on the Verilog(TM) Hardware Description Language* (IEEE Std 1364-1995). New York: The Institute of Electrical and Electronics Engineers, 1995.

5. THOMAS, D. E., AND P. R. MOORBY. *The Verilog Hardware Description Language* 4th ed. Boston: Kluwer Academic Publishers, 1998.

PROBLEMS

The plus (+) indicates a more advanced problem and the asterisk (*) indicates a solution is available on the Companion Website for the text.

7–1. Use manual or computer-based simulation to demonstrate that the clock gating function in Figure 7-1(c) works properly with the register in Figure 7-1(a). Use a positive-edge-triggered flip-flop with *Clock* as its clock input to generate *Load.* Be sure to use nonzero gate and flip-flop delays.

7–2. +Change the OR gate in Figure 7-1(c) into an AND gate, and remove the inverter on *Load.*
   (a) Perform the same simulation as in Problem 7–1 to demonstrate that the new clock gating circuitry does not work correctly. Explain what goes wrong.
   (b) Will the circuit work correctly if the flip-flop generating *Load* is triggered by the negative rather than the positive edge of *Clock?*

7–3. Assume that registers $R1$ and $R2$ in Figure 7-6 hold two unsigned numbers. When select input X is equal to 1, the adder–subtractor circuit performs the arithmetic operation "$R1$ + 2's complement of $R2$." This sum and the output carry C_n are transferred into $R1$ and C when $K_1 = 1$ and a positive edge occurs on the clock.
   (a) Show that if $C = 1$, then the value transferred to $R1$ is equal to $R1 - R2$, but if $C = 0$, the value transferred to $R1$ is the 2's complement of $R2 - R1$.
   (b) Indicate how the value in the C bit can be used to detect a borrow after the subtraction of two unsigned numbers.

7–4. *Perform the bitwise logic AND, OR, and XOR of the two 8-bit operands 10011001 and 11000011.

7–5. Given the 16-bit operand 00001111 10101010, what operation must be performed and what operand must be used
  (a) to clear all even bit positions to 0? (Assume bit positions are 15 through 0 from left to right.)
  (b) to set the leftmost 4 bits to 1?
  (c) to complement the center 8 bits?

7–6. *Starting from the 8-bit operand 01010011, show the values obtained after each shift microoperation given in Table 7-5.

7–7. *Modify the register of Figure 7-11 so that it will operate according to the following function table using mode selection inputs S_1 and S_0.

S_1	S_0	Register Operation
0	0	No change
0	1	Clear register to 0
1	0	Shift down
1	1	Load parallel data

7–8. *A ring counter is a shift register, as in Figure 7-9, with the serial output connected to the serial input.
  (a) Starting from an initial state of 1000, list the sequence of states of the four flip-flops after each shift.
  (b) Beginning in state 10...0, how many states are there in the count sequence of an n-bit ring counter?

7–9. A switch-tail ring counter (Johnson counter) uses the complement of the serial output of a right shift register as its serial input.
  (a) Starting from an initial state of 0000, list the sequence of states after each shift until the register returns to 0000.
  (b) Beginning in state 00...0, how many states are there in the count sequence of an n-bit switch-tail counter?

7–10. How many flip-flop values are complemented in an 8-bit binary ripple counter to reach the next count value after
  (a) 11101111?   **(b)** 01111111?

7–11. + For the CMOS logic family, the power consumption is proportional to the sum of the changes from 1-to-0 and 0-to-1 on all gate inputs and outputs in the circuit. When designing counters in very low power circuits, ripple counters are preferred over regular synchronous binary counters. Carefully count the numbers of changing inputs and outputs, including those related to the clock for a complete cycle of values in a 4-bit ripple counter versus a regular synchronous counter of the same length. Based on this examination, explain why the ripple counter is superior in terms of power consumption.

7–12. Construct a 16-bit serial-parallel counter, using four 4-bit parallel counters. Suppose all added logic is AND gates and serial connections are employed between the four counters. What is the maximum number of AND gates in a chain that a signal must propagate through in the 16-bit counter?

7–13. +A 64-bit synchronous parallel counter is to be designed.
  (a) Draw the logic diagram of a 64-bit parallel counter, using 8-bit parallel counter blocks and two levels of parallel gating connections between the blocks. In these blocks, CO is not driven by EN.
  (b) What is the ratio of the maximum frequency of operation of this counter to that of a 64-bit serial-parallel counter? Assume that the D flip-flop propagation time is twice the delay of an AND gate and that the flip-flop setup time is equal to the delay of an AND gate.

7–14. Using the synchronous binary counter of Figure 7-13 and an AND gate, construct a counter that counts from 0000 through 1010. Repeat for a count from 0000 to 1110. Minimize the number of inputs to the AND gate.

7–15. Using two binary counters of the type shown in Figure 7-13 and logic gates, construct a binary counter that counts from decimal 9 through decimal 129. Add an additional input to the counter that initializes it synchronously to 9 when the signal INIT is 1.

7–16. *Verify the flip-flop input equations of the synchronous BCD counter specified in Table 7–9. Draw the logic diagram of the BCD counter with a count enable input.

7–17. *Use D flip-flops and gates to design a binary counter with each of the following repeated binary sequences:
  (a) 0, 1, 2   **(b)** 0, 1, 2, 3, 4, 5

7–18. Use D-type flip-flops and gates to design a counter with the following repeated binary sequence: 0, 1, 3, 2, 4, 6.

7–19. Use only D-type flip-flops to design a counter with the following repeated binary sequence: 0, 1, 2, 4, 8.

7–20. Draw the logic diagram of a 4-bit register with mode selection inputs S_1 and S_0. The register is to be operated according to the following function table:

S_1	S_0	Register Operation
0	0	No change
0	1	Clear register to 0
1	0	Complement output
1	1	Load parallel data

7–21. *Show the diagram of the hardware that implements the register transfer statement

$$C_3: R2 \leftarrow R1, R1 \leftarrow R2$$

7–22. The outputs of registers $R0$, $R1$, $R2$, and $R3$ are connected through 4-to-1 multiplexers to the inputs of a fourth register R4. Each register is 8 bits long. The required transfers, as dictated by four control variables, are

$$C_0: R4 \leftarrow R0$$

$$C_1: R4 \leftarrow R1$$

$$C_2: R4 \leftarrow R2$$

$$C_3: R4 \leftarrow R3$$

The control variables are mutually exclusive (i.e., only one variable can be equal to 1 at any time) while the other three are equal to 0. Also, no transfer into $R4$ is to occur for all control variables equal to 0. (a) Using registers and a multiplexer, draw a detailed logic diagram of the hardware that implements a single bit of these register transfers. (b) Draw a logic diagram of the simple logic that maps the control variables as inputs to outputs that are the two select variables for the multiplexers to outputs that are the load signals for registers.

7–23. *Using two 4-bit registers $R1$ and $R2$, and AND gates, OR gates, and inverters, draw one bit slice of the logic diagram that implements all of the following statements:

$$C_0: R2 \leftarrow 0 \qquad \text{Clear } R2 \text{ synchronously with the clock}$$

$$C_1: R2 \leftarrow \overline{R2} \qquad \text{Complement } R2$$

$$C_2: R2 \leftarrow R1 \qquad \text{Transfer } R1 \text{ to } R2$$

The control variables are mutually exclusive (i.e., only one variable can be equal to 1 at any time) while the other two are equal to 0. Also, no transfer into $R2$ is to occur for all control variables equal to 0.

7–24. A register cell is to be designed for an 8-bit register A that has the following register transfer functions:

$$C_0: A \leftarrow A \wedge B$$

$$C_1: A \leftarrow A \vee \overline{B}$$

Find optimum logic using AND, OR, and NOT gates for the D input to the D flip-flop in the cell.

7–25. A register cell is to be designed for an 8-bit register $R0$ that has the following register transfer functions:

$$\overline{S_1} \cdot \overline{S_0}: R0 \leftarrow 0$$

$$\overline{S_1} \cdot S_0: R0 \leftarrow R0 \vee R1$$

$$S_1 \cdot \overline{S_0} : R0 \leftarrow R0 \oplus R1$$

$$S_1 \cdot S_0 : R0 \leftarrow R0 \wedge R1$$

Find optimum logic using AND, OR, and NOT gates for the D input to the D flip-flop in the cell.

7–26. A register cell is to be designed for register B, which has the following register transfers:

$$S_1 : B \leftarrow B + A$$

$$S_0 : B \leftarrow B + 1$$

Share the combinational logic between the two transfers as much as possible.

7–27. Logic to implement transfers among three registers, $R0$, $R1$, and $R2$, is to be implemented. Use the control variable assumptions given in Problem 7–2. The register transfers are as follows:

$$C_A : R1 \leftarrow R0$$

$$C_B : R0 \leftarrow R1, R2 \leftarrow R0$$

$$C_C : R1 \leftarrow R2, R0 \leftarrow R2$$

Using registers and dedicated multiplexers, draw a detailed logic diagram of the hardware that implements a single bit of these register transfers.

Draw a logic diagram of simple logic that converts the control variables C_A, C_B, and C_C as inputs to outputs that are the SELECT inputs for the multiplexers and LOAD signals for the registers.

7–28. *Two register transfer statements are given (otherwise, R1 is unchanged):

$$C_1 : R1 \leftarrow R1 + R2 \qquad \text{Add } R2 \text{ to } R1$$

$$\overline{C_1} C_2 : R1 \leftarrow R1 + 1 \qquad \text{Increment } R1$$

(a) Using a 4-bit counter with parallel load as in Figure 7-13 and a 4-bit adder as in Figure 5-5, draw the logic diagram that implements these register transfers.
(b) Repeat part (a) using a 4-bit adder as in Figure 5-5 plus external gates as needed. Compare with the implementation in part (a).

7–29. Repeat Problem 7–27 using one multiplexer-based bus and one direct connection from one register to another instead of dedicated multiplexers.

7–30. Draw a logic diagram of a bus system similar to the one shown in Figure 7-7, but use three-state buffers and a decoder instead of the multiplexers.

7–31. *A system is to have the following set of register transfers, implemented using buses:

$$C_a: R0 \leftarrow R1$$

$$C_b: R3 \leftarrow R1, R1 \leftarrow R4, R4 \leftarrow R0$$

$$C_c: R2 \leftarrow R3, R0 \leftarrow R2$$

$$C_d: R2 \leftarrow R4, R4 \leftarrow R2$$

(a) For each destination register, list all of the source registers.

(b) For each source register, list all of the destination registers.

(c) With consideration for which of the transfers must occur simultaneously, what is the minimum number of buses that can be used to implement the set of transfers? Assume that each register will have a single bus as its input.

(d) Draw a block diagram of the system, showing the registers and buses and the connections between them.

7–32. The following register transfers are to be executed in, at most, two clock cycles:

$$R0 \leftarrow R1$$
$$R2 \leftarrow R1$$
$$R4 \leftarrow R2$$
$$R6 \leftarrow R3$$
$$R8 \leftarrow R3$$
$$R9 \leftarrow R4$$
$$R10 \leftarrow R4$$
$$R11 \leftarrow R1$$

(a) What is the minimum number of buses required? Assume that only one bus can be attached to a register input and that any net connected to a register input is counted as a bus.

(b) Draw a block diagram connecting registers and multiplexers to implement the transfers.

7–33. What is the minimum number of clock cycles required to perform the following set of register transfers using two buses?

$R0 \leftarrow R1$	$R7 \leftarrow R1$
$R2 \leftarrow R3$	$R8 \leftarrow R4$
$R5 \leftarrow R6$	$R9 \leftarrow R3$

Assume that only one bus can be attached to a register input and that any net connected to a register input is counted as a bus.

7–34. *The content of a 4-bit register is initially 0000. The register is shifted eight times to the right, with the sequence 10110001 as the serial input. The leftmost bit of the sequence is applied first. What is the content of the register after each shift?

7–35. *The serial adder of Figure 7-22 uses two 4-bit registers. Register A holds the binary number 0111 and register B holds 0101. The carry flip-flop is initially reset to 0. List the binary values in register A and the carry flip-flop after each of four shifts.

All files referred to in the remaining problems are available in ASCII form for simulation and editing on the Companion Website for the text. A VHDL or Verilog compiler/simulator is necessary for the problems or portions of problems requesting simulation. Descriptions can still be written, however, for many problems without using compilation or simulation.

7–36. *Write a behavioral VHDL description for the 4-bit register in Figure 7-1 (a). Compile and simulate your description to demonstrate correctness.

7–37. Repeat problem 7-36 for the 4-bit register with parallel load in Figure 7-2.

7–38. Write a VHDL description for the 4-bit binary counter in Figure 7-13 using a register for the D flip-flops and Boolean equations for the logic. Compile and simulate your description to demonstrate correctness.

7–39. *Write a behavior Verilog description for the 4-bit register in Figure 7-1 (a). Compile and simulate your description to demonstrate correctness.

7–40. Repeat problem 5-39 for the 4-bit register with parallel load in Figure 7-2.

7–41. Write a Verilog description for the 4-bit binary counter in Figure 7-13 using a register for the D flip-flops and Boolean equations for the logic. Compile and simulate your description to demonstrate correctness.

8

SEQUENCING
AND CONTROL

I n Chapter 7, we introduced the concept of the datapath for processing data and
datapath implementation using registers and register transfers. In the current
chapter, our focus is the control unit also introduced in Chapter 7. Digital systems
can be classified as programmable or nonprogrammable systems depending on the
type of control unit. Nonprogrammable systems have inputs, but do not have any
mechanism for executing programs. Programmable systems are capable of executing
programs. In this chapter, our focus is on nonprogrammable systems, primarily using
a multiplier for illustration. Coverage of programmable systems begins in Chapter 10.

The algorithmic state machine (ASM), a more user-friendly version of the state
diagram for a sequential circuit, provides a representation of the behavior of the
control unit, as well as the controlled register transfers. By using register transfers in
the ASM, combined control unit and datapath behavior is represented.

The control units represented by ASMs in this chapter are implemented by using
hardwired control. Among the specialized techniques for hardwired control design, we
consider two approaches that simplify the design of larger control units when compared to the basic sequential circuit design approach in Chapter 6. Implementation of
control units using microprogrammed control is touched upon very briefly.

In this chapter, the main topics are algorithmic state machines, hardwired control, and
representation of algorithmic state machines using HDLs. Since the design
techniques are quite general, they have an impact on most of the electronic parts in
the generic computer diagram at the beginning of Chapter 1. Since the CPU and FPU
in the processor chip contain significant controls for activating and sequencing
register transfer operations, the processor is the major area impacted by the material
in this chapter.

8-1 THE CONTROL UNIT

The binary information stored in a digital computer can be classified as either data or control information. As we saw in the previous chapter, data is manipulated in a datapath by using microoperations implemented with register transfers. These operations are implemented with adder–subtractors, shifters, registers, multiplexers, and buses. The control unit provides signals that activate the various microoperations within the datapath to perform the specified processing tasks. The control unit also determines the sequence in which the various actions are performed. Because the logic design of a digital system is often treated in two distinct parts, the register and register transfer design for datapaths was covered in Chapter 7 and the design of the control unit is covered in this chapter.

Generally, the timing of all registers in a synchronous digital system is controlled by a master clock generator. The clock pulses are applied to all flip-flops and registers in the system, including those in the control unit. To prevent clock pulses from changing the state of all registers on every clock cycle, some registers have a load control signal that enables and disables the loading of new data into the register. The binary variables that control the selection inputs of multiplexers, buses, and processing logic and the load control inputs of registers are generated by the control unit.

The control unit that generates the signals for sequencing the microoperations is a sequential circuit with states that dictate the control signals for the system. At any given time, the state of the sequential circuit activates a prescribed set of microoperations. Using status conditions and control inputs, the sequential control unit determines the next state. The digital circuit that acts as the control unit provides a sequence of signals for activating the microoperations and also determines its own next state.

Based on the overall system design, there are two distinct types of control units used in digital systems, one for a programmable system and the other for a nonprogrammable system.

In a *programmable system,* a portion of the input to the processor consists of a sequence of *instructions.* Each instruction specifies the operation that the system is to perform, which operands to use, where to place the results of the operation, and, in some cases, which instruction to execute next. For programmable systems, the instructions are usually stored in memory, either in RAM or in ROM. To execute the instructions in sequence, it is necessary to provide the memory address of the instruction to be executed. This address comes from a register called the *program counter* (*PC*). As the name implies, the *PC* has logic that permits it to count. In addition, in order to change the sequence of operations using decisions based on status information from the datapath, the *PC* needs parallel load capability. So, in the case of a programmable system, the control unit contains a *PC* and associated decision logic, as well as the necessary logic to interpret the instruction. *Executing* an instruction means activating the necessary sequence of microoperations in the datapath required to perform the operation specified by the instruction.

For a *nonprogrammable system,* the control unit is not responsible for obtaining instructions from memory, nor is it responsible for sequencing the execution of those instructions. There is no *PC* or similar register in such a system. Instead, the control unit determines the operations to be performed and the sequence of those operations, based on its inputs and the status bits from the datapath.

This chapter focuses on nonprogrammable system design. It illustrates the use of algorithmic state machines (ASMs) for control unit design plus specialized techniques for ASM implementation. Programmable systems are covered in Chapters 10 and 12.

8-2 ALGORITHMIC STATE MACHINES

A processing task can be defined by register transfer microoperations controlled by a sequencing mechanism. Such a task can be specified as a hardware algorithm that consists of a finite number of procedural steps that perform the processing task. The most challenging and creative part of digital design is the formulation of hardware algorithms that achieve the required objectives. A hardware algorithm can be used as a basis for both the datapath and the control unit of a system.

A flowchart is a convenient way to specify a sequence of procedural steps and decision paths for an algorithm. A flowchart for a hardware algorithm must have special characteristics that tie it closely to the hardware implementation of the algorithm. As a consequence, we use a special flowchart called an *algorithmic state machine* (ASM) chart to define digital hardware algorithms. A *state machine* is just another term for a sequential circuit.

The ASM chart resembles a conventional flowchart, but is interpreted somewhat differently. A conventional flowchart describes procedural steps and decision paths without any concern for their relationship to time. By contrast, the ASM chart provides not only a sequence of events, but is distinguished by the fact that it describes the timing relationship between the states of the control unit and the datapath actions that occur in the states in response to clock pulses,

The ASM Chart

The ASM chart contains three basic elements: the state box, the scalar decision box, and the conditional output box, as illustrated in Figure 8-1. For convenience, a fourth element, the vector decision box, has been added. This additional component simplifies representation of multiway decisions and establishes a correspondence between HDL representations and ASM charts.

A state in the control sequence is indicated by a state box, as shown in Figure 8-1(a). The *state box* is a rectangle containing register transfer operations or output signals that are activated while the control unit is in the state. Implicitly, activation of an output signal means assigning a value of 1 to the signal. The symbolic name for the state is placed at the upper left corner of the box, and the binary code for the state, if assigned, is placed at the upper right corner of the box.

Figure 8-1(b) shows a specific example of a state box. The state has the symbolic name IDLE, and the binary code assigned to it is 000. Inside the box is the register transfer $R \leftarrow 0$ and the output *RUN*. The register transfer indicates that the

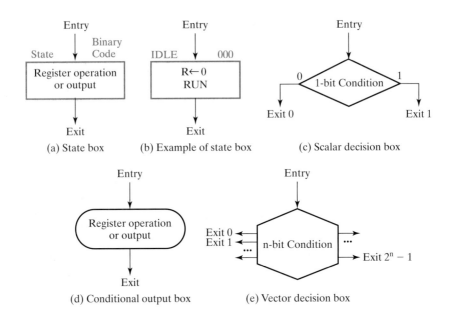

☐ **FIGURE 8-1**
ASM Chart Elements

register R is to be reset to 0 on any clock pulse that occurs while the control is in state IDLE. *RUN* indicates that the output signal *RUN* is to be 1 during the time that the control is in state IDLE. *RUN* is 1 for any state box in which it appears and is 0 for any state box in which it does not appear.

The *scalar decision box* describes the effect of an input on the control. It is a diamond-shaped box with two exit paths, as shown in Figure 8-1(c). The input condition is a single binary input variable or a single Boolean expression dependent only upon inputs. One exit path is taken if the input condition is true (1), and the other is taken if the input condition is false (0).

The third element, the *conditional output box*, is unique to the ASM chart. The oval shape of the box is shown in Figure 8-1(d). The rounded corners differentiate it from the state box. The entry path to a conditional output box from a state box must pass through one or more decision boxes. If the conditions specified on the path through the decision boxes leading from the state box to a conditional output box are satisfied, the register transfers or outputs listed inside the conditional output box are activated.

The *vector decision box* shown in Figure 8-1(e) describes the effect of a vector function of inputs on the control. It is a hexagon-shaped box with up to 2^n exit paths for an n-element binary vector. The input condition is a vector of $n > 1$ binary input variables or Boolean expressions dependent upon only the inputs. An exit path is taken if the vector value matches the label corresponding to the exit path.

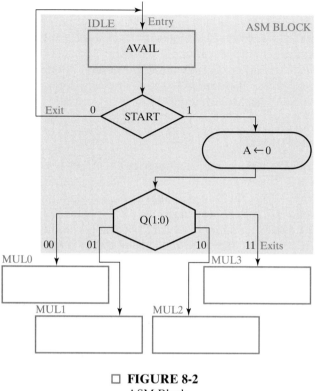

□ **FIGURE 8-2**
ASM Block

An *ASM block* consists of one state box and all of the decision and conditional output boxes connected between the state box exit and entry paths to the same or other state boxes. An example of an ASM block is shown in Figure 8-2. The block represents decisions and output actions that can take place in the state. Any outputs for which conditions are satisfied within the ASM block are activated in the block. Any register transfers for which conditions are satisfied within the ASM block will be executed when the clock event occurs. This same clock event will also transfer control to the next state as specified by decisions within the ASM block. For the block in Figure 8-2, the state is IDLE. While in the state IDLE, the output *AVAIL* is equal to 1. If *START* is 0, then the next state is IDLE. If *START* is 1, then at the clock event, *A* is cleared to all 0's, and, depending on the value of the vector $Q(1:0)$, the next state is MUL0, MUL1, MUL2, or MUL3. In the figure, the entry path and the five exit paths for the ASM block are labeled at the boundaries of the ASM block.

The ASM chart is really a form of state diagram for the sequential circuit part of the control unit. Each state box is equivalent to a node in the state diagram. The decision boxes are equivalent to input values on the lines that connect nodes in the diagram. The register transfers and outputs in the state boxes and the conditional output boxes correspond to the outputs of the sequential circuit. Outputs in a state box are those that would be specified on a state node in the state diagram with a Moore model dependency. Outputs in a conditional output box correspond to the input values on the lines connecting states in the state diagram. Since these depend on the inputs, a Mealy model dependency is present. If all dependencies in an ASM are Moore model dependencies (i.e., there are no conditional output boxes), the ASM is a Moore model. If there are one or more conditional boxes with Mealy dependency, the ASM is a Mealy model.

Timing Considerations

In order to clarify the timing considerations for the ASM, we use the sample ASM block in Figure 8-2. The timing of the events related to state IDLE is illustrated in Figure 8-3. In considering the timing of these events, recall that only positive-edge-triggered flip-flops are used. During clock cycle 1, the control unit is in present state IDLE, output *AVAIL* is 1, and input *START* is 0. Based on the ASM block, when a positive clock edge occurs, the state remains at IDLE, and *AVAIL* remains at 1. Also, the contents of register *A* remain unchanged. In clock cycle 2, *START* becomes 1. So when the next positive clock edge occurs, register *A* is cleared to 0. With *START* at 1, $Q(1:0)$ is examined and found to be 01. For this value, when the clock edge occurs, the next state becomes MUL1. The new state MUL1 and the new value of *A* both appear at the beginning of clock cycle 3. The value of *AVAIL* becomes 0, since *AVAIL* does not appear in the state box for state MUL1. Note that the output *AVAIL* = 1 appears concurrently with the present state IDLE, but the result of the register transfer for *A* appears concurrently with the next state MUL1. This is because outputs occur asynchronously in response to state and input values, but register transfers and state changes both wait until the next positive clock edge.

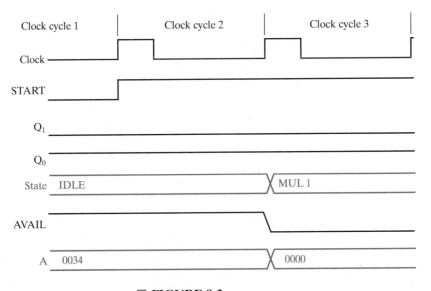

☐ FIGURE 8-3
ASM Timing Behavior

23	10111	Multiplicand
<u>19</u>	<u>10011</u>	Multiplier
	10111	
	10111	
	00000	
	00000	
	<u>10111 </u>	
437	110110101	Product

□ **FIGURE 8-4**
Hand Multiplication Example

8-3 ASM CHART EXAMPLES

A binary multiplier is used to illustrate ASM chart formulation. The multiplier multiplies two n-bit unsigned integers to produce a $2n$-bit integer result.

Binary Multiplier

In this example, we introduce a hardware algorithm for binary multiplication, propose a simple datapath and control unit for its implementation, and then describe its register transfers and control by use of an ASM. The system used for illustration multiplies two unsigned binary numbers. In Section 5-5, a hardware algorithm to execute this multiplication without using storage elements resulted in a combinational multiplier with many adders and AND gates. In contrast, the hardware algorithm developed here results in a sequential multiplier that uses only one adder and a long shift register. The algorithm is illustrated, the register transfer structure proposed, and the ASM chart formulated.

MULTIPLICATION ALGORITHM The multiplication of two unsigned binary numbers is done with paper and pencil by successive shifts of copies of the multiplicand to the left and an addition. The process is best illustrated with an actual example. Let us multiply the two binary numbers 10111 and 10011, as shown in Figure 8-4. To carry out the multiplication, we look at successive bits of the multiplier, least significant bit first. If the multiplier bit is 1, the multiplicand is copied down for use in the addition to follow. Otherwise 0's are copied down. The numbers copied in successive lines are shifted one position to the left from the previous number copied, to align them with the respective multiplier bit being processed. Finally, the numbers are added and their sum forms the product. Note that the product obtained by multiplying two n-bit binary numbers can have up to $2n$ bits for $n \geq 2$.

When the multiplication procedure is implemented with digital hardware, it is useful to change the process slightly. First, instead of having a digital circuit that adds n binary numbers simultaneously, it is less expensive to provide a circuit that adds just two numbers. Each time the multiplicand or 0's are copied, they are immediately added to a *partial product*. The partial product is stored in a register in preparation for the shift action to follow. Second, instead of shifting the copies of the multiplicand to the left, the partial product is shifted to the right. This leaves the partial product and the copy of the multiplicand in the same relative position as the left shift of the multiplicand did. But, more important, instead of a $2\,n$-bit adder, only an n-bit adder is needed. The addition always takes place in the same n positions, instead of moving to the left one bit position each time. Third, when the corresponding bit in the multiplier is 0, there is no need to add all 0's to the partial product, since this does not alter its resulting value.

The multiplication example is repeated in Figure 8-5 with these changes. Note that the initial partial product is 0. Each time the multiplier bit being processed is 1, an addition of the multiplicand, followed by a right shift, is performed. Each time the multiplier bit is a 0, only a right shift is performed. One of these two actions is performed for each bit of the multiplier, so in this case, five such actions occur. An unsigned overflow occurring during an addition is indicated in blue. This overflow is not a problem, however, since the right shift that immediately follows brings the extra partial product bit into the regular most significant bit position.

MULTIPLIER BLOCK DIAGRAM   The block diagram for the binary multiplier is shown in Figure 8-6. The multiplier datapath is first constructed from components covered in previous chapters. All but counter P are expanded to n bits; counter P requires $\lceil \log_2 n \rceil$ bits for counting the processing of the n bits of the multiplier. ($\lceil x \rceil$ denotes the smallest integer greater than or equal to x.) We use the parallel adder from Figure 5-5, a parallel-load register B similar to Figure 7-2, and parallel-load shift registers A and Q similar to Figure 7-10. Counter P is a version of the parallel-load

23	10111	Multiplicand
19	10011	Multiplier
	00000	Initial partial product
	10111	Add multiplicand, since multiplier bit is 1
	10111	Partial product after add and before shift
	010111	Partial product after shift
	10111	Add multiplicand, since multiplier bit is 1
	1000101	Partial product after add and before shift[a]
	1000101	Partial product after shift
	01000101	Partial product after shift
	001000101	Partial product after shift
	10111	Add multiplicand, since multiplier bit is 1
	110110101	Partial product after add and before shift
437	0110110101	Product after final shift

a. Note that overflow temporarily occurred.

☐ **FIGURE 8-5**
Hardware Multiplication Example

counter in Figure 7-14 that counts down instead of up, and C is a flip-flop that can be either synchronously cleared to 0 or loaded from C_{out}. These datapath components are connected as shown in Figure 8-6.

The multiplicand is loaded into register B from IN, the multiplier is loaded into register Q from IN, and the partial product is formed in register A and stored in registers A and Q. This dual use of Q is possible because we use a right shift of the multiplier in Q to examine each successive multiplier bit that appears in Q_0. The right shift vacates the most significant bit in register Q. This space accepts the least significant bit of the partial product from A as it is shifted. The n-bit binary adder is used for adding B to A. The C flip-flop stores the carry C_{out}, whether 0 or 1, from the addition and is reset to 0 during the right shift. In order to count the number of add-shift or shift actions that are to occur, counter P is provided. It is initially set to $n-1$ and counted down after the formation of each partial product. The value in P is checked just before it is decremented. So n operations occur, one operation for each value in P, $n-1$ down through 0. Each operation is either an add and shift or just a shift. When P contains 0, the final product is in the double register A and Q, and processing stops.

The control unit stays in an initial state until the Go signal G becomes 1. Then the system starts the multiplication. The sum of A and B forms the n most significant bits of the partial product, which is transferred back to A. C_{out} from the addition is transferred to C. Both the partial product and the multiplier in A and Q are shifted to the right. The carry from C shifts into the most significant bit of A,

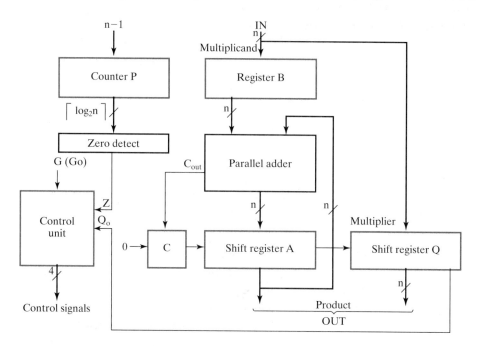

□ **FIGURE 8-6**
Block Diagram for Binary Multiplier

the least significant bit of A shifts into the most significant bit of Q, and the least significant bit of Q is discarded. After this right-shift operation, one (additional) bit of the partial product has transferred into Q, and the multiplier bits have shifted one position to the right. In this manner, the least significant bit of Q, Q_0, always holds the bit of the multiplier that the control unit examines next. The control unit "decides" whether to add, based on the value of this bit. It also checks signal Z, which is 1 for P equal to zero and 0 for P nonzero, to determine whether the multiplication is finished. Q_0 and Z are the status inputs for the control unit, with input G as the only external control input. The control signals from the control unit to the datapath activate the required microoperations.

MULTIPLIER ASM CHART   An ASM chart giving the sequence of operations in the binary multiplier is shown in Figure 8-7. Initially, the multiplicand is in B and the multiplier in Q. The loading of these two registers is not handled explicitly by the multiplier control unit. As long as the ASM is in state IDLE, and G is 0, no actions occur, and the ASM remains in IDLE. The multiplication process starts when G becomes 1. As the ASM moves from state IDLE to state MUL0, registers C and A are cleared to 0, and the counter P is loaded with the constant $n - 1$. In state MUL0, a decision is made based upon Q_0, the least significant bit of Q. If Q_0 is 1, the contents of B are added to those of A, with the result transferred to A and the carry transferred to C. If Q_0 is 0, register A and bit C are left unchanged. In both cases, the next state is MUL1.

In state MUL1, a right shift is performed on the combined contents of C, A, and Q. This shift can be expressed by the somewhat messy list of five simultaneous register transfers:

$$C \leftarrow 0, A(n-1) \leftarrow C, A \leftarrow \text{sr } A, Q(n-1) \leftarrow A(0), Q \leftarrow \text{sr } Q$$

To simplify representation of this operation, we add a bit of notation, using ‖ to define a *composite register* made up of other registers or pieces of other registers. This operation, ‖, is called *concatenation*. For example,

$$C \parallel A \parallel Q$$

represents a single register obtained by combining registers C, A, and Q from the most significant end to the least significant end. We can use this composite register to represent the right shift

$$C \parallel A \parallel Q \leftarrow \text{sr } C \parallel A \parallel Q$$

as shown in Figure 8-7. Recall that we are assuming that the leftmost bit of the result for a right shift takes on the value 0 unless otherwise specified, so C becomes 0. This is represented explicitly, however, in the ASM chart, since C is set to 0 in another state as well. The explicit listing allows $C \leftarrow 0$ to be performed by using a single control signal for both states.

Counter P is decremented in MUL1. The value in P is checked in state MUL1 before P is decremented. This illustrates a very important timing difference between a standard flowchart and an ASM chart. The decision on Z, which represents $P = 0$, follows the register transfer statement that updates P in the ASM

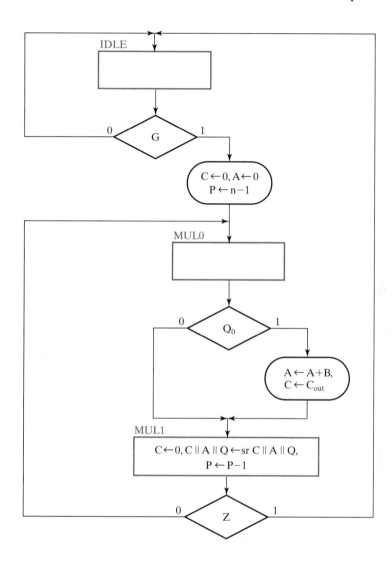

□ **FIGURE 8-7**
ASM Chart for Binary Multiplier

chart. Since the decision on P is performed asynchronously and the register transfer statement is synchronous with the next positive clock edge, the decision on P
precedes the update of P. At the next clock edge, when P is updated, the result of
this decision is available to determine the next state. The first $n - 1$ times that P is
checked, its content is nonzero, so status bit Z remains 0, and the loop, consisting
of states MUL0 and MUL1, is executed again. The nth time P is checked, the content of P is zero, so status bit Z is 1. This indicates that the multiplication is complete, causing the ASM to return to state IDLE. The final product is available in A
‖ Q, with A holding the n most significant bits and Q the n least significant bits of

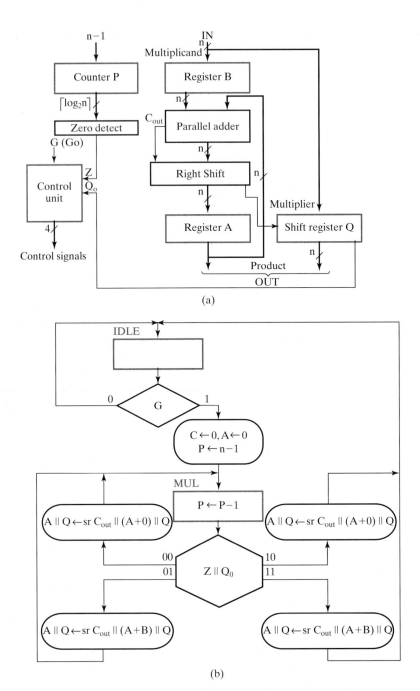

(a)

(b)

□ **FIGURE 8-8**
Alternative Binary Multiplier

the product. It is worthwhile to reexamine the hardware multiplication example for $n = 5$ in Figure 8-5, this time considering the relationship to the datapath and the flow of the ASM chart.

The type of registers selected for the datapath correspond to the microoperations listed in the ASM chart. Register A is a shift register with parallel load that accepts the sum from the adder. It also needs a synchronous clear to reset the register to 0. Register Q is a shift register. The C flip-flop needs to accept the input carry and also needs a synchronous clear. Registers B and Q also need parallel load in order to load the multiplicand and the multiplier prior to the start of the multiplication process.

Figure 8-8 shows an alternative multiplier design that uses the vector decision box in its ASM chart. In part (a) of the figure, Shift Register A has been replaced by a combinational right shifter that shifts 1 bit to the right, similar to Figure 5-13(c), and a register with load enable as shown in Figure 7-2. This combines the adder with a combinational right shift. This permits the number of states used in the ASM chart in Figure 8-8 to be reduced to just one. In order to represent the change in the multiplier datapath, it is necessary to write a register transfer statement that combines the addition with the shift. Also, the datapath change permits the flip-flop C to be deleted. Assuming that the combined delay of the adder and right shift (which consists only of wires) is no more than the adder, the reduction in states in the multiply loop substantially speeds up the multiplication operation.

To illustrate the vector decision box, we have used concatenation to combine Z and Q_0 into the vector (Z, Q_0) which is denoted as $Z \parallel Q_0$. The decision based on this vector is shown in the center of the ASM chart in Figure 8-8(b). There are four output combinations. For the combinations in which Z is 1, the next state is IDLE. For the combinations in which $Z = 0$, the next state is MUL. For the combinations in which $Q_0 = 1$, the output is an add-right shift with input operands A and B, and for the combinations in which $Q_0 = 0$, the output is an add-right shift with input operands A and 0. These are represented by the combined add and shift transfers in conditional output boxes for the four output combinations for $Z \parallel Q_0$.

8-4 HARDWIRED CONTROL

In implementing a control unit, two distinct aspects must be considered: the control of the microoperations and the sequencing of the control unit and microoperations. Very simply put, the first has to do with the part of the control that generates the control signals, and the second has to do with the part of the control that determines what happens next. Here, we separate these two aspects by dividing the original ASM specification into two parts: a table that defines the control signals in terms of states and inputs, and a simplified ASM chart that represents only transitions from state to state. Although we are separating these two aspects for design purposes, they can share logic.

The control signals are based on the ASM chart. The control signals needed for the multiplier datapath are listed in Table 8-1, where we have chosen to examine the datapath registers and tabulate the microoperations for each register. Based on the tabulated microoperations, the control signals are defined. A control

☐ **TABLE 8-1**
   Control Signals for Binary Multiplier

Block Diagram Module	Microoperation	Control Signal Name	Control Expression
Register A:	$A \leftarrow 0$	Initialize	IDLE $\cdot$ G
	$A \leftarrow A + B$	Load	MUL0 $\cdot$ Q
	$C \parallel A \parallel Q \leftarrow$ sr $C \parallel A \parallel Q$	Shift_dec	MUL1
Register B:	$B \leftarrow IN$	Load_B	*LOADB*
Flip-Flop C:	$C \leftarrow 0$	Clear_C	IDLE $\cdot$ G + MUL1
	$C \leftarrow C_{out}$	Load	—
Register Q:	$Q \leftarrow IN$	Load_Q	*LOADQ*
	$C \parallel A \parallel Q \leftarrow$ sr $C \parallel A \parallel Q$	Shift_dec	—
Counter P:	$P \leftarrow n - 1$	Initialize	—
	$P \leftarrow P - 1$	Shift_dec	—

signal can be used for activating microoperations in more than one register. This is reasonable, in this case, since the datapath is dedicated to only one operation, multiplication. Thus, the control signals do not need to be separated to provide the generality required for implementing additional, potentially unknown operations. Finally, the Boolean expression for each control signal is derived from the location or locations of the microoperation in the ASM chart. For example, for register A, there are three microoperations shown in Table 8-1: clear, add and load, and right shift. Since the clear operation always occurs at the same time as the clear for flip-flop C and the loading of counter P, all of these microoperations can be activated by the same control signal, named Initialize. Because C is cleared in state MUL1 as well, however, we choose to separate its control signal. So Initialize is used for clearing A and loading P. In the last column for Initialize, the Boolean expression for which Initialize is to be active, as determined from the ASM chart, is given in terms of the state IDLE and input G. Since Initialize is to be 1 when G is 1 in state IDLE, IDLE and G are ANDed. At this point, the name for the state is treated as a Boolean variable. Depending on the implementation, there may be such a signal representing the state, or the state may need to be expressed as a function of the state variables. The signal for clearing C, Clear_C, is to be active in state IDLE for G equal to 1, as well as in state MUL1. Thus, G is ANDed with IDLE, and the result is ORed with MUL1. The other two internal multiplier control signals, Load and Shift_Dec, are defined in a similar manner. The final two signals, Load_B and Load_Q, load the multiplicand and multiplier from outside the multiplier system. These signals will not be considered explicitly in the remainder of the design.

With the information on microoperations removed, we can redraw the ASM chart so that only the information on sequencing is represented. This modified ASM chart for the binary multiplier appears in Figure 8-9. Note that all of the conditional output boxes have been removed. In addition, any decision box not affecting the next state is removed. In particular, in Figure 8-7, the decision box Q_0

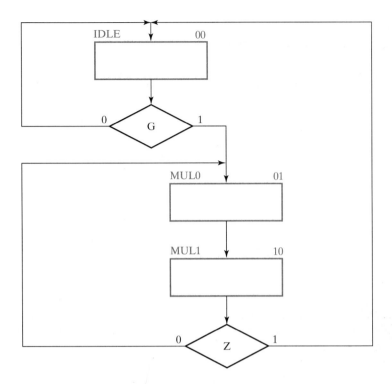

□ **FIGURE 8-9**

Sequencing Part of ASM Chart for the Binary Multiplier

affected only a conditional output box. Once that conditional output box is removed, the two exit paths from decision box Q_0 clearly go to the same state. So this decision box has no effect on the next state and is removed.

From this modified ASM chart, we can design the sequencing part of the control unit (i.e., the part that represents the next-state behavior). The division of control into next-state behavior in the form of the modified ASM chart and output behavior in the form of the control signal table shows how the ASM corresponds to the next-state and output parts of a sequential circuit. Figure 8-9 corresponds to the state diagram of a sequential circuit without the outputs specified, except that the representations used in the diagram for states and transitions are different. Because of this correspondence, we can treat the ASM chart as a state diagram and form a state table for the sequencing part of the control unit. Then the control unit can be designed by the sequential logic design procedure, as outlined in Chapter 4. However, in many cases, this method is difficult to carry out because of the large number of states for a typical control unit. As a consequence, we use specialized methods for control unit design that are variations of the classical sequential logic methods. We next present and illustrate two such design methods.

Sequence Register and Decoder

The sequence register and decoder method, as the name implies, uses a sequence register for the control states and a decoder to provide an output signal corresponding to each of the states. A register with n flip-flops can have up to 2^n states and an n-to-2^n decoder has up to 2^n outputs, one for each of the states. An n-bit sequence register is essentially n flip-flops, together with the associated gates that effect their state transitions. Additional logic may be required to produce the necessary control signal outputs.

   The sequencing part of the ASM chart for the binary multiplier has three states and two inputs. To implement the ASM chart with a sequence register and decoder, we need two flip-flops for the register and a 2-to-4-line decoder. Since there are three states, only three of the four decoder outputs are used. Although this is a simple example, the procedure to be outlined applies to more complex situations as well.

   The state table for the sequencing part of the control unit is shown in Table 8-2; it is derived directly from the ASM chart in Figure 8-9. We designate the two flip-flops as M_1 and M_0 and assign the binary states 00, 01, and 10 to IDLE, MUL0, and MUL1, respectively. Note that the input columns have unspecified entries ($\times$) whenever an input variable is not used to determine the next state. The outputs of the sequencing part of the control are designated by the state names. The binary code for the present state determines the particular output variable that is equal to 1 at any given time. Thus, when the present state is $M_1 M_0 = 00$, output IDLE equals 1, while the other outputs equal 0. Since these outputs depend on the present state only, they can be generated with the 2-to-4-line decoder having inputs M_1 and M_0 and outputs IDLE, MUL0, and MUL1.

   As mentioned earlier, the sequential circuit can be designed from the state table by means of the sequential logic design procedure presented in Chapter 4. This example has a small number of states and inputs, so we could use maps to simplify the Boolean functions. In most control logic applications, however, the number of states is much larger. The application of the conventional method

☐ **TABLE 8-2**
State Table for Sequence Register and Decoder Part
of Multiplier Control Unit

Present state			Inputs		Next state		Decoder Outputs		
Name	M_1	M_0	G	Z	M_1	M_0	IDLE	MUL0	MUL1
IDLE	0	0	0	$\times$	0	0	1	0	0
	0	0	1	$\times$	0	1	1	0	0
MUL0	0	1	$\times$	$\times$	1	0	0	1	0
MUL1	1	0	$\times$	0	0	1	0	0	1
	1	0	$\times$	1	0	0	0	0	1
—	1	1	$\times$	$\times$	$\times$	$\times$	$\times$	$\times$	$\times$

requires excessive work to obtain the simplified input equations for the flip-flops. Here, the design can be simplified if we take into consideration the fact that the decoder outputs are available for use in the design. Instead of using flip-flop outputs as the present state conditions, we might as well use the outputs of the decoder to obtain this information. These outputs supply a single signal representing each of the possible present states of the circuit. Moreover, instead of using maps to simplify the flip-flop equations, we can obtain them directly by inspection of the state table. For example, from the next-state conditions in the table, we find that the next state of M_0 is equal to 1 when the present state is IDLE and input G is equal to 1 or when the present state is MUL1 and input Z is equal to 0. These conditions give

$$D_{M_0} = \text{IDLE} \cdot G + \text{MUL1} \cdot \overline{Z}$$

for the D input of the M_0 flip-flop. Similarly, the D input of the M_1 flip-flop is

$$D_{M_1} = \text{MUL0}$$

Note that these equations derived by inspection from the state table use the state names rather than the state variable names, since the decoder producing the state symbols is present. In some cases, it may be possible to find simpler D flip-flop input equations by using the state variables directly instead of the states. We can remove redundancy and reduce cost by writing the Boolean equations for the decoder and applying a simplification program to the set of control equations.

The logic diagram for the control appears in Figure 8-10. It consists of a two-bit register with flip-flops M_1 and M_0 and a 2-to-4-line decoder. The three outputs

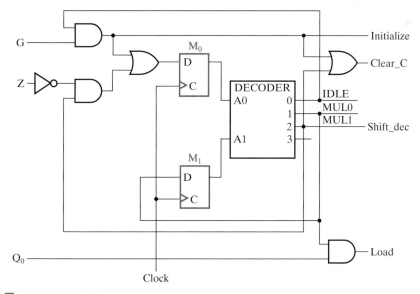

☐ **FIGURE 8-10**
Control Unit for Binary Multiplier Using a Sequence Register and a Decoder

of the decoder are used to generate the control outputs, as well as inputs to the next-state logic. The outputs Initialize, Clear_C, Shift_dec, and Load are determined from Table 8-1. Initialize and Shift_dec are already available as signals, so that only labeled output lines are added. However, as shown in the figure, we must add logic gates for Clear_C and Load. We complete the binary multiplier design by connecting the outputs of the control unit to the control inputs of the datapath.

One Flip-Flop per State

Another possible method of control logic design is the use of one flip-flop per state. A flip-flop is assigned to each of the states, and at any time, only one of the flip-flops contains a 1, with all the rest containing 0. When the 1 is in the flip-flop assigned to a particular state, the sequential circuit is in that same state. The single 1 propagates from one flip-flop to another under the control of decision logic. In such a configuration, each flip-flop represents a state that is present only when the single 1 is stored in the flip-flop.

It is obvious that, short of some error detection or correction techniques, this method uses the maximum number of flip-flops for the sequential circuit. For example, a sequential circuit with 12 states using minimum variable encoding needs four flip-flops. With one flip-flop per state, the circuit requires 12 flip-flops, one for each state. At first glance, it may seem that this method would increase the cost of the system, since more flip-flops are used. But the method offers some cost advantages that may not be apparent. One advantage is the simplicity with which the logic can be designed—merely by inspection of the ASM chart or state diagram. No state or excitation tables are needed if D flip-flops are employed. This offers a savings in design effort.

Figure 8-11 shows the symbol replacement rules for transforming an ASM chart into a sequential circuit with one flip-flop per state. These rules are most easily applied to an ASM chart representing only sequencing information, such as that of Figure 8-9. Each rule specifies the replacement of a component of an ASM chart with a logic circuit. As shown in Figure 8-11(a), the state box is replaced by a D flip-flop labeled with the name of the state. The entry to the state box corresponds to the D input to the flip-flop. The exit of the state box corresponds to the output of the flip-flop.

In Figure 8-11(b), the scalar decision box is replaced by a 2-way demultiplexer. The signal corresponding to the entry to the decision box is sent to one of two exit lines, depending on the value of signal X. If X is 0, the signal is sent to the exit 0 line; if X is 1, the signal is sent to the exit 1 line. So, for example, if the single 1 in the circuit is on the entry to the decision box, and X is 0, the 1 is passed to the exit 0 line. The demultiplexer acts like a switch that directs the 1 through the paths in the circuit corresponding to paths in the ASM chart.

In Figure 8-11(c), the vector decision box is replaced by an n-way demultiplexer. The signal corresponding to the entry to the decision box is sent to one of the $2^n - 1$ lines, depending on the value of the signal vector $X = X_0, ..., X_{n-1}$. If X is 0, the signal is sent to the exit 0 line; if X is 9, the signal is sent to the exit 9 line. So, for example, if the single 1 in the circuit is on the entry to the decision box, and X is 9,

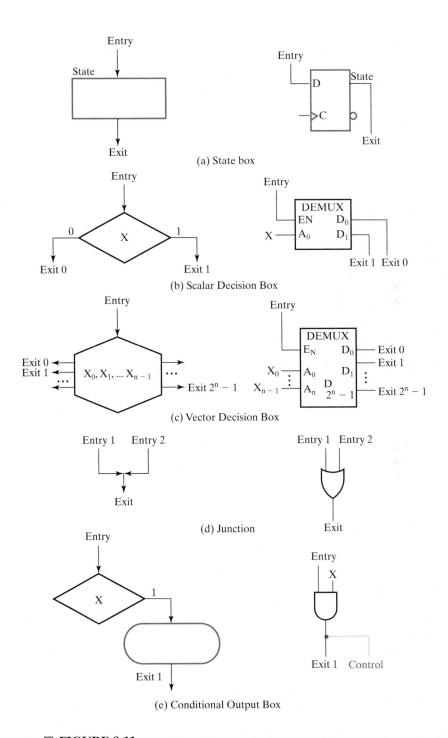

(a) State box

(b) Scalar Decision Box

(c) Vector Decision Box

(d) Junction

(e) Conditional Output Box

□ **FIGURE 8-11**
Transformation Rules for Control Unit with One Flip-Flop per State

the 1 is passed to the exit 9 line. The demultiplexer acts like a switch that directs the 1 through the paths in the circuit corresponding to paths in the ASM chart.

The junction in Figure 8-11(d) is any point at which two or more directed lines in the ASM chart join together. If a 1 is present in the circuit on any line corresponding to one of the entry paths, then it must appear on the line corresponding to the exit path, giving that line the value 1. If none of the lines corresponding to entry paths into the junction have the value 1, then the exit line must have the value 0. Thus, the junction is replaced by an OR gate.

With these four transformations, the sequencing part of the ASM chart can be replaced by a circuit with one flip-flop per state, just by inspection. In order to handle outputs, it is merely a matter of attaching control lines to the proper locations in the circuit or adding output logic. The outputs are based on the original ASM chart or the control signal table derived from the chart. Attaching a control line based on an ASM chart is illustrated by the conditional output box shown in Figure 8-11(e). The conditional output box in the ASM chart is just replaced by a connection in the circuit. But to cause the output actions to happen, a control line is tapped from the connection and labeled with the output variable. The transformation is shown in blue for clarity.

We now use these transformations to find the control unit with one flip-flop per state for the binary multiplier.

EXAMPLE 8-1 Binary Multiplier

The ASM chart in Figure 8-9 will be used for the sequencing part of the design. Note that the binary codes given are ignored, since they were for the former design approach. The resulting logic diagram is shown in Figure 8-12.

First, we replace each of the three state boxes by a *D* flip-flop labeled with the name of the state, as indicated by the circled 1's in the figure. Second, each of the decision boxes is replaced by a demultiplexer with the decision variable as its selection input, as indicated by the circled 2's in the figure. Third, each junction is replaced by an OR gate, as indicated by the circled 3's. Finally, the connections represented by the directed lines in the ASM chart are added from the outputs to the inputs of the corresponding components.

To handle the control outputs, we can use either Table 8-1 or the original ASM chart in Figure 8-7. From the table, we see that the Boolean function for Initialize is already available in the logic diagram, so we simply add the output labeled Initialize. Likewise, the output for Shift_dec can be added. For Clear_C and Load, however, logic gates are added. All of the output connections and logic added are designated by the circled 4's in Figure 8-12.

One final issue in the design of the control logic with one flip-flop per state is initialization to the state having a 1 in the IDLE flip-flop and a 0 in all of the others. This can be done by using an asynchronous PRESET input on the IDLE flip-flop and an asynchronous CLEAR on the other flip-flops. If only an asynchronous CLEAR is available, rather than both PRESET and CLEAR, a NOT gate can be placed just before the *D* input and another NOT gate just after the

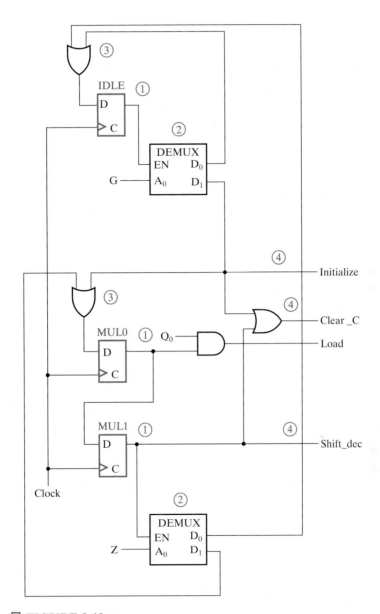

□ **FIGURE 8-12**
Control Unit with One Flip-Flop per State for the Binary Multiplier

output of the IDLE flip-flop. Then the IDLE flip-flop will actually contain a 0 when in state IDLE and a 1 at all other times. This permits the asynchronous CLEAR to be used to initialize all three flip-flops in the circuit. It should be noted that, other than for resetting the circuit, the use of asynchronous flip-flop inputs for implementing ASMs or other sequential circuits is generally poor design practice. ■

Once the basic design of the control logic with one flip-flop per state is completed, it may be desirable to refine the design. For example, if there are a number of junctions connected together by lines, the OR gates that resulted from the transformation may be combined. Also, demultiplexers cascaded with each other may be combined. Other logic reduction or and technology mapping may also be applied to the design.

8-5 HDL REPRESENTATION OF THE BINARY MULTIPLIER—VHDL

The binary multiplier just studied can be represented during the design process as a behavioral VHDL description. Such a description for a 4-bit version of the multiplier appears in Figures 8-13 and 8-14. This VHDL code represents the block diagram in Figure 8-6 and the ASM chart in Figure 8-7. The VHDL code consists of entity `binary_multiplier` and an architecture `behavior_4`. The architecture contains two assignment statements and three processes. The processes are similar to those used for the sequence recognizer in Chapter 6. The primary difference is that the output function process has been replaced by a process describing the datapath register transfers. Due to this change, the VHDL representation corresponds more closely to the description in Table 8-1 and the ASM chart in Figure 8-9 than to the ASM chart in Figure 8-7.

In the entity, multiplier inputs and outputs are defined. At the beginning of the architecture, a type declaration defines the three states. Internal signals, some of which will generate registers are declared next. Among these are `state` and `next_state` for the control, registers A, B, P and Q, and flip-flop C. Also, intermediate signal Z is declared for convenience. Next, an assignment is made which forces Z to be 1 whenever P contains value 0. Following this, the outputs of concatenated registers A and Q are assigned to the multiplier output `MULT_OUT`. This is necessary, rather than making A and Q circuit outputs, to permit A and Q to be used within the circuit.

The remainder of the description consists of the three processes. The first process describes the state register and includes a `RESET` as well as the clocking. The second process describes the next state function from Figure 8-8. Note that, since clocking and `RESET` are included in the state register, they do not appear here. In the sensitivity list, all signals that can affect the next state, G, Z, and `state` are included. Otherwise, this process resembles that for the `next_state` process in the sequence recognizer.

The final process in Figure 8-14 describes the datapath function. Since the conditions for performing an operation are defined in terms of the states and inputs, this process also implicitly defines the control signals given in Table 8-1. These control signals do not appear explicitly, however. Since the datapath function has registers as all assignment destinations, all transfers are controlled by `CLK`. Since contents will be loaded into these registers before the multiply operation is ever performed, it is unnecessary to provide a reset for these registers. The first **if** statement controls the loading of the multiplicand in register B and the second **if** statement controls the loading of the multiplier into register Q.

```vhdl
-- Binary Multiplier with n = 4: VHDL Description
-- See Figures 8-6 and 8-7 for block diagram and ASM Chart
library ieee;
use ieee.std_logic_1164.all;
use ieee.std_logic_unsigned.all;
entity binary_multiplier is
  port(CLK, RESET, G, LOADB, LOADQ: in std_logic;
    MULT_IN: in std_logic_vector(3 downto 0);
    MULT_OUT: out std_logic_vector(7 downto 0));
end binary_multiplier;

architecture behavior_4 of binary_multiplier is
  type state_type is (IDLE, MUL0, MUL1);
  signal state, next_state : state_type;
  signal A, B, Q: std_logic_vector(3 downto 0);
  signal P: std_logic_vector(1 downto 0);
  signal C, Z: std_logic;
begin
  Z <= P(1) NOR P(0);
  MULT_OUT <= A & Q;

  state_register: process (CLK, RESET)
  begin
    if (RESET = '1') then
      state <= IDLE;
    elsif (CLK'event and CLK = '1') then
      state <= next_state;
    end if;
  end process;

  next_state_func: process (G, Z, state)
  begin
    case state is
      when IDLE =>
        if G = '1' then
          next_state <= MUL0;
        else
          next_state <= IDLE;
        end if;
      when MUL0 =>
        next_state <= MUL1;
      when MUL1 =>
        if Z = '1' then
          next_state <= IDLE;
        else
          next_state <= MUL0;
        end if;
```

□ **FIGURE 8-13**
VHDL Description of a Binary Multiplier

```
      end case;
  end process;

datapath_func: process (CLK)
variable CA: std_logic_vector(4 downto 0);
begin
  if (CLK'event and CLK = '1') then
    if LOADB = '1' then
      B <= MULT_IN;
    end if;
    if LOADQ = '1' then
      Q <= MULT_IN;
    end if;
    case state is
      when IDLE =>
        if G = '1' then
          C <= '0';
          A <= "0000";
          P <= "11";
        end if;
      when MUL0 =>
        if Q(0) = '1' then
          CA := ('0' & A) + ('0' & B);
        else
          CA := C & A;
        end if;
          C <= CA(4);
          A <= CA(3 downto 0);
      when MUL1 =>
        C <= '0';
        A <= C & A(3 downto 1);
        Q <= A(0) & Q(3 downto 1);
        P <= P - "01";
    end case;
  end if;
end process;
end behavior_4;
```

☐ **FIGURE 8-14**
VHDL Description of a Binary Multiplier (Continued)

The register transfers directly involved in the multiplication are controlled by a **case** statement dependent upon the control state, input G, and internal signals Q(0) and Z. These transfers are outlined in Figure 8-7 and Table 8-1. Representation of the addition in state MUL0 requires some effort. First of all, to perform addition on std_logic vectors, a **use** statement appears just before the entity declaration for the package ieee.std_logic_unsigned.**all**. In addition to the sum from the addition, we also need to transfer the carry out, C_{out}, from the addition into C. To achieve this, we perform a 5-bit addition with 0's appended to the left of A and B and the result assigned to a 5-bit variable CA. The alternative would be to write C & A as the transfer destination, but use of concatenation & in destinations is not permitted in VHDL. Since CA is a variable, its value is assigned immediately and is available for assignment to C and A after the **if** statement. In state MUL1, the shift is performed by using concatenation, as was done in the example in Chapter 5. *P* is decremented by subtracting a 2-bit constant with value 1.

This description can be simulated to validate its correctness and synthesized to automatically produce the logic if desired.

8-6 HDL REPRESENTATION OF THE BINARY MULTIPLIER—VERILOG

The binary multiplier just studied can be represented during the design process as a behavioral Verilog description. Such a description for a 4-bit version of the multiplier appears in Figures 8-15 and 8-16. This Verilog code represents the block diagram in Figure 8-6 and the ASM chart in Figure 8-7. The Verilog code is contained in a module binary_multiplier_v. The description contains two assignment statements and three processes. The processes are similar to those used for the sequence recognizer in Chapter 4. The primary difference is that the output function process has been replaced by a process describing the datapath register transfer. Due to this change, the Verilog representation corresponds more closely to the description in Table 8-1 and the ASM chart in Figure 8-9 than to the ASM chart in Figure 8-7.

At the beginning of the description, multiplier inputs and outputs are defined. A parameter declaration defines the three states and their binary codes. Internal signals of type register are defined. Among these are the state and next_state for the control, registers A, B, P and Q, and flip-flop C. Based on clocking specifications, most of these will become actual positive-edge-triggered registers. The notable exception is next_state. Also, intermediate signal Z of type wire is declared for convenience. Next, an assignment is made which forces Z to be 1 whenever P contains value 0. This assignment uses the operation OR (|)as a *reduction operator*. Reduction is the application of an operator to a wire or register that combines the individual bits. In this case, the application of OR to P causes all bits of P to be ORed together. Since the OR is preceded by a ~, that overall operation performed is a NOR. Other operators may also be applied as reduction operators. The second assignment statement assigns the outputs of concatenated registers A and Q to the multiplier output MULT_OUT. This is done for convenience to make that output a single structure.

The remainder of the description consists of the three processes. The first process describes the state register and includes a RESET as well as the clocking. The second process describes the next state function from Figure 8-9. Note that since clocking and RESET are included in the state register, they do not appear here. In the event control statement, all signals that can affect the next state, G, Z, and state are included. Otherwise, this process resembles that for the next state process in the sequence recognizer.

```
// Binary Multiplier with n = 4: Verilog Description
// See Figures 8-6 and 8-7 for block diagram and ASM Chart

module binary_multiplier_v (CLK, RESET, G, LOADB, LOADQ,
        MULT_IN, MULT_OUT);
input CLK, RESET, G, LOADB, LOADQ;
input [3:0] MULT_IN;
output [7:0] MULT_OUT;
reg [1:0] state, next_state, P;
parameter IDLE = 2'b00, MUL0 = 2'b01, MUL1 = 2'b10;
reg [3:0] A, B, Q;
reg C;
wire Z;

assign Z = ~| P;
assign MULT_OUT = {A,Q};

//state register
always@(posedge CLK or posedge RESET)
begin
  if (RESET == 1)
    state <= IDLE;
  else
  state <= next_state;
end

//next state function
always@(G or Z or state)
begin
  case (state)
    IDLE:
      if (G == 1)
        next_state <= MUL0;
      else
        next_state <= IDLE;
    MUL0:
      next_state <= MUL1;
    MUL1:
      if (Z == 1)
        next_state <= IDLE;
      else
        next_state <= MUL0;
  endcase
end

//datapath function
always@(posedge CLK)
```

☐ **FIGURE 8-15**
Verilog Description of a Binary Multiplier

```verilog
begin
  if (LOADB == 1)
    B <= MULT_IN;
  if (LOADQ == 1)
    Q <= MULT_IN;
  case (state)
    IDLE:
      if (G == 1)
        begin
          C <= 0;
          A <= 4'b0000;
          P <= 2'b11;
        end
    MUL0:
      if (Q[0] == 1)
        {C, A} = A + B;
    MUL1:
      begin
        C <= 1'b0;
        A <= {C, A[3:1]};
        Q <= {A[0], Q[3:1]};
        P <= P - 2'b01;
      end
  endcase
end
endmodule
```

□ **FIGURE 8-16**

Verilog Description of a Binary Multiplier (Continued)

The final process describes the datapath function. Since the conditions for performing an operation are defined in terms of the states and inputs, this process also implicitly defines the control signals given in Table 8-1. These control signals do not appear explicitly, however. Since the datapath function has registers as all assignment destinations, all transfers are controlled by CLK. Since contents will be loaded into these registers before the multiply operation is ever performed, it is unnecessary to provide a reset for these registers. The first if statement controls the loading of the multiplicand into register B and the second if statement controls the loading of the multiplier into register Q.

The register transfers directly involved in the multiplication are controlled by a **case** statement dependent upon the control state, input G, and internal signals $Q(0)$ and Z. These transfers are outlined in Figure 8-7 and Table 8-1. Representation of the addition in state MUL0 uses concatenation of C and A to obtain the carry out, C_{out}, for loading into C. Verilog does permit the used of two 4-bit operands with a 5-bit result for the addition. In state MUL1, the shift is performed by using concatenation as was done in the example in Chapter 5. P is decremented by subtracting a 2-bit constant with value 1.

This description can be simulated to validate its correctness and synthesized to automatically produce the logic if desired.

8-7 MICROPROGRAMMED CONTROL

A control unit with its binary control values stored as words in memory is called a *microprogrammed control*. Each word in the control memory contains a *microinstruction* that specifies one or more microoperations for the system. A sequence of microinstructions constitutes a *microprogram*. The microprogram is usually fixed at the system design time and so is stored in ROM. Microprogramming involves placing representations for combinations of values of control variables in words of ROM. These representations are accessed via successive read operations for use by the rest of the control logic. The contents of a word in ROM at a given address specify the microoperations to be performed for both the datapath and the control unit. A microprogram can also be stored in RAM. In this case, it is loaded at system startup from some form of nonvolatile storage, such as a magnetic disk. With either ROM or RAM, the memory in the control unit is called *control memory*. If RAM is used, the memory is referred to as *writable control memory*.

Figure 8-17 shows the general configuration of a microprogrammed control. The control memory is assumed to be a ROM within which all control microprograms are permanently stored. The *control address register* (*CAR*) specifies the address of the microinstruction. The *control data register* (*CDR*), which is optional, may hold the microinstruction currently being executed by the datapath and the control unit. One function of the control word is to determine the address of the next microinstruction to be executed. This microinstruction may be the next one in sequence, or it may be located somewhere else in the control memory. Therefore, one or more bits that specify the method for determining the address of the next microinstruction are present in the current microinstruction. The next address may also be a function of status and external control inputs. When a microinstruction is executed, the *next-address generator* produces the next address. This address is transferred to the *CAR* on the next clock pulse and is used to read the next microinstruction to be executed from ROM. Thus, the microinstructions contain bits for activating microoperations in the datapath and bits that specify the sequence of microinstructions executed.

The next-address generator, in combination with the *CAR*, is sometimes called a microprogram *sequencer*, since it determines the sequence of instructions read from control memory. The address of the next microinstruction can be specified in several ways, depending on the sequencer inputs. Typical functions of a microprogram sequencer are incrementing the *CAR* by one and loading the *CAR*. Possible sources for the load operation include an address from control memory, an externally provided address, and an initial address to start control unit operation.

The *CDR* holds the present microinstruction while the next address is computed and the next microinstruction is read from memory. The *CDR* breaks up the long combinational delay paths through the control memory followed by the datapath. Its presence allows the system to use a higher clock frequency and process information faster. The inclusion of a *CDR* in a system, however, complicates the

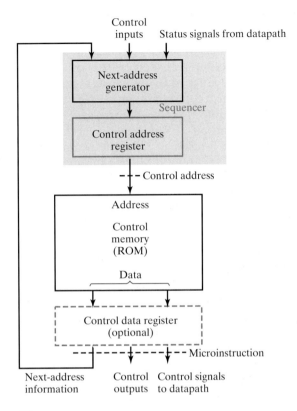

Control
inputs   Status signals from datapath

Next-address
generator

Sequencer

Control address
register

- -+- - Control address

Address

Control
memory
(ROM)

Data

Control data register
(optional)

- - -+ - - - - + - - + - - Microinstruction

Next-address   Control   Control signals
information   outputs   to datapath

□ **FIGURE 8-17**
Microprogrammed Control Unit Organization

sequencing of microinstructions, particularly when decisions are made based on status bits. For simplicity in our brief discussion, we omit the *CDR* and take the microinstructions directly from the ROM outputs. The ROM operates as a combinational circuit, with the address as the input and the corresponding microinstruction as the output. The contents of the specified word in ROM remain on the output lines as long as the address value is applied to the inputs. No read/write signal is needed, as it is with RAM. Each clock pulse executes the microoperations specified by the microinstruction and also transfers a new address to the *CAR*. In this case, the *CAR* is the only component in the control that receives clock pulses and stores state information. The next-address generator and the control memory are combinational circuits. Thus, the state of the control unit is given by the contents of the *CAR*.

Microprogrammed control has been a very popular alternative implementation technique for control units for both programmable and nonprogrammable systems. However, as systems have become more complex and performance specifications have increased the need for concurrent parallel sequences of activities, the lockstep nature of microprogramming has become less attractive for control unit implementation. Further, a large ROM or RAM tends to be much slower than the corresponding combinational logic. Finally, HDLs and synthesis tools facilitate the design of complex control units without the need for a lockstep programmable design approach. Overall, microprogrammed control for the design of control units, particularly direct datapath control in CPUs, has declined significantly. However, a new flavor of microprogrammed control has emerged, for implementing legacy computer architectures. These architectures have instruction sets that do not follow contemporary architecture principles. Nevertheless, such architectures must be implemented due to massive investments in software that uses them. Further, the contemporary architecture principles must be used in the implementations to meet performance goals. The control for these systems is hierarchical with microprogrammed control selectively used at the top level for complex instruction implementation and hardwired control at the lower level for implementing simple instructions and steps of complex instructions at a very rapid rate. This flavor of microprogramming is covered for a Complex Instruction Set Computer (CISC) in Chapter 12.

 Information on the more traditional flavor of microprogrammed control, derived from past editions of this text, is available in a supplement, Microprogrammed Control, on the Companion Website for the text.

8-8 CHAPTER SUMMARY

This chapter has examined the interaction between datapaths and control units and the difference between programmed and nonprogrammed systems. The algorithmic state machine (ASM) is a means for representing and specifying control functions. A binary multiplier was used to illustrate ASM chart formulation. Two implementation approaches to sequential circuit design, sequence register plus decoder and one flip-flop per state, were provided, in addition to the basic design procedure in Chapter 4. VHDL and Verilog alternatives for describing combinations of datapath and control were also illustrated. Finally, microprogrammed control was briefly discussed.

REFERENCES

1. MANO, M. M. *Computer Engineering: Hardware Design:* Englewood Cliffs, NJ: Prentice Hall, 1988.

2. MANO, M. M. *Digital Design,* 3rd Ed. Englewood Cliffs, NJ: Prentice Hall, 2002.

3. *IEEE Standard VHDL Language Reference Manual.* (ANSI/IEEE Std 1076-1993; revision of IEEE Std 1076-1987). New York: The Institute of Electrical and Electronics Engineers, 1994.

4. SMITH, D. J. *HDL Chip Design.* Madison, AL: Doone Publications, 1996.

5. *IEEE Standard Description Language Based on the Verilog(TM) Hardware Description Language* (IEEE Std 1364-1995). New York: The Institute of Electrical and Electronics Engineers, 1995.

6. PALNITKAR, S. *Verilog HDL: A Guide to Digital Design and Synthesis.* SunSoft Press (A Prentice Hall Title), 1996.

7. THOMAS, D. E., AND P. R. MOORBY. *The Verilog Hardware Description Language* 4th ed. Boston: Kluwer Academic Publishers, 1998.

PROBLEMS

The plus (+) indicates a more advanced problem and the asterisk (*) indicates a solution is available on the Companion Website for the text.

8–1. *A state diagram of a sequential circuit is given in Figure 8-18. Find the corresponding ASM chart. Minimize the chart complexity by using both vector and scalar decision boxes. The inputs to the circuit are X_1 and X_2, and the outputs are Z_1 and Z_2.

8–2. *Find the response for the ASM chart in Figure 8-19 to the following sequence of inputs (assume that the initial state is ST1):

A:	0	1	1	0	1	1	0	1
B:	1	1	0	1	0	1	0	1
C:	0	1	0	1	0	1	0	1

State: ST1

Z:

8–3. An ASM chart is given in Figure 8-19. Find the state table for the corresponding sequential circuit.

8–4. Find the ASM chart corresponding to the following description: There are two states, *A* and *B*. If in state *A* and input *X* is 1, then the next state is *A*. If in state *A* and input *X* is 0, then the next state is *B*. If in state *B* and input *Y* is 0, then the next state is *B*. If in state *B* and input *Y* is 1, then the next state is *A*. Output *Z* is equal to 1 while the circuit is in state *B*.

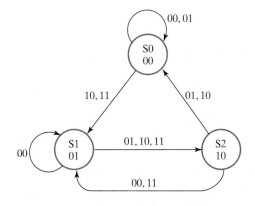

☐ **FIGURE 8-18**
State Diagram for Problem 8-1

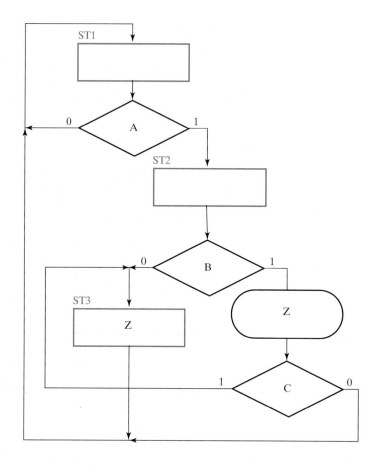

☐ **FIGURE 8-19**

ASM Chart for Problem 8-2 and Problem 8-3

8–5. *Find the ASM for a circuit that detects a difference in value in an input signal X at two successive positive clock edges. If X has different values at two successive positive clock edges, then output Z is equal to 1 for the next clock cycle. Otherwise, output Z is 0.

8–6. +The ASM chart for a synchronous circuit with clock CK for a washing machine is to be developed. The circuit has three external inputs, START, FULL, and EMPTY (which are 1 for at most a single clock cycle and are mutually exclusive), and external outputs, HOT, COLD, DRAIN, and TURN. The datapath for the control consists of a down-counter, which has three inputs, RESET, DEC, and LOAD. This counter synchronously decrements once each minute for DEC = 1, but can be loaded or synchronously reset on any cycle of clock CK. It has a single output, ZERO, which is 1 whenever the counter contains value zero and is 0 otherwise.

In its operation, the circuit goes through four distinct cycles, WASH, SPIN, RINSE, and SPIN, which are detailed as follows:

WASH: Assume that the circuit is in its power-up state IDLE. If START is 1 for a clock cycle, HOT becomes 1 and remains 1 until FULL = 1, filling the washer with hot water. Next, using LOAD, the down-counter is loaded with a value from a panel dial which indicates how many minutes the wash cycle is to last. DEC and TURN then become 1 and the washer washes its contents. When ZERO becomes 1, the wash is complete, and TURN and DEC become 0.

SPIN: Next, DRAIN becomes 1, draining the wash water. When EMPTY becomes 1, the down-counter is loaded with 7. DEC and TURN then become 1 and the remaining wash water is wrung from the contents. When ZERO becomes 1, DRAIN, DEC, and TURN return to 0.

RINSE: Next, COLD becomes 1 and remains 1 until FULL = 1, filling the washer with cold rinse water. Next, using LOAD, the down-counter is loaded with value 10. DEC and TURN then become 1 and the washer rinses its contents. When ZERO becomes 1, the rinse is complete, and TURN and DEC become 0.

SPIN: Next, DRAIN becomes 1, draining the rinse water. When EMPTY becomes 1, the down-counter is loaded with 8. DEC and TURN then become 1 and the remaining rinse water is wrung from the contents. When ZERO becomes 1, DRAIN, DEC, and TURN return to 0 and the circuit returns to state IDLE.

(a) Find the ASM chart for the washer circuit.

(b) Modify your design in part (a) assuming that there are two more inputs, PAUSE and STOP. PAUSE causes the circuit, including the counter, to halt and all outputs to go to 0. When START is pushed, the washer resumes operation at the point it paused. When STOP is pushed, all outputs are reset to 0 except for DRAIN which is set to 1. When EMPTY becomes 1, the state returns to IDLE.

8–7. Find an ASM chart for a traffic light controller that works as follows: A timing signal T is the input to the controller. T defines the yellow light interval, as well as the changes of the red and green lights. The outputs to the signals are defined by the following table:

Output	Light Controlled
GN	Green Light, North/South Signal
YN	Yellow Light, North/South Signal
RN	Red Light, North/South Signal
GE	Green Light, East/West Signal
YE	Yellow Light, East/West Signal
RE	Red Light, East/West Signal

While $T = 0$, the green light is on for one signal and the red light for the other. With $T = 1$, the yellow light is on for the signal that was previously green, and the signal that was previously red remains red. When T becomes 0, the signal that was previously yellow becomes red, and the signal that was previously red becomes green. This pattern of alternating changes in color continues. Assume that the controller is synchronous with a clock that changes much more frequently than input T.

8–8. *Implement the ASM chart in Figure 8-19 by using one flip-flop per state.

8–9. *Implement the ASM chart in Figure 8-19 by using a sequence register and decoder.

8–10. +Implement the ASM chart derived in Problem 8–6(a) by using one flip-flop per state.

8–11. *Multiply the two unsigned binary numbers 100110 (multiplicand) and 110101 (multiplier) by using both the hand method and the hardware method.

8–12. Manually simulate the process of multiplying the two unsigned binary numbers 1010 (multiplicand) and 1011 (multiplier). List the contents of registers A, Q, P, and C and the control state, using the system in Figure 8-6 with n equal to 4 and with the hardwired control in Figure 8-12.

8–13. Determine the time it takes to process the multiplication operation in the digital system described in Figure 8-6 and Figure 8-9. Assume that the Q register has n bits and the interval for a clock cycle is f nanoseconds.

8–14. Prove that the multiplication of two n-bit numbers gives a product of no more than $2n$ bits. Show that this condition implies that no overflow can occur in the final result in the multiplier circuit defined in Figure 8-6.

8–15. Consider the block diagram of the multiplier shown in Figure 8-6. Assume that the multiplier and multiplicand consist of 16 bits each.
   (a) How many bits can be expected in the product, and where is it available?
   (b) How many bits are in the P counter, and what is the binary number that must be loaded into it initially?
   (c) Design the combinational circuit that checks for zero in the P counter.

8–16. *Design a digital system with three 16-bit registers AR, BR, and CR and 16-bit data input IN to perform the following operations, assuming a two's complement representation and ignoring overflow:
   (a) Transfer two 16-bit signed numbers to AR and BR on successive clock cycles after a go signal G becomes 1.
   (b) If the number in AR is positive but nonzero, multiply the contents of BR by two and transfer the result to register CR.
   (c) If the number in AR is negative, multiply the contents of AR by two and transfer the result to register CR.
   (d) If the number in AR is zero, reset register CR to 0.

8–17. +Modify the multiplier design in Figure 8-6 and the ASM chart in Figure 8-7 to perform 2's complement signed-number multiplication using Booth's algorithm, which employs an adder–subtractor. The decision to add, to subtract, or to do nothing is made on the basis of the current least significant bit (LSB) in the Q register and on the previous LSB bit from the Q register before Q was shifted right. Thus, a flip-flop must be provided to store the previous LSB from the Q register. The initial value of the previous least significant bit is to be 0. The following table defines the decisions:

LSB of Q	Previous LSB of Q	Action
0	0	Leave partial product unchanged
0	1	Add multiplicand to partial product
1	0	Subtract multiplicand from partial product
1	1	Leave partial product unchanged

8–18. +Design a digital system that multiplies two unsigned binary numbers by the repeated addition method. For example, to multiply 5 by 4, the digital system adds the multiplicand four times: $5 + 5 + 5 + 5 = 20$. Let the multiplicand be in register BR, the multiplier in register AR, and the product in register PR. An adder circuit adds the contents of BR to PR, and AR is a down-counter. A zero-detection circuit Z checks when AR becomes zero after each time that it is decremented. Design the control by the flip-flop per state method.

8–19. *Write, compile, and simulate a VHDL description for the ASM shown in Figure 8-19. Use a simulation input that passes through all paths in the ASM chart, and include both the state and output Z as simulation outputs. Correct and resimulate your design if necessary.

8–20. *Write, compile, and simulate a Verilog description for the ASM in Figure 8-19. Use code 00 for state ST1, 01 for state ST2, and 10 for state ST3. Use a simulation input that passes through all paths in the ASM chart, and include both the state and Z as simulation outputs. Correct and resimulate your design if necessary.

8–21. Perform the design in Problem 8-5 using Verilog instead of an ASM chart. Use state names S0, S1, S2, … , and codes that are the binary equivalent of the integer in the state name. Compile and simulate your design using a simulation input that thoroughly validates the design and that provides both state and Z as simulation outputs. Correct and resimulate your design if necessary.

8–22. +Perform the design in Problem 8-7 using VHDL instead of an ASM chart. Compile and simulate your design by running the traffic light through two full cycles. Use realistic intervals for T and a slow clock. Adjust the clock intervals if necessary to avoid long simulation times.

8–23. +Perform the design in Problem 8-7 using Verilog instead of an ASM chart. Compile and simulate your design by running the traffic light through two full cycles. Use the state assignment method in Problem 8-21. Use realistic intervals for T and a slow clock. Adjust the clock intervals if necessary to avoid long simulation times.

CHAPTER

9

MEMORY BASICS

M emory is a major component of a digital computer and is present in a large proportion of all digital systems. Random-access memory (RAM) stores data temporarily, and read-only memory (ROM) stores data permanently. ROM is one form of a variety of components called programmable logic devices (PLDs) that use stored information to define logic circuits.

Our study of RAM begins by looking at it in terms of a model with inputs, outputs, and signal timing. We then use equivalent logical models to understand the internal workings of RAM chips. Both static RAM and dynamic RAM are considered. The various types of dynamic RAM used for movement of data at high speeds between the CPU and memory are surveyed. Finally, we put RAM chips together to build simple RAM systems.

In many of the previous chapters, the concepts presented were broad, pertaining to much of the generic computer at the beginning of Chapter 1. In this chapter, for the first time, we can be more precise and point to specific uses of memory and related components. Beginning with the processor, the internal cache is largely very fast RAM. Outside the CPU, the external cache is largely fast RAM. The RAM subsystem, by its very name, is a type of memory. In the I/O area, we find substantial memory for storing information about the screen image in the video adapter. RAM appears in disk cache in the disk controller, to speed up disk access. Aside from the highly central role of the RAM subsystem in storing data and programs, we find memory in various forms applied in most subsystems of the generic computer.

9-1 MEMORY DEFINITIONS

In digital systems, memory is a collection of cells capable of storing binary information. In addition to these cells, memory contains electronic circuits for storing and retrieving the information. As indicated in the discussion of the generic computer, memory is used in many different parts of a modern computer, providing temporary or permanent storage for substantial amounts of binary information. In order for this information to be processed, it is sent from the memory to processing hardware consisting of registers and combinational logic. The processed information is then returned to the same or to a different memory. Input and output devices also interact with memory. Information from an input device is placed in memory so that it can be used in processing. Output information from processing is placed in memory, and from there it is sent to an output device.

Two types of memories are used in various parts of a computer: *random-access memory* (RAM) and *read-only memory* (ROM). RAM accepts new information for storage to be available later for use. The process of storing new information in memory is referred to as a memory *write* operation. The process of transferring the stored information out of memory is referred to as a memory *read* operation. RAM can perform both the write and the read operations, whereas ROM as introduced in Chapter 3, performs only read operations. RAM sizes may range from hundreds to billions of bits.

9-2 RANDOM-ACCESS MEMORY

Memory is a collection of binary storage cells together with associated circuits needed to transfer information into and out of the cells. Memory cells can be accessed to transfer information to or from any desired location, with the access taking the same time regardless of the location, hence, the name *random-access memory*. In contrast, *serial memory,* such as is exhibited by a magnetic disk or tape unit, takes different lengths of time to access information, depending on where the desired location is relative to the current physical position of the disk or tape.

Binary information is stored in memory in groups of bits, each group of which is called a *word*. A word is an entity of bits that moves in and out of memory as a unit—a group of 1's and 0's that represents a number, an instruction, one or more alphanumeric characters, or other binary-coded information. A group of eight bits is called a *byte*. Most computer memories use words that are multiples of eight bits in length. Thus, a 16-bit word contains two bytes, and a 32-bit word is made up of four bytes. The capacity of a memory unit is usually stated as the total number of bytes that it can store. Communication between a memory and its environment is achieved through data input and output lines, address selection lines, and control lines that specify the direction of transfer of information. A block diagram of a memory is shown in Figure 9-1. The n data input lines provide the information to be stored in memory, and the n data output lines supply the information coming out of memory. The k address lines specify the particular word chosen among the many available. The two control inputs specify the direction of transfer desired: the Write input causes binary data to be transferred into memory, and the Read input causes binary data to be transferred out of memory.

The memory unit is specified by the number of words it contains and the number of bits in each word. The address lines select one particular word. Each word in memory is assigned an identification number called an *address*. Addresses

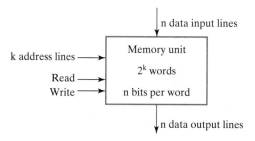

□ **FIGURE 9-1**
Block Diagram of Memory

range from 0 to $2^k - 1$, where k is the number of address lines. The selection of a specific word inside memory is done by applying the k-bit binary address to the address lines. A decoder accepts this address and opens the paths needed to select the word specified. Computer memory varies greatly in size. It is customary to refer to the number of words (or bytes) in memory with one of the letters K (kilo), M (mega), or G (giga). K is equal to 2^{10}, M is equal to 2^{20}, and G is equal to 2^{30}. Thus, $64K = 2^{16}, 2M = 2^{21}$, and $4G = 2^{32}$.

Consider, for example, a memory with a capacity of 1K words of 16 bits each. Since $1K = 1024 = 2^{10}$, and 16 bits constitute two bytes, we can say that the memory can accommodate 2048, or 2K, bytes. Figure 9-2 shows the possible contents of the first three and the last three words of this size of memory. Each word contains 16 bits that can be divided into two bytes. The words are recognized by their decimal addresses from 0 to 1023. An equivalent binary address consists of 10 bits. The first address is specified using ten 0's, and the last address is specified with ten 1's. This is because 1023 in binary is equal to 1111111111. A word in memory is selected by its binary address. When a word is read or written, the memory operates on all 16 bits as a single unit.

Memory address		
Binary	Decimal	Memory contents
0000000000	0	10110101 01011100
0000000001	1	10101011 10001001
0000000010	2	00001101 01000110
.	.	.
.	.	.
.	.	.
.	.	.
1111111101	1021	10011101 00010101
1111111110	1022	00001101 00011110
1111111111	1023	11011110 00100100

□ **FIGURE 9-2**
Contents of a 1024 × 16 Memory

The 1K × 16 memory of the figure has 10 bits in the address and 16 bits in each word. If, instead, we have a 64K × 10 memory, it is necessary to include 16 bits in the address, and each word will consist of 10 bits. The number of address bits needed in memory is dependent on the total number of words that can be stored there and is independent of the number of bits in each word. The number of bits in the address for a word is determined from the relationship $2^k \geq m$, where m is the total number of words and k is the minimum number of address bits satisfying the relationship.

Write and Read Operations

The two operations that a random-access memory can perform are write and read. A *write* is a transfer into memory of a new word to be stored. A *read* is a transfer of a copy of a stored word out of memory. A Write signal specifies the transfer-in operation, and a Read signal specifies the transfer-out operation. On accepting one of these control signals, the internal circuits inside memory provide the desired function.

The steps that must be taken for a write are as follows:

1. Apply the binary address of the desired word to the address lines.
2. Apply the data bits that must be stored in memory to the data input lines.
3. Activate the Write input.

The memory unit will then take the bits from the data input lines and store them in the word specified by the address lines.

The steps that must be taken for a read are as follows:

1. Apply the binary address of the desired word to the address lines.
2. Activate the Read input.

The memory will then take the bits from the word that has been selected by the address and apply them to the data output lines. The contents of the selected word are not changed by reading them.

Memory is made up of RAM integrated circuits (chips), plus additional logic circuits. RAM chips usually provide the two control inputs for the read and write operations in a somewhat different configuration from that just described. Instead of having separate Read and Write inputs to control the two operations, most integrated circuits provide at least a Chip Select that selects the chip to be read from or written to and a Read/$\overline{\text{Write}}$ that determines the particular operation. The memory operations that result from these control inputs are shown in Table 9-1.

The Chip Select is used to enable the particular RAM chip or chips containing the word to be accessed. When Chip Select is inactive, the memory chip or chips are not selected, and no operation is performed. When Chip Select is active, the Read/$\overline{\text{Write}}$ input determines the operation to be performed. While Chip Select accesses chips, a signal is also provided that accesses the entire memory. We will call this signal the Memory Enable.

Timing Waveforms

The operation of the memory unit is controlled by an external device, such as a CPU. The CPU is synchronized by its own clock pulses. The memory, however,

□ **TABLE 9-1**
Control Inputs to a Memory Chip

Chip select CS	Read/$\overline{\text{Write}}$ R/$\overline{\text{W}}$	Memory operation
0	×	None
1	0	Write to selected word
1	1	Read from selected word

does not employ the CPU clock. Instead, its read and write operations are timed by changes in values on the control inputs. The *access time* of a memory read operation is the maximum time from the application of the address to the appearance of the data at the Data Output. Similarly, the *write cycle time* is the maximum time from the application of the address to the completion of all internal memory operations required to store a word. Memory writes may be performed one after the other at the intervals of the cycle time. The CPU must provide the memory control signals in such a way as to synchronize its own internal clocked operations with the read and write operations of memory. This means that the access time and the write cycle time of the memory must be related within the CPU to a period equal to a fixed number of CPU clock pulse periods.

Assume, as an example, that a CPU operates with a clock frequency of 50 MHz, giving a period of 20 ns (1 ns = 10^{-9} s) for one clock pulse. Suppose now that the CPU communicates with a memory with an access time of 65 ns and a write cycle time of 75 ns. The number of clock pulses required for a memory request is the integer value greater than or equal to the larger of the access time and the write cycle time, divided by the clock period. Since the period of the CPU clock is 20 ns, and the larger of the access time and write cycle time is 75 ns, it will be necessary to devote at least four clock pulses to each memory request.

The memory cycle timing shown in Figure 9-3 is for a CPU with a 50 MHz clock and memory with a 75-ns write cycle time and a 65-ns access time. The write cycle in part (a) shows four pulses $T1$, $T2$, $T3$, and $T4$ with a cycle of 20 ns. For a write operation, the CPU must provide the address and input data to the memory. The address is applied, and Memory Enable is set to the high level at the positive edge of the $T1$ pulse. The data, needed somewhat later in the write cycle, is applied at the positive edge of $T2$. The two lines that cross each other in the address and data waveforms designate a possible change in value of the multiple lines. The shaded areas represent unspecified values. A change of the Read/$\overline{\text{Write}}$ signal to 0 to designate the write operation is also at the positive edge of $T2$. To avoid destroying data in other memory words, it is important that this change occur after the signals on the address lines have become fixed at the desired values. Otherwise, one or more other words might be momentarily addressed and accidentally written over with different data. The Read/$\overline{\text{Write}}$ signal must stay at 0 long enough after application of the address and Memory Enable to allow the write operation to complete. Finally, the address and data signals must remain stable for a short time after the Read/$\overline{\text{Write}}$ goes to 1, again to avoid destroying data in other memory

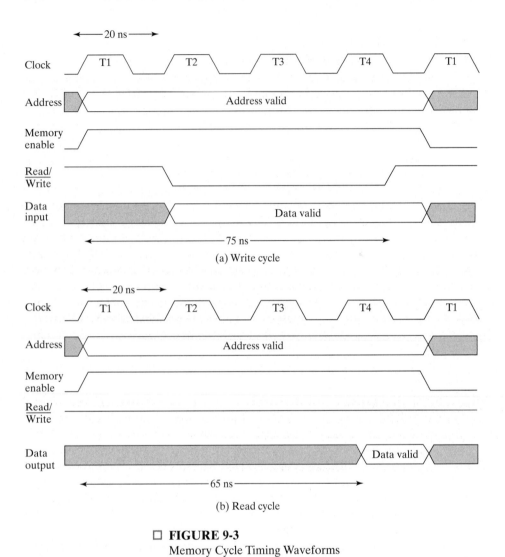

(a) Write cycle

(b) Read cycle

☐ **FIGURE 9-3**
Memory Cycle Timing Waveforms

words. At the completion of the fourth clock pulse, the memory write operation has ended with 5 ns to spare, and the CPU can apply the address and control signals for another memory request with the next $T1$ pulse.

The read cycle shown in Figure 9-3(b) has an address for the memory that is provided by the CPU. The CPU applies the address, sets the Memory Enable to 1, and sets Read/Write to 1 to designate a read operation, all at the positive edge of $T1$. The memory places the data of the word selected by the address onto the data output lines within 65 ns from the time that the address is applied and the memory enable is activated. Then, the CPU transfers the data into one of its internal registers during the positive transition of the next $T1$ pulse, which can also change the address and controls for the next memory request.

Properties of Memory

Integrated circuit RAM may be either static or dynamic. *Static* RAM (SRAM) consists of internal latches that store the binary information. The stored information remains valid as long as power is applied to the RAM. *Dynamic* RAM (DRAM) stores the binary information in the form of electric charges on capacitors. The capacitors are accessed inside the chip by *n*-channel MOS transistors. The stored charge on the capacitors tends to discharge with time, and the capacitors must be periodically recharged by *refreshing* the DRAM. This is done by cycling through the words every few milliseconds, reading and rewriting them to restore the decaying charge. DRAM offers reduced power consumption and larger storage capacity in a single memory chip, but SRAM is easier to use and has shorter read and write cycles. Also, no refresh is required for SRAM.

Memory units that lose stored information when power is turned off are said to be *volatile*. Integrated circuit RAMs, both static and dynamic, are of this category, since the binary cells need external power to maintain the stored information. In contrast, a *nonvolatile memory*, such as magnetic disk, retains its stored information after the removal of power. This is because the data stored on magnetic components is represented by the direction of magnetization, which is retained after power is turned off. Another nonvolatile memory is ROM, discussed in Section 3-9.

9–3   SRAM INTEGRATED CIRCUITS

As indicated earlier, memory consists of RAM chips plus additional logic. We will consider the internal structure of the RAM chip first. Then we will study combinations of RAM chips and additional logic used to construct memory. The internal structure of a RAM chip of m words with n bits per word consists of an array of mn binary storage cells and associated circuitry. The circuity is made up of decoders to select the word to be read or written, read circuits, write circuits, and output logic. The *RAM cell* is the basic binary storage cell used in the RAM chip, which is typically designed as an electronic circuit rather than a logic circuit. Nevertheless, it is possible and convenient to model the RAM chip using a logic model.

A static RAM chip serves as the basis for our discussion. We first present RAM cell logic for storing a single bit and then use the cell in a hierarchy to describe the RAM chip. Figure 9-4 shows the logic model of the RAM cell. The storage part of the cell is modeled by an SR latch. The inputs to the latch are enabled by a Select signal. For Select equal to 0, the stored content is held. For Select equal to 1, the stored content is determined by the values on B and $\overline{B}$. The outputs from the latch are gated by Select to produce cell outputs C and $\overline{C}$. For Select equal to 0, both C and $\overline{C}$ are 0, and for Select equal to 1, C is the stored value and $\overline{C}$ is its complement.

To obtain simplified static RAM diagrams, we interconnect a set of RAM cells and read and write circuits to form a *RAM bit slice* that contains all of the circuitry associated with a single bit position of a set of RAM words. The logic diagram for a RAM bit slice is shown in Figure 9-5(a). The portion of the model representing each RAM cell is highlighted in blue. The loading of a cell latch is

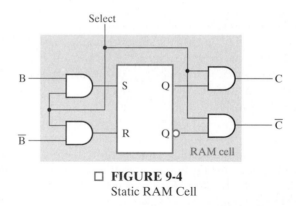

☐ **FIGURE 9-4**
Static RAM Cell

now controlled by a Word Select input. If this is 0, then both S and R are 0, and the cell latch contents remain unchanged. If the Word Select input is 1, then the value to be loaded into the latch is controlled by two signals B and $\overline{B}$ from the Write Logic. In order for either of these signals to be 1 and potentially change the stored value, Read/$\overline{\text{Write}}$ must be 0 and Bit Select must be 1. Then the Data In value and its complement are applied to B and $\overline{B}$, respectively, to set or reset the latch in the RAM cell selected. If Data In is 1 the latch is set to 1, and if Data In is 0 the latch is reset to 0, completing the write operation.

Only one word is written at a time. That is, only one Word Select line is 1, and all other Word Select lines are 0. Thus, only one RAM cell attached to B and $\overline{B}$ is written. The Word Select also controls the reading of the RAM cells, using shared Read Logic. If Word Select is 0, then the stored value in the SR latch is prevented by the AND gates from reaching the pair of OR gates in the Read Logic. But if Word Select is 1, the stored value passes through to the OR gates and is captured in the Read Logic SR latch. If Bit Select is also 1, the captured value appears on the Data Out line of the RAM bit slice. Note that for this particular Read Logic design, the read occurs regardless of the value of Read/$\overline{\text{Write}}$.

The symbol for the RAM bit slice given in Figure 9-5(b) is used to represent the internal structure of RAM chips. Each Word Select line extends beyond the bit slice, so that when multiple RAM bit slices are placed side by side, corresponding Word Select lines connect. The other signals in the lower portion of the symbol may be connected in various ways, depending on the structure of the RAM chip.

The symbol and block diagram for a 16×1 RAM chip are shown in Figure 9-6. Both have four address inputs for the 16 one-bit words stored in RAM. There are also Data Input, Data Output, and Read/$\overline{\text{Write}}$ signals. The Chip Select at the chip level corresponds to the Memory Enable at the level of a RAM consisting of multiple chips. The internal structure of the RAM chip consists of a RAM bit slice having 16 RAM cells. Since there are 16 Word Select lines to be controlled such that one and only one has the value logic 1 at a given time, a 4-to-16-line decoder is used to decode the four address bits into 16 Word Select bits.

The only additional logic in the figure is a triangular symbol with one normal input, one normal output, and a second input on the bottom of the symbol. This symbol is a three-state buffer that allows construction of a multiplexer with an

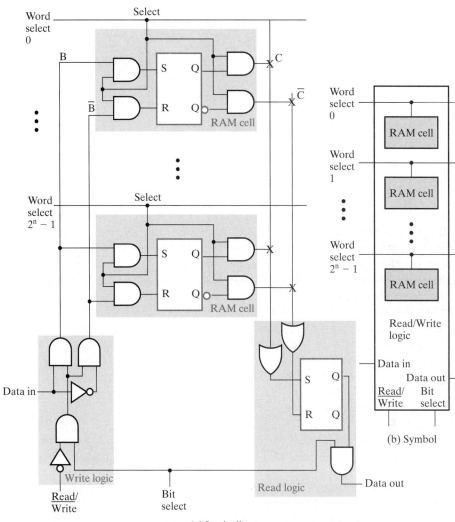

(a) Logic diagram

□ **FIGURE 9-5**
RAM Bit Slice Model

arbitrary number of inputs. Three-state outputs are connected together and properly controlled using the Chip Select inputs. By using three-state buffers on the outputs of RAM chips, these outputs can be connected together to provide the word from the chip being read on the bit lines attached to the RAM outputs. The enable signals in the preceding discussion correspond to the Chip Select inputs on the RAM chips. To read a word from a particular RAM chip, the Chip Select value for that chip must be 1, and for all other chips attached to the same output bit lines, the Chip Select must be 0. These combinations containing a single 1 can be obtained from a decoder.

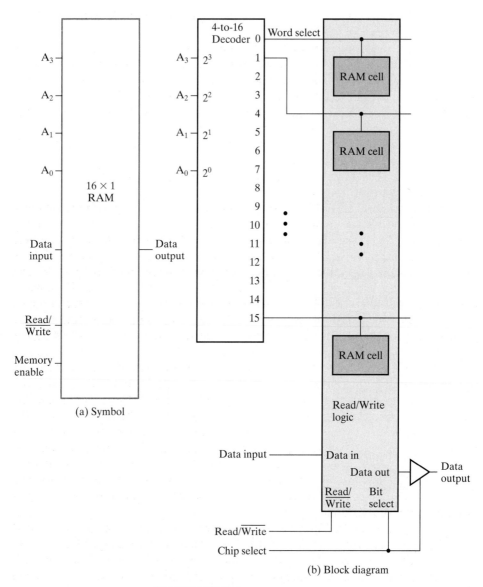

□ **FIGURE 9-6**
16-Word by 1-Bit RAM Chip

Coincident Selection

Inside a RAM chip, the decoder with k inputs and 2^k outputs requires 2^k AND gates with k inputs per gate if a straightforward design approach is used. In addition, if the number of words is large, and all bits for one bit position in the word are contained in a single RAM bit slice, the number of RAM cells sharing the read and write circuits is also large. The electrical properties resulting from both of these situations cause the access and write cycle times of the RAM to become long, which is undesirable.

The total number of decoder gates, the number of inputs per gate, and the number of RAM cells per bit slice can all be reduced by employing two decoders with a *coincident selection* scheme. In one possible configuration, two $k/2$-input decoders are used instead of one k-input decoder. One decoder controls the word select lines and the other controls the bit select lines. The result is a two-dimensional matrix selection scheme. If the RAM chip has m words with 1 bit per word, then the scheme selects the RAM cell at the intersection of the Word Select row and the Bit Select column. Since the Word Select is no longer strictly selecting words, its name is changed to *Row Select*. An output from the added decoder that selects one or more bit slices is referred to as a *Column Select*.

Coincident selection is illustrated for the 16×1 RAM chip with the structure shown in Figure 9-7. The chip consists of four RAM bit slices of four bits each and has a total of 16 RAM cells in a two-dimensional array. The two most significant address inputs go through the 2-to-4-line row decoder to select one of the four rows of the array. The two least significant address inputs go through the 2-to-4-line column decoder to select one of the four columns (RAM bit slices) of the array. The column decoder is enabled with the Chip Select input. When the Chip Select is 0, all outputs of the decoder are 0 and none of the cells is selected. This prevents writing into any RAM cell in the array. With Chip Select at 1, a single bit in the RAM is accessed. For example, for the address 1001, the first two address bits are decoded to select row 10 (2_{10}) of the RAM cell array. The second two address bits are decoded to select column $01(1_{10})$ of the array. The RAM cell accessed, in row 2 and column 1 of the array, is cell 9 ($10_2\ 01_2$). With a row and column selected, the $\overline{\text{Read}/\text{Write}}$ input determines the operation. During the read operation ($\overline{\text{Read}/\text{Write}}$ = 1), the selected bit of the selected row goes through the OR gate to the three-state buffer. Note that the gate is drawn according to the array logic established in Figure 3-22. Since the buffer is enabled by Chip Select, the value read appears at the Data Output. During the write operation ($\overline{\text{Read}/\text{Write}}$ = 0), the bit available on the Data Input line is transferred into the selected RAM cell. Those RAM cells not selected are disabled, and their previous binary values remain unchanged.

The same RAM cell array is used in Figure 9-8 to produce an 8×2 RAM chip (eight words of two bits each). The row decoding is unchanged from that in Figure 9-7; the only changes are in the column and output logic. Since there are just three address bits, and two are handled by the row decoder, the column decoder has only one address bit and Chip Select as inputs and produces just two Column Select lines. Since two bits at a time are to be written or read, the Column Select lines go to adjacent pairs of RAM bit slices. Two input lines, Data Input 0 and Data Input 1, each go to a different bit in all of the pairs. Finally, corresponding bits of the pairs share output OR gates and three-state buffers, giving output lines Data Output 0 and Data Output 1. The operation of this structure can be illustrated by the application of the address 3 (011_2). The first two bits of the address, 01, access row 1 of the array. The final bit, 1, accesses column 1, which consists of bit slices 2 (10_2) and 3 (11_2). So the word to be written or read lies in RAM cells 6 and 7 ($011\ 0_2$ and $011\ 1_2$), which contain bits 0 and 1, respectively, of word 3.

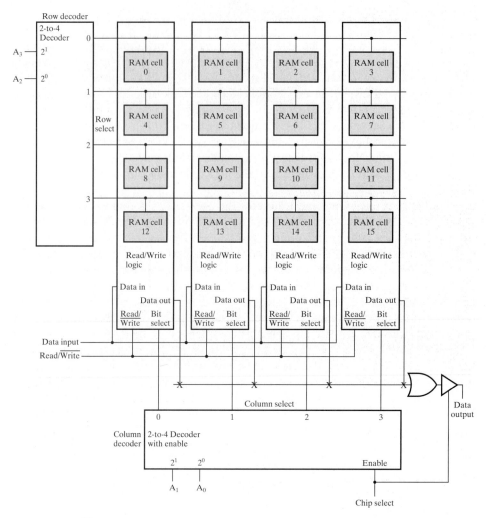

☐ **FIGURE 9-7**
Diagram of a 16×1 RAM Using a 4×4 RAM Cell Array

We can demonstrate the savings of the coincident selection scheme by considering a more realistic static RAM size, $32K \times 8$. This RAM chip contains a total of 256K bits. To make the number of rows and columns in the array equal, we take the square root of 256K, giving $512 = 2^9$. So the first nine bits of the address are fed to the row decoder and the remaining six bits to the column decoder. Without coincident selection, the single decoder would have 15 inputs and 32,768 outputs. With coincident selection, there is one 9-to-512-line decoder and one 6-to-64-line decoder. The number of gates for a straightforward design of the single decoder would be 32,800. For the two coincident decoders, the number of gates is 608, reducing the gate count by a factor of more than 50. In addition, although it appears that there are 64 times as many Read/Write circuits, the column selection

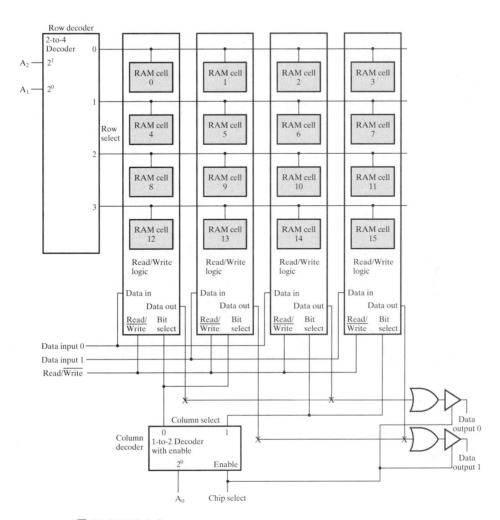

□ **FIGURE 9-8**
Block Diagram of an 8 × 2 RAM Using a 4 × 4 RAM Cell Array

can be done between the RAM cells and the Read/Write circuits, so that only the original eight circuits are required. Because of the reduced number of RAM cells attached to each Read/Write circuit at any time, the access time of the chip is also improved.

9-4 ARRAY OF SRAM ICs

Integrated circuit RAM chips are available in a variety of sizes. If the memory unit needed for an application is larger than the capacity of one chip, it is necessary to combine a number of chips in an array to form the required size of memory. The capacity of the memory depends on two parameters: the number of words and the number of bits per word. An increase in the number of words requires that we increase the address length. Every bit added to the length of the address doubles the number of words in memory. An increase in the number of bits per word requires that we increase the number of data input and output lines, but the address length remains the same.

To illustrate an array of RAM ICs, let us first introduce a RAM chip using the condensed representation for inputs and outputs shown in Figure 9-9. The capacity of this chip is 64K words of 8 bits each. The chip requires a 16-bit address and 8 input and output lines. Instead of 16 lines for the address and 8 lines each for data input and data output, each is shown in the block diagram by a single line. Each line has a slash across it with a number indicating the number of lines represented. The *CS* (Chip Select) input selects the particular RAM chip, and the *R/$\overline{W}$* (Read/Write) input specifies the read or write operation when the chip is selected. The small triangle shown at the outputs is the standard graphics symbol for three-state outputs. The *CS* input of the RAM controls the behavior of the data output lines. When *CS* = 0, the chip is not selected, and all its data outputs are in the high-impedance state. With *CS* = 1, the data output lines carry the eight bits of the selected word.

Suppose that we want to increase the number of words in the memory by using two or more RAM chips. Since every bit added to the address doubles the binary number that can be formed, it is natural to increase the number of words in factors of two. For example, two RAM chips will double the number of words and add one bit to the composite address. Four RAM chips multiply the number of words by four and add two bits to the composite address.

Consider the possibility of constructing a 256K × 8 RAM with four 64K × 8 RAM chips, as shown in Figure 9-10. The eight data input lines go to all the chips. The three-state outputs can be connected together to form the eight common data output lines. This type of output connection is possible only with three-state outputs. Just one chip select input will be active at any time, while the other three chips will be disabled. The eight outputs of the selected chip will contain 1's and 0's, and the other three will be in a high-impedance state, presenting only open circuits to the binary output signals of the selected chip.

The 256K-word memory requires an 18-bit address. The 16 least significant bits of the address are applied to the address inputs of all four chips. The two most

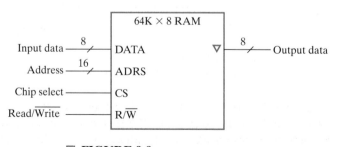

☐ **FIGURE 9-9**
Symbol for a 64K × 8 RAM Chip

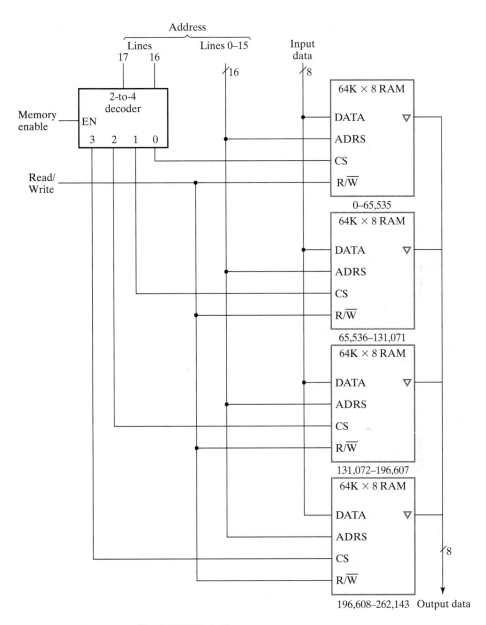

□ **FIGURE 9-10**
Block Diagram of a 256K × 8 RAM

significant bits are applied to a 2 × 4 decoder. The four outputs of the decoder are applied to the *CS* inputs of the four chips. The memory is disabled when the *EN* input of the decoder, Memory Enable, is equal to 0. All four outputs of the decoder are then 0, and none of the chips is selected. When the decoder is enabled, address bits 17 and 16 determine the particular chip that is selected. If these bits

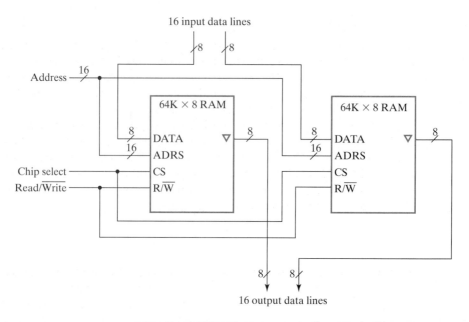

□ **FIGURE 9-11**
Block Diagram of a 64K × 16 RAM

are equal to 00, the first RAM chip is selected. The remaining 16 address bits then select a word within the chip in the range from 0 to 65,535. The next 65,536 words are selected from the second RAM chip with an 18-bit address that starts with 01 followed by the 16 bits from the common address lines. The address range for each chip is listed in decimal under its symbol in the figure.

It is also possible to combine two chips to form a composite memory containing the same number of words, but with twice as many bits in each word. Figure 9-11 shows the interconnection of two 64K × 8 chips to form a 64K × 16 memory. The 16 data input and data output lines are split between the two chips. Both receive the same 16-bit address and the common CS and $R/\overline{W}$ control inputs.

The two techniques just described may be combined to assemble an array of identical chips into a large-capacity memory. The composite memory will have a number of bits per word that is a multiple of that for one chip. The total number of words will increase in factors of two times the word capacity of one chip. An external decoder is needed to select the individual chips based on the additional address bits of the composite memory.

To reduce the number of pins on the chip package, many RAM ICs provide common terminals for the data input and data output. The common terminals are said to be *bidirectional*, which means that for the read operation they act as outputs, and for the write operation they act as inputs. Bidirectional lines are constructed with three-state buffers and are discussed further in Section 2-8. The use of bidirectional signals requires control of the three-state buffers by both Chip Select and Read/$\overline{Write}$.

9-5 DRAM ICs

Because of its ability to provide high storage capacity at low cost, dynamic RAM (DRAM) dominates the high-capacity memory applications, including the primary RAM in computers. Logically, DRAM in many ways is similar to SRAM. However, because of the electronic circuit used to implement the storage cell, its electronic design is considerably more challenging. Further, as the name "dynamic" implies, the storage of information is inherently only temporary. As a consequence, the information must be periodically "refreshed" to mimic the behavior of static storage. This need for refresh is the primary logical difference in the behavior of DRAM compared to SRAM. We explore this logical difference by examining the dynamic RAM cell, the logic required to perform the refresh operation, and the impact of the need for refresh on memory system operation.

DRAM Cell

The dynamic RAM cell circuit is shown in Figure 9-12(a). It consists of a capacitor C and a transistor T. The capacitor is used to store electrical charge. If there is sufficient charge stored on the capacitor, it can be viewed as storing a logical 1. If there is insufficient charge stored on the capacitor, it can be viewed as storing a logical 0. The transistor acts much like a switch, in the same manner as the transmission gate introduced in Chapter 2. When the switch is "open," the charge on the capacitor roughly remains fixed, in other words, is stored. But when the switch is "closed," charge can flow into and out of the capacitor from the external Bit (B) line. This charge flow allows the cell to be written with a 1 or 0 and to be read.

In order to understand the read and write operations for the cell, we will use a hydraulic analogy with charge replaced by water, the capacitor by a small storage tank, and the transistor by a valve. Since the bit line has a large capacitance, it is represented by a large tank and pumps which can fill and empty this tank rapidly. This analogy is given in Figures 9-12(b) and 9-12(c) with the valve closed. Note that in one case the small storage tank is full representing a stored 1 and in the other case, it is empty representing a stored 0. Suppose that a 1 is to be written into the cell. The valve is opened and the pumps fill up the large tank. Water flows through the valve, filling the small storage tank as shown in Figure 9-12(d). Then the valve is closed, leaving the small tank full which represents a 1. A 0 can be written using the same sort of operations, except that the pumps empty the large tank as shown in Figure 9-12(e).

Now, suppose we want to read a stored value and that the value is a 1 corresponding to a full storage tank. With the large tank at a known intermediate level, the valve is opened. Since the small storage tank is full, water flows from the small tank to the large tank increasing the level of the water surface in the large tank slightly as shown in Figure 9-12(f). This increase in level is observed as the reading of 1 from the storage tank. Correspondingly, if the storage tank is initially empty, there will be a slight decrease in the level in the large tank in Figure 9-12(g), which is observed as the reading of a 0 from the storage tank.

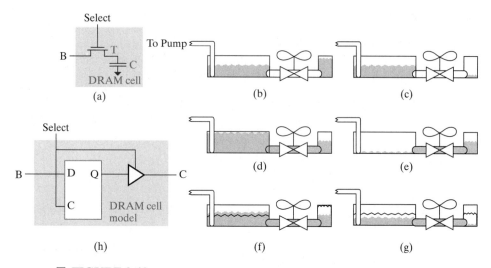

☐ **FIGURE 9-12**
Dynamic RAM cell, hydraulic analogy of cell operation, and cell model

In the read operation just described, Figures 9-12(f) and 9-12(g) show that, regardless of the initial stored value in the storage tank, it now contains an intermediate value which will not cause enough of a change in the level of the external tank to permit a 0 or 1 to be observed. So the read operation has destroyed the stored value; this is referred to as a *destructive read*. To be able to read the original stored value in the future, we must *restore* it (i.e., return the storage tank to its original level). To perform the restore for a stored 1 observed, the large tank is filled by the pumps and the small tank fills through the open valve. To perform the restore for a stored 0 observed, the large tank is emptied by the pumps and the small tank drains through the open valve.

In the actual storage cell, there are other paths present for charge flow. These paths are analogous to small leaks in the storage tank. Due to these leaks, a full small storage tank will eventually drain to a point at which the increase in the level of the large tank on a read cannot be observed as an increase. In fact, if the small tank is less than half full when read, it is possible that a decrease in the level of the large tank may be observed. To compensate for these leaks, the small storage tank storing a 1 must be periodically refilled. This is referred to as a refresh of the cell contents. Every storage cell must be refreshed before its level has declined to a point at which the stored value can no longer be properly observed.

Through the hydraulic analogy, the DRAM operation has been explained. Just as for the SRAM, we employ a logic model for the cell. The model shown in Figure 9-12(h) is a D latch. The *C* input to the D latch is Select and the *D* input to the D latch is *B*. In order to model the output of the DRAM cell, we use a three-state buffer with Select as its control input and *C* as its output. In the original electronic circuit for the DRAM cell in Figure 9-12(a), *B* and *C* are the same signal, but in the logical model they are separate. This is necessary in the modeling process to avoid connecting gate outputs together.

DRAM Bit Slice

Using the logic model for the DRAM cell, we can construct the DRAM bit-slice model shown in Figure 9-13. This model is similar to that for the SRAM bit-slice in Figure 9-5. It is apparent that, aside from the cell structure, the two RAM bit slices are logically similar. However, from the standpoint of cost per bit, they are quite different. The DRAM cell consists of a capacitor plus one transistor. The SRAM cell typically contains six transistors, giving a cell complexity roughly three times that of the DRAM. Therefore, the number of SRAM cells in a chip of a given size

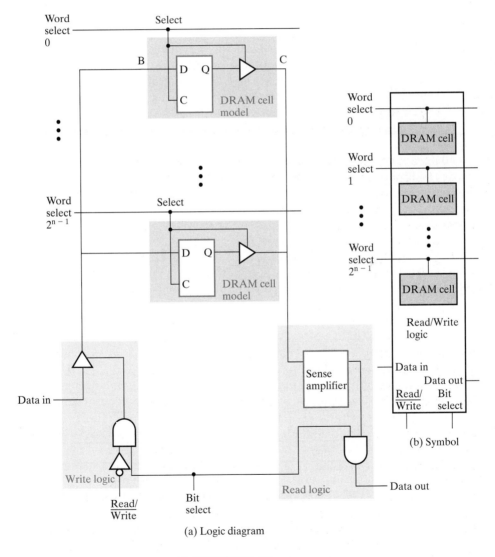

(a) Logic diagram

□ **FIGURE 9-13**
DRAM Bit Slice Model

is less than one-third of those in the DRAM. The DRAM cost per bit is less than 1/3 the SRAM cost per bit, which justifies the use of DRAM in large memories.

Refresh of the DRAM contents remains to be discussed. But first, we need to develop the typical structure used to handle addressing in DRAMs. Since many DRAM chips are used in a DRAM, we want to reduce the physical size of the DRAM chips. Large DRAMs require 20 or more address bits, which would require 20 address pins on each DRAM chip. To reduce the number of pins, the DRAM address is applied serially in two parts with the row address first and the column address second. This can be done since the row address, which performs the row selection, is actually needed before the column address, which reads out the data from the row selected. In order to hold the row address throughout the read or write cycle, it is stored in a register as shown in Figure 9-14. The column address is also stored in a register. The load signal for the row address register is $\overline{RAS}$ ($\overline{\text{Row Address Strobe}}$) and for the column addresses is $\overline{CAS}$ ($\overline{\text{Column Address Strobe}}$). Note that in addition to $\overline{RAS}$ and $\overline{CAS}$, control signals for the DRAM chip include R/$\overline{W}$ (Read/$\overline{\text{Write}}$), and $\overline{OE}$ ($\overline{\text{Output enable}}$). Note that this design uses signals active at the LOW (0) level.

The timing for DRAM write and read operation appears in Figure 9-15(a). The row address is applied to the address inputs, and then RAS changes from 1 to 0, loading the row address into the row address register. This address is applied to the row address decoder and selects a row of DRAM cells. Meanwhile, the column

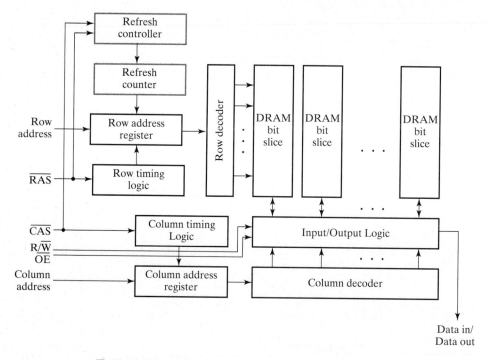

☐ **FIGURE 9-14**
Block Diagram of a DRAM Including Refresh Logic

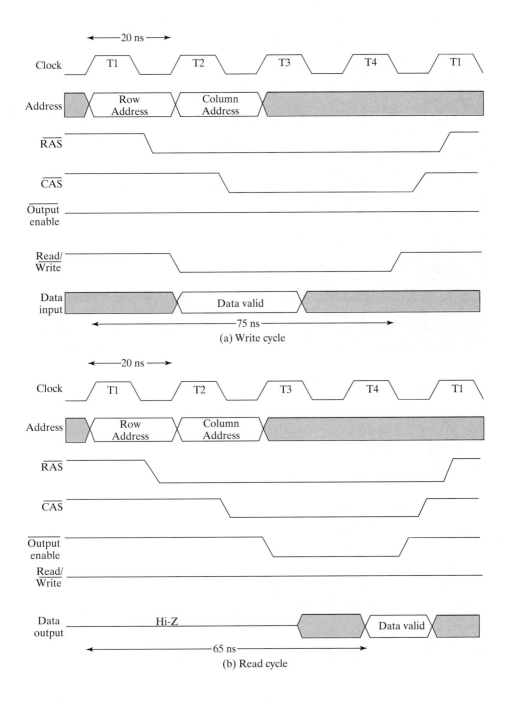

□ **FIGURE 9-15**
Timing for DRAM Write and Read Operations

address is applied, and then CAS changes from 1 to 0, loading the column address into the column address register. This address is applied to the column address decoder, which selects a set of columns of the RAM array of size equal to the number of RAM data bits. The input data with Read/$\overline{\text{Write}}$ = 0 is applied over a time interval similar to that for the column address. The data bits are applied to the set of bit lines selected by the column address decoder which, in turn apply the values to the DRAM cells in the selected row, writing the new data into the cells. When CAS and RAS return to 1, the write cycle is complete and the DRAM cells store the newly written data. Note that the stored data in all of the other cells in the addressed row has been restored.

The read operation timing shown in Figure 9-15(b) is similar. Timing of the address operations is the same. However, no data is applied and Read/$\overline{\text{Write}}$ is 1 instead of 0. Data values in the DRAM cells in the selected row are applied to the bit lines and sensed by the sense amplifiers. The column address decoder selects the values to be sent to the Data output, which is enabled by $\overline{\text{Output enable}}$. During the read operation, all values in the addressed row are restored.

To support refresh, additional logic shown in color is present in the block diagram in Figure 9-14. There is a Refresh counter and a Refresh controller. The Refresh counter is used to provide the address of the row of DRAM cells to be refreshed. It is essential for the refresh modes that require the address to be provided from within the DRAM chip. The refresh counter advances on each refresh cycle. Due to the number of bits in the counter, when it reaches $2^n - 1$, where n is the number of rows in the DRAM array, it advances to 0 on the next refresh. The standard ways in which a refresh cycle can be triggered and the corresponding refresh types are as follows:

1. **RAS only refresh.** A row address is placed on the address lines and RAS is changed to 0. In this case, the refresh addresses must be applied from outside the DRAM chip, typically by an IC called a DRAM controller.

2. **CAS before RAS refresh.** The CAS is changed from 1 to 0 followed by a change from 1 to 0 on RAS. Additional refresh cycles can be performed by changing RAS without changing CAS. The refresh addresses for this case come from the refresh counter, which is incremented after the refresh for each cycle.

3. **Hidden refresh.** Following a normal read or write, CAS is left at 0 and RAS is cycled, effectively performing a CAS before RAS refresh. During a hidden refresh, the output data from the prior read remains valid. Thus, the refresh is hidden. Unfortunately, the time taken by the hidden refresh is significant, so a subsequent read or write operation is delayed.

In all cases, note that the initiation of a refresh is controlled externally by using the $\overline{\text{RAS}}$ and $\overline{\text{CAS}}$ signals. Each row of a DRAM chip requires refreshing within a specified maximum refresh time, typically ranging from 16 to 64 milliseconds (ms). Refreshes may be performed at evenly spaced points in the refresh time, an approach called distributed refresh. Alternatively, all refreshes may be performed one after the other, an approach called burst refresh. For example, a 4M × 4 DRAM has a refresh time of 64 ms and has 4096 rows to be refreshed. The length of time to perform a single refresh is 60 ns, and the refresh interval for distributed refresh is 64 ms/4096 = 15.6 microseconds (µs). A total time out for refresh of 0.25 ms is used out of the 64 ms refresh interval. For the same DRAM, a burst refresh also takes 0.25 ms. The DRAM controller must initiate a refresh every 15.6 µs for distributed refresh and must initiate 4,096 refreshes sequentially every 64 ms for burst refresh. During any refresh cycle, no DRAM reads or writes can occur. Since use of burst refresh would halt computer operation for a fairly long period, distributed refresh is more commonly used.

9-6   DRAM TYPES

Over the last two decades, both the capacity and speed of DRAM has increased significantly. The quest for speed has resulted in the evolution of many types of DRAM. Several of the DRAM types are listed with brief descriptions in Table 9-2. Of the memory types listed, the first two have largely been replaced in the marketplace by the more advanced SDRAM and RDRAM approaches. Since we have chosen to provide a discussion of error-correcting codes (ECC) for memories on the text website, our discussion of memory types here will omit the ECC feature and focus on synchronous DRAM, double data rate synchronous DRAM, and Rambus® DRAM. Before considering these three types of DRAM, some of the underlying concepts are covered briefly.

First of all, all three of these DRAM types work well because of the particular environment in which they operate. In modern high-speed computer systems, the processor interacts with the DRAM within a memory hierarchy. Most of the instructions and data for the processor are fetched from two lower levels of the hierarchy, the L1 and L2 caches. These are comparatively smaller SRAM-based memory structures that are covered in detail in Chapter 14. For our purposes, the key issue is that most of the reads from the DRAM are not directly from the CPU, but instead are reads initiated to bring data and instructions into these caches. The reads are in the form of a *line* (i.e., some number of bytes in contiguous addresses in memory) that is brought into the cache. For example, in a given read, the 16 bytes in hexadecimal addresses 000000 through 00000F would be read. This is referred to as a *burst read*. For burst reads, the effective *rate* of reading bytes, which is dependent upon reading bursts from contiguous addresses, rather than the access time is the important measure. With this measure, the three DRAM types we are discussing provide very fast performance.

Second, the effectiveness of these three DRAM types depends upon a very fundamental principle involved in DRAM operation, the reading out of all of the bits in a row for each read operation. The implication of this principle is that all of the bits in a row are available after a read using that row if only they can be accessed. With these two concepts in mind, the synchronous DRAM can be introduced.

Synchronous DRAM (SDRAM)

The use of clocked transfers differentiates SDRAM from conventional DRAM. A block diagram of a 16-megabyte SDRAM IC appears in Figure 9-16. The inputs

☐ **TABLE 9-2**
DRAM Types

Type	Abbreviation	Description
Fast Page Mode DRAM	FPM DRAM	Takes advantage of the fact that, when a row is accessed, all of the row values are available to be read out. By changing the column address, data from different addresses can be read out without reapplying the row address and waiting for the delay associated with reading out the row cells to pass if the row portion of the addresses match.
Extended Data Output DRAM	EDO DRAM	Extends the length of time that the DRAM holds the data values on its output, permitting the CPU to perform other tasks during the access since it knows the data will still be available.
Synchronous DRAM	SDRAM	Operates with a clock rather than being asynchronous. This permits a tighter interaction between memory and CPU, since the CPU knows exactly when the data will be available. SDRAM also takes advantage of the row value availability and divides memory into distinct banks, permitting overlapped accesses.
Double Data Rate Synchronous DRAM	DDR SDRAM	The same as SDRAM except that data output is provided on both the negative and the positive clock edges.
Rambus® DRAM	RDRAM	A proprietary technology that provides very high memory access rates using a relatively narrow bus.
Error-Correcting Code	ECC	May be applied to most of the DRAM types above to correct single bit data errors and often detect double errors.

and outputs differ little from those for the DRAM block diagram in Figure 9-14 with the exception of the presence of the clock for synchronous operation. Internally, there are a number of differences. Since the SDRAM appears synchronous from the outside, there are synchronous registers on the address inputs and the data inputs and outputs. In addition, a column address counter has been added, which is key to the operation of the SDRAM. While the control logic may appear to be similar, the control in this case is much more complex since the SDRAM has a mode control word that can be loaded from the address bus. Considering a 16 MB memory, the memory array contains 134,217,728 bits and is almost square, with 8,192 rows and 16,384 columns. There are 13 row address bits. Since there are

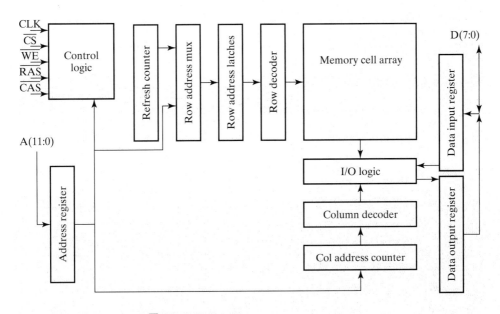

□ **FIGURE 9-16**
Block Diagram of a 16MB SDRAM

8 bits per byte, the number of column addresses is 16,384 divided by 8, which equals 2048. This requires 11 column address bits. Note that 13 plus 11 equals 24 giving the correct number of bits to address 16MB.

As with the regular DRAM, the SDRAM applies the row address first followed by the column address. The timing, however, is somewhat different, and some new terminology is used. Before performing an actual read operation from a specified column address, the entire row of 2048 bytes specified by the applied row address is read out internally and stored in the I/O logic. Internally, this step takes a few clock cycles. Next, the actual read step is performed with the column address applied. After an additional delay of a few clock cycles, the data bytes begin appearing on the output, one per clock period. The number of bytes that appear, the burst length, has been set by loading a mode control word into the control logic from the address input.

The timing of a burst read cycle with burst length equal to four is shown in Figure 9-17. The read begins with the application of the row address and the row address strobe (RAS), which causes the row address to be captured in the address register and the reading of the row to be initiated. During the next two clock periods, the reading of the row is taking place. During the third clock period, the column address and the column address strobe are applied, with the column address captured in the address register and the reading of the first data byte initiated. The data byte is then available to be read from the SDRAM at the positive clock edge occurring two cycles later. The second, third, and fourth bytes are available for reading on subsequent clock edges. In Figure 9-17, note that the bytes are presented in the order 1, 2, 3, 0. This is because, in the column address identifying the

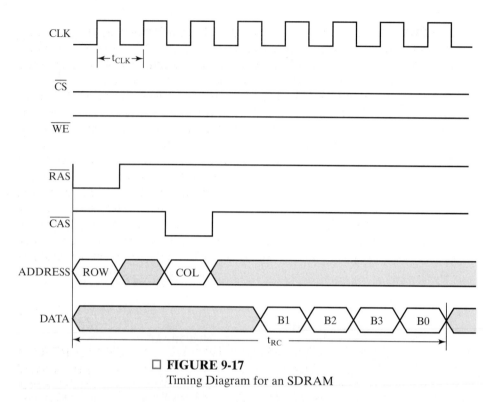

☐ **FIGURE 9-17**
Timing Diagram for an SDRAM

byte immediately needed by the CPU, the last two bits are 01. The subsequent bytes appear in the order of these two bits counted up modulo (burst length) by the column address counter, giving addresses ending in 01, 10, 11, and 00, with all other address bits fixed.

It is interesting to compare the byte rate for reading bytes from SDRAM to that of the basic DRAM. We assume that the read cycle time t_{RC} for the basic DRAM is 60 ns and that the clock period t_{CLK} for the SDRAM is 7.5 ns. The byte rate for the basic DRAM is one byte per 60 ns, or 16.67 MB/sec. For the SDRAM, from Figure 9-17, it requires 8.0 clock cycles, or 60 ns, to read four bytes, giving a byte rate of 66.67 MB/sec. If the burst is eight instead of four bytes, a read cycle time of 90 ns is required, giving a byte rate of 88.89 MB/sec. Finally, if the burst is the entire 2048-byte row of the SDRAM, the read cycle time becomes 60 + (2048 – 4) × 7.5 = 15,390 ns, giving a byte rate of 133.07 MB, which approaches the limit of one byte per 7.5 ns clock period.

Double Data Rate SDRAM (DDR SDRAM)

The second DRAM type, double data rate SDRAM (DDR SDRAM) overcomes the preceding limit without decreasing the clock period. Instead, it provides two bytes of data per clock period by using both the positive and negative clock edges. In Figure 9-17, four bytes are read, one per positive clock edge. By using both clock edges, eight bytes can be transferred in the same read cycle time t_{RC}. For a 7.5 ns clock period, the byte rate limit doubles in the example to 266.14 MB/sec.

Additional basic techniques can be applied to further increase the byte rate. For example, instead of having single byte data, an SDRAM IC can have the data I/O length of four bytes (32 bits). This gives a byte rate limit of 1.065 GB/sec with a 7.5 ns clock period. Eight bytes gives a byte rate limit of 2.130 GB/sec.

The byte rates achieved in the examples are upper limits. If the actual accesses needed are to different rows of the RAM, the delay from the application of the RAS pulse to read out the first byte of data is significant and leads to performance well below the limit. This can be partially offset by breaking up the memory into multiple banks where each of the banks performs the row read independently. Provided that the row and bank addresses are available early enough, row reads can be performed on one or more banks while data is still being transferred from the currently active row. When the column reads from the currently active row are complete, data can potentially be available immediately from other banks, permitting an uninterrupted flow of data from the memory. This permits the actual read rate to more closely approach the limit. Nevertheless, due to the fact that multiple row accesses to the same bank may occur in sequence, the maximum rate is not reached.

RAMBUS® DRAM (RDRAM)

The final DRAM type to be discussed is RAMBUS DRAM (RDRAM). RDRAM ICs are designed to be integrated into a memory system that uses a packet-based bus for the interaction between the RDRAM ICs and the memory bus to the processor. The primary components of the bus are a 3-bit path for the row address, a 5-bit path for the column address, and a 16-bit or an 18- bit path for data. The bus is synchronous and performs transfers on both clock edges, the same property possessed by the DDR SDRAM. Information on the three paths mentioned above is transferred in packets that are four clock cycles long, which means that there are eight transfers/packet. The number of bits per packet for each of the paths is 24 bits for the row address packet, 40 bits for the column address packet, and 128 bits or 144 bits for the data packet. The larger data packet includes 16 parity bits for implementing an error-correcting code. The RDRAM IC employs the concept of multiple memory banks mentioned earlier to provide capability for concurrent memory accesses with different row addresses. RDRAM uses the usual row activate technique in which the addressed row data of the memory is read. From this row data, the column address is used to select byte pairs in the order in which they are to be transmitted in the packet. A typical timing picture for an RDRAM read access is shown in Figure 9-18. Due to the sophisticated electronic design of the RAMBUS system, we can consider a clock period of 1.875 ns. Thus, the time for transmission of a packet is $t_{PACK} = 4 \times 1.875 = 7.5$ ns. The cycle time for accessing a single data packet of 8 byte pairs or 16 bytes is 32 clock cycles or 60 ns as shown in Figure 9-18. The corresponding byte rate is 266.67 MB/sec. If four of the byte packets are accessed from the same row, the rate increases to 1.067 GB/sec. By reading an entire RDRAM row of 2048 bytes, the cycle time increases to $60 + (2048 - 64) \times 1.875/4 = 990$ ns or a byte rate limit of $2048/(990 \times 10^{-9}) = 2.069$ MB/sec, approaching the ideal limit of 4/1.875 ns or 2.133 GB/sec.

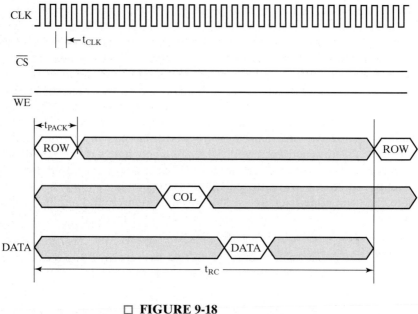

☐ **FIGURE 9-18**
Timing of a 16MB RDRAM

9-7 ARRAYS OF DYNAMIC RAM ICS

Many of the same design principles used for SRAM arrays in Section 9-4 apply to DRAM arrays. There are, however, a number of different requirements for the control and addressing of DRAM arrays. These requirements are typically handled by a *DRAM controller*. The functions performed by a DRAM controller include the following:

1. controlling separation of the address into a row address and a column address and providing these addresses at the required times,
2. providing the $\overline{RAS}$ and $\overline{CAS}$ signals at the required times for read, write, and refresh operations,
3. performing refresh operations at the necessary intervals, and
4. providing status signals to the rest of the system (e.g., indicating whether the memory is busy performing refresh).

The DRAM controller is a complex synchronous sequential circuit with the external CPU clock providing synchronization of its operation.

9-8 CHAPTER SUMMARY

Memory is of two types: random-access memory (RAM) and read-only memory (ROM). For both types, we apply an address to read from or write into a data word. Read and write operations have specific steps and associated timing parameters, including access time and write cycle time. Memory can be static or dynamic and volatile or nonvolatile. Internally, a RAM chip consists of an array of RAM cells, decoders, write circuits, read circuits, and output circuits. A combination of a write circuit, read circuit, and the associated RAM cells can be logically modeled as a RAM bit slice. RAM bit slices, in turn, can be combined to form two-dimensional RAM cell arrays, which, with decoders and output circuits added, form the basis for a RAM chip. Output circuits use three-state buffers in order to facilitate connecting together an array of RAM chips without significant additional logic. Due to the need for refresh, additional circuitry is required within DRAMs, as well as in arrays of DRAM chips. In a quest for faster memory access, a number of new DRAM types have been developed. The most recent forms of these high-speed DRAMs employ a synchronous interface that uses a clock to control memory accesses.

 Error detection and correction codes, often based on Hamming codes, are used to detect or correct errors in stored RAM data. Material from Edition 1 covering these codes is available on the Companion Website for the text.

 Material covering VHDL and Verilog for memories is available on the Companion Website for the text.

REFERENCES

1. WESTE, N. H. E., AND ESHRAGHIAN, K. *Principles of CMOS VLSI Design: A Systems Perspective*, 2nd ed. Reading, MA: Addison-Wesley, 1993.
2. Micron Technology, Inc. *Micron 256Mb: x4, x8, x16 SDRAM.* www.micron.com, 2002.
3. Micron Technology, Inc. *Micron 64Mb: x32 DDR SDRAM.* www.micron.com, 2001.
4. SOBELMAN, M., "Rambus Technology Basics," *Rambus Developer Forum.* Rambus, Inc., October 2001.
5. Rambus, Inc. *Rambus Direct RDRAM 128/144-Mbit (256x16/18x32s) - Preliminary Information*, Document DL0059 Version 1.11.

PROBLEMS

 The plus (+) indicates a more advanced problem and the asterisk (*) indicates a solution is available on the on the Companion Website for the text.

9–1. *The following memories are specified by the number of words times the number of bits per word. How many address lines and input-output data lines are needed in each case? (a) $16K \times 8$, (b) $256K \times 16$, (c) $64M \times 32$, and (d) $2G \times 8$.

9–2. Give the number of bytes stored in the memories listed in Problem 9–1.

9–3. *Word number $(835)_{10}$ in the memory shown in Figure 9-2 contains the binary equivalent of $(15,103)_{10}$. List the 10-bit address and the 16-bit memory contents of the word.

9–4. A 64K × 16 RAM chip uses coincident decoding by splitting the internal decoder into row select and column select. (a) Assuming that the RAM cell array is square, what is the size of each decoder, and how many AND gates are required for decoding an address? (b) Determine the row and column selection lines that are enabled when the input address is the binary equivalent of $(32000)_{10}$.

9–5. Assume that the largest decoder that can be used in an $m \times 1$ RAM chip has 13 address inputs and that coincident decoding is employed. In order to construct RAM chips that contain more 1-bit words than m, multiple RAM cell arrays, each with decoders and read/write circuits, are included in the chip.
  (a) With the decoder restrictions given, how many RAM cell arrays are required to construct a 512 M × 1 RAM chip?
  (b) Show the decoder required to select from among the different RAM arrays in the chip and its connections to address bits and column decoders.

9–6. A DRAM has 14 address pins and its row address is 1 bit longer than its column address. How many addresses, total, does the DRAM have?

9–7. A 256Mb DRAM uses 4-bit data and has equal length row and column addresses. How many address pins does the DRAM have?

9–8. A DRAM has a refresh interval of 128 ms and has 4096 rows. What is the interval between refreshes for distributed refresh? What is the minimum number of address pins on the DRAM?

9–9. *(a)How many 128K × 16 RAM chips are needed to provide a memory capacity of 1M bytes?
  (b) How many address lines are required to access 1M bytes? How many of these lines are connected to the address inputs of all chips?
  (c) How many lines must be decoded to produce the chip select inputs? Specify the size of the decoder.

9–10. Using the 64K × 8 RAM chip in Figure 9-9 plus a decoder, construct the block diagram for a 512K × 16 RAM.

9–11. Explain how SDRAM takes advantage of the two-dimensional storage array to provide a high data access rate.

9–12. Explain how a DDRAM achieves a data rate that is a factor of two higher than a comparable SDRAM.

COMPUTER DESIGN BASICS

I n Chapter 7, the separation of a design into a datapath that implements microoperations and a control unit that determines the sequence of microoperations was introduced. In this chapter, we define a generic computer datapath that implements register transfer microoperations and serves as a framework for the design of detailed processing logic. The concept of a control word provides a tie between the datapath and the control unit associated with it.

The generic datapath combined with a control unit and memory forms a programmable system, in this case, a simple computer. The concept of an instruction set architecture (ISA) is introduced as a means of specifying the computer. In order to implement the ISA, a control unit and the generic datapath are combined to form a CPU (Central Processing Unit). In addition, since this is a programmable system, memories are also present for storage of programs and data. Two different computers with two different control units are considered. The first computer has two memories, one for instructions and one for data, and performs all of its operations in a single clock cycle. The second computer has a single memory for both instructions and data and a more complex architecture requiring multiple clock cycles to perform its operations.

In the generic computer at the beginning of Chapter 1, register transfers, micro-operations, buses, datapaths, datapath components, and control words are used quite broadly. Likewise, control units appear in most of the digital parts of the generic computer. The design of processing units consisting of control units interacting with datapaths has its greatest impact within the generic computer in the CPU and FPU in the processor chip. These two components contain major datapaths that perform processing. The CPU and the FPU perform additions, subtractions, and most of the other operations specified by the instruction set.

10-1   INTRODUCTION

Computers and their design are introduced in this chapter. The specification for a computer consists of a description of its appearance to a programmer at the lowest level, its *instruction set architecture (ISA)*. From the ISA, a high-level description of the hardware to implement the computer, called the *computer architecture,* is formulated. This architecture, for a simple computer, is typically divided into a datapath and a control. The datapath is defined by three basic components:

1. a set of registers,

2. the microoperations that are performed on data stored in the registers, and

3. the control interface.

The control unit provides signals that control the microoperations performed in the datapath and in other components of the system, such as memories. In addition, the control unit controls its own operation, determining the sequence of events that occur. This sequence may depend upon the results of current and past microoperations executed. In a more complex computer, typically multiple control units and datapaths are present.

To build a foundation for considering computer designs, initially, we extend the ideas in Chapter 7 to the implementation of datapaths. Specifically, we consider a generic datapath, one that can be used, in some cases in modified form, in all of the computer designs considered in the remainder of the text. These future designs show how a given datapath can be used to implement different instruction set architectures by simply combining the datapath with different control units.

10-2   DATAPATHS

Instead of having each individual register perform its microoperations directly, computer systems often employ a number of storage registers in conjunction with a shared operation unit called an *arithmetic/logic unit*, abbreviated ALU. To perform a microoperation, the contents of specified source registers are applied to the inputs of the shared ALU. The ALU performs an operation, and the result of this operation is transferred to a destination register. With the ALU as a combinational circuit, the entire register transfer operation from the source registers, through the ALU, and into the destination register is performed during one clock cycle. The shift operations are often performed in a separate unit, but sometimes these operations are also implemented within the ALU.

Recall that the combination of a set of registers with a shared ALU and interconnecting paths is the datapath for the system. The rest of this chapter is concerned with the organization and design of datapaths and associated control units used to implement simple computers. The design of a particular ALU is undertaken to show the process involved in implementing a complex combinational circuit. We also design a shifter, combine control signals into control words, and then add control units to implement two different computers.

The datapath and the control unit are the two parts of the processor, or CPU, of a computer. In addition to the registers, the datapath contains the digital logic that implements the various microoperations. This digital logic consists of buses, multiplexers, decoders, and processing circuits. When a large number of registers is included in a datapath, the registers are most conveniently connected through one or more buses. Registers in a datapath interact by the direct transfer of data, as well as in the performance of the various types of microoperations. A simple bus-based datapath with four registers, an ALU, and a shifter is shown in Figure 10-1. The shading and blue signal names relate to Figure 10-10 and will be discussed in Section 10-5. The black signal names are used here to describe the details in Figure 10-1. Each register is connected to two multiplexers to form ALU and shifter input buses A and B. The select inputs on each multiplexer select one register for the corresponding bus. For Bus B, there is an additional multiplexer, MUX B, so that constants can be brought into the datapath from outside using Constant in. Bus B also connects to Data out, to send data outside the datapath to other components of the system, such as memory or input-output. Likewise, Bus A connects to Address out, to send address information outside of the datapath for memory or input-output.

Arithmetic and logic microoperations are performed on the operands on the A and B buses by the ALU. The G select inputs select the microoperation to be performed by the ALU. The shift microoperations are performed on data on Bus B by the shifter. The H select input either passes the operand on bus B directly through to the shifter output or selects a shift microoperation. MUX F selects the output of the ALU or the output of the shifter. MUX D selects the output of MUX F or external data applied to Data in to be applied to Bus D. The latter is connected to the inputs of all the registers. The destination select inputs determine which register is loaded with the data on Bus D. Since the select inputs are decoded, only one register Load signal is active for any transfer of data into a register from Bus D. A Load enable signal that can force all Load signals to 0 using AND gates is present for transfers that are not to change the contents of any of the four registers.

It is useful to have certain information, based on the results of an ALU operation, available for use by the control unit of the CPU to make decisions. Four status bits are shown with the ALU in Figure 10-1. The status bits, carry C and overflow V, were explained in conjunction with Figure 5-9. The zero status bit Z is 1 if the output of the ALU contains all zeros and is 0 otherwise. Thus, $Z = 1$ if the result of an operation is zero, and $Z = 0$ if the result is nonzero. The sign status bit N (for negative) is the leftmost bit of the ALU output, which is the sign bit for the result in signed-number representations. Status values from the shifter can also be incorporated into the status bits if desired.

The control unit for the datapath directs the information flow through the buses, the ALU, the shifter, and the registers by applying signals to the select inputs. For example, to perform the microoperation

$$R1 \leftarrow R2 + R3$$

the control unit must provide binary selection values to the following sets of control inputs:

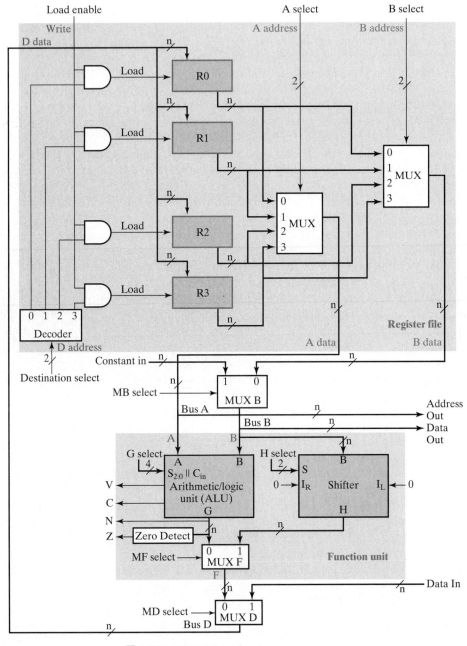

☐ **FIGURE 10-1**

Block Diagram of a Generic Datapath

1. *A* select, to place the contents of *R2* onto *A* data and, hence, Bus *A*.
2. *B* select, to place the contents of *R3* onto the 0 input of MUX *B*; and *MB* select, to put the 0 input of MUX *B* onto Bus *B*.
3. *G* select, to provide the arithmetic operation *A* + *B*.
4. *MF* select, to place the ALU output on the MUX *F* output.
5. *MD* select, to place the MUX *F* output onto Bus *D*.
6. Destination select, to select *R1* as the destination of the data on Bus *D*.
7. Load enable, to enable a register—in this case, *R1*—to be loaded.

The sets of values must be generated and must become available on the corresponding control lines early in the clock cycle. The binary data from the two source registers must propagate through the multiplexers and the ALU and on into the inputs of the destination register, all during the remainder of the same clock cycle. Then, when the next positive clock edge arrives, the binary data on Bus *D* is loaded into the destination register. To achieve fast operation, the ALU and shifter are constructed with combinational logic having a limited number of levels, such as a carry-lookahead adder.

10-3 THE ARITHMETIC/LOGIC UNIT

The ALU is a combinational circuit that performs a set of basic arithmetic and logic microoperations. The ALU has a number of selection lines used to determine the operation to be performed. The selection lines are decoded within the ALU, so that k selection lines can specify up to 2^k distinct operations.

Figure 10-2 shows the symbol for a typical *n*-bit ALU. The *n* data inputs from *A* are combined with the *n* data inputs from *B* to generate the result of an operation

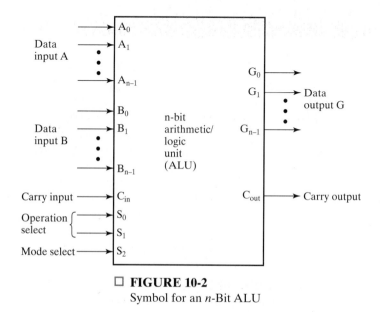

□ **FIGURE 10-2**
Symbol for an *n*-Bit ALU

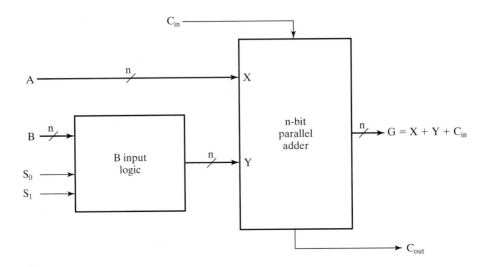

☐ **FIGURE 10-3**
Block Diagram of an Arithmetic Circuit

at the G outputs. The mode-select input S_2 distinguishes between arithmetic and logic operations. The two Operation select inputs S_1 and S_0 and the Carry input C_{in} specify the eight arithmetic operations with S_2 at 0. Operand select input S_0 and C_{in} specify the four logic operations with S_2 at 1.

We perform the design of this ALU in three stages. First, we design the arithmetic section. Then we design the logic section, and finally, we combine the two sections to form the ALU.

Arithmetic Circuit

The basic component of an arithmetic circuit is a parallel adder, which is constructed with a number of full-adder circuits connected in cascade, as shown in Figure 5-5. By controlling the data inputs to the parallel adder, it is possible to obtain different types of arithmetic operations. The block diagram in Figure 10-3 demonstrates a configuration in which one set of inputs to the parallel adder is controlled by the select lines S_1 and S_0. There are n bits in the arithmetic circuit, with two inputs A and B and output G. The n inputs from B go through the B input logic to the Y inputs of the parallel adder. The input carry C_{in} goes in the carry input of the full adder in the least-significant-bit position. The output carry C_{out} is from the full adder in the most-significant-bit position. The output of the parallel adder is calculated from the arithmetic sum as

$$G = X + Y + C_{in}$$

where X is the n-bit binary number from the inputs and Y is the n-bit binary number from the B input logic. C_{in} is the input carry, which equals 0 or 1. Note that the symbol + in the equation denotes arithmetic addition.

□ TABLE 10-1
Function Table for Arithmetic Circuit

Select		Input	$G = A + Y + C_{in}$	
S_1	S_0	Y	$C_{in} = 0$	$C_{in} = 1$
0	0	all 0's	$G = A$ (transfer)	$G = A + 1$ (increment)
0	1	B	$G = A + B$ (add)	$G = A + B + 1$
1	0	$\overline{B}$	$G = A + \overline{B}$	$G = A + \overline{B} + 1$ (subtract)
1	1	all 1's	$G = A - 1$ (decrement)	$G = A$ (transfer)

Table 10-1 shows the arithmetic operations that are obtainable by controlling the value of Y with the two selection inputs S_1 and S_0. If the inputs from B are ignored and we insert all 0's at the Y inputs, the output sum becomes $G = A + 0 + C_{in}$. This gives $G = A$ when $C_{in} = 0$ and $G = A + 1$ when $C_{in} = 1$. In the first case, we have a direct transfer from input A to output G. In the second case, the value of A is incremented by 1. For a straight arithmetic addition, it is necessary to apply the B inputs to the Y inputs of the parallel adder. This gives $G = A + B$ when $C_{in} = 0$. Arithmetic subtraction is achieved by applying the complement of inputs B to the Y inputs of the parallel adder, to obtain $G = A + \overline{B} + 1$ when $C_{in} = 1$. This gives A plus the 2's complement of B, which is equivalent to 2's complement subtraction. All 1's is the 2's complement representation for –1. Thus, applying all 1's to the Y inputs with $C_{in} = 0$ produces the decrement operation $G = A - 1$.

The B input logic in Figure 10-3 can be implemented with n multiplexers. The data inputs to each multiplexer in stage i for $i = 0, 1, \ldots, n - 1$ are $0, B_i, \overline{B}_i$, and 1, corresponding to selection values S_1S_0: 00, 01, 10, and 11, respectively. Thus, the arithmetic circuit can be constructed with n full adders and n 4-to-1 multiplexers.

The number of gates in the B input logic can be reduced if, instead of using 4-to-1 multiplexers, we go through the logic design of one stage (one bit) of the B input logic. This can be done as shown in Figure 10-4. The truth table for one typical

Inputs			Output	
S_1	S_0	B_i	Y_i	
0	0	0	0	$Y_i = 0$
0	0	1	0	
0	1	0	0	$Y_i = B_i$
0	1	1	1	
1	0	0	1	$Y_i = \overline{B}_i$
1	0	1	0	
1	1	0	1	$Y_i = 1$
1	1	1	1	

(a)Truth table

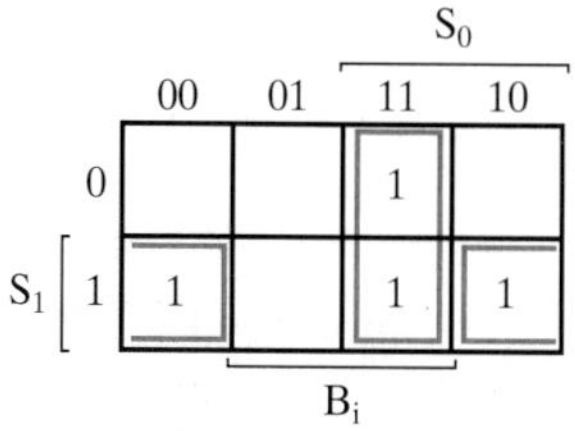

(b) Map Simplification:
$Y_i = B_iS_0 + \overline{B}_iS_1$

□ FIGURE 10-4
B Input Logic for One Stage of Arithmetic Circuit

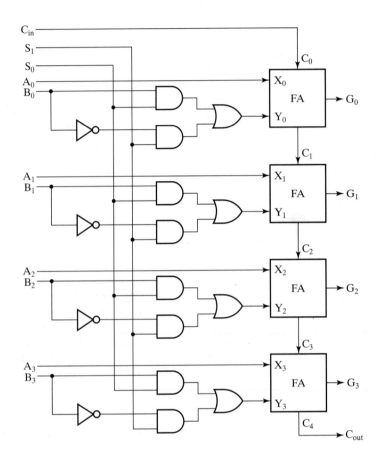

☐ **FIGURE 10-5**
Logic Diagram of a 4-bit Arithmetic Circuit

stage i of the logic is given in Figure 10-4(a). The inputs are S_1, S_0, and B_i, and the output is Y_i. Following the requirements specified in Table 10-1, we let $Y_i = 0$ when $S_1 S_0 = 00$, and similarly assign the other three values of Y_i for each of the combinations of the selection variables. Output Y_i is simplified in the map in Figure 10-4(b), to give

$$Y_i = B_i S_0 + \overline{B}_i S_1$$

where S_1 and S_0 are common to all n stages. Each stage i is associated with input B_i and output Y_i for $i = 0, 1, 2, ..., n - 1$. This logic corresponds to a 2-to-1 multiplexer with B_i on the select input and S_1 and S_0 on the data inputs.

Figure 10-5 shows the logic diagram of an arithmetic circuit for $n = 4$. The four full-adder (FA) circuits constitute the parallel adder. The carry into the first stage is the input carry C_{in}. All other carries are connected internally from one stage to the next. The selection variables are S_1, S_0, and C_{in}. Variables S_1 and S_0

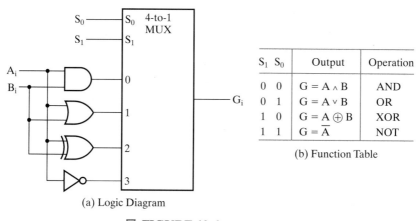

(a) Logic Diagram

S_1 S_0	Output	Operation
0 0	$G = A \wedge B$	AND
0 1	$G = A \vee B$	OR
1 0	$G = A \oplus B$	XOR
1 1	$G = \overline{A}$	NOT

(b) Function Table

□ **FIGURE 10-6**

One Stage of Logic Circuit

control all Y inputs of the full adders according to the Boolean function derived in Figure 10-4(b). Whenever C_{in} is $1, A + Y$ has 1 added. The eight arithmetic operations for the circuit as a function of S_1, S_0, and C_{in} are listed in Table 10-2. It is interesting to note that the operation $G = A$ appears twice in the table. This is a harmless by-product of using C_{in} as one of the control variables while implementing both increment and decrement instructions.

Logic Circuit

The logic microoperations manipulate the bits of the operands by treating each bit in a register as a binary variable, giving bitwise operations. There are four commonly used logic operations—AND, OR, XOR (exclusive-OR), and NOT—from which others can be conveniently derived.

Figure 10-6(a) shows one stage of the logic circuit. It consists of four gates and a 4-to-1 multiplexer, although simplification could yield less complex logic. Each of the four logic operations is generated through a gate that performs the required logic. The outputs of the gates are applied to the inputs of the multiplexer with two selection variables S_1 and S_0. These choose one of the data inputs of the multiplexer and direct its value to the output. The diagram shows a typical stage with subscript i. For the logic circuit with n bits, the diagram must be repeated n times, for $i = 0, 1, 2,..., n - 1$. The selection variables are applied to all stages. The function table in Figure 10-6(b) lists the logic operations obtained for each combination of the selection values.

Arithmetic/Logic Unit

The logic circuit can be combined with the arithmetic circuit to produce an ALU. Selection variables S_1 and S_0 can be common to both circuits, provided that we use a third selection variable to differentiate between the two. The configuration for one stage of the ALU is illustrated in Figure 10-7. The outputs of the arithmetic

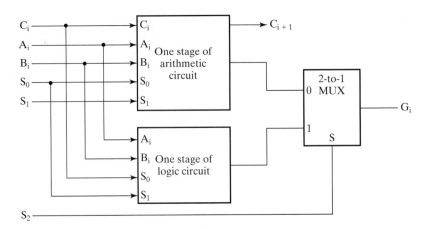

☐ **FIGURE 10-7**
One Stage of ALU

and logic circuits in each stage are applied to a 2-to-1 multiplexer with selection variable S_2. When $S_2 = 0$, the arithmetic output is selected, and when $S_2 = 1$, the logic output is selected. Note that the diagram shows just one typical stage of the ALU; the circuit must be repeated n times for an n-bit ALU. The output carry C_{i+1} of a given arithmetic stage must be connected to the input carry C_i of the next stage in sequence. The input carry to the first stage is the input carry C_{in}, which also acts as a selection variable for the arithmetic operations.

The ALU specified in Figure 10-7 provides eight arithmetic and four logic operations. Each operation is selected through the variables S_2, S_1, S_0, and C_{in}. Table 10-2 lists the 12 ALU operations. The first eight are arithmetic operations

☐ **TABLE 10-2**
Function Table for ALU

Operation Select					
S_2	S_1	S_0	C_{in}	Operation	Function
0	0	0	0	$G = A$	Transfer A
0	0	0	1	$G = A + 1$	Increment A
0	0	1	0	$G = A + B$	Addition
0	0	1	1	$G = A + B + 1$	Add with carry input of 1
0	1	0	0	$G = A + \overline{B}$	A plus 1's complement of B
0	1	0	1	$G = A + \overline{B} + 1$	Subtraction
0	1	1	0	$G = A - 1$	Decrement A
0	1	1	1	$G = A$	Transfer A
1	X	0	0	$G = A \wedge B$	AND
1	X	0	1	$G = A \vee B$	OR
1	X	1	0	$G = A \oplus B$	XOR
1	X	1	1	$G = \overline{A}$	NOT (1's complement)

and are selected with $S_2 = 0$. The next four are logic operations and are selected with $S_2 = 1$. To provide selection codes using as few bits as possible, S_0 and C_i are used to control the selection of the logic operations instead of S_2 and S_1. Selection input S_1 has no effect during the logic operations and is marked with X to indicate that its value may be either 0 or 1. Later in the design, it is assigned value 0 for logic operations.

The ALU logic we have designed is not as simple as it could be and has a fairly high number of logic levels, contributing to propagation delay in the circuit. With the use of logic simplification software, we can simplify this logic and reduce the delay. For example, it is quite easy to simplify the logic for a single stage of the ALU. For realistic n, a means of further reducing the carry propagation delay in the ALU, such as the carry lookahead adder from Section 5-2, is usually necessary.

10-4 THE SHIFTER

The shifter shifts the value on Bus B, placing the result on an input of MUX F. The basic shifter performs one of two main types of transformations on the data: right shift and left shift.

A seemingly obvious choice for a shifter would be a bidirectional shift register with parallel load. Data from Bus B can be transferred to the register in parallel and then shifted to the right, the left, or not at all. A clock pulse loads the output of Bus A into the shift register, and a second clock pulse performs the shift. Finally, a third clock pulse transfers the data from the shift register to the selected destination register.

Alternatively, the transfer from a source register to a destination register can be done using only one clock pulse if the shifter is implemented as a combinational circuit as done in Chapter 5. Because of the faster operation that results from the use of one clock pulse instead of three, this is the preferred method. In a combinational shifter, the signals propagate through the gates without the need for a clock pulse. Hence, the only clock needed for a shift in the datapath is for loading the data from Bus H into the selected destination register.

A combinational shifter can be constructed with multiplexers as shown in Figure 10-8. The selection variable S is applied to all four multiplexers to select the type of operation within the shifter. $S = 00$ causes B to be passed through the shifter unchanged. $S = 01$ causes a right-shift operation and $S = 10$ causes a left-shift operation. The right shift fills the position on the left with the value on serial input I_R. The left shift fills the position on the right with the value on serial input I_L. Serial outputs are available from serial output R and serial output L for right and left shifts, respectively.

The diagram of Figure 10-8 shows only four stages of the shifter, which has n stages in a system with n-bit operands. Additional selection variables may be employed to specify what goes into I_R and I_L during a single bit-position shift. Note that to shift an operand by $m > 1$ bit positions, this shifter must perform a series of m 1-bit position shifts, taking m clock cycles.

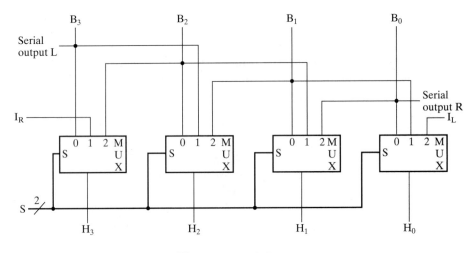

☐ **FIGURE 10-8**
4-Bit Basic Shifter

Barrel Shifter

In datapath applications, often the data must be shifted more than one bit position in a single clock cycle. A *barrel shifter* is one form of combinational circuit that shifts or rotates the input data bits by the number of bit positions specified by a binary value on a set of selection lines. The shift we consider here is a rotation to the left, which means that the binary data is shifted to the left, with the bits coming from the most significant part of the register rotated back into the least significant part of the register.

A 4-bit version of this kind of barrel shifter is shown in Figure 10-9. It consists of four multiplexers with common select lines S_1 and S_0. The selection variables determine the number of positions that the input data will be shifted to the left by rotation. When $S_1 S_0 = 00$, no shift occurs, and the input data has a direct path to the outputs. When $S_1 S_0 = 01$, the input data are rotated one position, with D_0 going to Y_1, D_1 going to Y_2, D_2 going to Y_3, and D_3 going to Y_0. When $S_1 S_0 = 10$, the input is rotated two positions, and when $S_1 S_0 = 11$, the rotation is by three bit positions. Table 10-3 gives the function table for the 4-bit barrel shifter. For each

☐ **TABLE 10-3**
Function Table for 4-Bit Barrel Shifter

Select		Output				
S_1	S_0	Y_3	Y_2	Y_1	Y_0	Operation
0	0	D_3	D_2	D_1	D_0	No rotation
0	1	D_2	D_1	D_0	D_3	Rotate one position
1	0	D_1	D_0	D_3	D_2	Rotate two positions
1	1	D_0	D_3	D_2	D_1	Rotate three positions

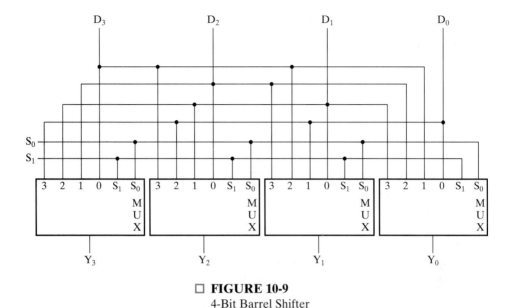

□ **FIGURE 10-9**
4-Bit Barrel Shifter

binary value of the selection variables, the table lists the inputs that go to the corresponding output. Thus, to rotate three positions, $S_1 S_0$ must be equal to 11, causing D_0 to go to Y_3, D_1 to go to Y_0, D_2 to go to Y_1, and D_3 to go to Y_2. Note that, by using this left-rotation barrel shifter, one can generate all desired right rotations as well. For example, a left rotation by three positions is the same as a right rotation by one position in this 4-bit barrel shifter. In general, in a 2^n-bit barrel shifter, i positions of left rotation is the same as $2^n - i$ bits of right rotation.

A barrel shifter with 2^n input and output lines requires 2^n multiplexers, each having 2^n data inputs and n selection inputs. The number of positions for the data to be rotated is specified by the selection variables and can be from 0 to $2^n - 1$ positions. For a large n, the fan-in to gates is too large, so larger barrel shifters consist of layers of multiplexers, as shown in Section 12-2, or of special structures designed at the transistor level.

10-5 DATAPATH REPRESENTATION

The datapath in Figure 10-1 includes the registers, selection logic for the registers, the ALU, the shifter, and three additional multiplexers. With a hierarchical structure, we can reduce the apparent complexity of the datapath. This reduction is important, since we frequently use this datapath. Also, as illustrated by the register file to be discussed next, the use of a hierarchy allows one implementation of a module to be replaced with another, so that we are not tied to specific logic implementations.

A typical datapath has more than four registers. Indeed, computers with 32 or more registers are common. The construction of a bus system with a large number of registers requires different techniques. A set of registers having common micro-operations performed on them may be organized into a *register file*. The typical register file is a special type of fast memory that permits one or more words to be read and one or more words to be written, all simultaneously. Functionally, a simple register file contains the equivalent of the logic shaded in blue in Figure 10-1. Due to the memory-like nature of register files, the *A* select, *B* select, and Destination select inputs in the figure, become three addresses. As shown in Figure 10-1 in blue and on the register file symbol in Figure 10-10, the *A* address accesses a word to be read onto *A* data, the *B* address accesses a second word to be read onto *B* data, and the *D* address accesses a word to be written into from *D* data. All of these accesses

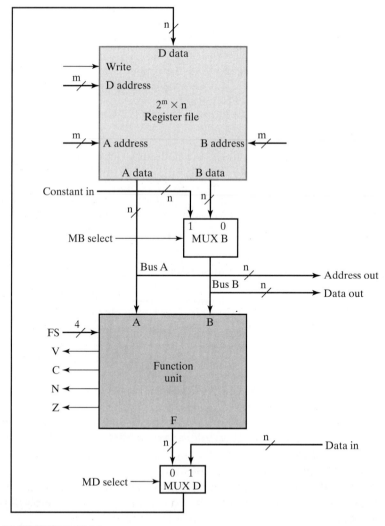

☐ **FIGURE 10-10**
Block Diagram of Datapath Using the Register File and Function Unit

□ **TABLE 10-4**
**G Select, H Select, and MF Select Codes Defined
in Terms of FS Codes**

FS(3:0)	MF Select	G Select(3:0)	H Select(3:0)	Microoperation
0000	0	0000	XX	$F = A$
0001	0	0001	XX	$F = A + 1$
0010	0	0010	XX	$F = A + B$
0011	0	0011	XX	$F = A + B + 1$
0100	0	0100	XX	$F = A + \overline{B}$
0101	0	0101	XX	$F = A + \overline{B} + 1$
0110	0	0110	XX	$F = A - 1$
0111	0	0111	XX	$F = A$
1000	0	1 X00	XX	$F = A \wedge B$
1001	0	1 X01	XX	$F = A \vee B$
1010	0	1 X10	XX	$F = A \oplus B$
1011	0	1 X11	XX	$F = \overline{A}$
1100	1	XXXX	0 0	$F = B$
1101	1	XXXX	0 1	$F = \text{sr } B$
1110	1	XXXX	1 0	$F = \text{sl } B$

occur in the same clock cycle. A Write input corresponding to the Load Enable signal is also provided. When at 1, the Write signal permits registers to be loaded, during the current clock cycle, and, when at 0, prevents register loading. The size of the register file is $2^m \times n$, where m is the number of register address bits and n is the number of bits per register. For the datapath in Figure 10-1, $m = 2$, giving four registers, and n is unspecified.

Since the ALU and the shifter are shared processing units with outputs that are selected by MUX F, it is convenient to group the two units and the MUX together to form a shared function unit. Gray shading in Figure 10-1 highlights the function unit, which can be represented by the symbol given in Figure 10-10. The inputs to the function unit are from Bus A and Bus B, and the output of the function unit goes to MUX D. The function unit also has the four status bits $V, C, N,$ and Z as added outputs.

In Figure 10-1, there are three sets of select inputs: the G select, H select, and MF select. In Figure 10-10, there is a single set of select inputs labeled FS, for "function select." To fully specify the function unit symbol in the figure, all of the codes for MF select, G select, and H select must be defined in terms of the codes for FS. Table 10-4 defines these code transformations. The codes for FS are given in the left column. From Table 10-4, it is apparent that MF is 1 for the leftmost two bits of FS both equal to 1. If MF select $= 0$, then the G select codes determine the function on the output of the function unit. If MF select $= 1$, then the H select codes determine the function on the output of the function unit. To show this dependency, the codes that determine the function unit outputs are highlighted in blue in the table. From Table 10-4, the code transformations can be implemented using the Boolean equations: $MF = F_3 \cdot F_2$, $G_3 = F_3$, $G_2 = F_2$, $G_1 = F_1$, $G_0 = F_0$, $H_1 = F_1$, and $H_0 = F_0$.

The status bits are assumed to be meaningless when the shifter is selected, although in a more complex system, shifter status bits can be designed to replace those for the ALU whenever a shifter microoperation is specified. Note that the status bit implementation depends on the specific implementation that has been used for the arithmetic circuit. Alternative implementations may not produce the same results.

10-6 THE CONTROL WORD

The selection variables for the datapath control the microoperations executed within the datapath for any given clock pulse. For the datapath in Section 10-5, the selection variables control the addresses for the data read from the register file, the function performed by the function unit, and the data loaded into the register file, as well as the selection of external data. We will now demonstrate how these control variables select the microoperations for the datapath. The choice of control variable values for typical microoperations will be discussed, and a simulation of the datapath will be illustrated.

A block diagram of a datapath that is a specific version of the datapath in Figure 10-10 is shown in Figure 10-11(a). It has a register file with eight registers, $R0$ through $R7$. The register file provides the inputs to the function unit through Bus A and Bus B. MUX B selects between constant values on Constant in and register values on B data. The ALU and zero-detection logic within the function unit generate the binary data for the four status bits: V (overflow), C (carry), N (sign), and Z (zero). MUX D selects the function unit output or the data on Data in as input for the register file.

There are 16 binary control inputs. Their combined values specify a *control word*. The 16-bit control word is defined in Figure 10-11(b). It consists of seven parts called *fields*, each designated by a pair of letters. The three register fields are three bits each. The remaining fields have one or four bits. The three bits of DA select one of eight destination registers for the result of the microoperation. The three bits of AA select one of eight source registers for the Bus A input to the ALU. The three bits of BA select a source register for the 0 input of the MUX B. The single MB bit determines whether Bus B carries the contents of the selected source register or a constant value. The 4-bit FS field controls the operation of the function unit. The FS field contains one of the 15 codes from Table 10-4. The single bit of MD selects the function unit output or the data on Data in as the input to Bus D. The final field, RW, determines whether a register is written or not. When applied to the control inputs, the 16-bit control word specifies a particular microoperation.

The functions of all meaningful control codes are specified in Table 10-5. For each of the fields, a binary code for each of the functions is given. The register selected by each of the fields DA, AA, and BA is the one with the decimal equivalent equal to the binary number for the code. MB selects either the register selected by the BA field or a constant from outside of the datapath on Constant in. The ALU operations, the shifter operations, and the selection of the ALU or

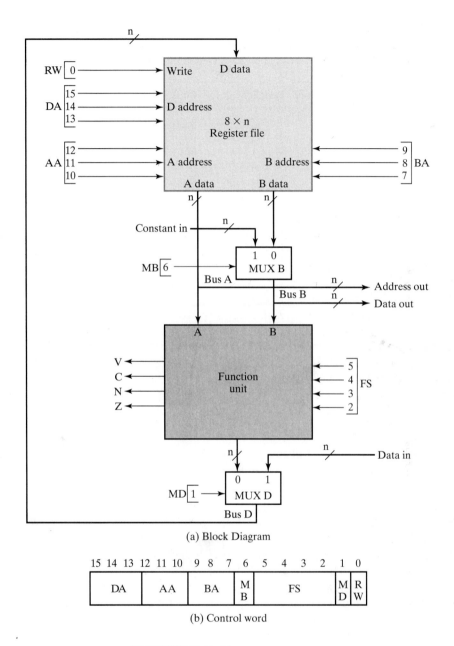

(a) Block Diagram

15 14 13	12 11 10	9 8 7	6	5 4 3 2	1	0
DA	AA	BA	M B	FS	M D	R W

(b) Control word

☐ **FIGURE 10-11**
Datapath with Control Variables

shifter outputs are all specified by the FS field. The field MD controls the information to be loaded into the register file. The final field, RW, has the functions No Write, to prevent writing to any registers, and Write, to signify writing to a register.

☐ **TABLE 10-5**
Encoding of Control Word for the Datapath

DA, AA, BA		MB		FS		MD		RW	
Function	Code	Function	Code	Function	Code	Function	Code	Function	Code
$R0$	000	Register	0	$F = A$	0000	Function	0	No write	0
$R1$	001	Constant	1	$F = A + 1$	0001	Data In	1	Write	1
$R2$	010			$F = A + B$	0010				
$R3$	011			$F = A + B + 1$	0011				
$R4$	100			$F = A + \overline{B}$	0100				
$R5$	101			$F = A + \overline{B} + 1$	0101				
$R6$	110			$F = A - 1$	0110				
$R7$	111			$F = A$	0111				
				$F = A \wedge B$	1000				
				$F = A \vee B$	1001				
				$F = A \oplus B$	1010				
				$F = \overline{A}$	1011				
				$F = B$	1100				
				$F = \text{sr } B$	1101				
				$F = \text{sl } B$	1110				

The control word for a given microoperation can be derived by specifying the value of each of the control fields. For example, a subtraction given by the statement

$$R1 \leftarrow R2 + \overline{R3} + 1$$

specifies $R2$ for the A input of the ALU and $R3$ for the B input of the ALU. It also specifies function unit operation $F = A + \overline{B} + 1$ and selection of the function unit output for input into the register file. Finally, the microoperation selects $R1$ as the destination register and sets RW to 1 to cause $R1$ to be written. The control word for this microinstruction is specified by its seven fields, with the binary value for each field obtained from the encoding listed in Table 10-5. The binary control word for this subtraction microoperation, 001_010_011_0_0101_0_1, (with underline "_" used for convenience to separate the fields) is obtained as follows:

Field:	DA	AA	BA	MB	FS	MD	RW
Symbolic:	$R1$	$R2$	$R3$	Register	$F = A + \overline{B} + 1$	Function	Write
Binary:	001	010	011	0	0101	0	1

The control word for the microoperation and those for several other microoperations are given in Table 10-6 using symbolic notation and in Table 10-7 using binary codes.

The second example in Table 10-6 is a shift microoperation given by the statement

$$R4 \leftarrow \text{sl } R6$$

□ **TABLE 10-6**
Examples of Microoperations for the Datapath, Using Symbolic Notation

Micro-operation	DA	AA	BA	MB	FS	MD	RW
$R1 \leftarrow R2 - R3$	$R1$	$R2$	$R3$	Register	$F = A + \overline{B} + 1$	Function	Write
$R4 \leftarrow$ sl $R6$	$R4$	—	$R6$	Register	$F = $ sl B	Function	Write
$R7 \leftarrow R7 + 1$	$R7$	$R7$	—	Register	$F = A + 1$	Function	Write
$R1 \leftarrow R0 + 2$	$R1$	$R0$	—	Constant	$F = A + B$	Function	Write
Data out $\leftarrow R3$	—	—	$R3$	Register	—	—	No Write
$R4 \leftarrow$ Data in	$R4$	—	—	—	—	Data in	Write
$R5 \leftarrow 0$	$R5$	$R0$	$R0$	Register	$F = A \oplus B$	Function	Write

This statement specifies a shift left for the shifter. The content of register $R6$, shifted to the left, is transferred to $R4$. Note that because the shifter is driven by the B bus, the source for the shift is specified in the BA field rather than the AA field. From the knowledge of the symbols in each field, the control word in binary is derived as shown in Table 10-7. For many microoperations, neither the A data nor the B data from the register file is used. In these cases, the respective symbolic field is marked with a dash. Since these values are unspecified, the corresponding binary values in Table 10-7 are Xs. Continuing with the last three examples in Table 10-6, to make the contents of a register available to an external destination only, we place the contents of the register on the B data output of the register file, with RW = No Write (0) to prevent the register file from being written. To place a small constant in a register or use a small constant as one of the operands, we place the constant on Constant in, set MB to Constant, and pass the value from Bus B through the ALU and Bus D to the destination register. To clear a register to 0, Bus D is set to all 0's by using the same register for both A data and B data with an XOR operation specified (FS = 1010) and MD = 0. The DA field is set to the code for the destination register, and RW is Write (1).

□ **TABLE 10-7**
Examples of Microoperations from Table 10-6, Using Binary Control Words

Micro-operation	DA	AA	BA	MB	FS	MD	RW
$R1 \leftarrow R2 - R3$	001	010	011	0	0101	0	1
$R4 \leftarrow$ sl $R6$	100	XXX	110	0	1110	0	1
$R7 \leftarrow R7 + 1$	111	111	XXX	0	0001	0	1
$R1 \leftarrow R0 + 2$	001	000	XXX	1	0010	0	1
Data out $\leftarrow R3$	XXX	XXX	011	0	XXXX	X	0
$R4 \leftarrow$ Data in	100	XXX	XXX	X	XXXX	1	1
$R5 \leftarrow 0$	101	000	000	0	1010	0	1

It is apparent from these examples that many microoperations can be performed by the same datapath. Sequences of such microoperations can be realized by providing a control unit that produces the appropriate sequences of control words.

To complete this section, we perform a simulation of the datapath in Figure 10-11. The number of bits in each register, n, is equal to 8. An unsigned decimal representation, which is most convenient for reading the simulation output, is used for all multiple bit signals. We assume that the microoperations in Table 10-7, executed in sequence, provide the inputs to the datapath and that the initial content of each register is its number in decimal (e.g., $R5$ contains $(0000\ 0101)_2 = (5)_{10}$). Figure 10-12 gives the result of this simulation. The first value displayed is the Clock with the clock cycles numbered for reference. The inputs, outputs, and state for the datapath are given roughly in the order of the flow of information through the path. The first four inputs are the primary control word fields, which specify the register addresses that determine the register file outputs and the function selection. Next are inputs Constant in and MB, which control the input to Bus B.

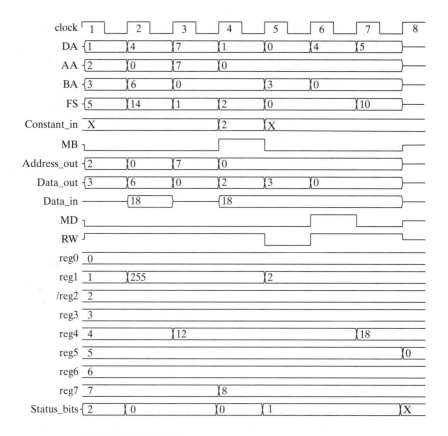

☐ **FIGURE 10-12**
Simulation of the Microoperation Sequence in Table 10-7

Following are the outputs Address out and Data out, which are the outputs from Bus A and Bus B, respectively. The next three variables—Data in, MD, and RW—are the final three inputs to the datapath. They are followed by the content of the eight registers and the Status bits, which are given as a vector (V, C, N, Z). The initial content of each register is its number in decimal. The value 2 is applied to Constant only in cycle 4 where MB equals 1. Otherwise, the value on Constant in is unknown as indicated by X. Finally, Data in has value 18. In the simulation, this value comes from a memory that is addressed by Address out and that has value 18 in location 0 with unknown values in all other locations. The resulting value, except when Address out is 0, is represented by a line midway between 0 and 1 indicating the value is unknown.

Of note in the simulation results is that changes in registers as a result of a particular microoperation appear in the clock cycle *after* that in which the microoperation is specified. For example, the result of the subtraction in clock cycle 1 appears in register $R1$ in clock cycle 2. This is because the result is loaded into flip-flops on the positive edge of the clock at the end of the clock cycle 1. On the other hand, the values on the Status bits, Address out, and Data out appear in the same clock cycle as the microoperation controlling them, since they do not depend on a positive clock edge occurring. Since there is no combinational delay specified in the simulation, these values change at the same time as the register values. Finally, note that eight clock cycles of simulation are used for seven microoperations so that the values in the registers that result from the last microoperation executed can be observed. Although Status bits appear for all microoperations, they are not always meaningful. For example, for the microoperations, R3 = Data out and R4 ← Data in, in clock cycles 5 and 6, respectively, the value of the status bits does not relate to the result since the Function unit is not used in these operations. Finally, for R5 ← R0 ⊕ R0 in clock cycle 7, the arithmetic unit is not used, so the values of V and C from that unit are irrelevant, but the values for N and Z do represent the status of the result as a signed 2's complement integer.

10-7   A SIMPLE COMPUTER ARCHITECTURE

We introduce a simple computer architecture to obtain a beginning understanding of computer design and to illustrate control designs for programmable systems. In a programmable system, a portion of the input to the processor consists of a sequence of *instructions*. Each instruction specifies the operation the system is to perform, which operands to use for the operation, where to place the results of the operation and, in some cases, which instruction to execute next. For the programmable system, the instructions are usually stored in memory, which is either RAM or ROM. To execute the instructions in sequence, it is necessary to provide the address in memory of the instruction to be executed. In a computer, this address comes from a register called the *program counter* (*PC*). As the name implies, the *PC* has logic that permits it to count. In addition, to change the sequence of operations using decisions based on status information, the *PC* needs parallel load capability. So, in the case of a programmable system, the control unit contains a *PC* and associated decision logic, as well as the necessary logic to inter-pret the instruction in order to execute it. *Executing* an instruction means activat-ing the necessary sequence of microoperations in the datapath (and elsewhere) required to perform the operation specified by the instruction. In contrast to the preceding, note that for a nonprogrammable system, the control unit is not responsible for obtaining instructions from memory, nor is it responsible for sequencing the execution of those instructions. There is no *PC* or similar register in such a system. Instead, the control unit determines the operations to be per-formed and the sequence of those operations, based on only its inputs and the sta-tus bits.

We show how the operations specified by instructions for the simple com-puter can be implemented by microoperations in the datapath, plus movement of information between the datapath and memory. We also show two different con-trol structures for implementing the sequences of operations necessary for control-ling program execution. The purpose here is to illustrate two different approaches to control design and the effects that such approaches have on datapath design and system performance. A more extensive study of the concepts associated with instruction sets for digital computers is presented in detail in the next chapter, and more complete CPU designs are undertaken in Chapter 12.

Instruction Set Architecture

The user specifies the operations to be performed and their sequence by the use of a *program*, which is a list of instructions that specifies the operations, the operands, and the sequence in which processing is to occur. The data processing performed by a computer can be altered by specifying a new program with different instruc-tions or by specifying the same instructions with different data. Instructions and data are usually stored together in the same memory. By means of the techniques discussed in Chapter 12, however, they may appear to be coming from different memories. The control unit reads an instruction from memory and decodes and executes the instruction by issuing a sequence of one or more microoperations. The ability to execute a program from memory is the most important single property of a general-purpose computer. Execution of a program from memory is in sharp contrast to the nonprogrammable multiplier control unit considered earlier, which executes only a single, fixed operation.

An *instruction* is a collection of bits that instructs the computer to perform a specific operation. We call the collection of instructions for a computer its *instruc-tion set* and a thorough description of the instruction set its *instruction set architec-ture (ISA)*. Simple instruction set architectures have three major components: the storage resources, the instruction formats, and the instruction specifications.

Storage Resources

The storage resources for the simple computer are represented by the diagram in Figure 10-13. The diagram depicts the computer structure as viewed by a user pro-gramming it in a language that directly specifies the instructions to be executed. It

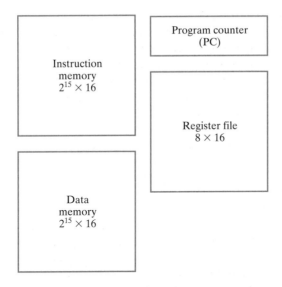

□ **FIGURE 10-13**
Storage Resource Diagram for a Simple Computer

gives the resources the user sees available for storing information. Note that the architecture includes two memories, one for storage of instructions and the other for storage of data. These may actually be different memories, or they may be the same memory, but viewed as different from the standpoint of the CPU as discussed in Chapter 12. Also visible to the programmer in the diagram is a register file with eight 16-bit registers and the 16-bit program counter.

Instruction Formats

The format of an instruction is usually depicted by a rectangular box symbolizing the bits of the instruction, as they appear in memory words or in a control register. The bits are divided into groups or parts called *fields*. Each field is assigned a specific item, such as the operation code, a constant value, or a register file address. The various fields specify different functions for the instruction and, when shown together, constitute an instruction format.

The *operation code* of an instruction, often shortened to "opcode," is a group of bits in the instruction that specifies an operation, such as add, subtract, shift, or complement. The number of bits required for the opcode of an instruction is a function of the total number of operations in the instruction set. It must consist of at least m bits for up to 2^m distinct operations. The designer assigns a bit combination (a code) to each operation. The computer is designed to accept this bit configuration at the proper time in the sequence of activities and to supply the proper control word sequence to execute the specified operation. As a specific example, consider a computer with a maximum of 128 distinct operations, one of them an addition operation. The opcode assigned to this operation consists of seven bits 0000010. When the opcode 0000010 is detected by the control unit, a sequence of control words is applied to the datapath to perform the intended addition.

The opcode of an instruction specifies the operation to be performed. The operation must be performed using data stored in computer registers or in memory (i.e., on the contents of the storage resources). An instruction, therefore, must specify not only the operation, but also the registers or memory words in which the operands are to be found and the result is to be placed. The operands may be specified by an instruction in two ways. An operand is said to be specified *explicitly* if the instruction contains special bits for its identification. For example, the instruction performing an addition may contain three binary numbers specifying the registers containing the two operands and the register that receives the result. An operand is said to be defined *implicitly* if it is included as a part of the definition of the operation itself, as represented by the opcode, rather than being given in the instruction. For example, in an Increment Register operation, one of the operands is implicitly $+1$.

The three instruction formats for the simple computer are illustrated in Figure 10-14. Suppose that the computer has a register file consisting of eight registers, $R0$ through $R7$. The instruction format in Figure 10-14(a) consists of an opcode that specifies the use of three or fewer registers, as needed. One of the registers is designated a destination for the result and two of the registers sources for operands. For convenience, the field names are abbreviated DR, for "Destination Register," SA for "Source Register A," and SB for "Source Register B." The number of register fields and registers actually used are determined by the specific opcode. The opcode also specifies the use of the registers. For example, for a subtraction operation, suppose that the three bits in SA are 010, specifying $R2$, the three bits in SB

15	9	8	6	5	3	2	0
Opcode		Destination register (DR)		Source register A (SA)		Source register B (SB)	

(a) Register

15	9	8	6	5	3	2	0
Opcode		Destination register (DR)		Source register A (SA)		Operand (OP)	

(b) Immediate

15	9	8	6	5	3	2	0
Opcode		Address (AD) (Left)		Source register A (SA)		Address (AD) (Right)	

(c) Jump and Branch

☐ **FIGURE 10-14**
Three Instruction Formats

are 011, specifying R3, and the three bits in DR are 001, specifying R1. Then the contents of R3 will be subtracted from the contents of R2, and the result will be placed in R1. As an additional example, suppose that the operation is a store (to memory). Suppose further, that the three bits in SA specify R4 and the three bits in SB specify R5. For this particular operation, it is assumed that the register specified in SA contains the address and the register specified in SB contains the operand to be stored. So the value in R5 is stored in the memory location given by the value in R4. The DR field has no effect, since the store operation prevents the register file from being written.

The instruction format in Figure 10-14(b), has an opcode, two register fields, and an operand. The operand is a constant called an *immediate operand*, since it is immediately available in the instruction. For example, for an add immediate operation with SA specified as R7, DR specified as R2, and operand OP equal to 011, the value 3 is added to the contents of R7, and the result of the addition is placed in R2. Since the operand is only three bits rather than a full 16 bits, the remaining 13 bits must be filled by using either zero fill or sign extension as discussed in Chapter 5. In this ISA, zero-fill is specified for the operand.

The instruction format in Figure 10-14(c), in contrast to the other two formats, does not change any register file or memory contents. Instead, it affects the order in which the instructions are fetched from memory. The location of an instruction to be fetched is determined by the program counter denoted by *PC*. Ordinarily, the program counter fetches the instructions from sequential addresses in memory as the program is executed. But much of the power of a processor comes from its ability to change the order of execution of the instructions based on results of the processing performed. These changes in the order of instruction execution are based on the use of instructions referred to as jumps and branches.

The example format given in Figure 10-14(c) for jump and branch instructions has an operation code, one register field SA, and a split address field AD. If a branch (possibly based on the contents of the register specified) is to occur, the new address is formed by adding the current PC contents and the contents of the 6-bit address field. This addressing method is called PC relative and the 6-bit address field, which is referred to as an *address offset* is treated as a signed two's complement number. To preserve the two's complement representation, *sign extension* is applied to the 6-bit address to form a 16-bit offset before the addition. If the leftmost bit of the address field AD is a 1, then the 10 bits to its left are filled with 1's to give a negative two's complement offset. If the leftmost bit of the address field is 0, then the 10 bits to its left are filled with 0's to give a positive two's complement offset. The offset is added to the contents of the *PC* to form the location from which the next instruction is to be fetched. For example, with the *PC* value equal to 55, suppose that a branch is to occur to location 35 if the contents of R6 is equal to zero. The opcode would specify a branch on zero instruction, SA would be specified as R6, and AD would be the 6-bit, two's complement representation of − 20. If R6 is zero, then *PC* contents becomes 55 + (− 20) = 35 and the next instruction would be fetched from address 35. Otherwise, if R6 is nonzero, the *PC* will count up to 56 and the next instruction will be fetched from address 56. This addressing method alone provides only branch addresses within a small range below and above the PC value. The jump provides a broader range of addresses by using the unsigned contents of a 16-bit register as the jump target.

The three formats in Figure 10-14 are used for the simple computer to be discussed in this chapter. In Chapter 11, we present and discuss more generally other instruction types and formats.

Instruction Specifications

Instruction specifications describe each of the distinct instructions that can be executed by the system. For each instruction, the opcode is given along with a shorthand name called a *mnemonic,* that can be used as a symbolic representation for the opcode. This mnemonic, along with a representation for each of the additional instruction fields in the format for the instruction, represents the notation to be used in specifying all of the fields of the instruction symbolically. This symbolic representation is then converted to the binary representation of the instruction by a program called an *assembler.* A description of the operation performed by the instruction execution is given, including the status bits that are affected by the instruction. This description may be text or may use a register transfer-like notation.

The instruction specifications for the simple computer are given in Table 10-8. The register transfer notation introduced in previous chapters is used to describe

☐ **TABLE 10-8**
Instruction Specifications for the Simple Computer

Instruction	Opcode	Mnemonic	Format	Description	Status Bits
Move A	0000000	MOVA	RD,RA	$R[DR] \leftarrow R[SA]$	N, Z
Increment	0000001	INC	RD,RA	$R[DR] \leftarrow R[SA] + 1$	N, Z
Add	0000010	ADD	RD,RA,RB	$R[DR] \leftarrow R[SA] + R[SB]$	N, Z
Subtract	0000101	SUB	RD,RA,RB	$R[DR] \leftarrow R[SA] - R[SB]$	N, Z
Decrement	0000110	DEC	RD,RA	$R[DR] \leftarrow R[SA] - 1$	N, Z
AND	0001000	AND	RD,RA,RB	$R[DR] \leftarrow R[SA] \wedge R[SB]$	N, Z
OR	0001001	OR	RD,RA,RB	$R[DR] \leftarrow R[SA] \vee R[SB]$	N, Z
Exclusive OR	0001010	XOR	RD,RA,RB	$R[DR] \leftarrow R[SA] \oplus R[SB]$	N, Z
NOT	0001011	NOT	RD,RA	$R[DR] \leftarrow \overline{R[SA]}$	N, Z
Move B	0001100	MOVB	RD,RB	$R[DR] \leftarrow R[SB]$	
Shift Right	0001101	SHR	RD,RB	$R[DR] \leftarrow sr\ R[SB]$	
Shift Left	0001110	SHL	RD,RB	$R[DR] \leftarrow sl\ R[SB]$	
Load Immediate	1001100	LDI	RD, OP	$R[DR] \leftarrow zf\ OP$	
Add Immediate	1000010	ADI	RD,RA,OP	$R[DR] \leftarrow R[SA] + zf\ OP$	
Load	0010000	LD	RD,RA	$R[DR] \leftarrow M[SA]$	
Store	0100000	ST	RA,RB	$M[SA] \leftarrow R[SB]$	
Branch on Zero	1100000	BRZ	RA,AD	if $(R[SA] = 0)$ $PC \leftarrow PC + se\ AD$	
Branch on Negative	1100001	BRN	RA,AD	if $(R[SA] < 0)$ $PC \leftarrow PC + se\ AD$	
Jump	1110000	JMP	RA	$PC \leftarrow R[SA]$	

the operation performed, and the status bits that are valid for each instruction are indicated. In order to illustrate the instructions, suppose that we have a memory with 16 bits per word with instructions having one of the formats in Figure 10-14. Instructions and data, in binary, are placed in memory as shown in Table 10-9. This stored information represents the four instructions illustrating the distinct formats. At address 25, we have a register format instruction that specifies an operation to subtract $R3$ from $R2$ and load the difference into $R1$. This operation is represented symbolically in the rightmost column of Table 10-9. Note that the 7-bit opcode for subtraction is 0000101, or decimal 5. The remaining bits of the instruction specify the three registers: 001 specifies the destination register as $R1$, 010 specifies the source register A as $R2$, and 011 specifies the source register B as $R3$.

In memory location 35 is a register format instruction to store the contents of $R5$ in the memory location specified by $R4$. The opcode is 0100000, or decimal 32, and the operation is given symbolically, again, in the rightmost column of the figure. Suppose $R4$ contains 70 and $R5$ contains 80. Then the execution of this instruction will store the value 80 in memory location 70, replacing the original value of 192 stored there.

At address 45, an immediate format instruction appears that adds 3 to the contents of $R7$ and loads the result into $R2$. The opcode for this instruction is 66, and the operand to be added is the value 3 (011) in the OP field, the last three bits of the instruction.

In location 55, the branch instruction previously described appears. The opcode for this instruction is 96, and source register A is specified as $R6$. Note that AD (Left) contains 101 and AD (Right) contains 100. Putting these two together

□ **TABLE 10-9**
Memory Representation of Instructions and Data

Decimal Address	Memory Contents	Decimal Opcode	Other Fields	Operation
25	0000101 001 010 011	5 (Subtract)	DR:1, SA:2, SB:3	$R1 \leftarrow R2 - R3$
35	0100000 000 100 101	32 (Store)	SA:4, SB:5	$M[R4] \leftarrow R5$
45	1000010 010 111 011	66 (Add Immediate)	DR:2, SA:7, OP:3	$R2 \leftarrow R7 + 3$
55	1100000 101 110 100	96 (Branch on Zero)	AD: 44, SA:6	If $R6 = 0$, $PC \leftarrow PC - 20$
70	0000000011000000	Data = 192. After execution of instruction in 35, Data = 80.		

and applying sign extension, we obtain 1111111111101100, which represents − 20 in two's complement. If register $R6$ is zero, then − 20 is added to the PC to give 35. If register $R6$ is nonzero, the new PC value will be 56. It should be noted that we have assumed that the addition to the PC content occurs before the PC has been incremented which would be the case in the simple computer. In real systems, however, the PC has sometimes been incremented to point to the next instruction in memory. In such a case, the value stored in AD needs to be adjusted accordingly to obtain the right branch address.

The placement of instructions in memory as shown in Table 10-9 is quite arbitrary. In many computers, the word length is from 32 to 64 bits, so the instruction formats can hold much larger immediate operands and addresses than those we have given. Depending on the computer architecture, some of the instruction formats may occupy two or more consecutive memory words. Also, the number of registers is often larger, so the register fields in the instructions must contain more bits.

At this point, it is vital to recognize the difference between a computer *operation* and a hardware *microoperation*. An operation is specified by an instruction stored in binary, in the computer's memory. The control unit in the computer uses the address or addresses provided by the program counter to retrieve the instruction from memory. It then decodes the opcode bits and other information in the instruction to perform the required microoperations for the execution of the instruction. In contrast, a microoperation is specified by the bits in a control word in the hardware which is decoded by the computer hardware to execute the microoperation. The execution of a computer operation often requires a sequence or program of microoperations, rather than a single microoperation.

10-8 SINGLE-CYCLE HARDWIRED CONTROL

The block diagram for a computer that has a hardwired control unit and that fetches and executes an instruction in a single clock cycle is shown in Figure 10-15. We refer to this computer as the single-cycle computer. The storage resources, instruction formats, and instruction specifications for this computer are given in the previous section. The datapath shown is the same as that in Figure 10-11 with $m = 3$ and $n = 16$. The data memory M is attached to the Address out, Data out, and Data in by connections to the datapath. It has a single control signal MW which is 1 to write the memory, and 0 otherwise.

The Control unit appears on the left in Figure 10-15. Although not usually thought of as part of the control unit, the instruction memory, together with its address inputs and instruction outputs, is shown for convenience with the control unit. We do not write to the instruction memory, in theory, making it a combinational rather than a sequential component. As previously discussed, the PC provides the instruction address to the instruction memory, and the instruction output from the instruction memory goes to the control logic, which, in this case, is the instruction decoder. The output from the instruction memory also goes to Extend and Zero fill, which provide the address offset to the PC and the constant input, Constant in, to the datapath, respectively. Extension appends the leftmost bit of the 6-bit address offset field AD to the left of AD, preserving its two's complement

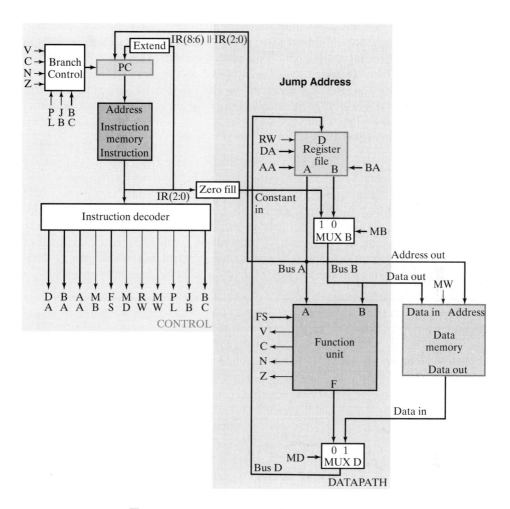

□ **FIGURE 10-15**

Block Diagram for a Single-Cycle Computer

representation. Zero fill appends 13 zeros to the left of the operand (OP) field of the instruction to form a 16-bit unsigned operand for use in the datapath. For example, operand value 110 becomes 0000000000000110 or +6.

The *PC* is updated in each clock cycle. The behavior of the *PC*, which is a complex register, is determined by the opcode, N, and Z, since C and V are not used in this control unit design. If a jump occurs, the new *PC* value becomes the value on Bus *A*. If a branch is taken, then the new *PC* value is the sum of the previous *PC* value and the sign-extended address offset, which in two's complement can be either positive or negative. Otherwise, the *PC* is incremented by 1. A jump occurs for bit 13 in the instruction equal to 1. For bit 13 equal to 0, a conditional branch occurs. The status bit that is the condition for the branch is selected by bit 9 of the instruction. For bit 9 equal to 1, N is selected and, for bit 9 equal to 0, Z is selected.

All parts of the computer that are sequential are shown in blue. Note that there is no sequential logic in the control part other than the *PC*. Thus, aside from providing the address to the instruction memory, the control logic is combinational in this case. That fact, combined with the structure of the datapath and the use of separate instruction and data memories, allows the single-cycle computer to obtain and execute an instruction from the instruction memory, all in a single clock cycle.

Instruction Decoder

The instruction decoder is a combinational circuit that provides all of the control words for the datapath, based on the contents of the fields of the instruction. A number of the fields of the control word can be obtained directly from the contents of the fields in the instruction. Looking at Figure 10-16, we see that the control word fields DA, AA, and BA are equal to the instruction fields DR, SA, and SB, respectively. Also, control field BC for selection of the branch condition status bits is taken directly from the last bit of Opcode. The remaining control word fields include datapath and data memory control bits MB, MD, RW, and MW. There are two added bits

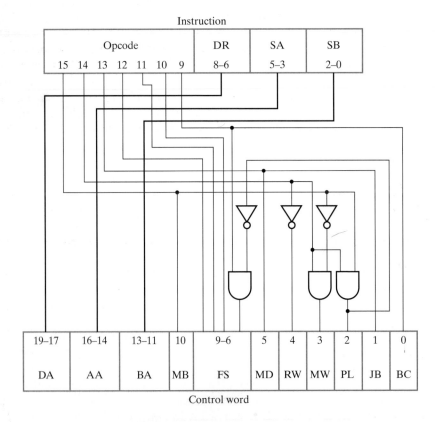

☐ **FIGURE 10-16**
Diagram of Instruction Decoder

for the control of the PC, PL, and JB. If there is to be a jump or branch, PL = 1, loading the *PC*. For PL = 0, the *PC* is incremented. With PL = 1, JB = 1 calls for a jump, and JB = 0 calls for a conditional branch. Some of the single bit control word fields require logic for their implementation. In order to design this logic, we divide the various instructions possible for the simple computer into different function types and then assign the first three bits of the opcode to the various types. The instruction function types shown in Table 10-10 are based on the use of specific hardware resources in the computer, such as MUX *B*, the Function unit, the Register file, Data memory, and the *PC*. For example, the first function type uses the ALU, sets MUX *B* to use the Register file source, sets MUX *D* to use the Function unit output, and writes to the Register file. Other instruction function types are defined as various combinations of use of a constant input instead of a register, Data memory reads and writes, and manipulation of the *PC* for jumps and branches.

By looking at the relationship between the instruction function types and the necessary control word values needed for their implementation, bits 15 through 13 and bit 9 were assigned as shown in Table 10-10. This assignment attempted to minimize the logic required to implement the decoder. To perform the design of the decoder, the values for all of the single bit fields in the control word were determined from the function types and entered into Table 10-10. Note that there are a number of don't care (X) entries. Treating Table 10-10 as a truth-table and optimizing the logic functions, the logic for the single bit outputs of the instruction decoder in Figure 10-16 results. In the optimization, the four unused codes for bits 15, 14, 13, and 9 were assumed to have X values for all of the single bit fields. This implies that if one of these codes occurs in a program, the effect is unknown. A more conservative design specifies RW, MW, and PL all zero for these four codes to insure that the storage resource state is unchanged for these

□ **TABLE 10-10**
Truth Table for Instruction Decoder Logic

Instruction Function Type	Instruction Bits				Control Word Bits						
	15	14	13	9	MB	MD	RW	MW	PL	JB	BC
Function unit operations using registers	0	0	0	X	0	0	1	0	0	X	X
Memory read	0	0	1	X	0	1	1	0	0	X	X
Memory write	0	1	0	X	0	X	0	1	0	X	X
Function unit operations using register and constant	1	0	0	X	1	0	1	0	0	X	X
Conditional branch on zero (Z)	1	1	0	0	X	X	0	0	1	0	0
Conditional branch on negative (N)	1	1	0	1	X	X	0	0	1	0	1
Unconditional Jump	1	1	1	X	X	X	0	0	1	1	X

unused codes. The optimization results in the logic in Figure 10-16 for implementing MB, MD, RW, MW, PL, and JB.

The remaining logic in the decoder deals with the FS field. For all but the conditional branch and unconditional jump instructions, bits 9 through 12 are fed directly through to form the FS field. During conditional branch operations, such as Branch on Zero, the value in source register A must be passed through the ALU so that the status bits N and Z can be evaluated. This requires FS = 0000. The use of bit 9, however, for status bit selection for conditional branches, requires at times that bit 9, which controls the rightmost bit of FS, be a 1. The contradiction in values between bit 9 and FS is resolved by adding an enable on bit 9 that forces FS_0 to zero whenever PL = 1 as shown in Figure 10-16.

Sample Instructions and Program

Six instructions for the single-cycle computer are listed in Table 10-11. The symbolic names associated with the instructions are useful for listing programs in symbolic form rather than in binary code. Because of the importance of instruction decoding, the rightmost six columns of the table show critical control signal values for each instruction, based on the values obtained using the logic in Figure 10-16.

Now suppose that the first instruction, "Add Immediate" (ADI), is present on the output of the instruction memory shown in Figure 10-15. Then, on the basis of the first three bits of the opcode, 100, the outputs of the instruction decoder have the values MB = 1, MD = 0, RW = 1, and MW = 0. The last three bits of the instruction, $OP_{2\text{-}0}$, are extended to 16 bits by zero fill. We denote this in a register transfer statement by zf. Since MB is 1, this zero-filled value is placed on Bus B. With MD equal to 0, the function unit output is selected, and since the last four bits of the opcode, 0010, specify field FS, the operation is $A + B$. So the zero-filled value on Bus B is added to the contents of register SA, with the result presented on Bus D. Since RW = 1, the value on Bus D is written into register DR. Finally, with MW = 0, no write into memory occurs. This entire operation takes place in a single clock cycle. At the beginning of the next cycle, the destination register is written and, since PL = 0, the PC is incremented to point to the next instruction.

The second instruction, LD, is a load from memory with opcode 0010000. The first three bits of this opcode, 001, give control values MD = 1, RW = 1, and MW = 0. These values, plus the register source field SA and register destination field DR, fully specify this instruction, which loads the contents of the memory address specified by register SA into register DR. Again, since PL = 0, the PC is incremented. Note that the values of JB and BC are ignored, since this is neither a jump nor a branch instruction.

The third instruction, ST, stores the contents of a register in memory. The first three bits of the opcode, 010, give control signal values MB = 0, RW = 0, and MW = 1. MW = 1 causes a memory write operation, with the address and data from the register file. RW = 0 prevents the register file from being written. The address for the memory write comes from the register selected by field SA, and the data for the memory write come from the register selected by SB, since MB = 0. The DR field, although present, is not used, since no write occurs to a register.

□ **TABLE 10-11**
Six Instructions for the Single-Cycle Computer

Operation code	Symbolic name	Format	Description	Function	MB	MD	RW	MW	PL	JB	BC
1000010	ADI	Immediate	Add immediate operand	$R[DR] \leftarrow R[SA] + \text{zf } I(2{:}0)$	1	0	1	0	0	0	0
0010000	LD	Register	Load memory content into register	$R[DR] \leftarrow M[R[SA]]$	0	1	1	0	0	0	0
0100000	ST	Register	Store register content in memory	$M[R[SA]] \leftarrow R[SB]$	0	1	0	1	0	0	0
0001110	SL	Register	Shift left	$R[DR] \leftarrow \text{sl} R[SB]$	0	0	1	0	0	0	0
0001011	NOT	Register	Complement register	$R[DR] \leftarrow \overline{R[SA]}$	0	0	1	0	0	0	1
1100000	BRZ	Jump/Branch	If $R[SA] = 0$, branch to PC + se AD	If $R[SA] = 0$, $PC \leftarrow PC + \text{se} AD$, If $R[SA] \neq 0, PC \leftarrow PC + 1$	1	0	0	0	1	0	0

Because this computer has load and store instructions and does not combine loading and storing of data operands with other operations, it is referred to as having a *load/store* architecture. The use of such an architecture simplifies the execution of instructions.

The next two instructions use the Function unit and write to the Register file without immediate operands. The last four bits of the opcode, the value for the FS field of the control word, specify Function unit operation. For these two instructions, only one source register, R[SA] for the NOT and R[SB] for the shift left, and a destination register are involved.

The final instruction is a conditional branch and manipulates the *PC* value. It has PL = 1, causing the program counter to be loaded instead of incremented, and JB = 0, causing a conditional branch rather than a jump. Since BC = 0, register R[SA] is tested for a value of zero. If R[SA] equals zero, the *PC* value becomes *PC* + se AD, where se stands for sign extend. Otherwise, *PC* is incremented. For this instruction, the DR and SB fields become the 6-bit address field AD, which is sign extended and added to the *PC*.

To demonstrate how instructions such as these can be used in a simple program, consider the arithmetic expression $83 - (2 + 3)$. The following program performs this computation, assuming that register $R3$ contains 248, location 248 in data memory contains 2, location 249 contains 83, and the result is to be placed in location 250:

LD	R1, R3	Load $R1$ with contents of location 248 in memory ($R1 = 2$)
ADI	R1, R1, 3	Add 3 to $R1$ ($R1 = 5$)
NOT	R1, R1	Complement $R1$
INC	R1, R1	Increment $R1$ ($R1 = -5$)
INC	R3, R3	Increment the contents of $R3$ ($R3 = 249$)
LD	R2, R3	Load $R2$ with contents of location 249 in memory ($R2 = 83$)
ADD	R2, R2, R1	Add contents of $R1$ to contents of $R2$ ($R2 = 78$)
INC	R3, R3	Increment the contents of $R3$ ($R3 = 250$)
ST	R3, R2	Store $R2$ in memory location 250 ($M[250] = 78$)

The subtraction in this case is done by taking the 2's complement of $(2 + 3)$ and adding it to 83; the subtraction operation SUB could have been used as well. If a register field is not used in executing an instruction, its symbolic value is omitted. The symbolic values for the register-type instructions, when the latter are present, are in the order DR, SA, and SB. For immediate types, the fields are in the order DR, SA, and OP. To store this program in the instruction memory, it is necessary to convert all of the symbolic names and decimal numbers used to their corresponding binary codes.

Single-Cycle Computer Issues

Although there may be instances in which single-cycle computer timing and control strategy is useful, it has a number of shortcomings. One shortcoming is in the area of performing complex operations. For example, suppose that an instruction is desired that executes unsigned binary multiplication using an add-and-shift algorithm. With the given datapath, this cannot be accomplished by a microoperation that can be executed in a single clock cycle. Thus, a control organization that provides multiple clock cycles for the execution of instructions is needed.

Also, the single-cycle computer has two distinct 16-bit memories, one for instructions and one for data. For a simple computer with instructions and data in the same 16-bit memory, two read accesses of memory are required to execute an instruction that loads a data word from memory into a register. The first access obtains the instruction, and the second access, if required, reads or writes the data word. Since two different addresses must be applied to the memory address inputs, at least two clock cycles, one for each address, are required for obtaining and executing the instruction. This can also be accomplished easily with multiple-cycle control.

Finally, the single-cycle computer has a lower limit on the clock period based on a long worst case delay path. This path is shown in blue in the simplified diagram of Figure 10-17. The total delay along the path is 17 ns. This limits the clock frequency to 58.8 MHz, which, although it may be adequate for some applications,

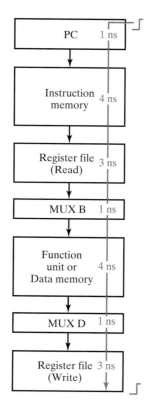

□ **FIGURE 10-17**
Worst Case Delay Path in Single-Cycle Computer

is too slow for a modern computer CPU. In order to have a higher clock frequency, either the delays of the components on the path or the number of components in the path must be reduced. If the delays of the components cannot be reduced, reducing the number of components in the path is the only alternative. In Chapter 12, pipelining of the datapath reduces the number of components in the longest combinational delay path and permits the clock frequency to be increased. A pipelined datapath and control given in Chapter 12, demonstrates the improved CPU performance that can be obtained.

10-9 MULTIPLE-CYCLE HARDWIRED CONTROL

To demonstrate multiple-cycle control, we use the architecture of the simple computer, but modify its datapath, memory, and control. The goal of the modifications is to demonstrate the use of a single memory for both data and instructions and to demonstrate how more complex instructions can be implemented by using multiple clock cycles per instruction. The block diagram in Figure 10-18 shows the modifications to the datapath, memory, and control.

The changes to the single-cycle computer can be observed by comparing Figures 10-15 and 10-18. The first modification, which is possible with, but not essential to, multiple-cycle operation, replaces the separate instruction memory and data memory in Figure 10-15 with the single Memory *M* in Figure 10-18. To fetch instructions, the *PC* is the address source for the memory, and to fetch data, Bus *A* is the address source. At the address input to memory, multiplexer MUX *M* selects between these two address sources. MUX *M* requires an additional control signal, MM, which is added to the control word format. Since instructions from Memory M are needed in the control unit, a path is added from its output to the instruction register IR in the control unit.

In executing an instruction across multiple clock cycles, data generated during the current cycle is often needed in a later cycle. This data can be temporarily stored in a register from the time it is generated until the time it is used. Registers used for such temporary storage during the execution of the instruction are usually not visible to the user (i.e., are not part of the storage resources). The second modification provides these temporary storage registers by doubling the number of registers in the register file. Registers 0 through 7 are storage resources and registers 8 through 15 are used only for temporary storage during instruction execution, so are not part of the storage resources visible to the user. The addressing of 16 registers requires 4 bits, and becomes more complex, since addressing of the first eight registers must be controlled from the instruction and the control unit, and the second eight registers are controlled from the control unit. This is handled by the Register address logic in Figure 10-18 and by modified DX, AX, and BX fields in the control word. The details of this change will be discussed later when the control is defined.

The *PC* is the only control unit component retained and it must also be modified. During the execution of a multiple-cycle instruction, the *PC* must be held at its current value for all but one of the cycles. To provide this hold capability, as well as an increment and two load operations, the *PC* is modified to be controlled by a 2-bit control word field, PS. Since the *PC* is controlled completely by the control word, the Branch control logic previously represented by BC is absorbed into the Control Logic block in Figure 10-18.

Because of the multiple cycles of the modified computer, the instruction needs to be held in a register for use during its execution since its values are likely to be needed for more than just the first cycle. The register used for this purpose is the *instruction register IR* in Figure 10-18. Since the *IR* loads only when an instruction is being read from memory, it has a load-enable signal IL that is added to the control word. Because of the multiple-cycle operation, a sequential control circuit, which can provide a sequence of control words for microoperations used

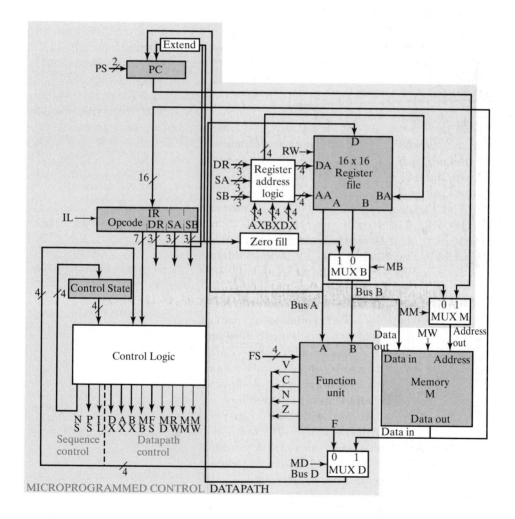

MICROPROGRAMMED CONTROL **DATAPATH**

□ **FIGURE 10-18**
Block Diagram for a Multiple-Cycle Computer

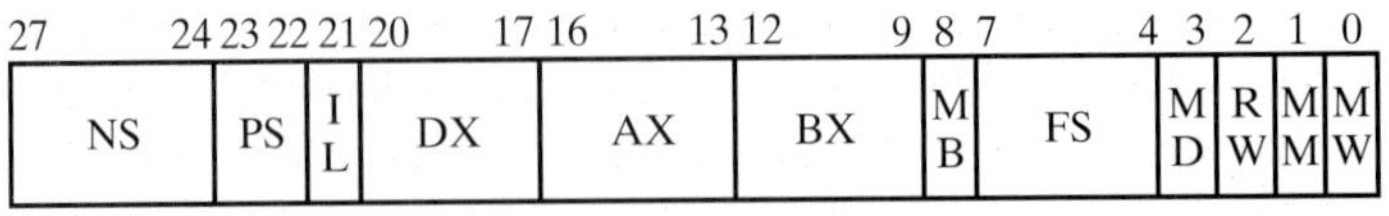

27	24 23 22 21 20		17 16	13 12	9 8 7		4 3 2 1 0

| NS | PS | I L | DX | AX | BX | M B | FS | M D | R W | M M | M W |

□ **FIGURE 10-19**
Control Word Format for Multiple-Cycle Computer

to interpret the instruction is required and replaces the Instruction decoder. The sequential control unit consists of the Control state register and the combinational Control logic. The Control logic has the state, the opcode, and the status bits as its inputs and produces the control word as its output. Conceptually, the control word is divided into two parts, one for Sequence control, which determines the next state of the overall control unit, and one for Datapath control, which controls the micro-operations executed by the Datapath and Memory M as shown in Figure 10-18.

The 28-bit modified control word is given in Figure 10-19 and the definitions of the fields of the control word are given in Table 10-12 and 10-13. In Table 10-12, the fields DX, AX, and BX control the register selection. If the MSB of one of these fields is 0, then the corresponding register address DA, AA, or BA is that given by $0 \parallel DR$, $0 \parallel SA$, and $0 \parallel SB$, respectively. If the MSB of one of these fields is 1, then the corresponding register address is the contents of the field DX, AX, or BX. This

□ **TABLE 10-12**
Control Word Information for Datapath

DX	AX	BX	Code	MB	Code	FS	Code	MD	RW	MM	MW	Code
$R[DR]$	$R[SA]$	$R[SB]$	0XXX	Register	0	$F = A$	0000	FnUt	No write	Address Out	No write	0
$R8$	$R8$	$R8$	1000	Constant	1	$F = A + 1$	0001	Data In	Write	PC	Write	1
$R9$	$R9$	$R9$	1001			$F = A + B$	0010					
$R10$	$R10$	$R10$	1010			Unused	0011					
$R11$	$R11$	$R11$	1011			Unused	0100					
$R12$	$R12$	$R12$	1100			$F = A + \overline{B} + 1$	0101					
$R13$	$R13$	$R13$	1101			$F = A - 1$	0110					
$R14$	$R14$	$R14$	1110			Unused	0111					
$R15$	$R15$	$R15$	1111			$F = A \wedge B$	1000					
						$F = A \vee B$	1001					
						$F = A \oplus B$	1010					
						$F = \overline{A}$	1011					
						$F = B$	1100					
						$F = sr\ B$	1101					
						$F = sl\ B$	1110					
						Unused	1111					

selection process is performed by the Register address logic, which contains three multiplexers, one for each of DA, AA, and BA, controlled by the MSB of DX, AX, and BX, respectively. Table 10-12 also gives the code values for the MM field, which determines whether Address out or PC serves as the Memory M address. The remaining fields in Table 10-12, MB, MD, RW, and MW, have the same functions as for the single-cycle computer.

In the sequential control circuit, the State control register has a set of states, just as a set of flip-flops in any other sequential circuit, has. At the level of our discussion, we assume that each state has an abstract name which can be used as both the state and the next state value. In the design process, a state assignment needs to be made to these abstract states. Referring to Table 10-13, the field NS in the control word provides the next state for the Control State register. We have assigned four bits for the state code, but this can be modified as necessary depending on the number of states needed and the state assignment used in the design. This particular field could be considered as integral to the control and sequential circuit and not part of the control word, but it will appear in the state table of the control in any case. The 2-bit PS field controls the program counter, PC. On a given clock cycle the PC holds its state (00), increments its state by 1 (01), conditionally loads PC plus sign-extended AD (10), or unconditionally loads the contents of $R[SA]$ (11). Finally, the instruction register is loaded only once during the execution of an instruction. Thus, on any given cycle, either a new instruction is loaded (IL = 1) or the instruction remains unchanged (IL = 0).

Sequential Control Design

The design of the sequential control circuit can be done using techniques from Chapter 6 and Chapter 8. However, compared to the examples there, even for this comparatively simple computer, the control is quite complex. Assuming there are four state variables, the combinational Control logic has 15 input variables and 28 output variables. It turns out that a condensed state table for the circuit is not too difficult to develop, but manual design of the detailed logic is very complex, making the use of a PLA or logic synthesis more viable options. As a consequence, we focus on state table development rather than detailed logic implementation

We begin by developing an ASM chart that represents the instructions that can be implemented with the minimum number of clock cycles. Extensions of this

☐ **TABLE 10-13**
Control Information for Sequence Control

NS	PS		IL	
Next State	Action	Code	Action	Code
Gives next state	Hold PC	00	No load	0
of Control State	Inc PC	01	Load instr.	1
Register	Branch	10		
	Jump	11		

chart can then be developed for implementation of instructions requiring more than the minimum number of clock cycles. The ASM charts provide the information needed to develop the state table entries for implementing the instruction set. For instructions requiring a memory access for data as well as for the instruction itself, at least two cycles are required. It is convenient to separate the cycles into two processing steps: *instruction fetch* and *instruction execution*. On the basis of this division, the ASM chart for the two-cycle instructions is given in Figure 10-20. The instruction fetch occurs in state INF at the top of the chart. The *PC* contains the address of the instruction in Memory *M*. This address is applied to the memory, and the word read from memory is loaded into the *IR* on the clock pulse that ends state INF. The same clock pulse causes the new state to become EX0. In state EX0, the instruction is decoded by use of a large vector decision box and the microoperations executing all or part of the instruction appears in a conditional output box. If the instruction can be completed in state EX0, the next state is INF in preparation for fetching of the next instruction. Further, for instructions that do not change *PC* contents during their execution, the *PC* is incremented. If additional states are required for instruction execution, the next state is EX1. In each of the execution states, there are 128 different input combinations possible, based on the opcode. When the status bits are used, typically one at a time, the output of the vector decision box feeds one or more scalar decision boxes as illustrated for the branch instructions on the lower right of Figure 10-20.

Next, we describe a sampling of the instruction executions specified by the ASM chart in Figure 10-20. The first opcode is 0000000 for the move A, (MOVA) instruction. This instruction involves a simple transfer from the source A register to the destination register, as specified by the register transfer shown in state EX0 for the instruction opcode. Although the status bits *N* and *Z* are valid, they are not used in the execution of this instruction. The *PC* is incremented on the clock edge ending state EX0, an action that occurs for all but branch and jump instructions in the ASM chart.

The third opcode is 0000010 for the ADD instruction with the register transfer for addition shown. In this case, status bits *V*, *C*, *N*, and *Z* are valid, although not used. The eleventh opcode, 0010000, is the load (LD) instruction, which uses the value in the register specified by SA for the address and loads the data word from Memory *M* into the register specified by DR. The twelfth opcode, 0100000, is for the store (ST) instruction, which stores the value in register SB into the location in Memory *M* specified by the address from register SA. The fourteenth opcode, 1001100, is add immediate (ADI), which adds the zero-filled value of the OP field, the rightmost three bits of the instruction, to the contents of register SA and places the result in the register DR.

The sixteenth opcode, 1100001, is the branch on negative (BRN) instruction. The decoding of this instruction causes the value in the register specified by SA to be passed through the Function unit in order to evaluate status bits N and Z. The values N and Z then propagate back to the Control logic. Based on the value of N, the branch is taken or not taken by adding the extended address AD from the instruction to the value in the *PC* or incrementing the *PC*, respectively. This is represented by the scalar decision box for N shown in Figure 10-20.

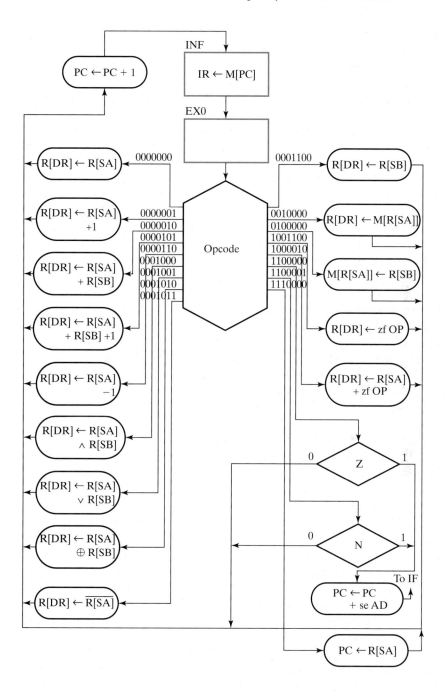

□ **FIGURE 10-20**
Basic ASM Chart for Multiple-Cycle Computer

□ **TABLE 10-14**
State Table for Two-Cycle Instructions

State	Inputs		Next state	Outputs											Comments	
	Opcode	VCNZ		IL	PS	DX	AX	BX	MB	FS	MD	RW	MM	MW		
INF	XXXXXXX	XXXX	EX0	1	00	XXXX	XXXX	XXXX	X	XXXX	X	0	1	0		$IR \leftarrow M[PC]$
EX0	0000000	XXXX	INF	0	01	0XXX	0XXX	XXXX	X	0000	0	1	X	0	MOVA	$R[DR] \leftarrow R[SA]$*
EX0	0000001	XXXX	INF	0	01	0XXX	0XXX	XXXX	X	0001	0	1	X	0	INC	$R[DR] \leftarrow R[SA] + 1$*
EX0	0000010	XXXX	INF	0	01	0XXX	0XXX	0XXX	0	0010	0	1	X	0	ADD	$R[DR] \leftarrow R[SA] + R[SB]$*
EX0	0000101	XXXX	INF	0	01	0XXX	0XXX	0XXX	0	0101	0	1	X	0	SUB	$R[DR] \leftarrow R[SA] + \overline{R[SB]} + 1$*
EX0	0000110	XXXX	INF	0	01	0XXX	0XXX	XXXX	X	0110	0	1	X	0	DEC	$R[DR] \leftarrow R[SA] + (-1)$*
EX0	0001000	XXXX	INF	0	01	0XXX	0XXX	0XXX	0	1000	0	1	X	0	AND	$R[DR] \leftarrow R[SA] \wedge R[SB]$*
EX0	0001001	XXXX	INF	0	01	0XXX	0XXX	0XXX	0	1001	0	1	X	0	OR	$R[DR] \leftarrow R[SA] \vee R[SB]$*
EX0	0001010	XXXX	INF	0	01	0XXX	0XXX	0XXX	0	1010	0	1	X	0	XOR	$R[DR] \leftarrow R[SA] \oplus R[SB]$*
EX0	0001011	XXXX	INF	0	01	0XXX	0XXX	XXXX	X	1011	0	1	X	0	NOT	$R[DR] \leftarrow \overline{R[SA]}$*
EX0	0001100	XXXX	INF	0	01	0XXX	0XXX	0XXX	0	1100	0	1	X	0	MOVB	$R[DR] \leftarrow R[SB]$*
EX0	0010000	XXXX	INF	0	01	0XXX	0XXX	XXXX	X	XXXX	1	1	0	0	LD	$R[DR] \leftarrow M[R[SA]]$*
EX0	0100000	XXXX	INF	0	01	XXXX	0XXX	0XXX	0	XXXX	X	0	0	1	ST	$M[R[SA]] \leftarrow R[SB]$*
EX0	1001100	XXXX	INF	0	01	0XXX	XXXX	XXXX	1	1100	0	1	0	0	LDI	$R[DR] \leftarrow zf\ OP$*
EX0	1000010	XXXX	INF	0	01	0XXX	0XXX	XXXX	1	0010	0	1	0	0	ADI	$R[DR] \leftarrow R[SA] + zf\ OP$*
EX0	1100000	XXX1	INF	0	10	XXXX	XXXX	XXXX	X	0000	X	0	0	0	BRZ	$PC \leftarrow PC + se\ AD$
EX0	1100000	XXX0	INF	0	01	XXXX	XXXX	XXXX	X	0000	X	0	0	0	BRZ	$PC \leftarrow PC + 1$
EX0	1100001	XX1X	INF	0	10	XXXX	XXXX	XXXX	X	0000	X	0	0	0	BRN	$PC \leftarrow PC + se\ AD$
EX0	1100001	XX0X	INF	0	01	XXXX	XXXX	XXXX	X	0000	X	0	0	0	BRN	$PC \leftarrow PC + 1$
EX0	1110000	XXXX	INF	0	11	XXXX	XXXX	XXXX	X	0000	X	0	0	0	JMP	$PC \leftarrow R[SA]$

* For this state and input combination, $PC \leftarrow PC + 1$ also occurs.

From this ASM chart, the state table for the sequential control circuit can be developed as shown in Table 10-14. The present states are given as abstract state names, and the opcodes and status bits serve as inputs. In the case of the status bits, only those bits that are used in the instruction are specified. By using combinations of bits and multiple status bit patterns, it is possible to specify functions of status bits. Note that many of the entries in Table 10-14 contain Xs, symbolizing "don't cares." For these entries, the input or resource is not used in the given microoperation or the specific bits of the code that are X are not used for controlling it. It is a useful exercise to determine how each of the entries in Table 10-14 is obtained, based on Table 10-12, Table 10-13, and Figure 10-20.

It is interesting to briefly compare the timing of the execution of instructions in this organization with that for the single-cycle computer. Each instruction requires two clock cycles to fetch and execute, compared with one clock cycle for the single-cycle computer. Because the very long delay path from the *PC* through the Instruction memory, Instruction decoder, datapath, and branch control is broken up by the instruction register, the clock periods are somewhat shorter. Nevertheless, due to setup time requirements for the added flip-flops in the *IR* and a potential imbalance in delays for the various paths through the circuit, the overall time taken to execute an instruction could be just as long as or longer than in the single-cycle computer. So what is the benefit of this organization other than ability to use a single memory? The next two instructions give the answer.

The first instruction to be added is a "load register indirect" (LRI), with opcode 0010001. In this instruction, the contents of register SA address a word in memory. The word, which is known as an *indirect address*, is then used to address the word in memory that is loaded into register DR. This can be represented symbolically as

$$R[DR] \leftarrow M[M[R[SA]]]$$

The ASM chart for the execution of this instruction is given in Figure 10-21. Following the instruction fetch, the state becomes EX0. In this state, R[SA] addresses the memory to obtain the indirect address, which is then placed in temporary register *R8*. In the next state, EX1, the next memory access occurs with the address from *R8*. The operand obtained is placed in R[DR] to complete the operation, and the *PC* is incremented. The ASM then returns to state INF to fetch the next instruction. The vector decision box for opcode is required for all states, since these same states are used by other instructions for their execution. Clearly, with two accesses to Memory *M*, this instruction could not be executed by the single-clock-cycle computer or using two clock cycles in the multiple-cycle computer. Also, to avoid disturbing the contents of registers *R0* through *R7* (except for R[SA]), the use of register *R8* for temporary storage is essential. The LRI instruction requires three clock cycles for its execution. To accomplish the same operation in the single-cycle computer requires two LD instructions, taking two clock cycles. In the multiple-cycle computer, due to two instruction fetches and two data accesses, it would require two LD instructions, but would take four clock cycles. So the LRI instruction gives an improvement in execution time in the latter case.

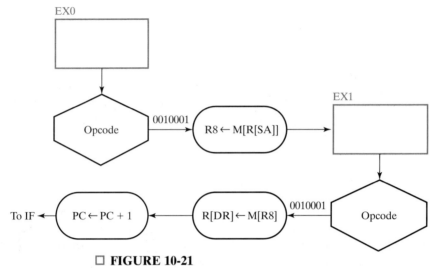

☐ **FIGURE 10-21**
ASM Chart for Register Indirect Instruction

The final two instructions to be added are "shift right multiple" (SRM) and "shift left multiple" (SLM), with opcodes 0001101 and 0001110, respectively. These two instructions can share most of the microinstruction sequence to be used. SRM specifies that the contents of register SA are to be shifted to the right by the number of positions given by the three bits of the OP field, with the result placed in register DR. The ASM chart for this operation (and for SLM) is given in Figure 10-22. Register $R9$ stores the number of bit positions remaining to be shifted, and the shifting is performed in register $R8$.

Initially, the contents of R[SA] to be shifted is placed in $R8$. As it is loaded into $R8$, it is checked to see if it is 0 and shifting is not needed. Likewise, the shift amount being loaded into $R9$ is checked to see whether it is 0, meaning that shifting is not needed. If either case is satisfied, the instruction execution is complete, and the ASM flow returns to state INF. Otherwise, a right-shift operation is performed on the contents of register $R8$. $R9$ is decremented and tested to see whether it will be 0. If $R9 \neq 0$, then the shift and decrement are repeated. If $R9 = 0$, then the contents of $R8$ have been shifted by the number of bit positions specified by OP, so the result is transferred to R[DR] to complete the instruction execution, and the ASM flow returns to state INF.

If both the operand and the shift amount are nonzero, SRM, including fetch, requires $2s + 4$ clock cycles, where s is the number of positions shifted. The range of clock cycles required, including the instruction fetch, is from 6 to 18. If the same operation were implemented by a program using the right-shift instruction plus increment and branching, then $3s + 3$ instructions would be required giving $6s + 6$ cycles. The improvement in the required number of clock cycles is $4s + 2$, so 6 to 30 clock cycles are saved in the multiple-cycle computer for a nonzero operand and shift amount. Also, five fewer memory locations are required for storage of the SRM instruction, in contrast to that for the program.

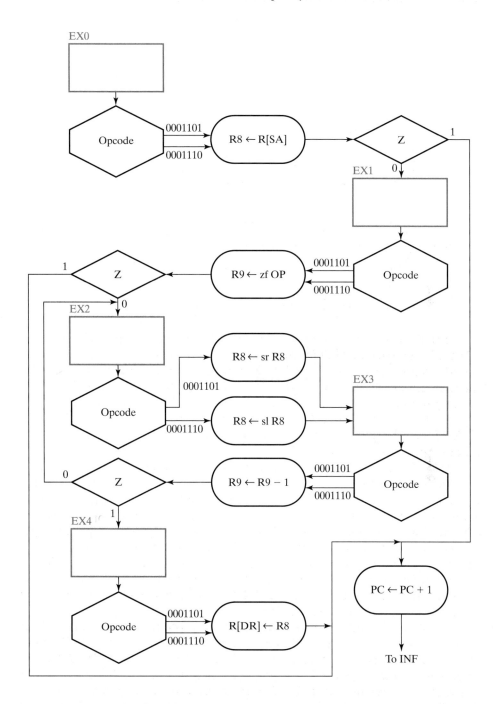

□ **FIGURE 10-22**
ASM Chart for Right-Shift Multiple Instruction

□ **TABLE 10-15**
State Table for Illustration of Instructions Having Three or More Cycles

State	Opcode	VCNZ	Next state	I L	PS	DX	AX	BX	MB	FS	MD	RW	MM	MW		Comments
EX0	0010001	XXXX	EX1	0	00	1000	0XXX	XXXX	X	0000	1	1	X	0	LRI	$R8 \leftarrow M[R[SA], \rightarrow EX1$
EX1	0010001	XXXX	INF	0	01	0XXX	1000	XXXX	X	0000	1	1	X	0	LRI	$R[DR] \leftarrow M[R8], \rightarrow INF^*$
EX0	0001101	XXX0	EX1	0	00	1000	0XXX	XXXX	X	0000	0	1	X	0	SRM	$R8 \leftarrow R[SA], \bar{Z}: \rightarrow EX1$
EX0	0001101	XXX1	INF	0	00	1000	0XXX	XXXX	X	0000	0	1	X	0	SRM	$R8 \leftarrow R[SA], Z: \rightarrow INF^*$
EX1	0001101	XXX0	EX2	0	01	1001	XXXX	XXXX	1	1100	0	1	X	0	SRM	$R9 \leftarrow zf\ OP, \bar{Z}: \rightarrow EX2$
EX1	0001101	XXX1	INF	0	01	1001	XXXX	XXXX	1	1100	0	1	X	0	SRM	$R9 \leftarrow zf\ OP, Z: \rightarrow INF^*$
EX2	0001101	XXXX	EX3	0	00	1000	XXXX	1000	0	1101	0	1	X	0	SRM	$R8 \leftarrow sr\ R8, \rightarrow EX3$
EX3	0001101	XXX0	EX2	0	00	1001	1001	XXXX	X	0110	0	1	X	0	SRM	$R9 \leftarrow R9 - 1, \bar{Z}: \rightarrow EX2$
EX3	0001101	XXX1	EX4	0	00	1001	1001	XXXX	X	0110	0	1	X	0	SRM	$R9 \leftarrow R9 - 1, Z: \rightarrow EX4$
EX4	0001101	XXXX	INF	0	01	0XXX	1000	XXXX	X	0000	0	1	X	0	SRM	$R[DR] \leftarrow R8, \rightarrow INF^*$
EX0	0001110	XXX0	EX1	0	00	1000	0XXX	XXXX	X	0000	0	1	X	0	SLM	$R8 \leftarrow R[SA], \bar{Z}: \rightarrow EX1$
EX0	0001110	XXX1	INF	0	00	1000	0XXX	XXXX	X	0000	0	1	X	0	SLM	$R8 \leftarrow R[SA], Z: \rightarrow INF^*$
EX1	0001110	XXX0	EX2	0	01	1001	XXXX	XXXX	1	1100	0	1	X	0	SLM	$R9 \leftarrow zf\ OP, \bar{Z}: \rightarrow EX2$
EX1	0001110	XXX1	INF	0	01	1001	XXXX	XXXX	1	1100	0	1	X	0	SLM	$R9 \leftarrow zf\ OP, Z: \rightarrow INF^*$
EX2	0001110	XXXX	EX3	0	00	1000	XXXX	1000	0	1110	0	1	X	0	SLM	$R8 \leftarrow sl\ R8, \rightarrow EX3$
EX3	0001110	XXX0	EX2	0	00	1001	1001	XXXX	X	0110	0	1	X	0	SLM	$R9 \leftarrow R9 - 1, \bar{Z}: \rightarrow EX2$
EX3	0001110	XXX1	EX4	0	00	1001	1001	XXXX	X	0110	0	1	X	0	SLM	$R9 \leftarrow R9 - 1, Z: \rightarrow EX4$
EX4	0001110	XXXX	INF	0	01	0XXX	1000	XXXX	X	0000	0	1	X	0	SLM	$R[DR] \leftarrow R8, \rightarrow IF^*$

*For this state and input combination, PC ← PC + 1 also occurs.

In the ASM chart in Figure 10-22, the states INF and EX0 (and EX1) are the same as those used for the two-cycle instructions in the ASM chart in Figure 10-20 and for the LRI instruction in Figure 10-21. Also, implementation of the left shift multiple operation is shown in Figure 10-22 in which, based on the opcode, the left shift of $R8$ replaces the right shift of $R8$. As a consequence, the logic implementing the states used for implementation of these two instructions can be shared. Further, the logic used for the sequencing of the states can be shared between the SRM and SLM instruction implementations.

The state table specification in Table 10-15 is derived by using the information from the ASM chart in Figure 10-22, and Tables 10-12 and 10-13. The codes are derived from the register transfer and sequencing action described in the comments on the right in the same way that Table 10-15 was derived.

Implementation of the LRI and SRM instructions illustrates the flexibility achieved using multiple-cycle control. Implementation of additional instructions is explored in the problems at the end of the chapter.

10-10 CHAPTER SUMMARY

In the first part of the chapter, the concept of datapaths for information processing in digital systems was introduced. Among the major components of datapaths are register files, buses, arithmetic/logic units (ALUs), and shifters. The control word provides a means of organizing the control of the microoperations performed by the datapath. These concepts were combined into the concept of a datapath, which serves as a basis for exploring computers in the remainder of the text.

In the second part of the chapter, control design for programmed systems was introduced by examining two different implementations of basic control units for a simple computer architecture. We introduced the concept of instruction set architectures and defined instruction formats and operations for the simple computer. The first implementation of this computer is capable of executing any instruction in a single clock cycle. Aside from having a program counter and its logic, the control unit of this computer consists of a combinational decoder circuit.

Among the shortcomings of the single-cycle computer are limitations on the complexity of the instructions that can be executed on it, problems with the interface to a single memory, and the relatively low clock frequencies attained. To deal with the first two of these shortcomings, we examined a multiple-cycle version of the simple computer in which a single memory is used and instructions are implemented using two distinct phases: instruction fetch and instruction execution. The remaining issue of long clock cycles is dealt with in Chapter 12 by introducing pipelined datapaths and control.

REFERENCES

1. MANO, M. M. *Computer Engineering: Hardware Design:* Englewood Cliffs, NJ: Prentice Hall, 1988.

2. MANO, M. M. *Computer System Architecture,* 3rd Ed. Englewood Cliffs, NY: Prentice Hall, 1993.

3. PATTERSON, D. A., AND J. L. HENNESSY. *Computer Organization and Design: The Hardware/Software Interface,* 2nd ed. San Francisco, CA: Morgan Kaufmann, 1998.

4. HENNESSY, J. L., AND D. A. PATTERSON. *Computer Architecture: A Quantitative Approach,* 2nd ed. San Francisco, CA: Morgan Kaufmann, 1996.

PROBLEMS

The plus (+) indicates a more advanced problem and the asterisk (*) indicates a solution is available on the Companion Website for the text.

10–1. A datapath similar to the one in Figure 10-1 has 128 registers. How many selection lines are needed for each set of multiplexers and for the decoder?

10–2. *Given an 8-bit ALU with outputs F_7 through F_0 and available carries C_8 and C_7, show the logic circuit for generating the signals for the four status bits N (sign), Z (zero), V (overflow), and C (carry).

10–3. *Design an arithmetic circuit with two selection variables S_1 and S_0 and two n-bit data inputs A and B. The circuit generates the following eight arithmetic operations in conjunction with carry C_{in}:

S_1	S_0	$C_{in} = 0$	$C_{in} = 1$
0	0	$F = A + B$ (add)	$F = A + \overline{B} + 1$ (subtract A − B)
0	1	$F = \overline{A} + B$	$F = \overline{A} + B + 1$ (subtract B − A)
1	0	$F = A - 1$ (decrement)	$F = A + 1$ (increment)
1	1	$F = \overline{A}$ (1's Complement)	$F = \overline{A} + 1$ (2's Complement)

Draw the logic diagram for the two least significant bits of the arithmetic circuit.

10–4. *Design a 4-bit arithmetic circuit, with two selection variables S_1 and S_0, that generates the following arithmetic operations:

$S_1 S_0$	$C_{in} = 0$	$C_{in} = 1$
0 0	$F = A + B$ (add)	$F = A + B + 1$
0 1	$F = A$ (transfer)	$F = A + 1$ (increment)
1 0	$F = \overline{B}$ (complement)	$F = \overline{B} + 1$ (negate)
1 1	$F = A + \overline{B}$	$F = A + \overline{B} + 1$ (subtract)

Draw the logic diagram for a single bit stage.

10–5. Inputs X_i and Y_i of each full adder in an arithmetic circuit have digital logic specified by the Boolean functions

$$X_i = A_i \qquad Y_i = \overline{B}_iS + B_i\overline{C}_{\text{in}}$$

where S is a selection variable, C_{in} is the input carry, and A_i and B_i are input data for stage i.

(a) Draw the logic diagram for the 4-bit circuit, using full adders and multiplexers.

(b) Determine the arithmetic operation performed for each of the four combinations of S and C_{in}: 00, 01, 10, and 11.

10–6. *Design one bit of a digital circuit that performs the four logic operations of exclusive-OR, exclusive-NOR, NOR, and NAND on register operands A and B with the result to be loaded into register A. Use two selection variables.

(a) Using a Karnaugh map, design minimum logic for one typical stage, and show the logic diagram.

(b) Repeat (a), trying different assignments of the selection codes to the four operations to see whether the logic for the stage can be simplified further.

10–7. +Design an ALU that performs the following operations:

$$\begin{array}{ll} A + B & \text{sl } A \\ A + \overline{B} + 1 & A \vee B \\ \overline{B} & A \oplus B \\ \overline{B} + 1 & A \wedge B \end{array}$$

Give the result of your design as the logic diagram for a single stage of the ALU. Your design should have only a single carry line between stages and three selection bits. If you have access to logic simplification software, apply it to the design to obtain reduced logic.

10–8. *Find the output Y of the 4-bit barrel shifter in Figure 10-9 for each of the following bit patterns applied to $S_1, S_0, D_3, D_2, D_1,$ and D_0:

(a) 000101   **(b)** 010011

(c) 101010   **(d)** 111100

10–9. Specify the 16-bit control word that must be applied to the datapath of Figure 10-11 to implement each of the following microoperations:

(a) $R0 \leftarrow R1 + R7$ **(b)** $R7 \leftarrow 0$

(c) $R6 \leftarrow \text{sl } R6$ **(d)** $R3 \leftarrow \text{sr} R4$

(e) $R1 \leftarrow R7 + 1$ **(f)** $R2 \leftarrow R4 - \text{Constant in}$

(g) $R1 \leftarrow R2 \oplus R3$ **(h)** $R5 \leftarrow \text{Data in}$

10–10. *Given the following 16-bit control words for the datapath of Figure 10-11, determine (a) the microoperation that is executed and (b) the change in the contents of the register for each control word (assume that the registers are 8-bit registers and that, before the execution of a control word, they contain the value of their number (e.g., register $R5$ contains 05 in hexadecimal)). Assume that Constant has value 6 and Data in has value 1B, both in hexadecimal.

(a) 101 100 101 0 1000 0 1

(b) 110 010 100 0 0101 0 1

(c) 101 110 000 0 1100 0 1

(d) 101 000 000 0 0000 0 1

(e) 100 100 000 1 1101 0 1

(f) 011 000 000 0 0000 1 1

10–11. Given the sequence of 16-bit control words below for the datapath in Figure 10-11 and the initial ASCII character codes in 8-bit registers, simulate the datapath to determine the alphanumeric characters in the registers after the execution of the sequence. The result is a scrambled word: what is it?

Control word	Register	Contents
011 011 001 0 0010 0 1	$R0$	00000000
100 100 001 0 1001 0 1	$R1$	00100000
101 101 001 0 1010 0 1	$R2$	01000100
001 001 000 0 1011 0 1	$R3$	01000111
001 001 000 0 0001 0 1	$R4$	01010100
110 110 001 0 0101 0 1	$R5$	01001100
111 111 001 0 0101 0 1	$R6$	01000001
001 111 000 0 0000 0 1	$R7$	01001001

10–12. A datapath has five major components, A through E, attached in a loop from register file to register file similar to that in Figure 10-17. The maximum delay of each of the components is A, 2 ns; B, 1 ns; C, 3 ns; D, 4 ns; and E, 4 ns.

(a) What is the maximum clock frequency that can be used for the datapath?

(b) The datapath is to be changed to one that is pipelined using three stages. How should the components be combined into stages, and what is the maximum clock frequency that can be achieved?

(c) Repeat (b) for four pipeline stages.

10–13. A computer has a 32-bit instruction word broken into fields as follows: opcode, 6 bits; two register fields, 6 bits each; and one immediate operand/register field, 14 bits.

(a) What is the maximum number of operations that can be specified?

(b) How many registers can be addressed?

(c) What is the range of unsigned immediate operands that can be provided?

(d) What is the range of signed immediate operands that can be provided, assuming that bit 13 is the sign bit?

10–14. *A digital computer has a memory unit with a 32-bit instruction and a register file with 32 registers. The instruction set consists of 110 different operations. There is only one type of instruction format, with an opcode part, a register file address, and an immediate operand part. Each instruction is stored in one word of memory.

(a) How many bits are needed for the opcode part of the instruction?

(b) How many bits are left for the immediate part of the instruction?

(c) If the immediate operand is used as an unsigned address to memory, what is the maximum number of words that can be addressed in memory?

(d) What are the largest and the smallest algebraic values of signed 2's complement binary numbers that can be accommodated as an immediate operand?

10–15. A digital computer has 32-bit instructions. There are a number of different instruction formats and the number of bits in each format used for opcodes varies depending on the bits needed for other fields. If the first bit of the opcode is 0, then there are 4 opcode bits. If the first bit of the opcode is 1 and the second bit of the opcode is 0, then there are 6 opcode bits. If the first bit of the opcode is 1 and the second bit of the opcode is 1, then there are 8 opcode bits. How many distinct opcodes are available for this computer?

10–16. The single-cycle computer in Figure 10-15 executes the five instructions described by the register transfers in the table that follows.

(a) Complete the following table, giving the binary instruction decoder outputs from Figure 10-16 during execution of each of the instructions:

Instruction—Register Transfer	DA	AA	BA	MB	FS	MD	RW	MW	PL	JB
$R[0] = R[7] \oplus R[3]$										
$R[1] \leftarrow M[R[4]]$										
$R[2] \leftarrow R[5] + 2$										
$R[3] \leftarrow \text{sl } R[6]$										
if $(R([4] = 0)$ $PC \leftarrow PC + \text{se } PC$ else $PC \leftarrow PC + 1$										

(b) Complete the following table, giving the instruction in binary for the single-cycle computer that executes the register transfer (if any field is not used, give it the value 0):

Instruction—Register Transfer	Opcode	DR	SA	SB or Operand
$R[0] = \text{sr}R[7]$				
$R[1] \leftarrow M[R[6]]$				
$R[2] \leftarrow R[5] + 4$				
$R[3] \leftarrow R[4] \oplus R[3]$				
$R[4] \leftarrow R[2] - R[1]$				

10–17. Using the information in the truth table in Table 10-10, verify that the design for the single-bit outputs in the decoder in Figure 10-16 is correct.

10–18. Manually simulate the single-cycle computer in Figure 10-15 for the following sequence of instructions, assuming that each register initially contains contents equal to its index (i.e., $R0$ contains 0, $R1$ contains 1, etc.):

SUB R0, R1, R2
SUB R3, R4, R5
SUB R6, R7, R0
SUB R0, R0, R3
SUB R0, R0, R6
ST R7, R0
LD R7, R6
ADI R0, R6, 0
ADI R3, R6, 3

Give (a) the binary value of the instruction on the current line of the results and (b) the contents of any register changed by the instruction, or the location and contents of any memory location changed by the instruction on the next line of the results. The results are positioned in this fashion because the new values do not appear in a register or memory, due to the execution of an instruction, until after a positive clock edge has occurred.

10–19. Give an instruction for the single-cycle computer that resets register $R4$ to 0 and updates the Z and N status bits based on the value 0 transferred to $R4$. (*Hint:* Try the exclusive-OR.) By examining the detailed ALU logic, determine the values of the V and C status bits.

10–20. List the control logic state table entries for the multiple-cycle computer (see Table 10-15) that implement the following register transfer statements. Assume that in all cases the present state is EX0 and the opcode is 0010001.

(a) $R3 \leftarrow R1 - R2$, $\rightarrow EX1$ Assume DR = 3, SA = 1, SB = 2.

(b) $R8 \leftarrow \text{sr } R8$, $\rightarrow INF$ Assume DR = 5, SB = 5

(c) if $(N = 0)$ then $(PC \rightarrow PC + \text{se}, \rightarrow INF)$ else $(PC \rightarrow PC + 1, \rightarrow INF)$

(d) $R6 \leftarrow R6, C \leftarrow 0, \rightarrow INF$ Assume DR = SA = 6.

10–21. Manually simulate the SRM instruction in the multiple-cycle computer for operand 0001001101111000 for OP = 6.

10–22. A new instruction is to be defined for the multiple-cycle computer with opcode 0010001. The instruction implements the register transfer

$$R[DR] \leftarrow R[SB] + M[R[SA]]$$

Find the ASM chart for implementing the instruction, assuming that 0010001 is the opcode. Form the part of the control state table that implements this instruction.

10–23. Repeat Problem 10-22 for the two instructions: Add and check OV (AOV), described by the register transfer

$$R[DR] \leftarrow R[SA] + R[SB], V{:}R8 \leftarrow 1, \overline{V}{:}R8 \leftarrow 0$$

and BRanch on oVerflow (BRV), described by the register transfer

$$R8 \leftarrow R8, V{:}PC \leftarrow PC + \text{se AD}, \overline{V}{:}PC \leftarrow PC + 1$$

The opcode for AOV is 1000101 and, for BRV, is 1000110. Note that register R8 is used as a "status" register that stores the overflow result V for the previous operation. All of the values N, Z, C and V could be stored in R8 to give a complete status on the prior arithmetic or logic operation.

10–24. +A new instruction is to be defined for the multiple-cycle computer. The instruction compares two unsigned integers stored in register R[SA] and R[SB]. If the integers are equal, then bit 0 of R[DR] is set to 1. If R[SA] is greater than R[SB], then bit 1 of R[DR] is set to 1. Otherwise, bits 0 and 1 are both 0. All other bits of R[DR] have value 0. Find the ASM chart for implementing the instruction, assuming that 0010001 is the opcode. Form the part of the control state table that implements this instruction.

CHAPTER

11

INSTRUCTION SET ARCHITECTURE

U p to this point, much of what we have studied has focused on digital system design, with computer components used as examples. In this chapter, the material studied becomes decidedly more specialized, dealing with instruction set architecture for general-purpose computers. We will examine the operations that the instructions perform and focus particularly on how the operands are obtained and where the results are stored. In our studies, we will contrast two distinct classes of architectures: reduced instruction set computers (RISCs) and complex instruction set computers (CISCs). We will classify elementary instructions into three categories: data transfer, data manipulation, and program control. In each of these categories, we will elaborate on typical elementary instructions.

In light of this change in focus, the general-purpose parts of the generic computer at the beginning of Chapter 1, including the central processing unit (CPU) and the accompanying floating-point unit (FPU), are heavily shaded. In addition, since a small general-purpose microprocessor may be present for controlling keyboard and monitor functions, we have lightly shaded these components. Aside from addressing used to access memory and I/O components, the concepts studied apply less to other areas of the computer. Increasingly, however, small CPUs are appearing more and more in the I/O components, giving a changing picture of the role of general-purpose instruction set architectures in the generic computer.

11-1 COMPUTER ARCHITECTURE CONCEPTS

The binary language in which instructions are defined and stored in memory is referred to as *machine language*. A symbolic language that replaces binary opcodes and addresses with symbolic names and that provides other features helpful to the programmer is referred to as *assembly language*. The logical structure of computers is normally described in assembly language reference manuals. Such manuals explain various internal elements of the computer that are of interest to the programmer, such as processor registers. The manuals list all hardware-implemented instructions, specify the symbolic names and binary code format of the instructions, and provide a precise definition of each instruction. In the past, this information represented the *architecture* of the computer. A computer was composed of its architecture, plus a specific *implementation* of that architecture. The implementation was separated into two parts: the organization and the hardware. The *organization* consists of structures such as datapaths, control units, memories, and the buses that interconnect them. *Hardware* refers to the logic, the electronic technologies employed, and the various physical design aspects of the computer. As computer designers pushed for higher and higher performance, and as increasingly more of the computer resided within a single IC, the relationships among architecture, organization, and hardware became so intertwined that a more integrated viewpoint became necessary. According to this new viewpoint, architecture as previously defined is more restrictively called *instruction set architecture* (*ISA*), and the term *architecture* is used to encompass the whole of the computer, including instruction set architecture, organization, and hardware. This unified view enables intelligent design trade-offs to be made that are apparent only in a tightly coupled design process. These trade-offs have the potential for producing better computer designs. In this chapter, we focus on instruction set architecture. In the next, we will look at two distinct instruction set architectures, with a focus on implementation using two very different organizations.

A computer usually has a variety of instructions and multiple instruction formats. It is the function of the control unit to decode each instruction and provide the control signals needed to process it. Simple examples of instructions and instruction formats were presented in Section 10-7. We will now expand this presentation by introducing typical instructions found in commercial general-purpose computers. We will also investigate the various instruction formats that may be encountered in a typical computer, with an emphasis on the addressing of operands. The format of an instruction is depicted in a rectangular box symbolizing the bits of the binary instruction. The bits are divided into groups called *fields*. The following are typical fields found in instruction formats:

1. An *opcode field*, which specifies the operation to be performed.
2. An *address field*, which provides either a memory address or an address for selecting a processor register.
3. A *mode field*, which specifies the way the address field is to be interpreted.

Other special fields are sometimes employed under certain circumstances—for example, a field that gives the number of positions to shift in a shift-type instruction or an operand field in an immediate operand instruction.

Basic Computer Operation Cycle

In order to comprehend the various addressing concepts to be presented in the next two sections, we need to understand the basic operation cycle of the computer. The control unit of a computer is designed to execute each instruction of a program in the following sequence of steps:

1. Fetch the instruction from memory into a control register.
2. Decode the instruction.
3. Locate the operands used by the instruction.
4. Fetch operands from memory (if necessary).
5. Execute the operation in processor registers.
6. Store the results in the proper place.
7. Go back to step 1 to fetch the next instruction.

As explained in Section 10-7, there is a register in the computer called the program counter (*PC*) that keeps track of the instructions in the program stored in memory. The *PC* holds the address of the instruction to be executed next and is incremented by one each time a word is read from the program in memory. The decoding done in Step 2 determines the operation to be performed and the addressing mode of the instruction. The operands in Step 3 are located from the addressing mode and the address field of the instruction. The computer executes the instruction, storing the result, and returns to Step 1 to fetch the next instruction in sequence.

Register Set

The *register set* consists of all registers in the CPU that are accessible to the programmer. These registers are typically those mentioned in assembly language programming reference manuals. In the simple CPUs we have dealt with so far, the register set has consisted of the programmer-accessible portion of the register file and the *PC*. The CPUs can also contain other registers, such as the instruction register, registers in the register file that are accessible only to microprograms, and pipeline registers. These registers, however, are not directly accessible to the programmer and, as a consequence, are not a part of the register set, which represents the stored information in the CPU that instructions can access. Thus, the register set has a considerable influence on instruction set architecture.

The register set for a realistic CPU can become quite complex. For the discussion in this chapter, we add two registers to the set we have used thus far: the *processor status register* (*PSR*) and the *stack pointer* (*SP*). The processor status register contains flip-flops that are selectively set by status values *C*, *N*, *V*, and *Z* from the ALU. These stored status bits are used to make decisions that determine the program flow, based on ALU results or the contents of registers. The stored status bits in the processor status register are also referred to as the *condition codes* or the *flags*. Additional bits in the *PSR* will be discussed when we cover associated concepts in this chapter.

11-2   OPERAND ADDRESSING

Consider an instruction such as ADD, which specifies the addition of two operands to produce a result. Suppose that the result of the addition is treated as just another operand. Then the ADD instruction has three operands: the addend, the augend, and the result. An operand residing in memory is specified by its address. An operand residing in a processor register is specified by a register address, a binary code of n bits that specifies one of 2^n registers in the register file. Thus, a computer with 16 processor registers, say, $R0$ through $R15$, has in its instructions one or more register address fields of four bits. The binary code 0101, for example, designates register $R5$.

Some operands, however, are not explicitly addressed, because their location is specified either by the opcode of the instruction or by an address assigned to one of the other operands. In such a case, we say that the operand has an *implied address*. If the address is implied, then there is no need for a memory or register address field for the operand in the instruction. On the other hand, if an operand has an address in the instruction, then we say that the operand is explicitly addressed or has an *explicit address*.

The number of operands explicitly addressed for a data manipulation operation such as ADD is an important factor in defining the instruction set architecture for a computer. An additional factor is the number of such operands that can be explicitly addressed in memory by the instruction. These two factors are so important in defining the nature of instructions that they act a means of distinguishing different instruction set architectures. They also govern the length of computer instructions.

We begin by illustrating simple programs with different numbers of explicitly addressed operands per instruction. Since each explicitly addressed operand has up to three memory or register addresses per instruction, we label the instructions as having three, two, one, or zero addresses. Note that, of the three operands needed for an instruction such as ADD, the addresses of all operands not having an address in the instruction are implied.

To illustrate the influence of the number of operands on computer programs, we will evaluate the arithmetic statement

$$X = (A + B)(C + D)$$

using three, two, one, and zero address instructions. We will assume that the operands are in memory addresses symbolized by the letters A, B, C, and D and must not be changed by the program. The result is to be stored in memory at a location with address X. The arithmetic operations to be used in the instructions are addition, subtraction, and multiplication, denoted by ADD, SUB, and MUL, respectively. Further, three operations needed to transfer data during the evaluation are move, load, and store, denoted by MOVE, LD, and ST, respectively. LD moves an operand from memory to a register and ST from a register to memory. Depending on the addresses permitted, MOVE can transfer data between registers, between memory locations, or from memory to register or register to memory.

Three-Address Instructions

A program that evaluates $X = (A + B)(C + D)$ using three-address instructions is as follows (a register transfer statement is shown for each instruction):

ADD T1, A, B	$M[T1] \leftarrow M[A] + M[B]$
ADD T2, C, D	$M[T2] \leftarrow M[C] + M[D]$
MUL X, T1, T2	$M[X] \leftarrow M[T1] \times M[T2]$

The symbol $M[A]$ denotes the operand stored in memory at the address symbolized by A. The symbol $\times$ designates multiplication. T1 and T2 are temporary storage locations in memory.

This same program can use registers as the temporary storage locations:

ADD R1, A, B	$R1 \leftarrow M[A] + M[B]$
ADD R2, C, D	$R2 \leftarrow M[C] + M[D]$
MUL X, R1, R2	$M[X] \leftarrow R1 \times R2$

Use of registers reduces the memory accesses required from nine to five. An advantage of the three-address format is that it results in short programs for evaluating expressions. A disadvantage is that the binary coded instructions require more bits to specify three addresses, particularly if they are memory addresses.

Two-Address Instructions

For two-address instructions, each address field can again specify either a possible register or a memory address. The first operand address listed in the symbolic instruction also serves as the implied address to which the result of the operation is transferred. The program is as follows:

MOVE T1, A	$M[T1] \leftarrow M[A]$
ADD T1, B	$M[T1] \leftarrow M[T1] + M[B]$
MOVE X, C	$M[X] \leftarrow M[C]$
ADD X, D	$M[X] \leftarrow M[X] + M[D]$
MUL X, T1	$M[X] \leftarrow M[X] \times M[T1]$

If a temporary storage register R1 is available, it can replace T1. Note that this program takes five instructions instead of the three used by the three-address instruction program.

One-Address Instructions

To perform instructions such as ADD, a computer with one-address instructions uses an implied address—such as a register called an *accumulator ACC*—for obtaining one of the operands and as the location of the result. The program to evaluate the arithmetic statement is as follows:

LD	A	$ACC \leftarrow M[A]$
ADD	B	$ACC \leftarrow ACC + M[B]$
ST	X	$M[X] \leftarrow ACC$
LD	C	$ACC \leftarrow M[C]$
ADD	D	$ACC \leftarrow ACC + M[D]$
MUL	X	$ACC \leftarrow ACC \times M[X]$
ST	X	$M[X] \leftarrow ACC$

All operations are done between the *ACC* register and a memory operand. In this case, the number of instructions in the program has increased to seven and the memory accesses is also seven.

Zero-Address Instructions

To perform an ADD instruction with zero addresses, all three addresses in the instruction must be implied. A conventional way of achieving this goal is to use a structure referred to as a *stack*, which is a mechanism or structure that stores information such that the item stored last is the first retrieved. Because of its "last in, first out" nature, a stack is also called a *last in, first out* (*LIFO*) queue. The operation of a computer stack is analogous to that of a stack of trays or plates in which the last tray placed on top of the stack is the first to be taken off. Data manipulation operations such as ADD are performed on the stack. The word at the top of the stack is referred to as TOS. The word below it in the stack is referred to as TOS_{-1}. When one or more words are used as operands for an operation, they are removed from the stack. The word below them then becomes the new TOS. When a resulting word is produced, it is placed on the stack and becomes the new TOS. Thus, TOS and a few locations below it are the implied addresses for operands, and TOS is the implied address for the result. For example, the instruction that specifies an addition is simply

ADD

The resulting register transfer action is $TOS \leftarrow TOS + TOS_{-1}$. Thus, there are no registers or register addresses used for data manipulation instructions in a stack architecture. Memory addressing, however, is used in such architectures for data transfers. For instance, the instruction

PUSH X

results in $TOS \leftarrow M[X]$, a transfer of the word in address X in memory to the top of the stack. A corresponding operation,

POP X

results in $M[X] \leftarrow TOS$, a transfer of the entry at the top of the stack to address X in memory.

The program for evaluating the sample arithmetic statement for the zero-address situation is as follows:

$$
\begin{array}{lll}
\text{PUSH} & \text{A} & TOS \leftarrow M[A] \\
\text{PUSH} & \text{B} & TOS \leftarrow M[B] \\
\text{ADD} & & TOS \leftarrow TOS + TOS_{-1} \\
\text{PUSH} & \text{C} & TOS \leftarrow M[C] \\
\text{PUSH} & \text{D} & TOS \leftarrow M[D] \\
\text{ADD} & & TOS \leftarrow TOS + TOS_{-1} \\
\text{MUL} & & TOS \leftarrow TOS \times TOS_{-1} \\
\text{POP} & \text{X} & M[X] \leftarrow TOS
\end{array}
$$

This program requires eight instructions—one more than the number required by the previous one-address program. However, it uses addressed memory locations or registers only for PUSH and POP and not to execute data manipulation instructions involving ADD and MUL.

Addressing Architectures

The programs just presented change if the number of addresses to the memory in the instructions is restricted or if the memory addresses are restricted to specific instructions. These restrictions, combined with the number of operands addressed, define addressing architectures. We can illustrate such architectures with the evaluation of an arithmetic statement in a three-address architecture that has all of the accesses to memory. Such an addressing scheme is called a *memory-to-memory architecture*. This architecture has only control registers, such as the program counter in the CPU. All operands come directly from memory, and all results are sent directly to memory. The formats of data transfer and manipulation instructions contain from one to three address fields, all of which are used for memory addresses. For the previous example, three instructions are required, but if an extra word must appear in the instruction for each memory address, then up to four memory reads are required to fetch each instruction. Including the fetching of operands and storing of results, the program to perform the addition would require 21 accesses to memory. If memory accesses take more than one clock cycle, the execution time would be in excess of 21 clock periods. Thus, even though the instruction count is low, the execution time is potentially high. Also, providing the capability for all operations to access memory increases the complexity of the control structures and may lengthen the clock cycle. Thus, this memory-to-memory architecture is typically not used in new designs.

In contrast, the three-address *register-to-register* or *load/store architecture*, which allows only one memory address and restricts its use to load and store types of instructions, is typical in modern processors. Such an architecture requires a sizeable register file, since all data manipulation instructions use register operands. With this architecture, the program to evaluate the sample arithmetic statement is as follows:

$$
\begin{array}{lll}
\text{LD} & \text{R1, A} & R1 \leftarrow M[A] \\
\text{LD} & \text{R2, B} & R2 \leftarrow M[B] \\
\text{ADD} & \text{R3, R1, R2} & R3 \leftarrow R1 + R2 \\
\text{LD} & \text{R1, C} & R1 \leftarrow M[C] \\
\text{LD} & \text{R2, D} & R2 \leftarrow M[D] \\
\text{ADD} & \text{R1, R1, R2} & R1 \leftarrow R1 + R2 \\
\text{MUL} & \text{R1, R1, R3} & R1 \leftarrow R1 \times R3 \\
\text{ST} & \text{X, R1} & M[X] \leftarrow R1
\end{array}
$$

Note that the instruction count increases to eight compared to three for the three-address, memory-to-memory case. Note also that the operations are the same as those for the stack case, except for the need for register addresses. By using registers, the number of accesses to memory for instructions, addresses, and operands is reduced from 21 to 18. If addresses can be obtained from registers instead of memory, as discussed in the next section, this number can be further reduced.

Variations on the previous two addressing architectures include three-address instructions and two-address instructions with one or two of the addresses to memory. The program lengths and number of memory accesses tend to be intermediate between the previous two architectures. An example of a two-address instruction with a single memory address allowed is

$$\text{ADD} \qquad \text{R1}, A \qquad R1 \leftarrow R1 + M[A]$$

This type of architecture is a *register-memory* architecture and remains prevalent among the current instruction set architectures, primarily to provide compatibility with older software using a specific architecture.

The program with one-address instructions illustrated previously gives the *single-accumulator architecture*. Since this architecture has no register file, its single address is for accessing memory. It requires 21 accesses to memory to evaluate the sample arithmetic statement. In more complex programs, significant additional memory accesses would be needed for temporary storage locations in memory. Because of its large number of memory accesses, this architecture is inefficient and, as a consequence, is restricted to use in CPUs for simple, low-cost applications that do not require high performance.

The zero-address instruction case using a stack supports the concept of a *stack architecture*. Data manipulation instructions such as ADD use no address, since they are performed on the top few elements of the stack. Single memory-address load and store operations, as shown in the program to evaluate the sample arithmetic statement, are used for data transfer. Since most of the stack is located in memory, one or more hidden memory accesses may be required for each stack operation. As register-register and load/store architectures have made strong performance advances, the high frequency of memory accesses in stack architectures has made them unattractive. However, recent stack architectures have begun to borrow technological advances from these other architectures. These new architectures store substantial numbers of stack locations in the processor chip and handle transfers between these locations and the memory transparently. Stack architectures are particularly useful for rapid interpretation of high-level language programs in which the intermediate code representation uses stack operations.

Stack architectures are compatible with a very efficient approach to expression processing which uses postfix notation rather than the traditional infix notation to which we are accustomed. The infix expression

$$(A + B) \times C + (D \times E)$$

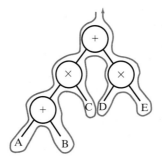

□ **FIGURE 11-1**
Graph for Example of Conversion from Infix to
RPN

with the operators between the operands can be written as postfix expression

$$A\ B + C \times D\ E \times +$$

Postfix notation is called reverse Polish notation (RPN), named for Polish mathematician Jan Lukasiewicz, who proposed prefix (the reverse of postfix) notation.

Conversion of $(A + B) \times C + (D \times E)$ to RPN can be achieved graphically as shown in Figure 11-1. When the path shown traversing the graph passes a variable, that variable is entered into the RPN expression. When the path passes an operation for the final time, the operation is entered into the RPN expression.

It is very easy to develop a program for an RPN expression. Whenever a variable is encountered, it is pushed onto the stack. Whenever an operation is encountered, it is executed on the implicit address TOS, or addresses TOS and TOS_{-1}, with the result placed in the new TOS. The program for the example RPN expression is

PUSH A
PUSH B
ADD
PUSH C
MUL
PUSH D
PUSH E
MUL
ADD

The execution of the program is illustrated by the successive stack states shown in Figure 11-2. As an operand is pushed on the stack, the stack contents are pushed down one stack location. When an operation is performed, the operand in the TOS is popped off and temporarily stored in a register. The operation is applied to the stored operand and the new TOS operand, and the result replaces the TOS operand.

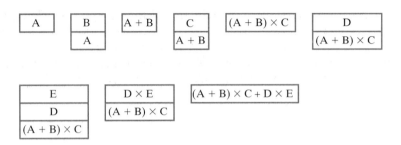

☐ **FIGURE 11-2**
Stack Activity for Execution of Example Stack Program

11-3 ADDRESSING MODES

The operation field of an instruction specifies the operation to be performed. This operation must be executed on data stored in computer registers or memory words. How the operands are selected during program execution is dependent on the addressing mode of the instruction. The addressing mode specifies a rule for interpreting or modifying the address field of the instruction before the operand is actually referenced. The address of the operand produced by the application of such a rule is called the *effective address*. Computers use addressing-mode techniques to accommodate one or both of the following provisions:

1. To give programming flexibility to the user via pointers to memory, counters for loop control, indexing of data, and relocation of programs.

2. To reduce the number of bits in the address fields of the instruction.

The availability of various addressing modes gives the experienced programmer the ability to write programs that require fewer instructions. The effect, however, on throughput and execution time must be carefully weighed. For example, the presence of more complex addressing modes may actually result in lower throughput and longer execution time. Also, most machine-executable programs are produced by compilers that often do not use complex addressing modes effectively.

In some computers, the addressing mode of the instruction is specified by a distinct binary code. Other computers use a common binary code that designates both the operation and the addressing mode of the instruction. Instructions may be defined with a variety of addressing modes, and sometimes two or more addressing modes are combined in one instruction.

An example of an instruction format with a distinct addressing-mode field is shown in Figure 11-3. The opcode specifies the operation to be performed. The

Opcode	Mode	Address or operand

☐ **FIGURE 11-3**
Instruction Format with Mode Field

mode field is used to locate the operands needed for the operation. There may or may not be an address field in the instruction. If there is an address field, it may designate a memory address or a processor register. Moreover, as discussed in the previous section, the instruction may have more than one address field. In that case, each address field is associated with its own particular addressing mode.

Implied Mode

Although most addressing modes modify the address field of the instruction, there is one mode that needs no address field at all: the implied mode. In this mode, the operand is specified implicitly in the definition of the opcode. It is the implied mode that provides the location for the two-operand-plus-result operations when fewer than three addresses are contained in the instruction. For example, the instruction "complement accumulator" is an implied-mode instruction because the operand in the accumulator register is implied in the definition of the instruction. In fact, any instruction that uses an accumulator without a second operand is an implied-mode instruction. For example, data manipulation instructions in a stack computer, such as ADD, are implied-mode instructions, since the operands are implied to be on top of stack.

Immediate Mode

In the immediate mode, the operand is specified in the instruction itself. In other words, an immediate-mode instruction has an operand field rather than an address field. The operand field contains the actual operand to be used in conjunction with the operation specified in the instruction. Immediate-mode instructions are useful, for example, for initializing registers to a constant value.

Register and Register-Indirect Modes

Earlier, we mentioned that the address field of the instruction may specify either a memory location or a processor register. When the address field specifies a processor register, the instruction is said to be in the register mode. In this mode, the operands are in registers that reside within the processor of the computer. The particular register is selected from a register address field in the instruction format.

In the register-indirect mode, the instruction specifies a register in the processor whose content gives the address of the operand in memory. In other words, the selected register contains the memory address of the operand, rather than the operand itself. Before using a register-indirect mode instruction, the programmer must ensure that the memory address is available in the processor register. A reference to the register is then equivalent to specifying a memory address. The advantage of register-indirect mode is that the address field of the instruction uses fewer bits to select a register than would have been required to specify a memory address directly.

An autoincrement or autodecrement mode is similar to the register-indirect mode, except that the register is incremented or decremented after (or before) its address value is used to access memory. When the address stored in the register refers to an array of data in memory, it is convenient to increment the register after each access to the array. This can be achieved by using a separate register-increment instruction. However, because it is such a common requirement, some computers incorporate an autoincrement mode that increments the content of the register containing the address after the memory data are accessed.

In the following instruction, an autoincrement mode is used to add the constant value 3 to the elements of an array addressed by register $R1$:

$$\text{ADD} \qquad (R1)+, 3 \qquad\qquad M[R1] \leftarrow M[R1] + 3, R1 \leftarrow R1 + 1$$

$R1$ is initialized to the address of the first element in the array. Then the ADD instruction is repeatedly executed until the addition of 3 to all elements of the array has occurred. The register transfer statement accompanying the instruction shows the addition of 3 to the memory location addressed by $R1$ and the incrementing of $R1$ in preparation for the next execution of the ADD on the next element in the array.

Direct Addressing Mode

In the direct addressing mode, the address field of the instruction gives the address of the operand in memory in a data transfer or data manipulation instruction. An example of a data transfer instruction is shown in Figure 11-4. The instruction in memory consists of two words. The first, at address 250, has the opcode for "load to ACC" and a mode field specifying a direct address. The second word of the

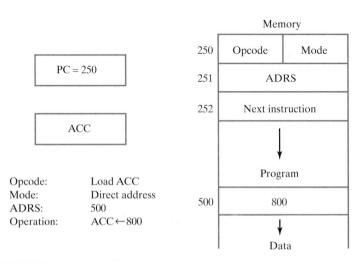

☐ **FIGURE 11-4**
Example Demonstrating Direct Addressing for a Data Transfer Instruction

instruction, at address 251, contains the address field, symbolized by ADRS, and is equal to 500. The *PC* holds the address of the instruction, which is brought from memory using two memory accesses. Simultaneously with or after the completion of the first access, the *PC* is incremented to 251. Then the second access for ADRS occurs and the *PC* is again incremented. The execution of the instruction results in the operation

$$ACC \leftarrow M[\text{ADRS}]$$

Since ADRS = 500 and M[500] = 800, the *ACC* receives the number 800. After the instruction is executed, the *PC* holds the number 252, which is the address of the next instruction in the program.

Now consider a branch-type instruction, as shown in Figure 11-5. If the contents of *ACC* equal 0, control branches to ADRS; otherwise, the program continues with the next instruction in sequence. When *ACC* = 0, the branch to address 500 is accomplished by loading the value of the address field ADRS into the *PC*. Control then continues with the instruction at address 500. When *ACC* ≠ 0, no branch occurs, and the *PC*, which was incremented twice during the fetch of the instruction, holds the address 302, the address of the next instruction in sequence.

Sometimes the value given in the address field is the address of the operand, but sometimes it is just an address from which the address of the operand is calculated. To differentiate among the various addressing modes, it is useful to distinguish between the address part of the instruction, as given in the address field, and the address used by the control when executing the instruction. Recall that we refer to the latter as the effective address.

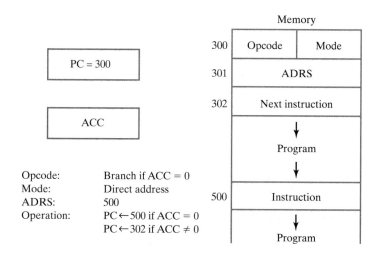

□ **FIGURE 11-5**
Example Demonstrating Direct Addressing in a Branch Instruction

Indirect Addressing Mode

In the indirect addressing mode, the address field of the instruction gives the address at which the effective address is stored in memory. The control unit fetches the instruction from memory and uses the address part to access memory again in order to read the effective address. Consider the instruction "load to *ACC*" given in Figure 11-4. If the mode specifies an indirect address, the effective address is stored in *M*[ADRS]. Since ADRS = 500 and *M*[ADRS] = 800, the effective address is 800. This means that the operand loaded into the *ACC* is the one found in memory at address 800 (not shown in the figure).

Relative Addressing Mode

Some addressing modes require that the address field of the instruction be added to the content of a specified register in the CPU in order to evaluate the effective address. Often, the register used is the *PC*. In the relative addressing mode, the effective address is calculated as follows:

Effective address = Address part of the instruction + Contents of *PC*

The address part of the instruction is considered to be a signed number that can be either positive or negative. When this number is added to the contents of the *PC*, the result produces an effective address whose position in memory is relative to the address of the next instruction in the program.

To clarify this with an example, let us assume that the *PC* contains the number 250 and the address part of the instruction contains the number 500, as in Figure 11-5, with the mode field specifying a relative address. The instruction at location 250 is read from memory during the fetch phase of the operation cycle, and the *PC* is incremented by 1 to 251. Since the instruction has a second word, the control unit reads the address field into a control register, and the *PC* is incremented to 252. The computation of the effective address for the relative addressing mode is 252 + 500 = 752. The result is that the operand associated with the instruction is 500 locations away, relative to the location of the next instruction.

Relative addressing is often used in branch-type instructions when the branch address is in a location close to the instruction word. Relative addressing produces more compact instructions, since the relative address can be specified with fewer bits than are required to designate the entire memory address.

Indexed Addressing Mode

In the indexed addressing mode, the content of an index register is added to the address part of the instruction to obtain the effective address. The index register may be a special CPU register or simply a register in a register file. We illustrate the use of indexed addressing by considering an array of data in memory. The address field of the instruction defines the beginning address of the array. Each operand in the array is stored in memory relative to the beginning address. The distance between the beginning address and the address of the operand is the index value stored in the register. Any operand in the array can be accessed with the same instruction, provided that the index register contains the correct index value. The index register can be incremented to facilitate access to consecutive operands.

Some computers dedicate one CPU register to function solely as an index register. This register is addressed implicitly when an index-mode instruction is used. In computers with many processor registers, any CPU register can be used as an index register. In such a case, the index register to be used must be specified with a register field within the instruction format.

A specialized variation of the index mode is the base-register mode. In this mode, the contents of a base register are added to the address part of the instruction to obtain the effective address. This is similar to indexed addressing, except that the register is called a base register instead of an index register. The difference between the two modes is in the way they are used rather than in the way addresses are computed: an index register is assumed to hold an index number that is relative to the address field of the instruction; a base register is assumed to hold a base address, and the address field of the instruction gives a displacement relative to the base address.

Summary of Addressing Modes

In order to show the differences among the various modes, we will investigate the effect of the addressing mode on the instruction shown in Figure 11-6. The instruction in addresses 250 and 251 is "load to *ACC*," with the address field ADRS (or an operand NBR) equal to 500. The *PC* has the number 250 for fetching this instruction. The contents of a processor register *R*1 are 400, and the *ACC* receives the result after the instruction is executed. In the direct mode, the effective address is 500, and the operand to be loaded into the *ACC* is 800. In the immediate mode, the operand 500 is loaded into the *ACC*. In the indirect mode, the effective address is 800, and the operand is 300. In the relative mode, the effective address is 500 + 252 = 752, and the operand is 600. In the index mode, the effective address is 500 + 400 = 900, assuming that *R*1 is the index register. In the register mode, the operand is in *R*1, and 400 is loaded into the *ACC*. In the register-indirect mode, the effective address is the contents of *R*1, and the operand loaded into the *ACC* is 700.

Table 11-1 lists the value of the effective address and the operand loaded into the *ACC* for seven addressing modes. The table also shows the operation with a register transfer statement and a symbolic convention for each addressing mode. LDA is the symbol for the load-to-accumulator opcode. In the direct mode, we use the symbol ADRS for the address part of the instruction. The # symbol precedes the operand NBR in the immediate mode. The symbol ADRS enclosed in square brackets symbolizes an indirect address, which some compilers or assemblers designate with the symbol @. The symbol $ before the address makes the effective address relative to the *PC*. An index-mode instruction is recognized by the symbol of a register placed in parentheses after the address symbol. The register mode is indicated by giving the name of the processor register following LDA. In the register-indirect mode, the name of the register that holds the effective address is enclosed in parentheses.

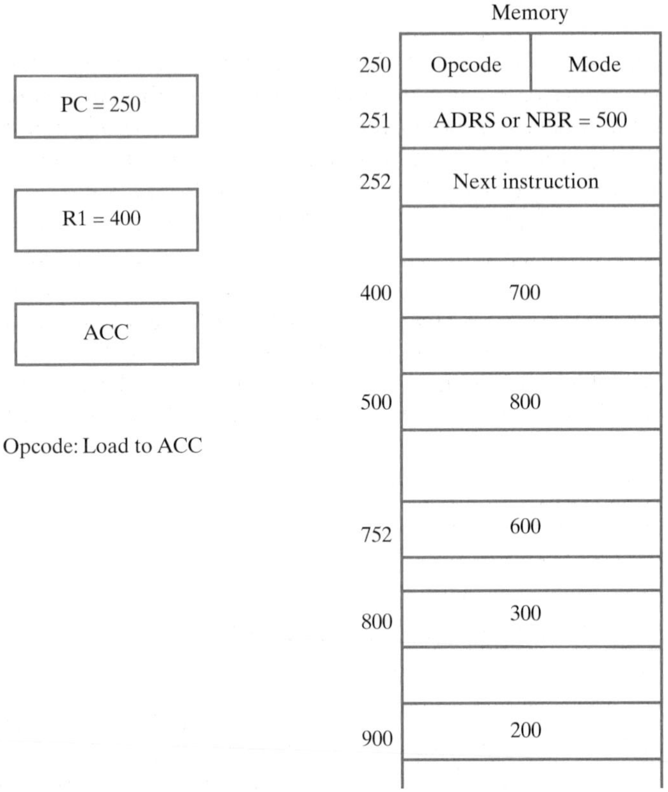

□ **FIGURE 11-6**
Numerical Example for Addressing Modes

□ **TABLE 11-1**
Symbolic Convention for Addressing Modes

Addressing mode	Symbolic convention	Register transfer	Refers to Figure 11-6	
			Effective address	Contents of *ACC*
Direct	LDA ADRS	$ACC \leftarrow M[ADRS]$	500	800
Immediate	LDA #NBR	$ACC \leftarrow NBR$	251	500
Indirect	LDA [ADRS]	$ACC \leftarrow M[M[ADRS]]$	800	300
Relative	LDA $ADRS	$ACC \leftarrow M[ADRS + PC]$	752	600
Index	LDA ADRS (R1)	$ACC \leftarrow M[ADRS + R1]$	900	200
Register	LDA R1	$ACC \leftarrow R1$	—	400
Register-indirect	LDA (R1)	$ACC \leftarrow M[R1]$	400	700

11-4 INSTRUCTION SET ARCHITECTURES

Computers provide a set of instructions to permit computational tasks to be carried out. The instruction sets of different computers differ in several ways from each other. For example, the binary code assigned to the opcode field varies widely for different computers. Likewise, although a standard exists (see Reference 7), the symbolic name given to instructions varies for different computers. In comparison to these minor differences, however, there are two major types of instruction set architectures that differ markedly in the relationship of hardware to software: *Complex instruction set computers* (CISCs) provide hardware support for high-level language operations and have compact programs; *Reduced instruction set computers* (RISCs) emphasize simple instructions and flexibility that, when combined, provide higher throughput and faster execution. These two architectures can be distinguished by considering the properties that characterize their instruction sets.

A *RISC architecture* has the following properties:

1. Memory accesses are restricted to load and store instructions, and data manipulation instructions are register-to-register.
2. Addressing modes are limited in number.
3. Instruction formats are all of the same length.
4. Instructions perform elementary operations.

The goal of a RISC architecture is high throughput and fast execution. To achieve these goals, accesses to memory, which typically take longer than other elementary operations, are to be avoided, except for fetching instructions. A result of this view is the need for a relatively large register file. Because of the fixed instruction length, limited addressing modes, and elementary operations, the control unit of a RISC is comparatively simple and is typically hardwired. In addition, the underlying organization is universally a pipelined design as covered in Chapter 12.

A purely *CISC architecture* has the following properties:

1. Memory access is directly available to most types of instructions.
2. Addressing modes are substantial in number.
3. Instruction formats are of different lengths.
4. Instructions perform both elementary and complex operations.

The goal of the CISC architecture is to match more closely the operations used in programming languages and to provide instructions that facilitate compact programs and conserve memory. In addition, efficiencies in performance may result through a reduction in the number of instruction fetches from memory, compared with the number of elementary operations performed. Because of the high memory accessibility, the register files in a CISC are smaller than in a RISC. Also, because of the complexity of the instructions and the variability of the instruction formats, microprogrammed control is often used. In the quest for speed, the microprogrammed control in newer designs is likely to be controlling a pipelined datapath. CISC instructions are converted to a sequence of RISC-like operations that are processed by the RISC-like pipeline as discussed in detail in Chapter 12.

Actual instruction set architectures range between those which are purely RISC and those which are purely CISC. Nevertheless, there is a basic set of elementary operations that most computers include among their instructions. In this chapter, we will focus primarily on elementary instructions that are included in both CISC and RISC instruction sets. Most elementary computer instructions can be classified into three major categories: (1) data transfer instructions, (2) data manipulation instructions, and (3) program control instructions.

Data transfer instructions cause transfer of data from one location to another without changing the binary information content. Data manipulation instructions perform arithmetic, logic, and shift operations. Program control instructions provide decision-making capabilities and change the path taken by the program when executed in the computer. In addition to the basic instruction set, a computer may have other instructions that provide special operations for particular applications.

11-5 DATA TRANSFER INSTRUCTIONS

Data transfer instructions move data from one place in the computer to another without changing the data. Typical transfers are between memory and processor registers, between processor registers and input and output registers, and among the processor registers themselves.

Table 11-2 gives a list of eight typical data transfer instructions used in many computers. Accompanying each instruction is a mnemonic symbol, the assembly language abbreviation recommended by an IEEE standard (Reference 6). Different computers, however, may use different mnemonics for the same instruction name. The load instruction is used to designate a transfer from memory to a processor register. The store instruction designates a transfer from a processor register into a memory word. The move instruction is used in computers with multiple processor registers to designate a transfer from one register to another. It is also used for data transfer between registers and memory and between two memory words.

☐ **TABLE 11-2**
Typical Data Transfer Instructions

Name	Mnemonic
Load	LD
Store	ST
Move	MOVE
Exchange	XCH
Push	PUSH
Pop	POP
Input	IN
Output	OUT

The exchange instruction exchanges information between two registers, between a register and a memory word, or between two memory words. The push and pop instructions are for stack operations described next.

Stack Instructions

The stack architecture introduced earlier possesses features that facilitate a number of data-processing and control tasks. A stack is used in some electronic calculators and computers for the evaluation of arithmetic expressions. Unfortunately, because of the negative effects on performance of having the stack reside primarily in memory, a stack in a computer typically handles only state information related to procedure calls and returns and interrupts, as explained in Section 11-8 and Section 11-9.

The stack instructions push and pop transfer data between a memory stack and a processor register or memory. The *push* operation places a new item onto the top of the stack. The *pop* operation removes one item from the stack so that the stack pops up. However, nothing is really physically pushed or popped in the stack. Rather, the memory stack is essentially a portion of a memory address space accessed by an address that is always incremented or decremented before or after the memory access. The register that holds the address for the stack is called a *stack pointer* (*SP*) because its value always points to TOS, the item at the top of the stack. Push and pop operations are implemented by decrementing or incrementing the stack pointer.

Figure 11-7 shows a portion of a memory organized as a stack that grows from higher to lower addresses. The stack pointer, *SP*, holds the binary address of the item that is currently on top of the stack. Three items are presently stored in the stack: *A*, *B*, and *C*, in consecutive addresses 103, 102, and 101, respectively. Item *C* is on top of the stack, so *SP* contains 101. To remove the top item, the stack is popped by reading the item at address 101 and incrementing *SP*. Item *B* is now on top of the stack, since

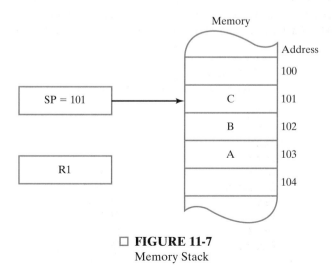

□ **FIGURE 11-7**
Memory Stack

SP contains address 102. To insert a new item, the stack is pushed by first decrementing *SP* and then writing the new item on top of the stack. Note that item *C* has been read out of the stack, but is not physically removed from it. This does not matter as far as the stack operation is concerned, because when the stack is pushed, a new item is written over it regardless of what was there before.

We assume that the items in the stack communicate with a data register *R*1 or a memory location X. A new item is placed on to the stack with the push operation as follows:

$$SP \leftarrow SP - 1$$

$$M[SP] \leftarrow R1$$

The stack pointer is decremented so that it points at the address of the next word. A memory write microoperation inserts the word from *R*1 onto the top of the stack. Note that *SP* holds the address of the top of the stack and that *M*[*SP*] denotes the memory word specified by the address presently in *SP*. An item is deleted from the stack with a pop operation as follows:

$$R1 \leftarrow M[SP]$$

$$SP \leftarrow SP + 1$$

The top item is read from the stack into *R*1. The stack pointer is then incremented to point at the next item in the stack, which is the new top of the stack.

The two microoperations needed for either the push or the pop operation are an access to memory through *SP* and an update of *SP*. Which microoperation is done first, and whether *SP* is updated by incrementing or decrementing it, depends on the organization of the stack. In Figure 11-7, the stack grows by decreasing the memory address. By contrast, a stack may be constructed to grow by increasing the memory address. In such a case, *SP* is incremented for the push operation and decremented for the pop operation. A stack may also be constructed so that *SP* points to the next empty location above the top of the stack. In that case, the sequence of microoperations must be interchanged.

A stack pointer is loaded with an initial value, which must be the bottom address of an assigned stack in memory. From then on, *SP* is automatically decremented or incremented with every push or pop operation. The advantage of a memory stack is that the processor can refer to it without having to specify an address, since the address is always available and automatically updated in the stack pointer.

The final pair of data transfer instructions, input and output, depend on the type of input-output used, as described next.

Independent versus Memory-Mapped I/O

Input and output (I/O) instructions transfer data between processor registers and input and output devices. These instructions are similar to load and store instructions, except that the transfers are to and from external registers instead of memory words. The computer is considered to have a certain number of input and output ports, with one or more ports dedicated to communication with a specific input or output device. A *port* is typically a register with input and/or output lines attached to the device. The particular port is chosen by an address, in a manner similar to the way an address selects a word in memory. Input and output instructions include an address field in their format, for specifying the particular port selected for the transfer of data.

Port addresses are assigned in two ways. In the *independent I/O system*, the address ranges assigned to memory and I/O ports are independent from each other. The computer has distinct input and output instructions, as listed in Table 11-2, containing a separate address field that is interpreted by the control and used to select a particular I/O port. Independent I/O addressing isolates memory and I/O selection, so that the memory address range is not affected by the port address assignment. For this reason, the method is also referred to as an i*solated I/O configuration.*

In contrast to independent I/O, *memory-mapped I/O,* assigns a subrange of the memory addresses for addressing I/O ports. There are no separate addresses for handling input and output transfers, since I/O ports are treated as memory locations in one common address range. Each I/O port is regarded as a memory location, similar to a memory word. Computers that adopt the memory-mapped scheme have no distinct input or output instructions, because the same instructions are used for manipulating both memory and I/O data. For example, the load and store instructions used for memory transfer are also used for I/O transfer, provided that the address associated with the instruction is assigned to an I/O port and not to a memory word. The advantage of this scheme is the simplicity that results with the same set of instructions serving for both memory and I/O access.

11-6 DATA MANIPULATION INSTRUCTIONS

Data manipulation instructions perform operations on data and provide the computational capabilities of the computer. In a typical computer, data manipulation instructions are usually divided into three basic types:

1. Arithmetic instructions.
2. Logical and bit manipulation instructions.
3. Shift instructions.

A list of elementary data manipulation instructions looks very much like the list of microoperations given in Chapter 10. However, an instruction is typically processed by executing a *sequence* of one or more microinstructions. A microoperation is an elementary operation executed by the hardware of the computer under the control of the control unit. In contrast, an instruction may involve several elementary operations that fetch the instruction, bring the operands from appropriate processor registers, and store the result in the specified location.

Arithmetic Instructions

The four basic arithmetic instructions are addition, subtraction, multiplication, and division. Most computers provide instructions for all four operations. Some small computers, however, have only addition and subtraction instructions; on such computers, multiplication and division must be carried out by means of programs. The four basic arithmetic operations are sufficient for formulating solutions to any numerical problem when they are used with numerical analysis methods.

A list of typical arithmetic instructions is given in Table 11-3. The increment instruction adds one to the value stored in a register or memory word. A common characteristic of the increment operation, when executed on a computer word, is that a binary number of all 1's produces a result of all 0's when incremented. The decrement instruction subtracts one from a value stored in a register or memory word. When decremented, a number of all 0's produces a number of all 1's.

The add, subtract, multiply, and divide instructions may be available for different types of data. The data type assumed to be in processor registers during the execution of these arithmetic operations is included in the definition of the opcode. An arithmetic instruction may specify unsigned or signed integers, binary or decimal numbers, or floating-point data. The arithmetic operations with binary integers were presented in Chapter 1 and Chapter 5. The floating-point representation is used for scientific calculations and is presented in the next section.

The number of bits in any register is finite; therefore, the results of arithmetic operations are of finite precision. Most computers provide special instructions to facilitate double-precision arithmetic. A carry flip-flop is used to store the carry from an operation. The instruction "add with carry" performs the addition with two operands plus the value of the carry from the previous computation. Similarly, the "subtract with borrow" instruction subtracts two operands and a borrow that may have resulted from a previous operation. The subtract reverse instruction reverses the order of the operands, performing $B - A$ instead of $A - B$. The negate instruction performs the 2's complement of a signed number, which is equivalent to multiplying the number by -1.

□ **TABLE 11-3**
Typical Arithmetic Instructions

Name	Mnemonic
Increment	INC
Decrement	DEC
Add	ADD
Subtract	SUB
Multiply	MUL
Divide	DIV
Add with carry	ADDC
Subtract with borrow	SUBB
Subtract reverse	SUBR
Negate	NEG

Logical and Bit Manipulation Instructions

Logical instructions perform binary operations on words stored in registers or memory words. They are useful for manipulating individual bits or a group of bits that represent binary-coded information. Logical instructions consider each bit of the operand separately and treat it as a Boolean variable. By proper application of the logical instructions, it is possible to change bit values, to clear a group of bits, or to insert new bit values into operands stored in registers or memory.

Some typical logical and bit manipulation instructions are listed in Table 11-4. The clear instruction causes the specific operand to be replaced by 0's. The set instruction causes the operand to be replaced by 1's. The complement instruction inverts all the bits of the operand. The AND, OR, and XOR instructions produce the corresponding logical operations on individual bits of the operand. Although logical instructions perform Boolean operations, when used on words they often are viewed as performing bit manipulation operations. There are three bit manipulation operations possible: A selected bit can be cleared to 0, set to 1, or complemented. The three logical instructions are usually applied to do just that.

The AND instruction is used to clear a bit or a selected group of bits of an operand to 0. For any Boolean variable X, the relationship $X \cdot 0 = 0$ dictates that a binary variable ANDed with a 0 produces a 0; and similarly, the relationship $X \cdot 1 = X$ dictates that the variable does not change when ANDed with a 1. Therefore, the AND instruction is used to selectively clear bits of an operand by ANDing the operand with a word that has 0's in the bit positions that must be cleared and 1's in the bit positions that must remain the same. The AND instruction is also called a *mask* because, by inserting 0's, it masks a selected portion of an operand. AND is also sometimes referred to as a *bit clear* instruction.

The OR instruction is used to set a bit or a selected group of bits of an operand to 1. For any Boolean variable X, the relationship $X + 1 = 1$ dictates that a binary variable ORed with a 1 produces a l; similarly, the relationship $X + 0 = X$ dictates that the variable does not change when ORed with a 0. Therefore, the OR instruction can be used to selectively set bits of an operand by ORing the operand with a word with 1's in the bit positions that must be set to 1. The OR instruction is sometimes called a *bit set* instruction.

□ **TABLE 11-4**
Typical Logical and Bit Manipulation Instructions

Name	Mnemonic
Clear	CLR
Set	SET
Complement	NOT
AND	AND
OR	OR
Exclusive-OR	XOR
Clear carry	CLRC
Set carry	SETC
Complement carry	COMC

The XOR instruction is used to selectively complement bits of an operand. This is because of the Boolean relationships $X \oplus 1 = \overline{X}$ and $X \oplus 0 = X$. A binary variable is complemented when XORed with a 1, but does not change value when XORed with a 0. The XOR instruction is sometimes called a *bit complement* instruction.

Other bit manipulation instructions included in Table 11-4 clear, set, or complement the carry bit. Additional instructions clear, set, or complement other status bits or flag bits in a similar manner.

Shift Instructions

Instructions to shift the content of an operand are provided in several varieties. Shifts are operations in which the bits of the operand are moved to the left or to the right. The incoming bit shifted in at the end of the word determines the type of shift. Instead of using just a 0, as for sl and sr in Chapter 10, here we add further possibilities. The shift instructions may specify either logical shifts, arithmetic shifts, or rotate-type operations.

Table 11-5 lists four types of shift instructions. The logical shift inserts 0 into the incoming bit position after the shift. Arithmetic shifts conform to the rules for shifting two's complement signed numbers. The arithmetic shift right instruction preserves the sign bit in the leftmost position. The value of the sign bit is shifted to the right together with the rest of the number, but the sign bit itself remains unchanged. The arithmetic shift left instruction inserts 0 into the incoming bit in the rightmost position and is identical to the logical shift left instruction. The two instructions may differ, however, in that an arithmetic shift left may set the overflow status bit *V*, while a logical shift left does not affect *V*.

The rotate instructions produce a circular shift: the values shifted out of the outgoing bit of the word are not lost, as in a logical shift, but are rotated back into the incoming bit. The rotate-with-carry instructions treat the carry bit as an extension of the register whose word is being rotated. Thus, a rotate left with carry transfers the carry bit into the incoming bit in the rightmost bit position of the register, transfers the outgoing bit from the leftmost bit of the register into the carry, and

☐ **TABLE 11-5**
Typical Shift Instructions

Name	Mnemonic
Logical shift right	SHR
Logical shift left	SHL
Arithmetic shift right	SHRA
Arithmetic shift left	SHLA
Rotate right	ROR
Rotate left	ROL
Rotate right with carry	RORC
Rotate left with carry	ROLC

shifts the entire register to the left. Some computers have a multiple-field format for the shift instruction. One field contains the opcode, and the others specify the type of shift and the number of positions that an operand is to be shifted. A shift instruction may include the following five fields:

OP   REG   TYPE   RL   COUNT

OP is the opcode field for specifying a shift, and REG is a register address that specifies the location of the operand. TYPE is a 2-bit field that specifies one of the four types of shifts (logical, arithmetic, rotate, and rotate with carry), while RL is a 1-bit field that specifies whether a shift is to the right or the left. COUNT is a k-bit field that specifies shifts of up to $2^k - 1$ positions. With such a format, it is possible to specify the type of shift, the direction of the shift, and the number of positions to be shifted, all in one instruction.

11-7 FLOATING-POINT COMPUTATIONS

In many scientific calculations, the range of numbers is very large. In a computer, the way to express such numbers is in floating-point notation. The floating-point number has two parts, one containing the sign of the number and a *fraction* (sometimes called a *mantissa*) and the other designating the position of the radix point in the number and called the *exponent*. For example, the decimal number +6132.789 is represented in floating-point notation as

Fraction	Exponent
+.6132789	+04

The value of the exponent indicates that the actual position of the decimal point is four positions to the right of the indicated decimal point in the fraction. This representation is equivalent to the scientific notation $+.6132789 \times 10^{+4}$. Decimal floating-point numbers are interpreted as representing a number in the form

$$F \times 10^E$$

where F is the fraction and E the exponent. Only the fraction and the exponent are physically represented in computer registers; radix 10 and the decimal point of the fraction are assumed and are not shown explicitly. A floating-point binary number is represented in a similar manner, except that it uses radix 2 for the exponent. For example, the binary number +1001.11 is represented with an 8-bit fraction and 6-bit exponent as

Fraction	Exponent
01001110	000100

The fraction has a 0 in the leftmost position to denote a plus. The binary point of the fraction follows the sign bit, but is not shown in the register. The exponent has the equivalent binary number $+4$. The floating-point number is equivalent to

$$F \times 2^E = +(0.1001110)_2 \times 2^{+4}$$

A floating-point number is said to be *normalized* if the most significant digit of the fraction is nonzero. For example, the decimal fraction 0.350 is normalized, but 0.0035 is not. Normalized numbers provide the maximum possible precision for the floating-point number. A zero cannot be normalized because it does not have a nonzero digit; it is usually represented in floating-point by all 0's in both the fraction and the exponent.

Floating-point representation increases the range of numbers that can be accommodated in a given register. Consider a computer with 48-bit registers. Since one bit must be reserved for the sign, the range of signed integers will be $\pm(2^{47} - 1)$, which is approximately $\pm 10^{14}$. The 48 bits can be used to represent a floating-point number, with one bit for the sign, 35 bits for the fraction, and 12 bits for the exponent. The largest positive or negative number that can be accommodated is thus

$$\pm(1 - 2^{-35}) \times 2^{+2047}$$

This number is derived from a fraction that contains 35 1's, and an exponent with a sign bit and 11 1's. The maximum exponent is $2^{11} - 1$, or 2047. The largest number that can be accommodated is approximately equivalent to decimal 10^{615}. Although a much larger range is represented, there are still only 48 bits in the representation. As a consequence, exactly the same number of numbers are represented. Hence, the range is traded for the precision of the numbers, which is reduced from 48 bits to 35 bits.

Arithmetic Operations

Arithmetic operations with floating-point numbers are more complicated than with integer numbers, and their execution takes longer and requires more complex hardware. Adding and subtracting two numbers requires that the radix points be aligned, since the exponent parts must be equal before adding or subtracting the fractions. The alignment is done by shifting one fraction and correspondingly adjusting its exponent until it is equal to the other exponent. Consider the sum of the following floating-point numbers:

$$.5372400 \times 10^2$$

$$+ .1580000 \times 10^{-1}$$

It is necessary that the two exponents be equal before the fractions can be added. We can either shift the first number three positions to the left or shift the second number three positions to the right. When the fractions are stored in registers, shifting to the left causes a loss of the most significant digits. Shifting to the right causes a loss of the least significant digits. The second method is preferable because it only reduces the precision, whereas the first method may cause an error. The usual alignment procedure is to shift the fraction with the smaller exponent to the right by a number of places equal to the difference between the exponents. After this is done, the fractions can be added:

$$.5372400 \times 10^2$$
$$+ .0001580 \times 10^2$$
$$\overline{.5373980 \times 10^2}$$

When two normalized fractions are added, the sum may contain an overflow digit. An overflow can be corrected by shifting the sum once to the right and incrementing the exponent. When two numbers are subtracted, the result may contain most significant zeros in the fraction, as shown in the following example:

$$.56780 \times 10^5$$
$$- .56430 \times 10^5$$
$$\overline{.00350 \times 10^5}$$

A floating-point number that has a 0 in the most significant position of the fraction is not normalized. To normalize the number, it is necessary to shift the fraction to the left and decrement the exponent until a nonzero digit appears in the first position. In the preceding example, it is necessary to shift left twice to obtain $.35000 \times 10^3$. In most computers, a normalization procedure is performed after each operation to ensure that all results are in normalized form.

Floating-point multiplication and division do not require an alignment of the fractions. Multiplication can be performed by multiplying the two fractions and adding the exponents. Division is accomplished by dividing the fractions and subtracting the exponents. In the examples shown, we used decimal numbers to demonstrate arithmetic operations on floating-point numbers. The same procedure applies to binary numbers, except that the base of the exponent is 2 instead of 10.

Biased Exponent

The sign and fraction part of a floating-point number is usually a signed-magnitude representation. The exponent representation employed in most computers is known as a *biased exponent*. The bias is an excess number added to the exponent so that, internally, all exponents become positive. As a consequence, the sign of the exponent is removed from being a separate entity.

Consider, for example, the range of decimal exponents from -99 to $+99$. This is represented by two digits and a sign. If we use an excess 99 bias, then the biased exponent e will be equal to $e = E + 99$, where E is the actual exponent. For $E = -99$, we have $e = -99 + 99 = 0$; and for $E = +99$, we have $e = 99 + 99 = 198$. In this way, the biased exponent is represented in a register as a positive number in the range from 000 to 198. Positive-biased exponents have a range of numbers from 099 to 198. Subtraction of the bias, 99, gives the positive values from 0 to $+99$. Negative-biased exponents have a range from 098 to 000. Subtraction of 99 gives the negative values from -1 to -99.

The advantage of biased exponents is that the resulting floating-point numbers contain only positive exponents. It is then simpler to compare the relative magnitude between two numbers without being concerned with the signs of their exponents. Another advantage is that the most negative exponent converts to a biased exponent with all 0's. The floating-point representation of zero is then a zero fraction and a zero biased exponent, which is the smallest possible exponent.

Standard Operand Format

Arithmetic instructions that perform operations with floating-point data often use the suffix F. Thus, ADDF is an add instruction with floating-point numbers. There are two standard formats for representing a floating-point operand: the single-precision data type, consisting of 32 bits, and the double-precision data type, consisting of 64 bits. When both types of data are available, the single-precision instruction mnemonic uses an FS suffix, and the double precision uses FL (for "floating-point long").

The format of the IEEE standard (see Reference 7) single-precision floating-point operand is shown in Figure 11-8. It consists of 32 bits. The sign bit s designates the sign for the fraction. The biased exponent e contains 8 bits and uses an excess 127 number. The fraction f consists of 23 bits. The binary point is assumed to be immediately to the left of the most significant bit of the f field. In addition, an implied 1 bit is inserted to the left of the binary point, which, in effect, expands the number to 24 bits representing a value from 1.0_2 to $1.11...1_2$. The component of the binary floating-point number that consists of a leading bit to the left of the implied binary point, together with the fraction in the field, is called the *significand*. Following are some examples of field values and the corresponding significands:

f Field	Significand	Decimal Equivalent
100 . . . 0	1.100 . . . 0	1.50
010 . . . 0	1.010 . . . 0	1.25
000 . . . 0	1.000 . . . 0*	1.00*

*Assuming the exponent is not equal to 00 . . . 0.

Even though the f field by itself may not be normalized, the significant is always normalized because it has a nonzero bit in the most significant position. Since normalized numbers must have a nonzero most significant bit, this 1 bit is not included explicitly in the format, but must be inserted by the hardware during arithmetic computations. The exponent field uses an excess 127 bias value for normalized numbers. The range of valid exponents is from -126 (represented as

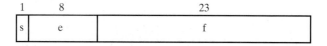

☐ **FIGURE 11-8**
IEEE Floating-Point Operand Format

00000001) through +127 (represented as 11111110). The maximum (11111111) and minimum (00000000) values for the e field are reserved to indicate exceptional conditions. Table 11-6 shows the biased and actual values of some exponents.

Normalized numbers are numbers that can be expressed as floating-point operands in which the e field is neither all 0's nor all 1's. The value of the number is derived from the three fields in the format of Figure 11-8 using the formula

$$(-1)^s 2^{e-127} \times (1.f)$$

The most positive normalized number that can be obtained has a 0 for the sign bit for a positive sign, a biased exponent equal to 254, and an f field with 23 1's. This gives an exponent $E = 254 - 127 = 127$. The significant is equal to $1 + 1 - 2^{-23} = 2 - 2^{-23}$. The maximum positive number that can be accommodated is

$$+2^{127} \times (2\ 2\ 2^{-23})$$

The smallest positive normalized number has a biased exponent equal to 00000001 and a fraction of all 0's. The exponent is $E = 1 - 127 = -126$, and the significant is equal to 1.0. The smallest positive number that can be accommodated is $+2^{-126}$. The corresponding negative numbers are the same, except that the sign bit is negative. As mentioned before, exponents with all 0's or all 1's (decimal 255) are reserved for the following special conditions:

1. When $e = 255$ and $f = 0$, the number represents plus or minus infinity. The sign is determined from the sign bit s.

2. When $e = 255$ and $f \neq 0$, the representation is considered to be *not a number*, or NaN, regardless of the sign value. NaNs are used to signify invalid operations, such as the multiplication of zero by infinity.

3. When $e = 0$ and $f = 0$, the number denotes plus or minus zero.

4. When $e = 0$, and $f \neq 0$, the number is said to be denormalized. This is the name given to numbers with a magnitude less than the minimum value that is represented in the normalized format.

□ **TABLE 11-6**
Evaluating Biased Exponents

Exponent E in decimal	Biased exponent $e = E + 127$	
	Decimal	**Binary**
−126	−126 + 127 = 1	00000001
−001	−001 + 127 = 126	01111110
000	000 + 127 = 127	01111111
+001	001 + 127 = 128	10000000
+126	126 + 127 = 253	11111101
+127	127 + 127 = 254	11111110

11-8 PROGRAM CONTROL INSTRUCTIONS

The instructions of a program are stored in successive memory locations. When processed by the control, the instructions are read from consecutive memory locations and executed one by one. Each time an instruction is fetched from memory, the *PC* is incremented so that it contains the address of the next instruction in sequence. In contrast, a program control instruction, when executed, may change the address value in the *PC* and cause the flow of control to be altered. The change in the *PC* as a result of the execution of a program control instruction causes a break in the sequence of execution of instructions. This is an important feature of digital computers, since it provides control over the flow of program execution and a capability of branching to different program segments, depending on previous computations.

Some typical program control instructions are listed in Table 11-7. The branch and jump instructions are often used interchangeably to mean the same thing, although sometimes they are used to denote different addressing modes. For example, the jump may use direct or indirect addressing, whereas the branch uses relative addressing. The branch (or jump) is usually a one-address instruction. When executed, the branch instruction causes a transfer of the effective address into the *PC*. Since the *PC* contains the address of the instruction to be executed next, the next instruction will be fetched from the location specified by the effective address.

Branch and jump instructions may be conditional or unconditional. An unconditional branch instruction causes a branch to the specified effective address without any conditions. The conditional branch instruction specifies a condition that must be met in order for the branch to occur, such as the value in a specified register being negative. If the condition is met, the *PC* is loaded with the effective address, and the next instruction is taken from this address. If the condition is not met, the *PC* is not changed, and the next instruction is taken from the next location in sequence.

The skip instruction does not need an address field. A conditional skip instruction will skip the next instruction if the specified condition is met. This is

□ **TABLE 11-7**
Typical Program Control Instructions

Name	Mnemonic
Branch	BR
Jump	JMP
Skip next instruction	SKP
Call procedure	CALL
Return from procedure	RET
Compare (by subtraction)	CMP
Test (by ANDing)	TEST

accomplished by incrementing the *PC* during the execute phase of the instruction, in addition to incrementing it during the fetch phase. If the condition is not met, control proceeds to the next instruction in sequence, at which point the programmer may insert an unconditional branch instruction. Thus, a conditional skip instruction followed by an unconditional branch instruction causes a branch if the condition is not met. This contrasts with a single conditional branch instruction, which causes a branch if the condition *is* met. Since the skip involves the execution of two instructions, it is slower and uses more instruction memory.

The call and return instructions are used in conjunction with procedures. Their performance and implementation are discussed later in this section.

The compare instruction performs a comparison via a subtraction, with the difference not retained. Instead, the comparison causes a conditional branch, changes the contents of a register, or sets or resets stored status bits. Similarly, the test instruction performs the logical AND of two operands without retaining the result and executes one of the actions listed for the compare instruction.

Based on their three possible actions, compare and test instructions are viewed to be of three distinct types, depending upon the way in which conditional decisions are handled. The first type executes the entire decision as a single instruction. For example, the contents of two registers can be compared and a branch or jump taken if the contents are equal. Since there are two register addresses and a memory address involved, such an instruction requires three addresses. The second type of compare and test instruction also uses three addresses, all of which are register addresses. Considering the same example, if the contents of the first two registers are equal, a 1 is placed in the third register. If the contents are not equal, then a 0 is placed in the third register. These two types of instruction avoid the use of stored status bits. In the first case, no such bit is required, and in the second case, a register is used to simulate the presence of a status bit. The third type of compare and test, with the most complex structure, has compare and test operations that set or reset stored status bits. Branch or jump instructions are then used to conditionally change the program sequence. This third type of compare and test instruction is the focus of discussion in the next subsection.

Conditional Branch Instructions

A conditional branch instruction is a branch instruction that may or may not cause a transfer of control, depending on the value of stored bits in the *PSR*. Each conditional branch instruction tests a different combination of status bits for a condition. If the condition is true, control is transferred to the effective address. If the condition is false, the program continues with the next instruction.

Table 11-8 gives a list of conditional branch instructions that depend directly on the bits in the *PSR*. In most cases, the instruction mnemonic is constructed with the letter B (for "branch") and a letter for the name of the status bit. The letter N (for "not") is included if the status bit is tested for a 0 condition. Thus, BC is a branch if carry = 1, and BNC is branch if carry = 0.

The zero status bit *Z* is used to check whether the result of an ALU operation is equal to zero. The carry bit *C* is used to check the carry after the addition

☐ **TABLE 11-8**
Conditional Branch Instructions Relating to Status Bits in the PSR

Branch condition	Mnemonic	Test condition
Branch if zero	BZ	$Z = 1$
Branch if not zero	BNZ	$Z = 0$
Branch if carry	BC	$C = 1$
Branch if no carry	BNC	$C = 0$
Branch if minus	BN	$N = 1$
Branch if plus	BNN	$N = 0$
Branch if overflow	BV	$V = 1$
Branch if no overflow	BNV	$V = 0$

or the borrow after the subtraction of two operands in the ALU. It is also used in conjunction with shift instructions to check the value of the outgoing bit. The sign bit N reflects the state of the leftmost bit of the output from the ALU. $N = 0$ denotes a positive sign and $N = 1$ a negative sign. These instructions can be used to check the value of the leftmost bit, whether it represents a sign or not. The overflow bit V is used in conjunction with arithmetic operations with signed numbers.

As stated previously, the compare instruction performs a subtraction of two operands, say, $A - B$. The result of the operation is not transferred into a destination register, but the status bits are affected. The status bits provide information about the relative magnitude between A and B. Some computers provide special branch instructions that can be applied after the execution of a compare instruction. The specific conditions to be tested depend on whether the two numbers are considered to be unsigned or signed.

The relative magnitude between two unsigned binary numbers A and B can be determined by subtracting $A - B$ and checking the C and Z status bits. Most commercial computers consider the C status bit as a carry after addition and a borrow after subtraction. A borrow occurs when $A < B$ because the most significant position must borrow a bit to complete the subtraction. A borrow does not occur if $A \geq B$, because the difference $A - B$ is positive. The condition for borrowing is the inverse of the condition for carrying when the subtraction is done by taking the 2's complement of B. Computers that use the C status bit as a borrow after a subtraction complement the output carry after adding the 2's complement of the subtrahend and call this bit a borrow. The technique is typically applied to all instructions that use subtraction within the functional unit, not just the subtract instruction. For example, it applies to compare instructions.

The conditional branch instructions for unsigned numbers are listed in Table 11-9. It is assumed that a previous instruction has updated status bits C and Z after a subtraction $A - B$ or some other similar instruction. The words "higher," "lower," and "equal" are used to denote the relative magnitude between two unsigned numbers. The two numbers are equal if $A = B$. This is determined from

□ **TABLE 11-9**
Conditional Branch Instructions for Unsigned Numbers

Branch condition	Mnemonic	Condition	Status bits*
Branch if higher	BH	$A > B$	$C + Z = 0$
Branch if higher or equal	BHE	$A \geq B$	$C = 0$
Branch if lower	BL	$A < B$	$C = 1$
Branch if lower or equal	BLE	$A \leq B$	$C + Z = 1$
Branch if equal	BE	$A = B$	$Z = 1$
Branch if not equal	BNE	$A \neq B$	$Z = 0$

*Note that C here is a borrow bit.

the zero status bit Z, which is equal to 1 because $A - B = 0$. A is lower than B and the borrow $C = 1$ when $A < B$. For A to be lower than or equal to B $(A \leq B)$, we must have $C = 1$ or $Z = 1$. The relationship $A > B$, is the inverse of $A \leq B$ and is detected from the complemented condition of the status bits. Similarly, $A \geq B$ is the inverse of $A < B$, and $A \neq B$ is the inverse of $A = B$.

The conditional branch instructions for signed numbers are listed in Table 11-10. Again, it is assumed that a previous instruction has updated the status bits N, V, and Z after a subtraction $A - B$. The words "greater," "less," and "equal" are used to denote the relative magnitude between two signed numbers. If $N = 0$, the sign of the difference is positive, and A must be greater than or equal to B, provided that $V = 0$, indicating that no overflow occurred. An overflow causes a sign reversal, as discussed in Section 5-4. This means that if $N = 1$ and $V = 1$, there was a sign reversal, and the result should have been positive, which makes A greater than or equal to B. Therefore, the condition $A \geq B$ is true if both N and V are equal to 0 or both are equal to 1. This is the complement of the exclusive-OR operation.

For A to be greater than but not equal to B $(A > B)$, the result must be positive and nonzero. Since a zero result gives a positive sign, we must ensure that the Z bit is 0 to exclude the possibility that $A = B$. Note that the condition $(N \oplus V) + Z = 0$ means that both the exclusive-OR operation and the Z bit must be equal to 0. The other two conditions in the table can be derived in a similar manner. The conditions BE (branch on equal) and BNE (branch on not equal) given for unsigned numbers apply to signed numbers as well and can be determined from $Z = 1$ and $Z = 0$, respectively.

□ **TABLE 11-10**
Conditional Branch Instructions for Signed Numbers

Branch condition	Mnemonic	Condition	Status bits
Branch if greater	BG	$A > B$	$(N \oplus V) + Z = 0$
Branch if greater or equal	BGE	$A \geq B$	$N \oplus V = 0$
Branch if less	BL	$A < B$	$N \oplus V = 1$
Branch if less or equal	BLE	$A \leq B$	$(N \oplus V) + Z = 1$

Procedure Call and Return Instructions

A *procedure* is a self-contained sequence of instructions that performs a given computational task. During the execution of a program, a procedure may be called to perform its function many times at various points in the program. Each time the procedure is called, a branch is made to the beginning of the procedure to start executing its set of instructions. After the procedure has been executed, a branch is made again to return to the main program. A procedure is also called a *subroutine*.

The instruction that transfers control to a procedure is known by different names, including call procedure, call subroutine, jump to subroutine, branch to subroutine, and branch and link. We will refer to the routine containing the procedure call as the calling procedure. The call procedure instruction has a one-address field and performs two operations. First, it stores the value of the *PC*, which is the address following the call procedure instruction, in a temporary location. This address is called the *return address*, and the corresponding instruction is the *continuation point* in the calling procedure. Second, the address in the call procedure instruction—the address of the first instruction in the procedure—is loaded into the *PC*. When the next instruction is fetched, it comes from the called procedure.

The final instruction in every procedure must be a return to the calling procedure. The return instruction takes the address that was stored by the call procedure instruction and places it in the *PC*. This results in a transfer of program execution back to the continuation point in the calling procedure.

Different computers use different temporary locations for storing the return address. Some computers store it in a fixed location in memory, some store it in a processor register, and some store it in a memory stack. The advantage of using a stack for the return address is that, when a succession of procedures are called, the sequential return address can be pushed onto the stack. The return instruction causes the stack to pop, and the contents of the top of the stack are then transferred to the *PC*. In this way, a return is always to the program that last called the procedure. A procedure call instruction using a stack is implemented with the following microoperations:

$$SP \leftarrow SP - 1 \qquad \text{Decrement stack pointer}$$
$$M[SP] \leftarrow PC \qquad \text{Store return address on stack}$$
$$PC \leftarrow \text{Effective address} \qquad \text{Transfer control to procedure}$$

The return instruction is implemented by popping the stack and transferring the return address to the *PC*:

$$PC \leftarrow M[SP] \qquad \text{Transfer return address to } PC$$
$$SP \leftarrow SP + 1 \qquad \text{Increment stack pointer}$$

By using a procedure stack, all return addresses are automatically stored by the hardware in the memory stack. Thus, the programmer does not have to be concerned about managing the return addresses for procedures called from within procedures.

11-9 PROGRAM INTERRUPT

A program interrupt is used to handle a variety of situations that require a departure from the normal program sequence. A program interrupt transfers control from a program that is currently running to another service program as a result of an externally or internally generated request. Control returns to the original program after the service program is executed. In principle, the interrupt procedure is similar to a call procedure, except in three respects:

1. The interrupt is usually initiated at an unpredictable point in the program by an external or internal signal, rather than the execution of an instruction.

2. The address of the service program that processes the interrupt request is determined by a hardware procedure, rather than the address field of an instruction.

3. In response to an interrupt, it is necessary to store information that defines all or part of the contents of the register set, rather than storing only the program counter.

After the computer has been interrupted and the appropriate service program executed, the computer must return to exactly the same state that it was in before the interrupt occurred. Only if this happens will the interrupted program be able to resume exactly as if nothing happened. The state of the computer at the end of an execution of an instruction is determined from the contents of the register set. In addition to containing the condition codes, the *PSR* can specify what interrupts are allowed to occur and whether the computer is operating in user or system mode. Most computers have a resident operating system that controls and supervises all other programs. When the computer is executing a program that is part of the operating system, the computer is placed in system mode, in which certain instructions are privileged and can be executed in the system mode only. The computer is in user mode when it executes user programs, in which case it cannot execute the privileged instructions. The mode of the computer at any given time is determined from a special status bit or bits in the *PSR*.

Some computers store only the program counter when responding to an interrupt. In such computers, the program that performs the data processing for servicing the interrupt must include instructions to store the essential contents of the register set. Other computers store the entire register set automatically in response to an interrupt. Some computers have two sets of processor registers, so that when the program switches from user to system mode in response to an interrupt, it is not necessary to store the contents of processor registers because each computer mode employs its own set of registers.

The hardware procedure for processing interrupts is very similar to the execution of a procedure call instruction. The contents of the register set of the processor are temporarily stored in memory, typically by being pushed onto a memory stack, and the address of the first instruction of the interrupt service program is loaded into the *PC*. The address of the service program is chosen by the hardware. Some computers assign one memory location for the beginning address of the service program:

the service program must then determine the source of the interrupt and proceed to service it. Other computers assign a separate memory location for each possible interrupt source. Sometimes, the interrupt source hardware itself supplies the address of the service routine. In any case, the computer must possess some form of hardware procedure for selecting a branch address for servicing the interrupt.

Most computers will not respond to an interrupt until the instruction that is in the process of being executed is completed. Then, just before going to fetch the next instruction, the control checks for any interrupt signals. If an interrupt has occurred, control goes to a hardware interrupt cycle. During this cycle, the contents of some part or all of the register set are pushed onto the stack. The branch address for the particular interrupt is then transferred to the *PC*, and the control goes to fetch the next instruction, which is the beginning of the interrupt service routine. The last instruction in the service routine is a return from the interrupt instruction. When this return is executed, the stack is popped to retrieve the return address, which is transferred to the *PC* as well as any stored contents of the rest of the register set, which are transferred back to the appropriate registers.

Types of Interrupts

The three major types of interrupts that cause a break in the normal execution of a program are as follows:

1. External interrupts.
2. Internal interrupts.
3. Software interrupts.

External interrupts come from input or output devices, from timing devices, from a circuit monitoring the power supply, or from any other external source. Conditions that cause external interrupts are an input or output device requesting a transfer of data, an external device completing a transfer of data, the time-out of an event, or an impending power failure. A time-out interrupt may result from a program that is in an endless loop and thus exceeds its time allocation. A power failure interrupt may have as its service program a few instructions that transfer the complete contents of the register set of the processor into a nondestructive memory such as a disk in the few milliseconds before power ceases.

Internal interrupts arise from the invalid or erroneous use of an instruction or data. Internal interrupts are also called *traps*. Examples of interrupts caused by internal conditions are an arithmetic overflow, an attempt to divide by zero, an invalid opcode, a memory stack overflow, and a protection violation. A *protection violation* is an attempt to address an area of memory that is not supposed to be accessed by the currently executing program. The service programs that process internal interrupts determine the corrective measure to be taken in each case.

External and internal interrupts are initiated by the hardware of the computer. By contrast, a *software interrupt* is initiated by executing an instruction. The software interrupt is a special call instruction that behaves like an interrupt rather than a procedure call. It can be used by the programmer to initiate an interrupt procedure at any desired point in the program. Typical use of the software interrupt is associated with a system call instruction. This instruction provides a means for switching from user mode to system mode. Certain operations in the computer may be performed by the operating system only in system mode. For example, a complex input or output procedure is done in system mode. In contrast, a program written by a user must run in user mode. When an input or output transfer is required, the user program causes a software interrupt, which stores the contents of the *PSR* (with the mode bit set to "user"), loads new *PSR* contents (with the mode bit set to "system"), and initiates the execution of a system program. The calling program must pass information to the operating system in order to specify the particular task that is being requested.

An alternative term for an interrupt is an *exception*, which may apply only to internal interrupts or to all interrupts, depending on the particular computer manufacturer. As an illustration of the use of the two terms, what one programmer calls interrupt-handling routines may be referred to as exception-handling routines by another programmer.

Processing External Interrupts

External interrupts may have single or multiple interrupt input lines. If there are more interrupt sources than there are interrupt inputs in the computer, two or more sources are ORed to form a common line. An interrupt signal may originate at any time during program execution. To ensure that no information is lost, the computer usually acknowledges the interrupt only after the execution of the current instruction is completed and only if the state of the processor warrants it.

Figure 11-9 shows a simplified external interrupt configuration. Four external interrupt sources are ORed to form a single interrupt input signal. Within the CPU is an enable-interrupt flip-flop (*EI*) that can be set or reset with two program instructions: enable interrupt (ENI) and disable interrupt (DSI). When *EI* is 0, the interrupt signal is neglected. When *EI* is 1 and the CPU is at the end of executing an instruction, the computer acknowledges the interrupt by enabling the interrupt acknowledge output *INTACK*. The interrupt source responds to *INTACK* by providing an interrupt vector address *IVAD* to the CPU. The program-controlled *EI* flip-flop allows the programmer to decide whether to use the interrupt facility. If a DSI instruction to reset *EI* has been inserted in the program, it means that the programmer does not want the program to be interrupted. The execution of an ENI instruction to set *EI* indicates that the interrupt facility will be active while the program is running.

The computer responds to an interrupt request signal if $EI = 1$ and execution of the present instruction is completed. Typical microinstructions that implement the interrupt are as follows:

$$SP \leftarrow SP - 1 \qquad \text{Decrement stack pointer}$$
$$M[SP] \leftarrow PC \qquad \text{Store return address on stack}$$
$$SP \leftarrow SP - 1 \qquad \text{Decrement stack pointer}$$

$M[SP] \leftarrow PSR$	Store processor status word on stack
$EI \leftarrow 0$	Reset enable-interrupt flip-flop
$INTACK \leftarrow 1$	Enable interrupt acknowledge
$PC \leftarrow IVAD$	Transfer interrupt vector address to PC
	Go to fetch phase.

The return address available in the PC is pushed onto the stack, and the PSR contents are pushed onto the stack. EI is reset to disable further interrupts. The program that services the interrupt can set EI with an instruction whenever it is appropriate to enable other interrupts. The CPU assumes that the external source will provide an $IVAD$ in response to an $INTACK$. The $IVAD$ is taken as the address of the first instruction of the program that services the interrupt. Obviously, a program must be written for that purpose and stored in memory.

The return from an interrupt is done with an instruction at the end of the service program that is similar to a return from a procedure. The stack is popped, and the return address is transferred to the PC. Since the EI flip-flop is usually included in the PSR, the value of EI for the original program is returned to EI when the old value of the PSR is returned. Thus, the interrupt system is enabled or disabled for the original program, as it was before the interrupt occurred.

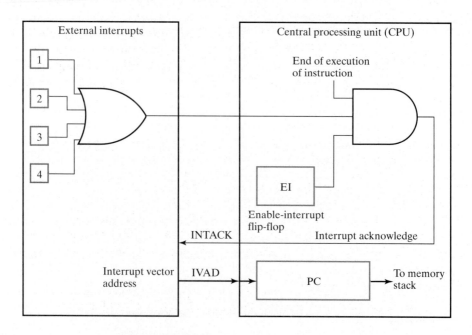

☐ **FIGURE 11-9**
Example of External Interrupt Configuration

11-10 CHAPTER SUMMARY

In this chapter, we defined the concepts of instruction set architecture and the components of an instruction and explored the effects on programs of the maximum address count per instruction, using both memory addresses and register addresses. This led to the definitions of four types of addressing architecture: memory-to-memory, register-to-register, single-accumulator, and stack. Addressing modes specify how the information in an instruction is interpreted in determining the effective address of an operand.

Reduced instruction set computers (RISCs) and complex instruction set computers (CISCs) are two broad categories of instruction set architecture. A RISC has as its goals high throughput and fast execution of instructions. In contrast, a CISC attempts to closely match the operations used in programming languages and facilitates compact programs.

Three categories of elementary instructions are data transfer, data manipulation, and program control. In elaborating data transfer instructions, the concept of the memory stack appears. Transfers between the CPU and I/O are addressed by two different methods: independent I/O, with a separate address space, and memory-mapped I/O, which uses part of the memory address space. Data manipulation instructions fall into three classes: arithmetic, logical, and shift. Floating-point formats and operations handle broader ranges of operand values for arithmetic operations.

Program control instructions include basic unconditional and conditional transfers of control, the latter of which may or may not use condition codes. Procedure calls and returns permit programs to be broken up into procedures that perform useful tasks. Interruption of the normal sequence of program execution is based on three types of interrupts: external, internal, and software. Also referred to as exceptions, interrupts require special processing actions upon the initiation of routines to service them and upon returns to execution of the interrupted programs.

REFERENCES

1. MANO, M. M. *Computer Engineering: Hardware Design.* Englewood Cliffs, NJ: Prentice Hall, 1988.

2. GOODMAN, J., AND K. MILLER *A Programmer's View of Computer Architecture.* Fort Worth, TX: Saunders College Publishing, 1993.

3. HENNESSY, J. L., AND D. A. PATTERSON *Computer Architecture: A Quantitative Approach,* 2nd Ed. San Francisco, CA: Morgan Kaufmann, 1996.

4. MANO, M. M. *Computer System Architecture,* 3rd Ed. Englewood Cliffs, NJ: Prentice Hall, 1993.

5. PATTERSON, D. A., AND J. L. HENNESSY *Computer Organization and Design: The Hardware/Software Interface,* 2nd Ed. San Mateo, CA: Morgan Kaufmann, 1998.

6. *IEEE Standard for Microprocessor Assembly Language.* (IEEE Std 694-1985.) New York, NY: The Institute of Electrical and Electronics Engineers.

7. *IEEE Standard for Binary Floating-Point Arithmetic.* (ANSI/IEEE Std 754-1985.) New York, NY: The Institute of Electrical and Electronics Engineers.

PROBLEMS

The plus (+) indicates a more advanced problem and the asterisk (*) indicates a solution is available on the Companion Website for the text.

11–1. Based on operations illustrated in Section 11-1, write a program to evaluate the arithmetic expression

$$X = (A - B) \times (A + C) \times (B - D)$$

Make effective use of the registers to minimize the number of MOV or LD instructions where possible.

(a) Assume a register-to-register architecture with three-address instructions.

(b) Assume a memory-to-memory architecture with two-address instructions.

(c) Assume a single-accumulator computer with one-address instructions.

11–2. *Repeat Problem 11-1 for

$$Y = (A + B) \times C \div (D - E \times F)$$

All operands are initially in memory and DIV represents divide.

11–3. *A program is to be written for a stack architecture for the arithmetic expression

$$X = (A - B) \times (A + C) \times (B - D)$$

(a) Find the corresponding RPN expression.

(b) Write the program using PUSH, POP, ADD, MUL, SUB, and DIV instructions.

(c) Show the contents of the stack after the execution of each of the instructions.

11–4. Repeat Problem 11-3 for the arithmetic expression

$$(A + B) \times C \div (D - (E \times F))$$

11–5. A two-word instruction is stored in memory at an address designated by the symbol *W*. The address field of the instruction (stored at *W* + 1) is designated by the symbol *Y*. The operand used during the execution of the instruction is stored at an address symbolized by *Z*. An index register contains the value *X*. State how *Z* is calculated from the other addresses if the addressing mode of the instruction is **(a)** direct; **(b)** indirect; **(c)** relative; **(d)** indexed.

11–6. *A two-word relative mode branch-type instruction is stored in memory at location 207 and 208 (decimal). The branch is made to an address equivalent to decimal 195. Let the address field of the instruction (stored at address 208) be designated by X.

(a) Determine the value of X in decimal.

(b) Determine the value of X in binary, using 16 bits. (Note that the number is negative and must be in 2's complement notation. Why?)

11–7. Repeat Problem 11-6 for a branch instruction in locations 143 and 144 and a branch address equivalent to 1000. All values are in decimal.

11–8. How many times does the control unit refer to memory when it fetches and executes a two-word indirect addressing-mode instruction if the instruction is (a) a computational type requiring one operand from a memory location with the return of the result to the same memory location; (b) a branch type?

11–9. An instruction is stored at location 300 with its address field at location 301. The address field has the value 211. A processor register $R1$ contains the number 189. Evaluate the effective address if the addressing mode of the instruction is (a) direct; (b) immediate; (c) relative; (d) register indirect; (e) indexed with $R1$ as the index register.

11–10. *A computer has a 32-bit word length, and all instructions are one word in length. The register file of the computer has 16 registers.

(a) For a format with no mode fields and three register addresses, what is the maximum number of opcodes possible?

(b) For a format with two register address fields, one memory field, and a maximum of 100 opcodes, what is the maximum number of memory address bits available?

11–11. A computer with a register file, but without PUSH and POP instructions, is to be used to implement a stack. The computer does have the following register indirect modes:

Register indirect + increment:

$$\begin{array}{ll} \text{LD R2 R1} & R2 \leftarrow M[R1] \\ & R1 \leftarrow R1 + 1 \\ \text{ST R2 R1} & M[R1] \leftarrow R2 \\ & R1 \leftarrow R1 + 1 \end{array}$$

Decrement + register indirect:

$$\begin{array}{ll} \text{LD R2 R1} & R1 \leftarrow R1 - 1 \\ & R2 \leftarrow M[R1] \\ \text{ST R2 R1} & R1 \leftarrow R1 - 1 \\ & M[R1] \leftarrow R2 \end{array}$$

Show how these instructions can be used to provide the equivalent of PUSH and POP by using the instructions and register $R6$ as the stack pointer.

11–12. A complex instruction, push registers (PSHR), pushes the contents of all of the registers onto the stack. There are eight registers, $R0$ through $R7$, in the CPU. A corresponding instruction, POPR, pops the saved contents of the registers back from the stack into the registers.

(a) Write a register transfer description for the execution of PSHR.

(b) Write a register transfer description for the execution of POPR.

11–13. A computer with an independent I/O system has the input and output instructions

IN R[DR] ADRS

OUT ADRS R[SB]

where ADRS is the address of an I/O register port. Give the equivalent instructions for a computer with memory-mapped I/O.

11–14. *Assume a computer with 8-bit words for the multiple-precision addition of two 32-bit unsigned numbers,

1F C6 24 7B + 00 57 ED 4B

(a) Write a program to execute the addition, using add and add with carry instructions.

(b) Execute the program for the given operands. Each byte is expressed as a 2-digit hexadecimal number.

11–15. Perform the logic AND, OR, and XOR with the two bytes 00110101 and 10111001.

11–16. Given the 16-bit value 1010 1001 0111 1100, what operation must be performed, and what operand is needed, in order to

(a) set the least significant 8 bits to 1's?

(b) complement the bits in odd positions (The leftmost bit is 15 and the rightmost bit is 0)?

(c) clear the bits in odd positions to 0's?

11–17. *An 8-bit register contains the value 01101001, and the carry bit is equal to 1. Perform the eight shift operations given by the instructions listed in Table 11-5 as a sequence of operations on this register.

11–18. Show how the following two floating-point numbers are to be added to get a normalized result:

$$(-.12345 \times 10^{+3}) + (+.71234 \times 10^{-1})$$

11–19. *A 36-bit floating-point number consists of 26 bits plus sign for the fraction and 8 bits plus sign for the exponent. What are the largest and smallest positive nonzero quantities for normalized numbers?

11–20. *A 4-bit exponent uses an excess 7 number for the bias. List all biased binary exponents from +8 through −7.

11–21. The IEEE standard double-precision floating-point operand format consists of 64 bits. The sign occupies 1 bit, the exponent has 11 bits, and the fraction occupies 52 bits. The exponent bias is 1023 and the base is 2. There is an implied bit to the left of the binary point in the fraction. Infinity is represented with a biased exponent equal to 2047 and a fraction of 0.

    (a) Give the formula for finding the decimal value of a normalized number.

    (b) List a few biased exponents in binary, as is done in Table 11-6.

    (c) Calculate the largest and smallest positive normalized numbers that can be accommodated.

11–22. Prove that if the equality $2^x = 10^y$ holds, then $y = 0.3x$. Using this relationship, calculate the largest and smallest normalized floating-point numbers in decimal that can be accommodated in the single-precision IEEE format.

11–23. *It is necessary to branch to ADRS if the bit in the least significant position of the operand in a 16-bit register is equal to 1. Show how this can be done with the TEST (Table 11-7) and BNZ (Table 11-8) instructions.

11–24. Consider the two 8-bit numbers $A = 00101101$ and $B = 01101001$.

    (a) Give the decimal equivalent of each number, assuming that (1) they are unsigned and (2) they are signed 2's complement.

    (b) Add the two binary numbers and interpret the sum, assuming that the numbers are (1) unsigned and (2) signed two's complement.

    (c) Determine the values of the C (carry), Z (zero), N (sign), and V (overflow) status bits after the addition.

    (d) List the conditional branch instructions from Table 11-8 that will have a true condition.

11–25. *The program in a computer compares two unsigned numbers A and B by performing a subtraction $A - B$ and updating the status bits.
Let $A = 01011101$ and $B = 01011100$.

    (a) Evaluate the difference and interpret the binary result.

    (b) Determine the values of status bits C (borrow) and Z (zero).

    (c) List the conditional branch instructions from Table 11-9 that will have a true condition.

11–26. The program in a computer compares two signed 2's complement numbers A and B by performing subtraction $A - B$ and updating the status bits.
Let $A = 11011110$ and $B = 11010110$.

    (a) Evaluate the difference and interpret the binary result.

    (b) Determine the value of status bits N (sign), Z (zero), and V (overflow).

    (c) List the conditional branch instructions from Table 11-10 that will have a true condition.

11-27. *The top of a memory stack contains 3000. The stack pointer *SP* contains 2000. A two-word procedure call instruction is located in memory at address 2000, followed by the address field of 0301 at location 2001. What are the contents of *PC*, *SP*, and the top of the stack

(a) before the call instruction is fetched from memory?

(b) after the call instruction is executed?

(c) after the return from the procedure?

11-28. A computer has no stack, but instead uses register *R7* as a link register (i.e., the computer stores the return address in *R7*).

(a) Show the register transfers for a branch and link instruction.

(b) Assuming that another branch and link is present in the procedure called, what action must be taken by software before the branch and link occurs?

11-29. What are the basic differences between a branch, a procedure call, and a program interrupt?

11-30. *Give five examples of external interrupts and five examples of internal interrupts. What is the difference between a software interrupt and a procedure call?

11-31. A computer responds to an interrupt request signal by pushing onto the stack the contents of the *PC* and the current *PSR*. The computer then reads new *PSR* contents from memory from the location given by the interrupt vector address (*IVAD*). The first address of the service program is taken from memory at location *IVAD* + 1.

(a) List the sequence of microoperations implementing the interrupt.

(b) List the sequence of microoperations implementing the return from interrupt.

12

RISC AND CISC CENTRAL PROCESSING UNITS

T he central processing Unit (CPU) is the key component of a digital computer. Its purpose is to decode instructions received from memory and perform transfer, arithmetic, logic, and control operations with data stored in internal registers, memory, or I/O interface units. Externally, the CPU provides one or more buses for transferring instructions, data, and control information to and from components connected to it.

In the generic computer at the beginning of Chapter 1, the CPU is a part of the processor and is heavily shaded. CPUs, however, may also appear elsewhere in computers. Small, relatively simple computers called microcontrollers are used in computers and in other digital systems to perform limited or specialized tasks. For example, a microcontroller is present in the keyboard and in the monitor in the generic computer; thus, these components are also shaded. In such microcontrollers, the CPU may be quite different from those discussed in this chapter. The word lengths may be short (e. g., eight bits), the number of registers small, and the instruction sets limited. Performance, relatively speaking, is low, but adequate. Most important, the cost of these microcontrollers is very low, making their use cost effective.

The approach in this chapter builds upon and parallels that in Chapter 10. It begins by converting the datapath in Chapter 10 to a pipelined datapath. A pipelined control unit is added to form a reduced instruction set computer (RISC) that is analogous to the single-cycle computer. Problems that arise due to the use of pipelining are introduced and solutions are offered in the context of the RISC design. Next, the control unit is augmented to provide a complex instruction set computer (CISC) that is analogous to the multiple-cycle computer. A brief overview of some of the methods used to enhance pipelined processor performance is presented. Finally, we relate the design ideas discussed to general digital system design.

12-1 PIPELINED DATAPATH

Figure 10-17 was used to illustrate the long delay path present in the single-cycle computer and the resultant clock frequency limit. With a narrower focus, Figure 12-1(a) illustrates maximum delay values for each of the components of a typical datapath. A maximum of 4 ns (3 ns + 1 ns) is required to read two operands from the register file or to read one operand from the register file and obtain a constant from MUX *B*. A maximum of 4 ns is also required to execute an operation in the functional unit. Finally, a maximum of 4 ns (1 ns + 3 ns) is required to write the result back into the register file, including the delay of MUX *D*. Adding these delays, we find that 12 ns are required to perform a single microoperation. The maximum rate at which the microoperations can be performed is the inverse of 12 ns (i.e., 83.3 MHz). This is the maximum frequency at which the clock can be operated, since 12 ns is the smallest clock period that will allow each microoperation to be completed with certainty. As illustrated in Figure 10-17, delay paths that

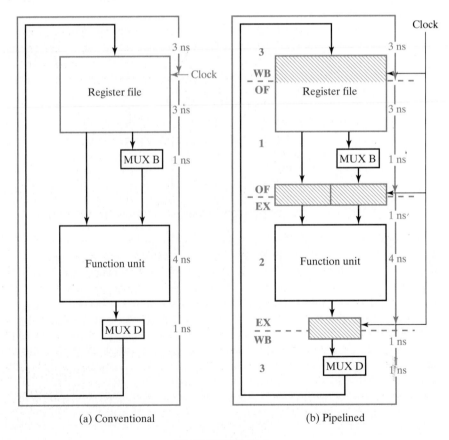

(a) Conventional   (b) Pipelined

☐ **FIGURE 12-1**
Datapath Timing

pass through both the datapath and the control unit limit the clock frequency to an even smaller value. For the datapath alone and for the combination of the datapath and control unit in the single-cycle computer, the execution of a microoperation constitutes the execution of an instruction. Thus, the rate of execution of instructions equals the clock frequency.

Now suppose that the datapath execution rate is not adequate for a particular application, and that there are no faster components available with which to reduce the 12 ns required to complete a microoperation. Still, it may be possible to reduce the clock period and increase the clock frequency. This can be done by breaking up the 12-ns delay path with registers. The resulting datapath, sketched in Figure 12-1(b), is referred to as a *pipelined datapath*, or just a *pipeline*.

Three sets of registers break the delay of the original datapath into three parts. These registers are shown crosshatched in blue. The register file contains the first set of registers. Cross-hatching covers only the top half of the register file, since the lower half is viewed as the combinational logic that selects the two registers to be read. The two registers that store the *A* data from the register file and the output of MUX *B* constitute the second set of registers. The third set of registers stores the inputs to MUX *D*.

The term "pipeline," unfortunately, does not provide the best analogy for the corresponding datapath structure. A better analogy for the datapath pipeline is a production line. A common illustration of a production line is an automated car wash in which cars are pulled through a series of stations at which a particular step of car washing is performed:

1. Wash - Flush with hot, soapy water,
2. Rinse - Flush with plain warm water, and
3. Dry - Blow air over the surface.

In this example, the processing of a vehicle through the car wash consists of three steps and requires a certain amount of time to complete. Using this analogy, the processing of an instruction by a pipeline consists of $n > 2$ steps and requires a certain amount of time to complete. The length of time required to process an instruction is called the *latency time*. Using the car wash analogy, the latency time is the length of time it takes for a car to pass through the three stations performing the three steps of the process. This time remains the same regardless of whether there is a single car or there are three cars in the car wash at a given time.

Continuing this analogy, with the pipeline datapath corresponding to the car wash, what corresponds to the nonpipelined datapath? A car wash with all of the steps available at a single station, with the steps performed serially. We now can compare the analogies, thereby comparing the pipelined and nonpipelined datapath. For the multiple station car wash and the single station car wash, the latencies are approximately the same. So by going to the multiple station car wash, there is no decrease in the time required to wash a car. However, suppose that we consider the frequency at which a washed car emerges from the two types of car washes. For the single station car wash, this frequency is the inverse of the latency time. In contrast, for the multiple station car wash with three stages, a washed car emerges at a frequency of three times the inverse of the latency time. Thus, there is a factor of three improvement in the frequency or rate of delivery of washed cars. Based on the analogy to pipelined datapaths with n stages and nonpipelined datapaths, the former has a processing rate or *throughput* for instructions that is n times that of the latter.

The desired structure, based on the nonpipelined, conventional datapath described in Chapter 10, is sketched in Figure 12-1(b). The operand fetch (OF) is stage 1, the execution (EX) is stage 2, and the write-back (WB) is stage 3. These stages are labeled at their boundaries with appropriate abbreviations. At this point, the analogy breaks down somewhat because the cars move smoothly through the car wash while the data within the pipeline moves synchronously with a clock controlling the movement from stage to stage. This has some interesting implications. First of all, the movement of the data through the pipeline is in discrete steps rather than continuous. Second, the length of time in each of the stages must be the clock period and is the same for all stages. To provide the mechanism separating the stages in the pipeline, registers are placed between the stages of the pipeline. These registers provide temporary storage for data passing through the pipeline and are called *pipeline platforms*.

Returning to the pipelined datapath example in Figure 12-1(b), Stage 1 of the pipeline has the delay required for reading the register file followed by selection by MUX *B*. This delay is 3 plus 1 ns, or 4 ns. Stage 2 of the pipeline has the 1 ns delay of the platform plus the 4 ns delay of the functional unit, giving 5 ns. Stage 3 has the 1 ns delay of the platform, the delay for the selection by MUX *D*, and the delay for writing back into the register file. This delay is $1 + 1 + 3$, for a total of 5 ns. Thus, all flip-flop–to–flip-flop delays are at most 5 ns, allowing a minimum clock period of 5 ns (assuming that the setup times for the flip-flops are zero) and a maximum clock frequency of 200 MHz, compared with the 83.3 MHz for the single state datapath. This clock frequency corresponds to the maximum throughput of the pipeline which is 200 million instructions per second, about 2.4 times that of the nonpipelined datapath. Even though there are three stages, the improvement factor is not three. This is due to two factors: (1) the delay contributed by the pipeline platforms and (2) the differences between the delay of the logic assigned to each stage. The clock period is governed by the longest delay, rather than the average delay assigned to any stage.

A more detailed diagram of the pipelined datapath appears in Figure 12-2. In this diagram, rather than showing the path from the output of MUX *D* to the register file input, the register file is shown *twice*—once in the OF stage, where it is read, and once in the WB stage, where it is written.

The first stage, OF, is the operand fetch stage. The operand fetch consists of reading register values to be used from the register file and, for Bus B, selecting between a register value or a constant by using MUX *B*. Following the OF stage is the first pipeline platform. The pipeline registers store the operand or operands for use in the next stage during the next clock cycle.

The second stage of the pipeline is the execute stage, denoted EX. In this stage, a function unit operation occurs for most microoperations. The results produced from this stage are captured by the second pipeline platform.

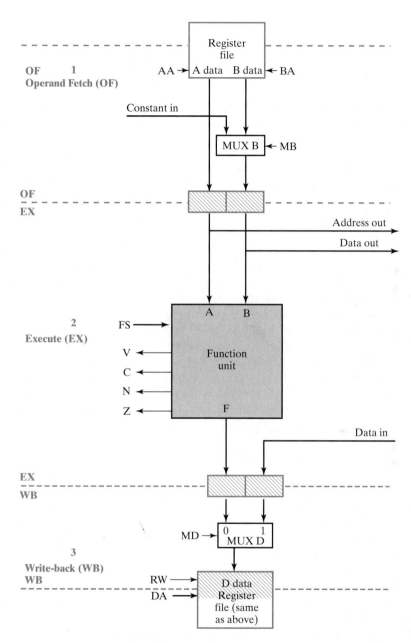

☐ **FIGURE 12-2**

Block Diagram of Pipelined Datapath

The third and final stage of the pipeline is the write-back stage, denoted WB. In this stage, the result saved from the EX stage, or the value on Data in, is selected by MUX D and written back into the register file at the end of the stage. In this case, the write part of the register file is the pipeline platform. The WB stage completes the execution of each microoperation that requires writing to a register.

Before leaving the car wash analogy, we examine the cost of the single-stage car wash versus that of the three-stage car wash. First, even though the three-stage car wash washes vehicles three times as fast as the single-stage car wash does, it costs three times as much in terms of space. Plus, it has the overhead of the mechanism to move the cars along through the stages. So it appears that it is not very cost effective compared with having three single-stage assembly stations operating in parallel. Nevertheless, from a business standpoint, it has proven to be cost effective. In terms of the car wash, can you figure out why? In contrast, for the pipelined datapath, pipeline platforms cut a single datapath into three pieces. Thus, a first order estimate of the cost increase is mainly that of the pipeline platforms.

Execution of Pipeline Microoperations

There are up to three operations at some stage of completion in the car wash at any given time. By analogy, we should be able to have three microoperations at some stage of completion in the pipelined datapath at any given time.

We now examine the execution of this sequence of microoperations with respect to the stages of the pipeline in Figure 12-2. In clock period 1, microoperation 1 is in the OF stage. In clock period 2, microoperation 1 is in the EX stage, and microoperation 2 is in the OF stage. In clock period 3, microoperation 1 is in the WB stage, microoperation 2 is in the EX stage, and microoperation 3 is in the OF stage. So at the end of the third clock period, microoperation 1 has completed execution, microoperation 2 is two-thirds finished, and microoperation 3 is one-third finished. So we have completed $1 + 2/3 + 1/3 = 2.0$ microoperations in three clock periods, or 15 ns. In the conventional datapath, we would have completed execution of microoperation 1 only. So, indeed, the pipelined datapath performance is superior in this example.

The procedure we have been using to analyze the sequence of microoperations so far is somewhat tedious. So to finish the analysis of the timing of the sequence, we will use a *pipeline execution pattern* diagram, as shown in Figure 12-3. Each vertical position in this diagram represents a microoperation to be performed, and each horizontal position represents a clock cycle. An entry in the diagram represents the stage of processing of the microoperation. So, for example, the execution (EX) stage of microoperation 4, which adds the constant 2 to $R0$, occurs in clock cycle 5.

We can see from the overall diagram that the sequence of seven microoperations requires nine clock cycles to execute completely. The time required for execution is $9 \times 5 = 45$ ns, compared to $7 \times 12 = 84$ ns for the conventional datapath. Thus, the sequence of microoperations is executed about 1.9 times faster.

Now let us examine the pipeline execution pattern carefully. In the first two clock cycles, not all of the pipeline stages are active, since the pipeline is *filling*. In the next five clock cycles, all stages of the pipeline are active, as indicated in blue, and the pipeline is fully utilized. In the last two clock cycles, not all stages of the

Clock cycle

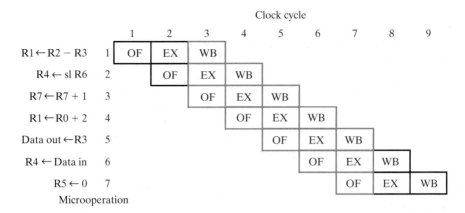

□ **FIGURE 12-3**
Pipeline Execution Pattern for Microoperation Sequence

pipeline are active, since the pipeline is *emptying*. If we want to find the maximum possible improvement of the pipelined datapath over the conventional one, we compare the two when the pipeline is fully utilized. Over these five clock cycles, 3 through 7, the pipeline executes $(5 \times 3) \div 3 = 5$ microoperations in 25 ns. In the same time, the conventional datapath executes $25 \div 12 = 2.083$ microoperations. So the pipelined datapath executes at best $5 \div 2.083 = 2.4$ times as many microoperations in a given time as the conventional datapath. In this ideal situation, we say that the throughput of the pipelined datapath is 2.4 times that of the conventional one. Note that filling and emptying reduce the pipeline speed below the maximum of 2.4. Additional topics associated with pipelines—in particular, providing a control unit for a pipelined datapath and dealing with pipeline hazards—are covered in the next two sections.

12-2 PIPELINED CONTROL

In this section, a control unit is specified to produce a CPU by using the datapath from the last section. Since the instruction must be fetched from a memory as well as executed, we add a stage to the analogous car wash used for illustration in that section. Analogous to the instruction fetch from the instruction memory, the operations in the car wash are specified by order sheets, produced by an attendant, that permit the functions performed in the stages of the car wash to vary. The order sheet, which is analogous to an instruction, accompanies the car as it moves down the line.

Figure 12-4 shows the block diagram of a pipelined computer based on the single-cycle computer. The datapath is that of Figure 12-2. The control has an added stage for instruction fetch that includes the *PC* and instruction memory. This becomes stage 1 of the combined pipeline. The instruction decoder and register file read are now in stage 2, the function unit and data memory read and write are in stage 3, and the register file write is in stage 4. These stages are labeled at their

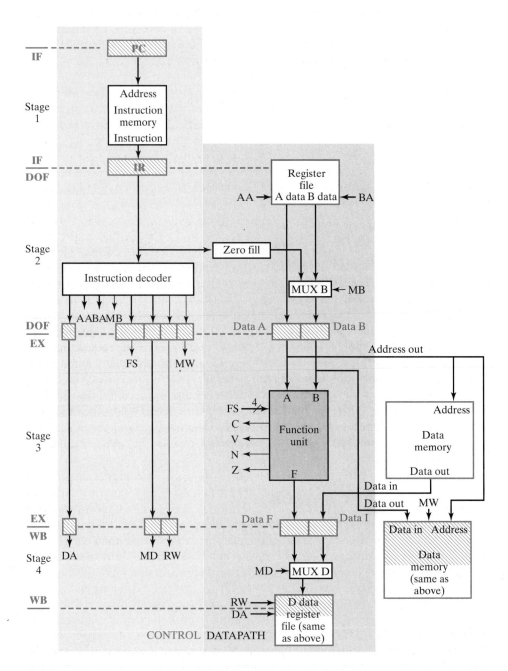

☐ **FIGURE 12-4**
Block Diagram of Pipelined Computer

boundaries with appropriate abbreviations. In the figure, we have added registers to the pipeline platforms between stages, as necessary to pass the decoded instruction information through the pipeline along with the data being processed. These additional registers serve to pass along the instruction information, just as order information was passed along in the car wash.

The added first stage is the instruction fetch stage, denoted by IF, which lies wholly in the control. In this stage, the instruction is fetched from the instruction memory, and the value in the *PC* is updated. Due to additional complexities of handling jumps and branches in a pipelined design, *PC* update is restricted here to an increment, with a more complete treatment provided in the next section. Between the first stage and the second stage is an interstage pipeline platform that plays the role of instruction register, so it has been labeled *IR*.

In the second stage, DOF for decode and operand fetch, decoding of the *IR* into control signals takes place. Among the decoded signals, the register file addresses AA and BA and the multiplexer control signal MB are used in this stage for operand fetch. All other decoded control signals are passed on to the next pipeline platform, to be used later. Following the DOF stage is the second pipeline platform, whose registers store control signals to be used later. The third stage of the pipeline is the execution stage, denoted EX. In this stage, an ALU operation, a shift operation, or a memory operation is executed for most instructions. Thus, the control signals used in this stage are FS and MW. The read part of the data memory *M* is considered a part of the stage. For a memory read, the value of the word addressed is read to Data out from the data memory. All of the results produced from this stage, plus the control signals for the last stage, are captured by the third pipeline platform. The write part of data memory *M* is considered a part of this platform, so a memory write may occur here. The control information held in the final pipeline platform consists of DA, MD, and RW, which are used in the final writeback stage, WB.

The location of the pipeline platforms has balanced the partitioning of the delays, so that the delays per stage are no more that 5 ns. This gives a potential maximum clock frequency of 200 MHz, about 3.4 times that of the single-cycle computer. Note, however, that an instruction takes $4 \times 5 = 20$ ns to execute. This latency of 20 ns compares to that of 17 ns for the single-cycle computer. So if only one instruction at a time is being executed, even fewer instructions are executed per second than for the single-cycle computer.

Pipeline Programming and Performance

If our hypothetical car wash is extended to four stages, there are up to four operations at some stage of completion at any given time. By analogy, then, we should be able to have four instructions at some stage of completion in the pipeline of our computer at any given time. Suppose we consider a simple calculation: Load the constants 1 through 7 into the seven registers $R1$ through $R7$, respectively. The program to do this is as follows (the number on the left is a number to identify the instruction):

1	LDI R1, 1
2	LDI R2, 2
3	LDI R3, 3
4	LDI R4, 4
5	LDI R5, 5
6	LDI R6, 6
7	LDI R7, 7

Let us examine the execution of this program with respect to the stages of the pipeline in Figure 12-4. We employ the pipeline execution pattern diagram shown in Figure 12-5. In clock period 1, instruction 1 is in the IF stage of the pipeline. In clock period 2, instruction 1 is in the DOF stage and instruction 2 is the IF stage. In clock period 3, instruction 1 is in the EX stage, instruction 2 is in the DOF stage, and instruction 3 is in the IF stage. In clock period 4, instruction 1 is in the WB stage, instruction 2 is in the EX stage, instruction 3 is in the DOF stage, and instruction 4 is in the IF stage. So at the end of the fourth clock period, instruction 1 has completed execution, instruction 2 is three-fourths finished, instruction 3 is half finished, and instruction 4 is one-fourth finished. So we have completed 1 + 3/4 + 1/2 + 1/4 = 2.5 instructions in four clock periods, or 20 ns. We can see from the overall diagram that the complete program of seven instructions requires 10 clock cycles to execute. Thus, the time required is 50 ns, compared to 119 ns for the single-cycle computer, and the program is executed about 2.4 times faster.

Now suppose that we examine the pipeline execution pattern carefully. In the first three clock cycles, not all of the pipeline stages are active, since the pipeline is *filling*. In the next four clock cycles, all stages of the pipeline are active, as indicated in blue, and the pipeline is fully utilized. In the last three clock cycles, not all stages of the pipeline are active, since the pipeline is *emptying*. If we want to find the

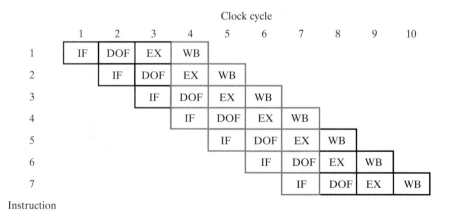

☐ **FIGURE 12-5**
Pipeline Execution Pattern of Register Number Program

maximum possible improvement of the pipelined computer over the single-cycle computer, we compare the two in the situation in which the pipeline is fully utilized. Over these four clock cycles, or 20 ns, the pipeline executes $4 \times 4 \div 4 = 4.0$ instructions. In the same time, the single-cycle computer executes $20 \div 17 = 1.18$ instructions. So in the best case, the pipelined computer executes $4 \div 1.18 = 3.4$ times as many instructions in a given time as the single-cycle computer does. In this ideal situation, we say that the throughput of the pipelined computer is 3.4 times that of the single-cycle computer. Note that even though the pipeline has four stages, the pipelined computer is not four times as fast as the single-cycle computer, because the delays of the latter cannot be divided exactly into four equal pieces and the delays of the added pipeline platforms. Also, filling and emptying the pipeline reduces its speed enough that the speed of the pipelined computer is less than the ideal maximum speed of 3.4 times as fast as the single-cycle computer.

The study of the pipelined computer here, along with the single-cycle computer and multiple-cycle computer in Chapter 10, completes our examination of three computer control organizations. Both the pipelined datapaths and the controls we have studied here are simplified and missing elements. Next we present two CPU designs that illustrate combinations of architectural characteristics of the instruction set, the datapath, and the control unit. The designs are top down, but reuse prior component designs, illustrating the influence of the instruction set architecture on the datapath and control units, and the influence of the datapath on the control unit. The material makes extensive use of tables and diagrams. Although we reuse and modify component designs from Chapter 10, background information from these chapters is not repeated here. Pointers, however, are given to earlier sections of the book, where detailed information can be found.

The two CPUs presented are for a RISC using a pipelined datapath with a hardwired pipelined control unit and a CISC based on the RISC using an auxiliary microprogrammed control unit. These two designs represent two distinct instruction set architectures with architectures using a common pipelined core that contributes enhanced performance.

12-3 THE REDUCED INSTRUCTION SET COMPUTER

The first design we examine is for a reduced instruction set computer with a pipelined datapath and control unit. We begin by describing the RISC instruction set architecture, which is characterized by load/store memory access, four addressing modes, a single instruction format length, and instructions that require only elementary operations. The operations, resembling those that can be performed by the single-cycle computer, can be performed by a single pass through the pipeline. The datapath for implementing the ISA is based on the single-cycle datapath initially described in Figure 10-11 and converted to a pipeline in Figure 12-2. In order to implement the RISC instruction set architecture, modifications are made to the register file and the function unit. These modifications represent the effects of a longer instruction word length and the desire to include multiple position shifts among the elementary operations. The control unit is based on the pipelined control unit in Figure 12-4. Modifications include support for the 32-bit instruction word and a more extensive program counter structure for dealing with branches in the pipeline environment. In response to data and control hazards associated with pipelined designs, additional changes will be made to both the control and datapath to sustain the performance gain achieved by using a pipeline.

Instruction Set Architecture

Figure 12-6 shows the CPU registers accessible to the programmer in this RISC. All registers are 32 bits. The register file has 32 registers, $R0$ through $R31$. $R0$ is a special register that supplies the value zero when used as a source and discards the result when used as a destination. The size of the programmer-accessible register file is comparatively large in the RISC because of the load/store instruction set architecture. Since the data manipulation operations can use only register operands, many active operands need to be present in the register file. Otherwise, numerous stores and loads would be needed to temporarily save operands in the data memory between data manipulation operations. In addition, in many real pipelines, these stores and loads require more than one clock cycle for their execution. To prevent these factors from degrading RISC performance, a larger register file is required.

In addition to the register file, only a program counter, PC, is provided. If stack pointer-based or processor status register-based operations are required, they are simply implemented by sequences of instructions using registers.

Figure 12-7 gives the three instruction formats for the RISC CPU. The formats use a single word of 32 bits. This longer word length is needed to hold realistic address values, since additional instruction words for holding addresses are difficult to accommodate in the RISC CPU. The first format specifies three registers. The two registers addressed by the 5-bit source register fields SA and SB contain the two operands. The third register, addressed by a 5-bit destination register field DR, specifies the register location for the result. A 7-bit OPCODE provides for a maximum of 128 operations.

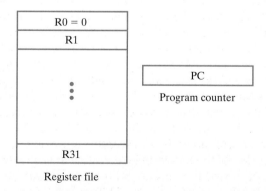

Register file

□ **FIGURE 12-6**
CPU Register Set Diagram for RISC

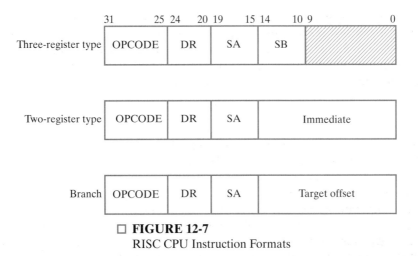

□ **FIGURE 12-7**
RISC CPU Instruction Formats

The remaining two formats replace the second register with a 15-bit constant. In the two-register format, the constant acts as an immediate operand and, in the branch format, the constant is a *target offset*. The *target address* is another name for the effective address, particularly if the address is used in a branch instruction. The target address is formed by adding the target offset to the contents of the *PC*. Thus, branching uses relative addressing based on the updated value of the *PC*. In order to branch backward from the current *PC* location, the offset, regarded as a 2's complement number with sign extension is added to the *PC*. The branch instructions specify source register SA. Whether the branch or jump is taken is based on whether the source register contains zero. The DR field is used to specify the register in which to store the return address for the procedure call. Finally, the rightmost 5 bits of the 15-bit constant are also used as the shift amount SH for multiple bit shifts.

Table 12-1 contains the 27 operations to be performed by the instructions. A mnemonic, an opcode, and a register transfer description are given for each operation. All of the operations are elementary and can be described by a single register transfer statement. The only operations that can access memory are Load and Store. A significant number of immediate instructions help to reduce data memory accesses and speed up execution when constants are employed. Since the immediate field of the instruction is only 15 bits, the leftmost 17 bits must be filled to form a 32-bit operand. In addition to using zero fill for logical operations, a second method used is called *sign extension*. The most significant bit of the immediate operand, bit 14 of the instruction, is viewed as a sign bit. To form a 32-bit 2's-complement operand, this bit is copied into the 17 bits. In Table 12-1, the sign extension of the immediate field is denoted by se *IM*. The same notation, se *IM*, also represents the sign extension of the target offset field discussed previously.

The absence of stored versions of status bits is handled by the use of three instructions: Branch if Zero (BZ), Branch if Nonzero (BNZ), and Set if Less Than (SLT). BZ and BNZ are single instructions that determine whether a register operand is zero or nonzero and branch accordingly. SLT stores a value in register

□ **TABLE 12-1**
RISC Instruction Operations

Operation	Symbolic Notation	Opcode	Action
No Operation	NOP	0000000	None
Move A	MOVA	1000000	$R[DR] \leftarrow R[SA]$
Add	ADD	0000010	$R[DR] \leftarrow R[SA] + R[SB]$
Subtract	SUB	0000101	$R[DR] \leftarrow R[SA] + \overline{R[SB]} + 1$
AND	AND	0001000	$R[DR] \leftarrow R[SA] \wedge R[SB]$
OR	OR	0001001	$R[DR] \leftarrow R[SA] \vee R[SB]$
Exclusive-OR	XOR	0001010	$R[DR] \leftarrow R[SA] \oplus R[SB]$
Complement	NOT	0001011	$R[DR] \leftarrow \overline{R[SA]}$
Add Immediate	ADI	0100010	$R[DR] \leftarrow R[SA] + se\ IM$
Subtract Immediate	SBI	0100101	$R[DR] \leftarrow R[SA] + \overline{(se\ IM)} + 1$
AND Immediate	ANI	0101000	$R[DR] \leftarrow R[SA] \wedge (0 \parallel IM)$
OR Immediate	ORI	0101001	$R[DR] \leftarrow R[SA] \vee (0 \parallel IM)$
Exclusive-OR Immediate	XRI	0101010	$R[DR] \leftarrow R[SA] \oplus (0 \parallel IM)$
Add Immediate Unsigned	AIU	1000010	$R[DR] \leftarrow R[SA] + (0 \parallel IM)$
Subtract Immediate Unsigned	SIU	1000101	$R[DR] \leftarrow R[SA] + \overline{(0 \parallel IM)} + 1$
Move B	MOVB	0001100	$R[DR] \leftarrow R[SB]$
Logical Right Shift by SH Bits	LSR	0001101	$R[DR] \leftarrow lsr\ R[SA]\ by\ SH$
Logical Left Shift by SH Bits	LSL	0001110	$R[DR] \leftarrow lsl\ R[SA]\ by\ SH$
Load	LD	0010000	$R[DR] \leftarrow M[R[SA]]$
Store	ST	0100000	$M[R[SA]] \leftarrow R[SB]$
Jump Register	JMR	1110000	$PC \leftarrow R[SA]$
Set if Less Than	SLT	1100101	If $R[SA] < R[SB]$ then $R[DR] = 1$
Branch on Zero	BZ	1100000	If $R[SA] = 0$, then $PC \leftarrow PC + 1 + se\ IM$
Branch on Nonzero	BNZ	1010000	If $R[SA] \neq 0$, then $PC \leftarrow PC + 1 + se\ IM$
Jump	JMP	1101000	$PC \leftarrow PC + 1 + se\ IM$
Jump and Link	JML	0110000	$PC \leftarrow PC + 1 + se\ IM, R[DR] \leftarrow PC + 1$

$R[DR]$ that acts like a negative status bit. If $R[SA]$ is less than $R[SB]$, a 1 is placed in register $R[DR]$; if $R[SA]$ is greater than or equal to $R[SB]$, a 0 is placed in $R[DR]$. The register $R[DR]$ can then be examined by a subsequent instruction to see whether it is zero (0) or nonzero (1). Thus, using two instructions, the relative values of two operands or the sign of one operand (by letting $R[SB]$ equal $R0$) can be determined.

The Jump and Link (JML) instruction provides a mechanism for implementing procedures. The value in the PC after updating is stored in register $R[DR]$, and then the sum of the PC and the sign-extended target offset from the instruction is placed in the PC. The return from a called procedure can use the Jump Register instruction with SA equal to DR for the calling procedure. If a procedure is to be called from within a called procedure, then each successive procedure that is called will need its own register for storing the return value. A software stack that moves return addresses from $R[DR]$ to memory at the beginning of a called procedure and restores them to $R[SA]$ before the return can also be used.

Addressing Modes

The four addressing modes in the RISC are register, register indirect, immediate, and relative. The mode is specified by the operation code, rather than by a separate mode field. As a consequence, the mode for a given operation is fixed and cannot be varied. The three-operand data manipulation instructions use register mode addressing. Register indirect, however, applies only to the load and store instructions, the only instructions that access data memory. Instructions using the two-register format have an immediate value that replaces register address SB. Relative addressing applies exclusively to branch and jump instructions and so produces addresses only for the instruction memory.

When programmers want to use an addressing mode not provided by the instruction set architecture, such as indexed addressing, they must use a sequence of RISC instructions. For example, for an indexed address for a load operation, the desired transfer is

$$R15 \leftarrow M[R5 + 0 \parallel I]$$

This transfer can be accomplished by executing two instructions:

AIU R9, R5, I

LD   R15, R9

The first instruction, Add Immediate Unsigned, forms the address by appending 17 0's to the left of I and adding the result to $R5$. The resulting effective address is then temporarily stored in $R9$. Next, the Load instruction uses the contents of $R9$ as the address at which to fetch the operand and places the operand in the destination register $R15$. Since, for indexed addressing, I is regarded as a positive offset in memory, the use of unsigned addition is appropriate. Sequences of operations for implementing addressing modes is the primary justification for having unsigned immediate addition available.

Datapath Organization

The pipelined datapath in Figure 12-2 serves as the basis for the datapath here, and we deal only with modifications. These modifications affect the register file, the function unit, and the bus structure. The reader should also refer to the datapath in Figure 12-2 and the new datapath shown in Figure 12-8 in order to understand fully the discussion that follows. We treat each modification in turn, beginning with the register file.

In Figure 12-2, there are 16 16-bit registers, and all registers are identical in function. In the new datapath, there are 32 32-bit registers. Also, reading register

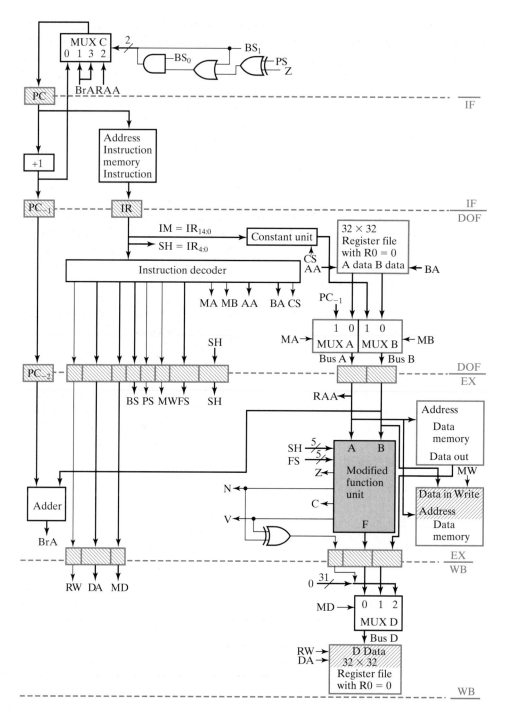

☐ **FIGURE 12-8**
Pipelined RISC CPU

$R0$ gives a constant value of zero. If a write is attempted into $R0$, the data will be lost. These changes are implemented in the new register file in Figure 12-8. All data inputs and the data output are 32 bits. To correspond to the 32 registers, the address inputs are five bits. The fixed value of 0 in $R0$ is implemented by replacing the storage elements for $R0$ with open circuits on the lines that were their inputs, and with constant zero values on the lines that were their outputs.

A second major modification to the datapath is the replacement of the single-bit position shifter with a barrel shifter to permit multiple-position shifting. This barrel shifter can perform a logical right or logical left shift of from 0 to 31 positions. A block diagram for the barrel shifter appears in Figure 12-9. The data input is 32-bit operand A, and the output is 32-bit result G. Left/$\overline{\text{right}}$, a control signal decoded from OPCODE, selects a left or right shift. The shift amount field $SH = IR(4:0)$ specifies the number of bit positions to shift the data input and takes on values from 0 through 31. A logical shift of p bit positions involves inserting p zeros into the result. In order to provide these zeros and simplify the design of the shifter, we will perform both the left and right shift by using a right rotate. The input to this rotate will be the input data A with 32 zeros concatenated to its left. A right shift is performed by rotating the input p positions to the right; a left shift is performed by rotating $64 - p$ positions to the right. This number of positions can be obtained by taking the 2's complement of the 6-bit value of $0 \parallel SH$.

The 63 different rotates can be obtained by using three levels of 4-to-1 multiplexers, as shown in Figure 12-8. The first level shifts by 0, 16, 32, or 48 positions, the second level by 0, 4, 8, or 12 positions, and the third level by 0, 1, 2, or 3 positions. The number of positions for A to be shifted, 0 through 63, can be implemented by representing $0 \parallel SH$ as a three-digit base-4 integer. From left to right, the digits have weights $4^2 = 16$, $4^1 = 4$, and $4^0 = 1$. The digit values in each of the positions are 0, 1, 2, and 3. Each digit controls a level of the 4-to-1 multiplexers, the

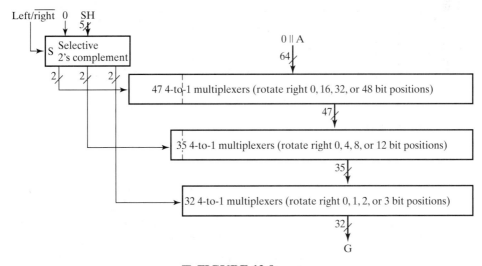

□ **FIGURE 12-9**
32-bit Barrel Shifter

most significant digit controlling the first level, the least significant the third level. Due to the presence of 32 zeros in the 64-bit input, fewer than 64 multiplexers can be used in each level. A level requires the number of multiplexers to be 32 plus the total number of positions its output can be shifted by subsequent levels. The output of the first level can be shifted at most $12 + 3 = 15$ positions to the right. Thus, this level requires $32 + 15 = 47$ multiplexers. The output of the second level can be shifted at most 3 positions, giving $32 + 3 = 35$ multiplexers. The final level cannot be shifted further and so needs just 32 multiplexers.

In the function unit, the ALU is expanded to 32 bits, and the barrel shifter replaces the single position shifter. The resulting modified function unit uses the same function codes as in Chapter 10, except that the two codes for shifts are now labeled as logical shifts, and some codes are not used. The shift amount SH is a new 5-bit input to the modified function unit in Figure 12-8.

The remaining datapath changes are shown in Figure 12-8. Beginning at the top of the datapath, zero fill has been replaced by the constant unit. The constant unit performs zero fill for $CS = 0$ and sign extension for $CS = 1$. MUX A is added to provide a path for the updated PC, PC_{-1}, to the register file for implementation of the Jump and Link (JML) instruction.

One other change in the figure helps implement the Set if Less Than (SLT) instruction. This logic provides a 1 to be loaded into $R[DA]$ if $R[AA] - R[BA] < 0$ and a 0 to be loaded into $R[DA]$ if $R[AA] - R[BA] \geq 0$. It is implemented by adding an additional input to MUX D. The leftmost 31 bits of the input are 0; the rightmost bit is 1 if N is 1 and V is 0 (i.e., if the result of the subtraction is negative and there is no overflow). It is also 1 if N is 0 and V is 1 (i.e., if the result of the subtraction is positive and there is an overflow). These represent all cases in which $R[AA]$ is greater than $R[BA]$ and can be implemented using an exclusive-OR of N and V.

A final difference in the datapath is that the register file is no longer edge triggered and is no longer a part of a pipeline platform at the end of the write-back (WB) stage. Instead, the register file uses latches and is written much earlier than the positive clock edge. Special timing signals are provided that permit the register file to be written in the first half and to be read in the last half of the cycle. In particular, in the second half of the cycle, it is possible to read data written into the register file during the first half of the same clock cycle. This is called a *read-after-write* register file, and it both avoids added complexity in the logic used for handling hazards and reduces the cost of the register file.

Control Organization

The control organization in the RISC is modified from that in Figure 12-4. The modified instruction decoder is essential to deal with the new instruction set. In Figure 12-8, SH is added as an *IR* field, a 1-bit *CS* field is added to the instruction decoder, and MD is expanded to two bits. There is a new pipeline platform for SH, and expanded 2-bit platforms for MD.

The remaining control signals are included to handle the new control logic for the *PC*. This logic permits the loading of addresses into the *PC* for implementing branches and jumps. MUX C selects from three different sources for the next value of *PC*. The updated *PC* is used to move sequentially through a program. The branch target address *BrA* is formed from the sum of the updated *PC* value for the branch instruction and the sign-extended target offset. The value in $R[AA]$ is used for the register jump. The selection of these values is controlled by the field BS. The effects of BS are summarized in Table 12-2. If $BS_0 = 0$, then the updated *PC* is selected by $BS_1 = 0$, and $R[AA]$ is selected by $BS_1 = 1$. If $BS_0 = 1$ and $BS_1 = 1$, then *BrA* is selected unconditionally. If $BS_0 = 1$ and $BS_1 = 0$, then, for $PS = 0$, a branch to *BrA* occurs for $Z = 1$, and for $PS = 1$, a branch to *BrA* occurs for $Z = 0$. This implements the two conditional branch instructions BZ and BNZ.

In order to have the value of the updated *PC* for the branch and jump instructions when they reach the execution stage, two pipeline registers, PC_{-1} and PC_{-2}, are added. PC_{-2} and the value from the constant unit are inputs to the dedicated adder that forms *BrA* in the execution stage. Note that MUX *C* and the attached control logic are in the EX stage, although shown above the *PC*. The related clock cycle difference causes problems with instructions following branches that we will deal with in later subsections.

The heart of the control unit is the instruction decoder. This is combinational circuitry that converts the operation code in the *IR* into the control signals necessary for the datapath and control unit. In Table 12-3, each instruction is identified by its mnemonic. A register transfer statement and the opcode are given for the instruction. The opcodes are selected such that the least significant four of the seven bits match the bits in the control field FS whenever it is used. This leads to simpler decoding. The register file addresses AA, BA, and DA come directly from SA, SB, and DR, respectively, in the *IR*.

Otherwise, to determine the control codes, the CPU is viewed much as is the single-cycle CPU in Figure 10-15. The pipeline platforms can be ignored in this determination; however, it is important to examine the timing carefully to be sure that various parts of the register transfer statement for the operation take place in the right stage of the pipeline. For example, note that the adder for the *PC* is in stage EX. This adder is connected to MUX *C* and its attached control logic, and to the incrementer +1 for the *PC*. Thus, all of this logic is in the EX stage, and the loading

□ **TABLE 12-2**
Definition of Control Fields BS and PS

Register Transfer	BS Code	PS Code	Comments
$PC \leftarrow PC + 1$	00	X	Increment *PC*
$Z: PC \leftarrow BrA, \overline{Z}: PC \leftarrow PC + 1$	01	0	Branch on Zero
$\overline{Z}: PC \leftarrow BrA, Z: PC \leftarrow PC + 1$		1	Branch on Nonzero
$PC \leftarrow R[AA]$	10	X	Jump to Contents of $R[AA]$
$PC \leftarrow BrA$	11	X	Unconditional Branch

☐ **TABLE 12-3**
Control Words for Instructions

Symbolic Notation	Action	Op Code	RW	MD	BS	PS	MW	FS	MB	MA	CS
								Control Word Values			
NOP	None	0000000	0	XX	00	X	0	XXXX	X	X	X
MOVA	$R[DR] \leftarrow R[SA]$	1000000	1	00	00	X	0	0000	X	0	X
ADD	$R[DR] \leftarrow R[SA] + R[SB]$	0000010	1	00	00	X	0	0010	0	0	X
SUB	$R[DR] \leftarrow R[SA] + \overline{R[SB]} + 1$	0000101	1	00	00	X	0	0101	0	0	X
AND	$R[DR] \leftarrow R[SA] \wedge R[SB]$	0001000	1	00	00	X	0	1000	0	0	X
OR	$R[DR] \leftarrow R[SA] \vee R[SB]$	0001001	1	00	00	X	0	1001	0	0	X
XOR	$R[DR] \leftarrow R[SA] \oplus R[SB]$	0001010	1	00	00	X	0	1010	0	0	X
NOT	$R[DR] \leftarrow \overline{R[SA]}$	0001011	1	00	00	X	0	1011	X	0	X
ADI	$R[DR] \leftarrow R[SA] + se\,IM$	0100010	1	00	00	X	0	0010	1	0	1
SBI	$R[DR] \leftarrow R[SA] + \overline{(se\,IM)} + 1$	0100101	1	00	00	X	0	0101	1	0	1
ANI	$R[DR] \leftarrow R[SA] \wedge zf\,IM$	0101000	1	00	00	X	0	1000	1	0	0
ORI	$R[DR] \leftarrow R[SA] \vee zf\,IM$	0101001	1	00	00	X	0	1001	1	0	0
XRI	$R[DR] \leftarrow R[SA] \oplus zf\,IM$	0101010	1	00	00	X	0	1010	1	0	0
AIU	$R[DR] \leftarrow R[SA] + zf\,IM$	1000010	1	00	00	X	0	0010	1	0	0
SIU	$R[DR] \leftarrow R[SA] + \overline{(zf\,IM)} + 1$	1000101	1	00	00	X	0	0101	1	0	0
MOVB	$R[DR] \leftarrow R[SB]$	0001100	1	00	00	X	0	1100	0	X	X
LSR	$R[DR] \leftarrow lsr\,R[SA]\,by\,SH$	0001101	1	00	00	X	0	1101	X	0	X
LSL	$R[DR] \leftarrow lsl\,R[SA]\,by\,SH$	0001110	1	00	00	X	0	1110	X	0	X
LD	$R[DR] \leftarrow M[R[SA]]$	0010000	1	01	00	X	0	XXXX	X	0	X
ST	$M[R[SA]] \leftarrow R[SB]$	0100000	0	XX	00	X	1	XXXX	0	0	X
JMR	$PC \leftarrow R[SA]$	1110000	0	XX	10	X	0	XXXX	X	0	X
SLT	If $R[SA] < R[SB]$ then $R[DR] = 1$	1100101	1	10	00	X	0	0101	0	0	X
BZ	If $R[SA] = 0$, then $PC \leftarrow PC + 1 + se\,IM$	1100000	0	XX	01	0	0	0000	1	0	1
BNZ	If $R[SA] \neq 0$, then $PC \leftarrow PC + 1 + se\,IM$	1010000	0	XX	01	1	0	0000	1	0	1
JMP	$PC \leftarrow PC + 1 + se\,IM$	1101000	0	XX	11	X	0	XXXX	1	X	1
JML	$PC \leftarrow PC + 1 + se\,IM, R[DR] \leftarrow PC + 1$	0110000	1	00	11	X	0	0000	1	1	1

of the *PC* that begins the IF stage is controlled from the EX stage. Likewise, the input $R[AA]$ is in the same combinational block of logic and comes not from the *A* Data output of the register file, but from Bus *A* in the EX stage, as shown.

Table 12-3 can serve as the basis for the design of the instruction decoder. It contains the values for all control signals, except the register addresses from *IR*. In contrast to the instruction decoder in Section 10-8, the logic is complex and is most easily designed by using a computer-based logic synthesis program.

Data Hazards

In Section 12-1, we examined a pipeline execution diagram and found that filling and flushing of the pipeline reduced the throughput below the maximum level achievable. Unfortunately, there are other problems with pipeline operation that reduce throughput. In this and the next subsection, we will examine two such problems: data hazards and control hazards. Hazards are timing problems that arise because the execution of an operation in a pipeline is delayed by one or more clock cycles from the time at which the instruction containing the operation was fetched. If a subsequent instruction tries to use the result of the operation as an operand before the result is available, it uses the old or stale value, which is very likely to give a wrong result. To deal with data hazards, we present two solutions, one that uses software and another that uses hardware.

Two data hazards are illustrated by examining the execution of the following program:

  1 MOVA R1, R5

  2 ADD   R2, R1, R6

  3 ADD   R3, R1, R2

The execution diagram of this program appears in Figure 12-10(a). The MOVA instruction places the contents of $R5$ into $R1$ in the first half of WB in cycle 4. But,

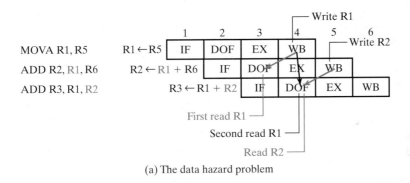

(a) The data hazard problem

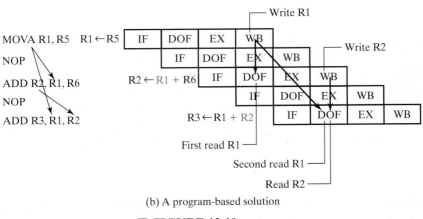

(b) A program-based solution

□ **FIGURE 12-10**
Example of Data Hazard

as shown by the blue arrow, the first ADD instruction reads $R1$ in the last half of DOF in cycle 3, one cycle before it is written. Thus, the ADD instruction uses the stale value in $R1$. The result of this operation is placed in $R2$ in the first half of WB in cycle 5. The second ADD instruction, however, reads both $R1$ and $R2$ in the second half of DOF in cycle 4. In the case of $R1$, the value read was written in the first half of WB in cycle 4. So the value read in the second half of cycle 4 is the new value. The write-back of $R2$, however, occurs in the first half of cycle 5, after it is read by the next instruction during cycle 4. So $R2$ has not been updated to the new value at the time it is read. This gives two data hazards, as indicated by the large blue arrows in the figure. The registers that are not properly updated to new values are highlighted in blue in the program and in the register transfer statements in the figure. In each of these cases, the read of the involved register occurs one clock cycle too soon with respect to the write of that register.

One possible remedy for data hazards is to have the compiler or programmer generate the machine code to delay instructions so that new values are available. The program is written so that any pending write to a register occurs in the same or an earlier clock cycle than a subsequent read from the register. To accomplish this, the programmer or compiler needs to have detailed information on how the pipeline operates. Figure 12-10(b) illustrates a modification of the simple three-line program that solves the problem. No-operation (NOP) instructions are inserted between the first and second instructions, and between the second and third instructions to delay the respective reads relative to the writes by one clock cycle. The execution diagram shows that, at worst, this approach has writes and subsequent reads in the same clock cycle. This is indicated by the pairs consisting of a register write and a subsequent register read connected by a black arrow in the diagram. Because of the read-after-write assumption for the register file, the timing shown permits the program to be executed on correct operands.

This approach solves the problem, but what is the cost? First of all, the program is obviously longer, although it may be possible to place other, unrelated instructions in the NOP positions instead of just wasting them. Also, the program takes two clock cycles longer and reduces the throughput from 0.5 instruction per cycle to 0.375 instruction per cycle with the NOPs in place.

Figure 12-11 illustrates an alternative solution involving added hardware. Instead of the programmer or compiler putting NOPs in the program, the hardware inserts the NOPs automatically. When an operand is found at the DOF stage that has not been written back yet, the associated execution and write-back are delayed by stalling the pipeline flow in IF and DOF for one clock cycle. Then the flow resumes with completion of the instruction when the operand becomes available, and a new instruction is fetched as usual. The delay of one cycle is enough to permit the result to be written before it is read as an operand.

When the actions associated with an instruction flowing through the pipe are prevented from happening at a given point, the pipeline is said to contain a *bubble* in subsequent clock cycles and stages for that instruction. In Figure 12-11, when the flow for the first ADD instruction is prevented beyond the DOF stage, in the next two clock cycles a bubble passes through the EX and the WB stages, respectively. The holding of the pipeline flow in the IF and DOF stages delays the

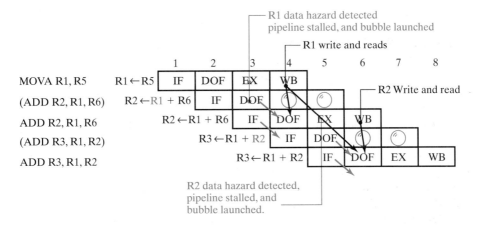

□ **FIGURE 12-11**

Example of Data Hazard Stall

microoperations taking place in these stages for one clock cycle. In the figure, this delay is represented by two diagonal blue arrows from the initial location in which the completion of the microoperation is prevented to the location one clock cycle later in which the microoperation is performed. When the pipeline flow is held in IF and DOF for an extra clock cycle, the pipeline is said to be *stalled*, and if the cause of the stall is a data hazard, then the stall is referred to as a *data hazard stall*.

An implementation of data hazard handling for the pipelined RISC that uses data hazard stalls is presented in Figure 12-12. The added or modified hardware is shown in the areas shaded in light blue. For this particular pipeline stage arrangement, a data hazard will occur for a register file read if there is a destination register at the execution stage that is to be written back in the next clock cycle and that is to be read at the current DOF stage as either of the two operands. So we have to determine whether such a register exists. This is done by evaluating the Boolean equations

$$HA = \overline{MA_{\mathrm{DOF}}} \cdot (DA_{\mathrm{EX}} = AA_{\mathrm{DOF}}) \cdot RW_{\mathrm{EX}} \cdot \sum_{i=0}^{4} (DA_{\mathrm{EX}})_i$$

$$HB = \overline{MB_{\mathrm{DOF}}} \cdot (DA_{\mathrm{EX}} = BA_{\mathrm{DOF}}) \cdot RW_{\mathrm{EX}} \cdot \sum_{i=0}^{4} (DA_{\mathrm{EX}})_i$$

and

$$DHS = HA + HB$$

The following events must all occur for *HA*, which represents a hazard for the *A* data, to equal 1:

1. *MA* in the DOF stage must be 0, meaning that the *A* operand is coming from the register file.

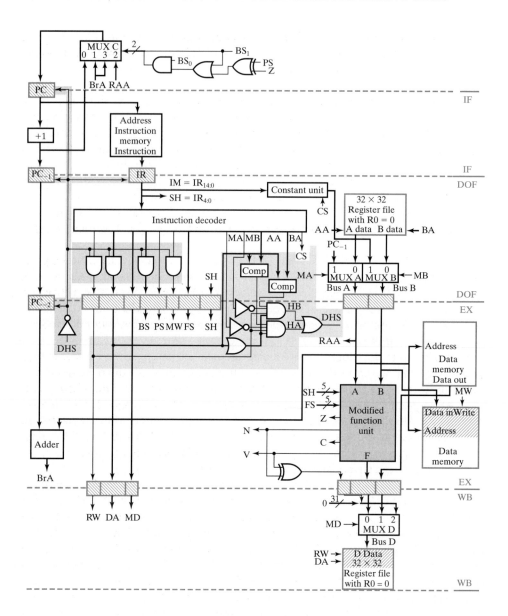

☐ **FIGURE 12-12**
Pipelined RISC: Data Hazard Stall

2. *AA* in the DOF stage equals *DA* in the EX stage, meaning that there is potentially a register being read in the DOF stage that is to be written in the next clock cycle.

3. *RW* in the EX stage is 1, meaning that register *DA* in the EX stage will definitely be written in WB during the next clock cycle.

4. The OR (Σ) of all bits of DA is 1, meaning that the register to be written is not $R0$ and so is a register that must be written before being read. ($R0$ has the same value 0 regardless of any writes to it.)

If all these conditions hold, there is a write pending for the next clock cycle to a register that is the same as one being read and used on Bus A. Thus, a data hazard exists for the A operand from the register file. HB represents the same combination of events for the B data. If either of the HA or HB terms equals 1, there is a data hazard and DHS is 1, meaning that a data hazard stall is required.

The logic implementing the preceding equations is shown in the shaded area in the center of Figure 12-12. The blocks marked "Comp" are equality comparators that have output 1 if and only if the two 5-bit inputs are equal. The OR gate with DA entering it ORs together the five bits of DA and has output 1 as long as DA is not 00000 ($R0$).

DHS is inverted and the inverted signal is used to initiate a bubble in the pipeline for the instruction currently in the IR, as well as to stop the PC and IR from changing. The bubble, which prevents actions from occurring as the instruction passes through the EX and WB stages, is produced by using AND gates to force RW and MW to 0. These 0s prevent the instruction from writing the register file and the memory. AND gates also force BS to 0 causing the PC to be incremented instead of loaded during the EX stage for a jump register or branch instruction affected by a data hazard. Finally, to prevent the data stall from continuing for the next and subsequent clock cycles, AND gates force DA to 0 so that it appears that $R0$ is being written, giving a condition which does not cause a stall. The registers to remain unchanged in the stall are the PC, the PC_{-1}, PC_{-2}, and the IR. These registers are replaced with registers with load control signals driven by $\overline{DHS}$. When $\overline{DHS}$ goes to 0, requesting a stall, the load signals become 0 and these pipeline platform registers hold their contents unchanged for the next clock cycle.

Returning to Figure 12-12, we see that in cycle 3 the data hazard for $R1$ is detected, so that $\overline{DHS}$ goes to 0 before the next clock edge. RW, MW, BS, and DA are set to 0, and at the clock edge, a bubble is launched into the EX stage for the ADD. At the same clock edge, the IF and DOF stages are stalled, so the information in them now is associated with clock cycle 4 instead of 3. In clock cycle 4, since DA_{EX} is 0, there is no stall, so the execution of the stalled ADD instruction proceeds. The same sequence of events occurs for the next ADD. Note that the execution diagram is identical to that in Figure 12-10(b), except that the NOPs are replaced by stalled instructions, shown in parentheses. Thus, although it removes the need for programming NOPs into the software, the data hazard stall solution has the same throughput penalty as the program with the NOPs.

A second hardware solution, *data forwarding*, does not have this penalty. Data forwarding is based on the answer to the following question: When a data hazard is detected, is the result available somewhere else in the pipeline, so that it can be used immediately in the operation having the data hazard? The answer is "almost." The result will be on Bus D, but it is not available until the next clock cycle. The result is to be written into the destination register during that clock cycle. The information needed to form the result, however, is available on the

inputs to the pipeline platform that provides the inputs to MUX D. All that is
needed to form the result during the current clock cycle is a multiplexer to select
from the three values, just as MUX D does. MUX D' is accordingly added to produce the result on Bus D'. In Figure 12-13, instead of reading the operand from
the register file, we use data forwarding to replace the operand with the value on
Bus D'. This replacement is implemented with an additional input to MUX A and
to MUX B from Bus D' as shown. Essentially the same logic as before is used to
detect the data hazard, except that the separate detection signals HA and HB are

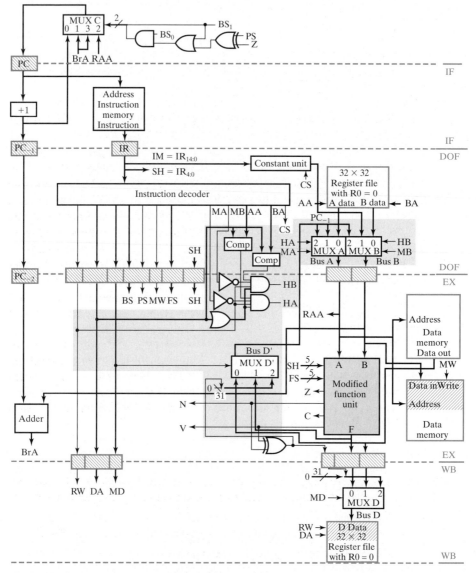

□ **FIGURE 12-13**
Pipeline RISC: Data Forwarding

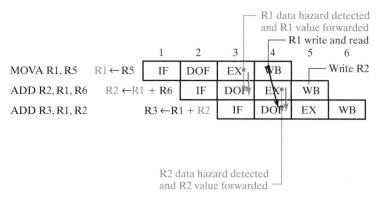

□ **FIGURE 12-14**
Example of Data Forwarding

used directly for *A* data and *B* data, respectively, so that the replacement occurs for the operand that has the data hazard.

The data-forwarding execution diagram for the three-instruction example appears in Figure 12-14. The data hazard for *R*1 is detected in cycle 3. This causes the value to go into *R*1 in the next cycle, to be forwarded from the EX stage of the first instruction in cycle 3. The correct value of *R*1 enters the DOF/EX platform at the next clock edge so that execution of the first ADD can proceed normally. The data hazard for *R*2 is detected in cycle 4, and the correct value is forwarded from the EX stage of the second instruction in that cycle. This gives the correct value in the DOF/EX platform needed for the second ADD to proceed normally. In contrast to the data hazard stall method, data forwarding does not increase the number of clock cycles required to execute the program and hence does not affect the throughput in terms of the number of clock cycles required. It may, however, add combinational delay, causing the clock period to be somewhat longer.

Data hazards can also occur with memory access, as well as with register access. For the ST and LD instructions, it is not likely that a data memory read can be performed after a write in a single clock cycle. Further, some memory reads may take more than one clock cycle, in contrast to what we have assumed here. Thus, the reduction in throughput for a data hazard may be increased due to a longer delay before the data is available.

Control Hazards

Control hazards are associated with branches in the control flow of the program. The following program containing a conditional branch illustrates a control hazard:

1	BZ	R1, 18
2	MOVA	R2, R3
3	MOVA	R1, R2
4	MOVA	R4, R2
20	MOVA	R5, R6

The execution diagram for this program is given in Figure 12-15(a). If $R1$ is zero, then a branch to the instruction in location 20 (recall that addressing is PC relative) is to occur, skipping the instructions in locations 2 and 3. If $R1$ is nonzero, then the instructions in locations 2 and 3 are to be executed in sequence. Assume that the branch is taken to location 20 because $R1$ is equal to zero. The fact that $R1$ equals 0 is not detected until EX in cycle 3 of the first instruction in Figure 12-15(a). So the PC is set to 20 on the clock edge at the end of cycle 3. But the MOVA instructions in locations 2 and 3 are into the EX and DOF stages, respectively, after the clock edge. Thus, unless corrective action is taken, these instructions will complete execution, even though the programmer's intention was for them to be skipped. This situation is one form of a *control hazard*.

NOP instructions can be used to deal with control hazards just as they were used with data hazards. The insertion of NOPs is performed by the programmer or compiler generating the machine language program. The program must be written so that only operations intended to be performed, regardless of whether the branch is taken, are introduced into the pipeline before the branch execution actually occurs. Figure 12-15(b) illustrates a modification of the simple three-line program that satisfies this condition. Two NOPs are inserted after the branch instruction BZ. These two NOPs can be performed regardless of whether the branch is taken in the

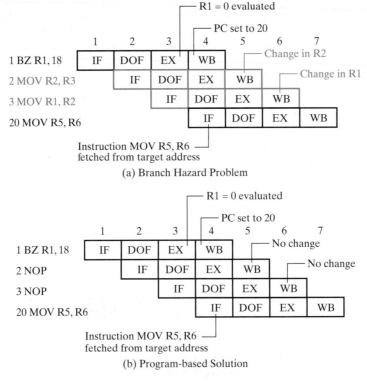

(a) Branch Hazard Problem

(b) Program-based Solution

☐ **FIGURE 12-15**
Example of Control Hazard

EX stage of BZ in cycle 3 with no adverse effects on the correctness of the program. When control hazards in the CPU are handled in this manner by programming, the branch hazard dealt with by the NOPs is referred to as a *delayed branch*. Branch execution is delayed by two clock cycles in this CPU.

The NOP solution in Figure 12-15(b) increases the time required to process the simple program by two clock cycles, regardless of whether the branch is taken. Note, however, that these wasted cycles can sometimes be avoided by rearranging the order of instructions. Suppose that those instructions to be executed regardless of whether the branch is taken can be placed in the two locations following the branch instruction. In this situation, the lost throughput is completely recovered.

Just as in the case of the data hazard, a stall can be used to deal with the control hazard. But, also as in the case of the data hazard, the reduction in throughput will be the same as with the insertion of NOPs. This solution is referred to as a *branch hazard stall* and will not be presented here.

A second hardware solution is to use *branch prediction*. In its simplest form, this method predicts that branches will never be taken. Thus, instructions will be fetched and decoded and operands fetched on the basis of the addition of 1 to the value of the *PC*. These actions occur until it is known during the execution cycle whether the branch in question will be taken. If the branch is not taken, the instructions already in the pipeline due to the prediction will be allowed to proceed. If the branch is taken, the instructions following the branch instruction need to be cancelled. Usually, the cancellation is done by inserting bubbles into the execution and write-back stages for these instructions. This is illustrated for the four-instruction program in Figure 12-16. On the basis of the prediction that the branch will not be taken, the two MOVA instructions after BZ are fetched, the first one is decoded, and its operands are fetched. These actions take place in cycles 2 and 3. In cycle 3, the condition upon which the branch is based has been evaluated, and it is found that $R1 = 0$. Thus, the branch is to be taken. At the end of cycle 3, the *PC* is set to 20, and the instruction fetch in cycle 4 is performed using the new value of the *PC*. In cycle 3, the fact that the branch is taken has been detected, and bubbles

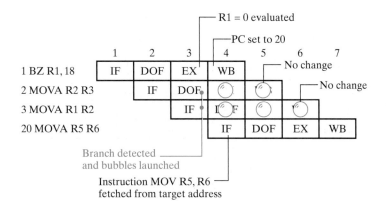

□ **FIGURE 12-16**
Example of Branch Prediction with Branch Taken

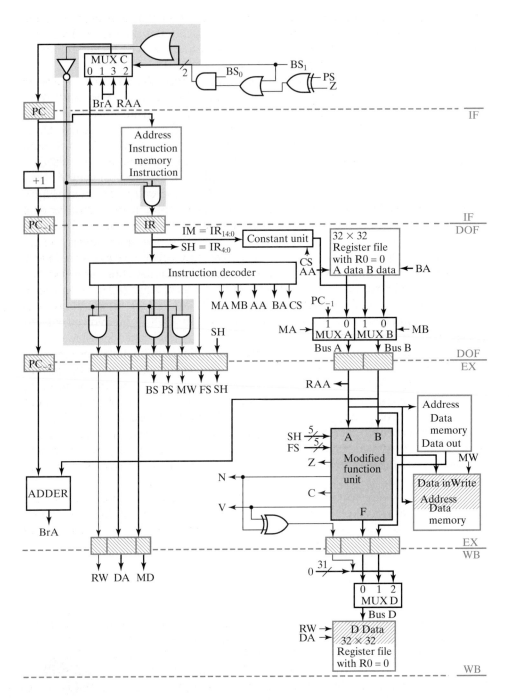

□ **FIGURE 12-17**
Pipelined RISC: Branch Prediction

are inserted into the pipeline for instructions 2 and 3. Proceeding through the pipeline, these bubbles have the same effect as two NOP instructions. However, because the NOPs are not present in the program, there is no delay or performance penalty when the branch is not taken.

The branch prediction hardware is shown in Figure 12-17. Whether a branch is taken is determined by looking at the selection values on the inputs to MUX C. If the pair of inputs is 01, then a conditional branch is being taken. If the pair is 10, then an unconditional JMR is occurring. If the pair is 11, then an unconditional JMP or JML is taking place. On the other hand, if the pair of inputs is 00, then no branch is occurring. Thus, a branch occurs for all combinations other than 00 (i.e., for at least one 1) on the pair of lines. Logically, this corresponds to the OR of the lines, as shown in the figure. The output of the OR is inverted and then ANDed with the *RW* and *MW* fields, so that the register file and the data memory cannot be written for the instruction following the branch instruction if the branch is taken. The inverted output is also ANDed with the *BS* field, so that a branch in the next instruction is not executed. In order to cancel the second instruction following the branch, the inverted OR output is ANDed with the *IR* output. This gives an instruction of all 0's, for which the OPCODE field is defined as NOP. If the branch is not taken, however, the inverted OR output is 1, and the *IR* and the three control fields remain unchanged, giving normal execution of the two instructions following the branch.

Branch prediction can also be done on the assumption that the branch is taken. In this case, the instructions and operands must be fetched down the path of the branch target. Thus, the branch target address must be computed and used for fetching the instruction in the branch target location. In case the branch does not take place, however, the updated value of the *PC* must also be saved. As a consequence, this solution will require additional hardware to compute and store the branch target address. Nevertheless, if branches are more likely to be taken than not, the "branch taken" prediction may yield a more favorable cost–performance trade-off than the "branch not taken" prediction.

For simplicity of presentation, we have treated the hardware solutions for dealing with hazards one at a time. In an actual CPU, these solutions need to be combined. In addition, other hazards, such as those associated with writing and reading memory locations, need to be handled.

12-4 THE COMPLEX INSTRUCTION SET COMPUTER

CISC instruction set architectures are characterized by complex instructions that are, at worst, impossible, and, at best, difficult to implement using a single cycle computer or a single pass through a pipeline. A CISC ISA often employs a sizable number of addressing modes. Further, the ISA often employs variable length instructions. The support for decision making via conditional branching is also more sophisticated than the simple concepts of branch on zero register contents and setting a register bit to 1 based on a comparison of two registers. In this section, a basic architecture for a CISC is developed with the high-performance of a RISC for simple instructions and most of the characteristics of a CISC ISA as just described.

Suppose that we are to implement a CISC architecture, but we are interested in approaching a throughput of one instruction per short RISC clock cycle for simple, frequently used instructions. To accomplish this goal, we use a pipelined datapath and a combination of pipelined and microprogrammed control as shown in Figure 12-18. An instruction is fetched into the IR and enters the Decode and Operand Fetch stage. If it is a simple instruction that executes completely in a single pass through the normal RISC pipeline, it is decoded and operand fetch occurs as usual. On the other hand, if the instruction requires multiple microoperations or multiple memory accesses in sequence, the decode stage produces a microcode address for the microcode ROM and replaces the usual decoder outputs with control values from the microcode ROM. Execution of microinstructions from the ROM, selected by the microprogram counter, continues until the execution of the instruction is completed.

Recall that to execute a sequence of microinstructions, it is often necessary to have temporary registers in which to store information. An organization of this type will frequently supply temporary registers with a convenient mechanism for switching between temporary registers and the usual programmer-accessible register resources.

The preceding organization supports an architecture that has combined CISC-RISC properties. It illustrate that pipelines and microprograms can be compatible and need not be viewed as mutually exclusive. The most frequent use of such a combined architecture allows existing software designed for a CISC to take advantage of a RISC architecture while preserving the existing ISA. The CISC-RISC architecture is a combination of concepts from the multiple-cycle computer in Chapter 10, the RISC CPU in the previous section, and the microprogramming concept introduced briefly in Chapter 10. This combination of concepts makes sense, since the CISC CPU executes instructions using multiple passes through the RISC datapath pipeline. To sequence these multiple-pass

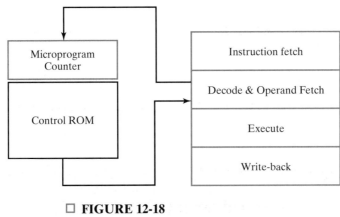

☐ **FIGURE 12-18**
Combined CISC-RISC Organization

instruction implementations, a sequential control of considerable complexity is needed, so microprogrammed control is chosen.

The development of the architecture begins with some minor modifications to the RISC ISA to obtain some capabilities desirable in the CISC ISA. Next, the datapath is modified to support the ISA changes. These include modification of the Constant Unit, addition of a Condition Code register *CC*, and deletion of the hardware for supporting the SLT instruction. Further, the Register file addressing logic is modified to provide addressing for 16 temporary registers for multiple-pass use of the datapath with 16 registers remaining in the storage resources. This is in contrast to the 32 registers in the storage resources for the RISC. The next step is to adapt the RISC control to work with the microprogrammed control in implementing the multiple pass instructions. Finally, the microprogrammed control itself is developed and its operation is illustrated by the implementation of three CISC instructions that characterize a CISC ISA.

ISA Modifications

The first modification to the RISC ISA is the addition of a new format for branch instructions. In terms of the instructions provided in the CISC, it is desirable to have the capability to compare the contents of two source registers and branch, indicating the relationship between the contents of the two registers. To perform such a comparison, a format with two source register fields SA and SB and a target offset are required. Referring to Figure 12-7, addition of the SB field to the branch format reduces the length of the target offset from 15 bits to 10 bits. The resulting Branch 2 format added for the CISC instructions is shown in Figure 12-19.

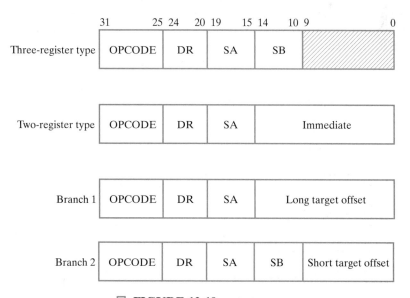

□ **FIGURE 12-19**
CISC CPU Instruction Formats

The second modification is to partition the Register file to provide addressing for 16 temporary registers for multiple-pass use of the datapath. With the partition, there are only 16 registers remaining in the storage resources. Rather than modify all of the register address fields in the instruction formats, we will simply ignore the most significant bit of these fields. For example, only the rightmost four bits of the field DR will be used. DR_4 will be ignored.

The third modification to the RISC ISA is the addition of condition codes (also called flags) as discussed in Chapter 11. The condition codes provided are designed specifically to be used in combination with branch on zero and branch on nonzero in implementing instructions that will provide a wide spectrum of decisions, such as greater than, less than, less than or equal to, etc. for both signed and unsigned integers. The codes are zero (Z), negative (N), carry (C), overflow (V), and L (less than). The first four are stored versions of the status outputs of the Function Unit. The less than (L) bit is the exclusive OR of Z and V which is useful in easily implementing particular decisions. The inclusion of the L bit in the condition codes eliminates the need for the SLT instruction.

To make the most effective use of these condition codes, it is useful to control whether or not they are modified for a particular microoperation execution from the instructions. Examination of the RISC instruction codes in Table 12-1 shows that bit 4 (third from the left) of the opcode is 0 for the operations down through instruction LSL. This bit can be used for these instructions to control whether the condition codes are affected by the instruction. If the bit is 1, then the condition code values are affected by the execution of the instruction. If it is 0, then the condition codes will not be affected. This permits flexible use of the condition codes in making decisions at both the ISA level and in the microcode.

Datapath Modifications

Several changes to the datapath are required to support the ISA modifications. These changes will be covered beginning with the datapath components in the DOF stage in Figure 12-20.

First, modifications are made to the Constant unit to handle the change in the length of the target offset. Logic added to the Constant unit extracts a constant, $IM_S = IR_{9:0}$, from constant IM. Sign extension is applied to IM_S to obtain a 32-bit word. Also, for use in comparisons with condition code values, an 8-bit constant CA is provided from the microinstruction register, MIR, in the microprogrammed control. This constant is zero-filled to form a 32-bit word. The CS control field for the Constant unit is expanded to two bits to perform selection from among the four possible constant sources.

Second, the Register address logic from the multiple-cycle computer in Chapter 10 is added to the address inputs of the Register file. The purpose of this change is to support the ISA modification that provides 16 temporary registers and 16 registers that are a part of the storage resources. An additional mode supports the use of DX as a register file source address with BX as the corresponding register file destination address. This is necessary to capture the contents for $R[DR]$ for use in destination address mode calculations.

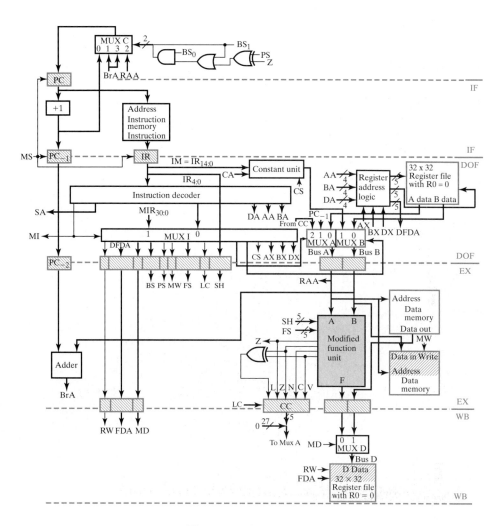

☐ **FIGURE 12-20**
Pipelined CISC CPU

Third, a number of changes are made to support the modification adding condition codes. In the DOF stage, an additional port is added on MUX *A* in order to provide access to *CC*, the stored condition codes, for storage in temporary registers or comparison to constant values. In the EX stage, the condition code bit L (less than) is implemented and the condition code register *CC* is added to the pipeline platform. The new control signal LC determines whether *CC* is loaded for the execution of a specific microoperation using a function unit operation. In the WB stage, the logic for support of the SLT instruction is replaced by a zero-filled *CC* value, which is passed to the new port on MUX *A*. Since the new condition code structure provides support for the same decision making as SLT did and more, support for SLT is no longer needed.

Control Unit Modifications

The addition of a microprogrammed control to the control unit to support instruction implementation using multiple passes through the pipeline causes significant changes to the existing control as shown in Figure 12-20. The microprogrammed control is a part of the instruction decoding hardware in the DOF stage, but it interacts with other parts of the control as well. For convenience, it will be described separately.

A quick overview of the execution of a multiple-pass instruction provides a perspective for the control unit changes. The PC points to the instruction in the Instruction memory. The instruction is fetched in the IF stage, and on the next clock edge, it is loaded into the IR and the PC is updated. The instruction is identified as a multiple-pass instruction from its opcode. Decoding of the opcode changes signal MI to 1 to indicate that this instruction is to use the microprogrammed control. The decoder also produces an 8-bit starting address, SA, that identifies the beginning of the microprogram in the Microcode ROM. Since multiple passes through the pipeline are needed to implement the instruction, the loading of subsequent instructions into the IR and further updating of the PC must be prevented. A signal MS produced by the microprogrammed control logic becomes 1 and stalls the PC and the IR. This prevents the PC from incrementing, but permits $PC + 1$ to continue down the pipeline into PC_{-1} and PC_{-2} for use in a branch. This stall remains until the multiple pass instruction has been executed or until there is branch or jump action on the PC. Also, when MI = 1, most of the fields of the decoded instruction are replaced with fields of the current microinstruction, which is a decoded NOP (no operation). This 31-bit field replacement, performed by MUX I, prevents the instruction itself from causing any direct actions. Some changes have been made to the control word to control modified datapath resources. Fields CS and MA have been expanded to two bits each, and field LC has been added. At this point, the microprogrammed control is now controlling the pipeline and supplies a series of microinstructions (control words) to implement the instruction execution. The control word format follows that for the multiple-cycle computer and includes fields such as SH, AX, BX, and DX. DX is modified to match the register address changes described for the datapath. In addition, the microprogrammed control has to interact with the datapath in order to perform decisions. This interaction includes application of the constant CA, use of the condition codes CC, and use of the zero detect signal Z.

To support the operations just discussed, the following changes are made to the control unit:

1. the addition of the stall signal MS to the PC, PC_{-1}, and IR,
2. changes in the instruction decoder to produce MI and SA,
3. expansion of the fields CS and MA to two bits,
4. addition of MUX I, and
5. addition of control fields AX, BX, and DX, and LC.

The definitions of new and modified control fields are given in Table 12-4.

□ **TABLE 12-4**
Added or Modified Control Word (Microinstruction) Fields for CISC

Control Fields			Register Fields		CS		MA		LC	
MZ 2b	CA 8h	BS P 2b S	Action	Code 5h	Action	Code 2b	Action	Code 2b	Action	Code
See Table 12-3	Next Address or Constant	See Table 12-2	**AX, BX**		zf IM	00	A Data	00	Hold CC	0
					se IM	01	PC_{-1}	01	Load CC	1
			R[SA], R[SB]	0X	se IM_S	10	$0 \| CC$	10		
			R_{16}	10	zf CA	11				
			...	...						
			R_{31}	1F						
			DX							
			Source R[DR] and Dest. R[SB]	00						
			Dest R[DR] with X ≠ 0	0X						
			R_{16}	10						
			...	...						
			R_{31}	1F						

Except for the addition of the microprogrammed control discussed in the next section, this completes the changes to the control unit.

Microprogrammed Control

A block diagram for the microprogrammed control and the format for microinstructions appear in Figure 12-21. The control is centered about the Microcode ROM, which has an 8-bit address and stores up to 256 41-bit microinstructions. The microprogram counter MC stores the address corresponding to the current microinstruction stored in the microinstruction register, MIR. The address for the ROM is provided by MUX E, which selects from the incremented MC, the jump address obtained from the microinstruction, CA, the prior value of the jump address, CA_{-1}, and the starting address from the instruction decoder in the control unit, SA. Table 12-5 defines the 2-bit select input ME for MUX E and stall bit, MS, in terms of the new control field MZ plus other variables. This function is implemented by the Microaddress Control logic. To set the context for the discussion, in location 0 of the ROM, the IDLE state 0 for the microprogrammed control contains a microinstruction that is a NOP consisting of all zeros. This microinstruction has MZ = 0 and CA = 0. From Table 12-5, with MI = 0, the microprogram address is CA = 0, causing the control to remain in this state until

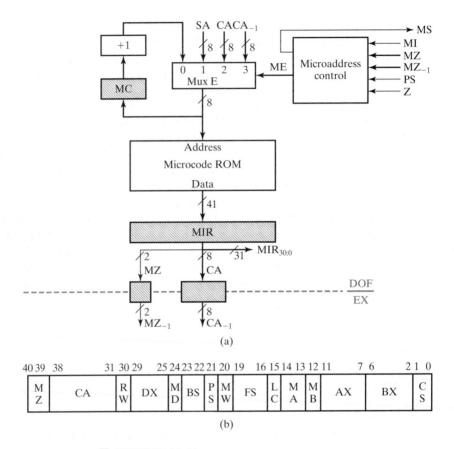

(a)

40 39 38		31 30 29	25 24 23 22 21 20 19	16 15 14 13 12 11	7 6	2 1 0
M Z	CA	R W	DX	M D	BS	P S	M W	FS	L C	M A	M B	AX	BX	C S

(b)

☐ **FIGURE 12-21**
Pipelined CISC CPU: Microprogrammed Control

MI = 1. With MI = 1, starting address SA is applied to fetch the first microinstruction of the microprogram for the complex instruction being held in *IR*. In the
control unit, MI = 1 also switches MUX *I* from the normal control word coming
from the decoder to the 31-bit *MIR* portion that is a NOP instruction. In addition, the output MS from the Microaddress control becomes 1, stalling the *PC*,
PC_{-1}, and the *IR* in the main control. At the next clock edge, the microinstruction fetched from the starting address SA enters the *MIR*, and the pipeline is
now controlled by the microprogram.

   In Figure 12-21, two pipeline registers are required as a part of the microprogrammed control. The stored pipeline values, MZ_{-1} and CA_{-1}, are required for the
execution of a conditional microbranch since the value of *Z* to be tested occurs
during the execution cycle for the microbranch instruction, one clock cycle after it
enters the *MIR*.

   During the execution of the microprogram, the microaddress is controlled
by MZ, MZ_{-1}, MI, PS, and Z. For $MZ_{-1} = 11$, MZ = 01 since the microinstruction

following a conditional microbranch must be a NOP. Under these conditions, the ME values are controlled by PS and Z with $MS = 1$. For PS and Z having opposite values, a conditional branch to the microaddress value from CA_{-1} occurs. Otherwise, for $MZ_{-1} = 11$ and $MZ = 01$, the next microaddress becomes the incremented value of MC.

For $MZ_{-1} \neq 11$, MZ, MI, and PS control the microaddress. For $MZ = 00$, the values of ME and MS are controlled by MI. For $MI = 0$, the next microaddress is CA and $MS = 0$, corresponding to the idle state for the microprogrammed control. For $MI = 1$, the next microaddress is SA and $MS = 1$, selecting the next microinstruction from the Microcode ROM and stalling the first two pipeline platforms. For $MZ = 01$, the next microaddress is the incremented value of MC, advancing execution to the next microinstruction in sequence. For $MZ = 10$, an unconditional jump is performed in the microcode control and the value of MS is controlled by PS. PS = 1 causes $MS = 1$, continuing microprogram execution. PS = 0 forces $MS = 0$, removing the stall, and returning control to the pipeline. This causes MI to become 0 (if the new instruction is not also a complex one). If CA = 0, the microprogrammed control is locked the IDLE state until MI = 1. In order for this to happen, the final instruction in the microprogram must have MZ = 10, PS = 0, and CA = 0.

□ **TABLE 12-5**
  Address Control

	Inputs				Outputs			
MZ_{-1}	MZ	MI	PS	Z	ME_1	ME_0	MS	Register Transfer Due to ME
11	01	X	0	0	0	0	1	$\overline{PS}\cdot\overline{Z}: MC \leftarrow MC + 1$
11	01	X	0	1	0	1	1	$\overline{PS}\cdot Z: MC \leftarrow CA_{-1}$
11	01	X	1	0	0	1	1	$PS\cdot\overline{Z}: MC \leftarrow CA_{-1}$
11	01	X	1	1	0	0	0	$\overline{PS}\cdot Z: MC \leftarrow MC + 1$
0X	01	X	X	X	0	0	1	$MC \leftarrow MC + 1$
X0	01	X	X	X	0	0	1	$MC \leftarrow MC + 1$
XX	00	0	X	X	1	0	0	$MC \leftarrow CA$
XX	00	1	X	X	0	1	1	$MC \leftarrow SA$
XX	10	X	0	X	1	0	0	$\overline{PS}: MC \leftarrow CA$
XX	10	X	1	X	1	0	1	$PS: MC \leftarrow CA$
XX	11	X	X	X	0	0	1	$MC \leftarrow MC + 1$

Microprograms for Complex Instructions

Three examples illustrate complex instructions implemented by using the CISC capabilities provided by the design just completed. The resulting microprograms are given in Table 12-6.

EXAMPLE 12-1 LD Instruction with Indirect Indexed Addressing (LII)

The LII instruction adds the target offset to the contents of a register that is being used as an index register. In the indirection step, the indexed address formed is then used to fetch the effective address from memory. Finally, the effective address is used to fetch the operand from memory. The opcode for this instruction is 0110001, and the instruction uses the Immediate format with the SA register field and a 15-bit target offset. When the LII instruction is fetched and appears in the *IR*, the instruction decoder sets MI equal to 1 and provides the microcode address symbolically represented by LII0 in Table 12-6. The first microinstruction to be executed is the one appearing in the IDLE address. This microoperation executes a NOP in the datapath and memory, but in the presence of MI = 1, the address control selects SA as the next microinstruction address, thereby leaving the IDLE state. The LII0 microinstruction forms the indexed address and increments the address in *MC* to fetch the next microinstruction LII1. This causes the NOP microinstruction in address LII1 to be fetched for execution in the pipeline. This NOP has been inserted, since the result of the microinstruction in LII0 is not placed in R_{16} until the WB stage. The next microinstruction in LII2 fetches the effective address from memory. A NOP is required next, due the clock cycle delay in writing the effective address to R_{17}. The microinstruction in LII4 applies the effective address to the memory to obtain the operand and place it in the destination register *R*[DR]. Since this completes the LII implementation, the microprogrammed control state in *MC* returns to IDLE and the next instruction following LII is fetched from the instruction memory by using the address in the *PC*. ∎

In Table 12-6, this sequence of microinstructions is described in the Action column by register transfer statements, and symbolic names are provided for the addresses of the microinstructions in the Microcode ROM. The remainder of the columns in the table provide the coding of the microinstruction fields. These codes are selected from Tables 10-12, 12-2, 12-3, and 12-5, to implement the register transfers. Of particular note is the appearance of MC = 10, PS = 0, and CA = IDLE (00) in microinstruction LII4 causing the microprogram control to return to IDLE and program control to return to the pipeline control.

EXAMPLE 12-2 Branch on Less Than or Equal to (BLE)

The BLE instruction compares the contents of registers *R*[SA] and *R*[SB]. If *R*[SA] is less than or equal to *R*[SB], then the *PC* branches to *PC* + 1 plus the sign-extended Short Target Offset (IM_S). Otherwise, the incremented *PC* is used. The opcode for the instruction is 1100101.

□ **TABLE 12-6**
Example Microprograms for CISC Architecture

				Microinstructions												
				R	M		P	M		L			M			
Action	Address	MZ	CA	W	DX	D	BS	S	W	FS	C	MA	B	AX	BX	CS
Shared Microinstructions																
$MI: MC \leftarrow SA, \overline{MI}:MC \leftarrow 00$	IDLE	00	00	0	00	0	00	0	0	0	0	00	0	00	00	00
$MC \leftarrow MC + 1$ (NOP)	Arbitrary	01	XX	0	00	0	00	0	0	0	0	00	0	00	00	00
Load Indirect Indexed (LII)																
$R_{16} \leftarrow R[SA] + \text{zf } IM_L$	LII0	01	00	1	10	0	00	0	0	2	0	00	1	00	00	00
$MC \leftarrow MC + 1$ (NOP)	LII1	01	00	0	00	0	00	0	0	0	0	00	0	00	00	00
$R_{17} \leftarrow M[R_{16}]$	LII2	01	00	1	11	1	00	0	0	0	0	00	0	10	00	00
$MC \leftarrow MC + 1$ (NOP)	LII3	01	00	0	00	0	00	0	0	0	0	00	0	00	00	00
$R[DR] \leftarrow M[R_{17}]$	LII4	10	IDLE	1	01	1	00	0	0	0	0	00	0	11	00	00
Compare Less Than or Equal To (BLE)																
$R[SA] - R[SB],$ $CC \leftarrow L\|Z\|N\|C\|V$	BLE0	01	00	0	01	0	00	0	0	5	1	00	0	00	00	00
$MC \leftarrow MC + 1$ (NOP)	BLE1	01	00	0	00	0	00	0	0	0	0	00	0	00	00	00
$R_{31} \leftarrow CC \wedge 11000$	BLE2	01	18	1	1F	0	00	0	0	8	0	10	1	00	00	11
$MC \leftarrow MC + 1$ (NOP)	BLE3	01	00	0	00	0	00	0	0	0	0	00	0	00	00	00
if $(R_{31} \neq 0)MC \leftarrow BLE7$ else MC $\leftarrow MC + 1$	BLE4	11	BLE7	0	00	0	00	1	0	0	0	00	0	1F	00	00
$MC \leftarrow MC + 1$ (NOP)	BLE5	01	00	0	00	0	00	0	0	0	0	00	0	00	00	00
$MC \leftarrow IDLE$	BLE6	00	IDLE	0	00	0	00	0	0	0	0	00	0	00	00	00
$PC \leftarrow (PC_{-1}) + \text{se } IM_L,$ $MC \leftarrow IDLE$	BLE7	10	IDLE	0	00	0	11	0	0	0	0	01	1	00	00	10
Move Memory Block (MMB)																
$R_{16} \leftarrow R[SB]$	MMB0	01	00	1	10	0	00	0	0	C	0	00	0	00	00	00
$MC \leftarrow MC + 1$ (NOP)	MMB1	01	00	0	00	0	00	0	0	0	0	00	0	00	00	00
$R_{16} \leftarrow R_{16} - 1$	MMB2	01	01	1	10	0	00	0	0	5	0	00	1	00	00	11
$R_{17} \leftarrow R[DR]$	MMB3	01	00	1	00	0	00	0	0	C	0	00	0	00	11	00
$R_{18} \leftarrow R[SA] + R_{16}$	MMB4	01	00	1	12	0	00	0	0	2	0	00	0	00	10	00
$R_{19} \leftarrow R_{17} + R_{16}$	MMB5	01	00	1	13	0	00	0	0	2	0	00	0	11	10	00
$R_{20} \leftarrow M[R_{18}]$	MMB6	01	00	1	14	1	00	0	0	0	0	00	0	12	00	00
$MC \leftarrow MC + 1$ (NOP)	MMB7	01	00	0	00	0	00	0	0	0	0	00	0	00	00	00
$M[R_{19}] \leftarrow R_{20}$	MMB8	01	00	0	00	0	00	0	1	0	0	00	0	13	14	00
if $(R_{16} \neq 0)MC \leftarrow MMB2$	MMB9	11	MMB2	0	00	0	00	1	0	0	1	00	0	10	00	00
$MC \leftarrow MC + 1$ (NOP)	MMB10	01	00	0	00	0	00	0	0	0	0	00	0	00	00	00
$MC \leftarrow IDLE$	MMB11	10	IDLE	0	00	0	00	0	0	0	0	00	0	00	00	00

The register transfers for the instruction are given in the Action column of Table 12-6. In microinstruction BLE0, $R[SB]$ is subtracted from $R[SA]$ and the condition codes L through V are captured in register CC. Due to the one-cycle delay in writing to CC, a NOP is required in microinstruction BLE1. $R[SA]$ is less than or equal to $R[SB]$ if (L + Z) = 1 (+ is OR in this expression). Thus, of the five condition code bits, only L and Z are of interest. So in microinstruction BLE2, the least significant three bits of CC are masked out using the mask 11000 ANDed with CC. The result is placed in register R_{31}, and, in BLE3, another NOP is required waiting for R_{31} to be written. In BLE4, a microbranch on R_{31} nonzero occurs. If R_{31} is nonzero, then L + Z = 1 giving $R[SA]$ less than or equal to $R[SB]$. Otherwise, both L and Z are 0 indicating $R[SA]$ is not less than or equal to $R[SB]$. Due to the microbranch, a NOP is required in BLE5. The connections to MUX E require only one NOP after a microbranch instead of the two NOPs needed for the conditional branch in the main control. If the branch is not taken, the next microinstruction BLE6 executes, returning MC to IDLE and reactivating the pipeline control to execute the next instruction. If the branch is taken, microinstruction BLE7 is executed, placing $PC + 1 + BrA$ into the PC for fetching the next instruction when the microinstruction reaches the EX stage. Note that such a branch on the PC can take place only after MS becomes 0 and the pipeline is reactivated. In this regard, a control hazard exists for this instruction in the main control, so it must be followed by a NOP. The codes for the microinstruction fields appear in Table 12-6 ∎

EXAMPLE 12-3 Move Memory Block (MMB)

The MMB instruction copies a block of information from one set of contiguous locations in memory to another. It has opcode 0100011 and uses the three-register type format. Register $R[SA]$ specifies address A, the beginning location of the source block in memory, and register $R[DR]$ specifies address B, the beginning location of the destination block. $R[SB]$ gives the number n of words in the block.

The register transfers for the instruction are given in the Action column of Table 12-6. In microinstruction MMB0, $R[SB]$ is loaded into R_{16}. MMB1 contains a NOP waiting for $R16$ to be written. In MMB2, R_{16} is decremented, providing an index with n values, $n - 1$ down to 0, for use in addressing the copying of n words. Since $R[DR]$ is a destination register, it is ordinarily not available as a source. But to do address manipulation for the destination locations, it is necessary for its value be placed in a register that can act as a source. Thus, in MMB3, the value of $R[DR]$ is copied to register R_{17} by using the register code DX = 00000, which treats $R[DR]$ as the source and the register specified in the BX field, R_{17}, as the destination. In microinstructions MMB4 and MMB5, R_{16} is added to $R[SA]$ and to $R[SB]$ to serve as pointers to the addresses in the blocks. Due to these operations, the words in the blocks are transferred from the highest location first. In MMB6, the first word is transferred from the first source address in memory to temporary register R_{20}. In MMB7, a NOP appears to permit the writing of the value in R_{20} by MMB6 before the use of the value by MMB8. In MMB8, the first word is transferred from R_{20} to the first destination address in memory. In MMB9, a branch on zero is done on the contents of R_{16} to determine if all of the words in the block have been transferred. If not, then MM2 is the next microaddress in which the next word transfer begins. If R_{16} equals zero, the next microinstruction is the NOP placed in MMB10 due to the branch. The final microinstruction in MMB11 returns the *MC* to IDLE and returns execution back to the pipeline control.

The codes for the microinstructions appear in Table 12-6. The code consists of simple register and memory transfers with a single branch to provide the looping capability and NOPs to deal with data and control hazards. ■

12-5 MORE ON DESIGN

The two designs considered in this chapter represent two different ISAs and two different supporting CPU organizations. The RISC architecture matches well with the pipelined control organization because of the simplicity of the instructions. Due to the need for high performance, the modern CISC architecture presented is built upon the RISC foundation. In this section, we will deal with additional features for speeding up the fundamental RISC pipeline. Finally, we relate the two organizations to more general digital systems design.

High-Performance CPU Concepts

Among the various methods used to design high-speed CPUs are multiple units organized as a pipeline-parallel structure, superpipelines, and superscalar architectures.

Consider the case in which an operation takes multiple clock cycles to execute, but the instruction fetch and write-back operations can be handled in a single cycle. Then it is possible to initiate an instruction every clock cycle, but not possible to complete the execution of an instruction every cycle. In such a situation, the performance of the CPU can be substantially improved by having multiple execution units in parallel. A high-level block diagram for this kind of system is shown in Figure 12-22. The instruction fetch, decoding, and operand fetch are carried out in the I-unit pipeline. In addition, the I-unit handles branches. When decoding of a nonbranch instruction has been completed, the instruction and operands are *issued* to the appropriate E-unit. When execution of the instruction is completed by the E-unit, the write-back to the register file occurs. If a memory access is required, then the D-unit is used to execute the memory write. If the operation is a store, it goes immediately to the D-unit. Note that the actual execution units may be microprogrammed and may also have internal pipelines.

Suppose that a sequence of three instructions—say, a multiplication, a 16-bit shift, and an addition—has no data hazards. Suppose further that there is a single pipelined E-unit that performs all of these operations, which take 17, 8, and 2 clock cycles, respectively, and that both the multiplication and the shift require multiple passes through portions of the E-unit pipeline. This situation allows only one clock cycle of overlap between pairs of the three instructions. Thus, the fastest that the sequence of operations executes in the E-unit is $17 + 8 + 2 - 2 = 25$ clock cycles.

But with an E-unit for each operation, these operations can be executed in max(17, 1 + 8, 2 + 2) clock cycles, which equals 17 clock cycles. The additional 1 and 2 are due to the issuing of one instruction per clock cycle to the E-unit set. The resulting execution throughput is improved by a factor of 25/17 = 1.5.

In all of the methods considered thus far, the peak throughput possible is one instruction per clock cycle. With this limitation, it is desirable to maximize the clock rate by minimizing the maximum pipeline stage delay. If, as a consequence, a large number of pipeline stages is used, the CPU is said to be *superpipelined*. A superpipelined CPU will generally have a very high clock frequency, in the range of a GHz. In such an organization, however, handling hazards effectively is critical, since any stalling or reinitialization of the pipeline will degrade the performance of the CPU significantly. Also, as more pipeline stages are added, further dividing up the combinational logic, the setup and propagation delay times of the flip-flops begin to dominate the platform-to-platform delay and the speed of the clock. The improvement

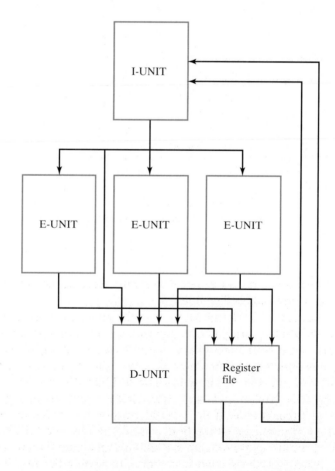

□ **FIGURE 12-22**
Multiple Execution Unit Organization

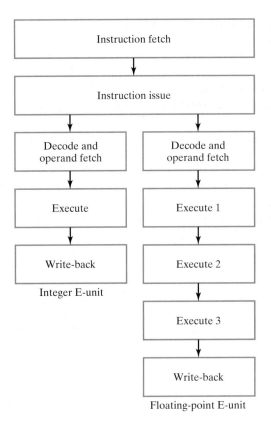

□ **FIGURE 12-23**
Superscalar Organization

achieved is less, and when hazards are taken into account, the performance may actually become worse rather than better.

For fast execution, an alternative to superpipelining is the use of a *superscalar* organization. The goal of this kind of organization is to have a peak rate of initiating instructions in excess of one instruction per clock cycle. A superscalar CPU that fetches a pair of instructions simultaneously by using a double-word wide path from instruction memory is illustrated in Figure 12-23. The processor checks for hazards among the instructions, as well as available execution units in the instruction issue stage of the pipeline. If there are hazards or busy execution units corresponding to the first instruction, then both instructions are held for later issuing. If the first instruction has no hazard and its E-unit is available, but there is a hazard or no available E-unit for the second instruction, then only the first instruction is issued. Otherwise, both instructions are issued in parallel. If a given superscalar architecture has the ability to issue up to four instructions simultaneously, then its peak execution rate is four instructions per clock cycle. If the clock cycle is 5 ns, then such a CPU has a peak execution rate of 800 MIPS. Note that the hazard checking for instructions in the execution stages and those in the issue stage become very complex as the maximum number of instructions issued simultaneously is increased. The resulting hardware complexity has the potential to increase the clock cycle length, so the trade-offs in such a design need to be examined very carefully.

We close this section with two observations. First, as the quest for better performance causes us to design increasingly complex organizations, hazards cause the order of the instructions to play a more important role in the throughput that is achievable. Also, improved performance can be achieved by reducing the number of hazard-producing instructions, such as branches. As a consequence, to fully exploit the performance capabilities of the hardware, the assembly language programmer and the compiler writer need to be very knowledgeable about the behavior of not only the instruction set architecture, but also the underlying organization of the hardware of the CPU.

When multiple execution units are involved, very often the CPU design we have been considering here actually becomes the design for the entire processor, as is shown for the generic computer. This is apparent in the superscalar organization in Figure 12-23, which contains the floating-point unit (FPU). The FPU, the MMU, and the portion of the internal cache that handles data are effectively types of E-units. The portion of the internal cache that handles the instructions can be viewed as a part of the I-Unit that fetches instructions. Thus, in the quest for higher and higher throughput, the realm of the CPU becomes that of the processor, as in the generic computer.

Recent Architectural Innovations

Beyond the concepts presented in the previous section, two general trends have become apparent in one of the most recent high-performance architectures. The first trend is the development of compilers and hardware architectures that permit the compiler to explicitly identify to the hardware instructions that can be executed in parallel. In this approach, the identification of parallelism typically done in hardware in the superscalar architecture has now been moved to a fair degree into the compiler. This releases hardware for other uses, notably more execution units and larger register files. The second trend is the use of techniques that allow the processor to avoid waiting for branches to be taken and for data values to become available. Three techniques that support this trend will be discussed in the remainder of this section.

Instead of waiting for a branch to be taken, the processor will execute both sides of the branch and produce results for both sides. When the results of the branch becomes available, the right result is selected and the computation proceeds. Thus, there is no delay waiting for a branch, significantly improving performance for long pipelines. This simple approach is referred to as *predication* and uses special 1-bit registers referred to as predicate registers that determine which result is used when the branch outcome is known.

Instead of waiting to load data from memory until it is known that the data is needed, *speculative loading* of data from memory is performed before it is known for sure whether or not the data is needed. The reason for use of this technique is to avoid the relatively long delay required to fetch an operand from memory. If the data that is speculatively fetched turns out to be the data needed, then the data will be available and the computation can proceed immediately without having to wait for a memory access to get the data.

Instead of waiting for data to become available, *data speculation* uses methods to predict data values and proceeds to compute using these values. When the actual value becomes known and matches the predicted value, then the result produced from the predicted value can be used to carry forward the computation. If the actual value and the predicted value differ, then the result based on the predicted value is discarded and the actual value is used to continue computation. An example of data speculation is permitting a value to be loaded from memory before a store into the same memory location occurring earlier in the program has been executed. In this case, it is predicted that the store will not change the value of the data in memory, so that the value loaded before the store will be valid. If, at the time the store occurs, the loaded value is not valid, the result of computation using it is discarded.

All of these techniques perform operations or sequences of operations for which results are discarded with some frequency. Thus, there is "wasted" computation. To be able to do large amounts of useful computation, as well as the wasted computation, more parallel resources, as well as specialized hardware for implementing the techniques, are required. The payoff in return for the cost of these resources is potentially higher performance.

Digital Systems

The two sizable digital system designs we have examined in this chapter are general-purpose CPUs. How does their design relate to that of other digital systems? First of all, each digital system has an architecture. Although that architecture may not in any way deal with instructions to be executed, it is likely that it still can be described by using register transfer descriptions and, possibly, one or more algorithmic state machines. On the other hand, it might have instructions, but they may be quite different from those for a CPU. The system may have no datapath at all or may have several datapaths. There is likely to be some form of control unit, and there may be multiple control units that interact. The system may or may not include memories. Thus, the total spectrum of digital systems has a very wide range of architectural possibilities.

So, what is the connection of the general digital system to the content of this chapter? Simply stated, the connection is design techniques. To illustrate, consider that we have shown in detail how a system with instructions can be implemented using a datapath and a control unit. From here, it is relatively easy to implement a simpler system without instructions. We have shown how high speeds can be achieved by using pipelines or parallel execution units. Thus, if the goal of a system is high speed, then pipelining or parallel units are techniques to consider. For example, one of the authors, in an example design of a system for implementing a portion of a USB transmitter (see section 13-4), used a pipelined datapath with a control that involved both pipeline and conventional sequential control. We have shown how microprogramming has been used to implement controls for complex functions carried out in a pipeline. If a system has one or more very complex functions, whether pipelined, programmable, or not, then a microprogrammed control is a possibility.

12-6 CHAPTER SUMMARY

This chapter has covered the design of two processors—one for a reduced instruction set computer (RISC) and one for a complex instruction set computer CISC. As a prelude to the design of these processors, the chapter began with an illustration of a pipelined datapath. The pipeline concept enables operations to be performed with clock frequencies and throughput not achievable with the same processing components in a conventional datapath. The pipeline execution pattern diagram was introduced for visualizing the behavior of a pipeline and estimating its peak performance. The problem of the low clock frequency of the single-cycle computer was addressed by adding a pipelined control unit to the datapath.

Next, we examined a RISC design with a pipelined datapath and control unit. Based on the single-cycle computer in Chapter 10, the RISC ISA is characterized by a single instruction length, a limited number of instructions with only a few addressing modes, and memory access restricted to load and store operations. Most RISC operations are simple in the sense that, in a conventional architecture, they can be executed using a single microoperation.

The RISC ISA is implemented by using a modified version of the pipelined datapath in Figure 12-2. Modifications include an increase of the word length to 32 bits, doubling of the number of registers in the register file, and replacement of the shifter in the function unit with a barrel shifter. Likewise, a modified version of the control unit in Figure 12-4 is used. Control changes were performed to accommodate the datapath changes and to handle branches and jumps in a pipeline environment. After completion of the basic design, consideration was given to data hazard and control hazard problems. We examined each type of hazard, as well as software and hardware solutions for each.

The ISA of the CISC has the potential for performing many distinct operations, with memory access supported by several addressing modes. The CISC also has operations that are complex in the sense that they require many clock cycles for their execution. The CISC permits many of the instructions to perform memory accesses and is characterized by complex conditional branching supported by condition codes (status bits). Although, in general, a CISC ISA permits multiple instruction lengths, this feature is not provided by the example architecture.

To provide high throughput, the RISC architecture serves as the core of the CISC architecture. Simple instructions can be executed at the RISC throughput, with complex instructions, executed by multiple passes through the RISC pipeline, reducing overall throughput. RISC datapath modification provided registers for temporary operand storage and condition code storage. Changes to the control unit were required to support these datapath changes. The primary control unit modification, however, was the addition of the microprogram control for execution of complex instructions. Added changes to the RISC control unit were required to integrate the microprogram control into the control pipeline. Examples of micro-programs for three complex instructions were provided.

After completing the CISC and RISC designs, we touched on some advanced concepts, including parallel execution units, superpipelined CPUs, superscalar CPUs, and predictive and speculative techniques for high performance. Finally, we related the design techniques in this chapter to more general digital system design.

REFERENCES

1. MANO, M. M. *Computer System Architecture,* 3rd Ed. Englewood Cliffs, NY: Prentice Hall, 1993.

2. PATTERSON, D. A., AND J. L. HENNESSY *Computer Organization and Design: The Hardware/Software Interface,* 2nd ed. San Francisco, CA: Morgan Kaufmann, 1998.

3. HENNESSY, J. L., AND D. A. PATTERSON *Computer Architecture: A Quantitative Approach,* 2nd ed. San Francisco, CA: Morgan Kaufmann, 1996.

4. DIETMEYER, D. L., *Logic Design of Digital Systems,* 3rd ed. Boston, MA: Allyn-Bacon, 1988.

5. KANE, G., AND J. HEINRICH *MIPS RISC Architecture.* Englewood Cliffs, NJ: Prentice Hall, 1992.

6. SPARC INTERNATIONAL, INC. *The SPARC Architecture Manual: Version 8.* Englewood Cliffs, NJ: Prentice Hall, 1992.

7. WEISS, S., AND J. E. SMITH *POWER and PowerPC.* San Mateo, CA: Morgan Kaufmann, 1994.

8. WYANT, G., AND T. HAMMERSTROM *How Microprocessors Work.* Emeryville, CA: Ziff-Davis Press, 1994.

9. HEURING, V., AND H. JORDAN *Computer Systems Design and Architecture.* Upper Saddle River, NJ: Prentice-Hall, 1997.

PROBLEMS

The plus (+) indicates a more advanced problem and the asterisk (*) indicates a solution is available on the Companion Website for the text.

12–1. A pipelined datapath is similar to that in Figure 12-1(b), but with the delays from the top to the bottom replaced by the following values: 1.0 ns, 1.0 ns, 0.1 ns, 0.2 ns, 1.3 ns, 0.2 ns, and 0.1 ns. Determine (a) the maximum clock frequency, (b) the latency time, and (c) the maximum throughput for this datapath.

12–2. *A program consisting of a sequence of 12 instructions without branch or jump instructions is to be executed in a six-stage pipelined computer with a clock period of 1.25 ns. Determine (a) the latency time for the pipeline, (b) the maximum throughput for the pipeline, and (c) the time required for executing the program.

12–3. The sequence of seven LDI instructions in the register number program with the pipeline execution pattern given below Figure 12-5 is fetched and executed. Manually simulate the execution by giving, for each clock cycle, the values in pipeline registers PC, IR, Data A, Data B, Data F, Data I, and in the register file having its value changed for each clock cycle. Assume that all file registers initially contain -1 (all 1's).

12–4. For each of the RISC operations in Table 12-1, list the addressing mode or modes used.

12–5. Simulate the operation of the barrel shifter in Figure 12-8 for each of the following shifts and $A = 7E93C2A1_{16}$. List the hexadecimal values on the 47 lines, 35 lines, and 32 lines out of the three levels of the shifter.
(a) Left, $SH = 11$
(b) Right, $SH = 13$
(c) Left, $SH = 30$

12–6. *For the RISC CPU in Figure 12-9, manually simulate, in hexadecimal, the processing of the instruction ADI R1 R16 2F01 located in $PC = 10F$. Assume that $R16$ contains 0000001F. Show the contents of each of the pipeline platforms and of the register file (the latter only when a change occurs) for each of the clock cycles.

12–7. Repeat Problem 12–6 for the instruction SLT R31 R10 R16 with $R10$ containing 0000100F and R16 containing 00001022.

12–8. Repeat Problem 12–6 for the instruction LSL R1 R16 000F.

12–9. +Use a computer-based logic minimization program to design the instruction decoder for a RISC from Table 12-3. The field FS need not be done, since it can be wired directly from OPCODE.

12–10. *For the RISC design, draw the execution diagram for the following RISC program, and indicate any data hazards that are present:

```
1 MOVA    R7, R6
2 SUB     R8, R8, R6
3 AND     R8, R8, R7
```

12–11. For the RISC design, draw the execution diagram for the following RISC program (with the contents of $R7$ nonzero after the subtraction), and indicate any data or control hazards that are present:

1	SUB	R7, R7, R6
2	BNZ	R7, 000F
3	AND	R8, R7, R6
4	OR	R5, R8, R5

12–12. *Rewrite the RISC programs in Problem 12–10 and Problem 12–11 using NOPs to avoid all data and control hazards and draw the new execution diagrams.

12–13. Draw the execution diagrams for the program in Problem 12–10, assuming
(a) the RISC CPU with data stall given in Figure 12-12.
(b) the RISC CPU with data forwarding in Figure 12-13.

12–14. Simulate the processing of the program in Problem 12–11 using the RISC CPU with data hazard stall in Figure 12-12. Give the contents of each pipeline platform and the register file (the latter only whenever a change occurs) for each clock cycle. Initially, $R6$ contains 00000010_{16}, $R7$ contains 00000020_{16}, $R8$ contains 00000030_{16}, and the PC contains 00000001_{16}. Is the data hazard avoided?

12–15. *Repeat Problem 12–14 using the RISC CPU with data forwarding in Figure 12-13.

12–16. Draw the execution diagram for the program in Problem 12–11, assuming the combination of the RISC CPU with branch prediction in Figure 12-17 and the RISC CPU with data forwarding in Figure 12-13.

12–17. Design the Constant Unit in the Pipelined CISC CPU by using the information given in Table 12-5 and multiple -bit multiplexers, AND gates, OR gates, and inverters.

12–18. *Design the Register Address Logic in the Pipelined CISC CPU by using information given in the register fields of Table 12-5 plus multiple-bit multiplexers, AND gates, OR gates, and inverters.

12–19. Design the Address Control logic described by Table 12-4 by using AND gates, OR gates, and inverters.

12–20. Write microcode for the execution part of each of the following CISC instructions. Give both a register transfer description and binary or hexadecimal representations similar to those shown in Table 12-6 for the binary code for each microinstruction.
(a) Compare Greater Than
(b) Branch if less than zero (CC bit N = 1)
(c) Branch if overflow (CC bit V = 1)

12–21. Repeat problem 12-20 for the following CISC instructions that are specified by register transfer statements.

(a) Push: $R[SA] \leftarrow R[SA] + 1$ followed by $M[R[SA]] \leftarrow R[SB]$

(b) Pop: $R[DR] \leftarrow M[R[SA]]$ followed by $R[SA] \leftarrow R[SA] - 1$

12–22. *Repeat problem 12-21 for the following CISC instructions.

(a) Add with carry: $R[DR] \leftarrow R[SA] + R[SB] + C$

(b) Subtract with borrow: $R[DR] \leftarrow R[SA] - R[SB] - B$

Borrow B is defined as the complement of the carry out, C.

12–23. Repeat problem 12-21 for the following CISC instructions.

(a) Add Memory Indirect: $R[DR] \leftarrow R[SA] + M[M[R[SB]]]$

(b) Add to Memory: $M[R[DR]] \leftarrow M[R[SA]] + R[SB]$

12–24. *Repeat problem 12-20 for the CISC instruction, Memory Scalar Add. This instruction uses the contents of $R[SB]$ as the vector length. It adds the elements of the vector with its least significant element in memory pointed to by $R[SA]$ and places the result in the memory location pointed to by $R[DR]$.

12–25. Repeat problem 12-20 for the CISC instruction, Memory Vector Add. This instruction uses the contents of $R[SB]$ as the vector length. It adds the vector with its least significant element in memory pointed to by $R[SA]$ to the vector with its least significant element in memory pointed to by $R[DR]$. The result of the addition replaces the vector with its least significant element pointed to by $R[DR]$.

13

INPUT–OUTPUT
AND COMMUNICATION

I n this chapter, we give an overview of selected aspects of computer input-output (I/O) and communication between the CPU and I/O devices, I/O interfaces, and I/O processors. Because of the wide variety of different I/O devices and the quest for faster handling of programs and data, I/O is one of the most complex areas of computer design. As a consequence, we are able to present only selected pieces of the I/O puzzle. We illustrate just three devices in detail: a keyboard, a hard disk, and a graphics display. We then introduce the I/O bus and the I/O interfaces that connect to I/O devices. We consider serial communication and use the I/O structure for the keyboard as an illustration. We then look at the Universal Serial Bus (USB), an alternative solution to the problem of accessing I/O devices. Finally, we discuss four modes for performing data transfers: program-controlled transfer, interrupt-initiated transfer, direct memory access, and the use of an I/O processor.

In terms of the generic computer at the beginning of Chapter 1, it is apparent that I/O involves a very large part of the computer. Only the processor, external cache, and RAM are not as highly involved, although they, too, are used extensively in directing and performing I/O transfers. Even the generic computer, which has fewer I/O devices than most PC systems, has a diverse set of such devices requiring significant digital electronic hardware for support.

13-1 COMPUTER I/O

The input and output subsystem of a computer provides an efficient mode of communication between the CPU and the outside environment. Programs and data must be entered into the memory for processing, and results obtained from computations must be recorded or displayed. Among the input and output devices that are commonly found in computer systems are keyboards, monitors, printers, magnetic disks, and compact disk read-only memory (CD-ROM) drives. Other input and output devices frequently encountered are modems or other communication interfaces, scanners, and sound cards with speakers and microphones. Significant numbers of computers, such as those used in automobiles, have analog-to-digital converters, digital-to-analog converters, and other data acquisition and control components.

The I/O facility of a computer is a function of its intended application. This results in a wide diversity of attached devices and corresponding differences in the needs for interacting with them. Since each device behaves differently, it would be time consuming to dwell on the detailed interconnections needed between the computer and each peripheral. We will, therefore, examine just three peripherals that appear in most computers: the keyboard, the hard disk, and the graphics display. These represent typical points in the range of data transfer rates required for peripherals. In addition, we present some of the common characteristics found in the I/O subsystem of computers, as well as the various techniques available for transferring data either in parallel, using many conducting paths, or serially, through communication lines.

13-2 SAMPLE PERIPHERALS

Devices that the CPU controls directly are said to be connected *on-line*. These devices communicate directly with the CPU or transfer binary information into or out of the memory upon command from the CPU. Input or output devices attached to the computer on-line are called *peripherals*. In this section, we examine three peripheral devices: a keyboard, a hard disk, and a graphics display. We also use the keyboard as an example to illustrate I/O concepts in a later section. We introduce the hard disk both to motivate the need for direct memory access and to provide background for the role of the device in Chapter 14 as a component in a memory hierarchy. We include the graphics display to illustrate the very high potential transfer rate requirements of contemporary applications.

Keyboard

The keyboard is among the simplest of the electromechanical devices attached to the typical computer. Since it is manually controlled, it has one of the slowest data rates of any peripheral.

The keyboard consists of a collection of keys that can be depressed by the user. It is necessary to detect which of the keys have been depressed. To do this, a *scan matrix* that lies beneath the keys is used, as shown in Figure 13-1. This two-dimensional matrix is conceptually similar to the matrix used in RAM. The matrix shown in the figure is 8×16, giving 128 intersections, so it can handle up to 128 keys. A decoder drives the X lines of the matrix, which are analogous to the word lines of a RAM. A multiplexer is attached to the Y lines of the matrix, which are analogous to the bit lines of a RAM. The decoder and the multiplexer are controlled by a

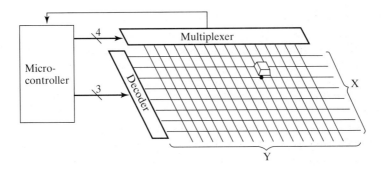

□ **FIGURE 13-1**
Keyboard Scan Matrix

microcontroller, a tiny computer that contains RAM, ROM, a timer, and simple I/O interfaces.

The microcontroller is programmed to periodically scan all intersections in the matrix by manipulating the control inputs of the decoder and multiplexer. If the key is depressed at an intersection, a signal path is closed from an output of the X decoder to an input of the Y multiplexer. The existence of this path is sensed at an input to the microcontroller. The 7-bit control code applied to the decoder and multiplexer at the time identifies the key. To allow for "rollover" in typing, in which multiple keys are depressed before any of them is released, the microcontroller actually identifies the depressing and release of the keys. Whether a key is depressed or released, the control code at the time of the event is sensed and is translated by the microcontroller into a *K-scan code*. When a key is depressed, a *make code* is produced; when a key is released, a *break code* is produced. Thus, there are two codes for each key, one for when the key is depressed and one for when the key is released. Note that the scanning of the entire keyboard occurs hundreds of times per second, so there is no danger of missing any depression or release of a key.

After presenting a number of I/O interface concepts, we will revisit the keyboard to see what happens to the K-scan codes before they are finally translated to ASCII characters.

Hard Disk

The hard disk is the primary intermediate-speed, nonvolatile, writable storage medium for most computers. The typical hard drive stores information serially on a nonremovable disk with a few to many platters, as shown in the upper right of the generic computer at the beginning of Chapter 1. Each platter is magnetizable on one or both surfaces. There are one or more read/write *heads* per recording surface; for the remainder of our discussion, we will assume a single head per surface. Each platter is divided into concentric *tracks*, as illustrated in Figure 13-2. The set of tracks that are at the same distance from the center of the disk on all platter surfaces is referred to as a *cylinder*. Each track is divided into *sectors* containing a fixed

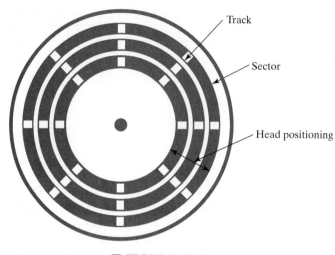

Track

Sector

Head positioning

☐ **FIGURE 13-2**
Hard Disk Format

number of bytes. The number of bytes per sector typically ranges from 256 to 4K. The typical byte address includes the cylinder number, head number, sector number, and word offset within the sector. The addressing assumes that the number of sectors per track is fixed. In modern, high-capacity disks, more sectors are included in the longer outer tracks than in the shorter inner tracks. In addition, a number of spare sectors are reserved to take the place of defective sectors. As a consequence of these design choices, the actual physical address of a sector on the disk is likely to be different from the address of the sector sent to the disk controller. The mapping from this address to the physical address is typically accomplished in the disk controller or drive electronics.

To enable information to be accessed, the set of heads is mounted on an actuator that can move the heads radially over the disk, as shown in the generic computer drawing. The time required to move the heads from the current cylinder to the desired cylinder is called the *seek time*. The time required to rotate the disk from its current position to that having the desired sector under the heads is called the *rotational delay*. In addition, a certain amount of time is required by the disk controller to access and output information. This time is the *controller time*. The time required to locate a word on the disk is the *disk access time*, which is the sum of the controller time, the seek time, and the rotational delay. Average values over all possibilities are used for these four parameters. Words may be transferred singly, but as we will see in Chapter 14, they are often accessed in blocks. The transfer rate for a block of words, once the block has been located, is the *disk transfer rate*, typically specified in megabytes/second (MB/s). The transfer rate required by the CPU-memory bus to transfer a sector from disk is the number of bytes in the sector divided by the length of time taken to read a sector from the disk. The length of time required to read a sector is equal to the proportion of the cylinder occupied by the sector divided by the rotational speed of the disk. For example, with 63 sectors,

512 B per sector, a rotational speed of 5400 rpm, and allowance for the gap between sectors, this time is about 0.15 ms, giving a transfer rate of 512/0.15 ms = 3.4 MB/s. The controller will store the information read from the sector in its memory. The sum of the disk access time and the disk transfer rate times the number of bytes per sector gives an estimate of the time required to transfer the information in a sector to or from the hard disk. Typical values in the mid-1990s are a seek time of 10 ms, a rotational delay of 6 ms, a sector transfer time of 0.15 ms, and a negligible controller time, giving an access time for an isolated sector of 16.15 ms.

Graphics Display

The graphics display is the primary output device for the interactive use of computers. Displays use a number of different technologies, the most prevalent of which is currently the cathode-ray tube (CRT), illustrated in Figure 13-3. The most modern versions of the CRT display are based on analog signals, which are generated on the display adapter board. The display is defined in terms of picture elements called *pixels*. The color display has three locations associated with each pixel on the screen. These locations correspond to the primary colors red, green, and blue (RGB). At each location, there is the corresponding colored phosphor. A phosphor emits light of its color when excited by a beam of electrons. In order to excite the three phosphors simultaneously, three electron guns are used, one for red, one for green, and one for blue—hence the RGB electron guns shown in the figure. The color that results for a given pixel is determined by the intensity of the electron beams striking the phosphors within the pixel.

The electron beams are scanned across the screen to form a set of horizontal lines called *scan lines*. This set of lines is referred to as a *raster*. The lines are scanned from top to bottom, beginning at the upper left and ending at the lower right. The electron guns remain at zero intensity as they scan from right to left in preparation for drawing the next scan line. The resolution of the information displayed is given in terms of the number of pixels per scan line and the number of

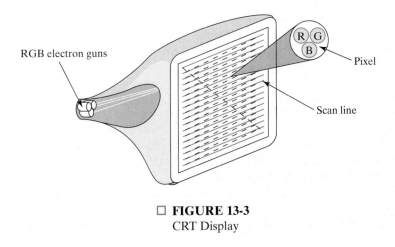

□ **FIGURE 13-3**
CRT Display

scan lines in the raster. A high-resolution super video graphics array (SVGA) display may have as many as 1280 pixels per scan line and 1024 lines in the raster. The electron beams scan the entire raster in 1/60 of a second.

Each of the pixels is controlled by the display adapter. A typical adapter uses a byte to define the color of a pixel. Since the byte contains 8 bits, it can define 256 colors at any given time. The byte does not directly drive the display, but instead selects 1 out of 256 registers in the graphics adapter to define the color. Each register is 20 bits or more, so the 256 colors can be selected from over 1 million colors by defining the contents of the registers.

Typically, the display adapter has video RAM that stores all of the bytes which control the display pixels. For a high-resolution display with 1280 pixels per scan line and 1024 scan lines, the number of pixels is $1280 \times 1024 = 1,310,720$. So, for 256 colors, a single screen of information requires at least 1.25 MB of video RAM.

I/O Transfer Rates

An indicated earlier, the three peripheral devices discussed in this section give a sense of the range of peak I/O transfer rates. The keyboard data transfer rate is less than 10 bytes/s. For the hard disk, while the disk controller is capturing the data arriving rapidly from the disk in the sector buffer, the transfer of data from the buffer to main memory is impossible. Thus, in the case in which the next sector is to be read immediately, all of the data from the sector buffer needs to be stored in main memory during the time the gap on the disk between the sectors passes under the disk head. For 63 sectors and a rotational speed of 5400 rpm, this time is about 25 μs. Thus, the peak transfer rate required is 512B/25 ms = 20 MB/s. For a display with 256 colors, if a screen is to be changed entirely every 1/60 of a second, 1.25 MB of data must be delivered to the video RAM from the CPU in that amount of time. This requires a data rate of $1.25 \text{ MB} \times 60 = 75$ MB/s.

Based on the preceding, we can conclude that the peak data rates required by the particular peripherals we have considered have a wide range. The rates for the hard disk and the display are high enough compared to the maximum rate of transfer on the computer buses to provide a challenge to designers. Attempts to meet this challenge use techniques in the disk controller and the graphics adapter to reduce the peak transfer rates required and use fast bus designs between the peripheral interfaces and memory.

13-3   I/O INTERFACES

Peripherals connected to a computer need special communication links to interface them with the CPU. The purpose of these links is to resolve the differences in the properties of the CPU and memory and the properties of each peripheral. The major differences are as follows:

1. Peripherals are often electromechanical devices whose manner of operation is different from that of the CPU and memory, which are electronic devices. Therefore, a conversion of signal values may be required.

2. The data transfer rate of peripherals is usually different from the clock rate of the CPU. Consequently, a synchronization mechanism may be needed.

3. Data codes and formats in peripherals differ from the word format in the CPU and memory.

4. The operating modes of peripherals differ from each other, and each must be controlled in a way that does not disturb the operation of other peripherals connected to the CPU.

To resolve these differences, computer systems include special hardware components between the CPU and the peripherals to supervise and synchronize all input and output transfers. These components are called *interface units*, because they interface between the bus from the CPU and the peripheral device. In addition, each device has its own controller to supervise the operations of the particular mechanism of that peripheral. For example, the controller in a printer attached to a computer controls the motion of the paper, the timing of the printing, and the selection of the characters to be printed.

I/O Bus and Interface Unit

A typical communication structure between the CPU and several peripherals is shown in Figure 13-4. Each peripheral has an interface unit associated with it. The common bus from the CPU is attached to all peripheral interfaces. To communicate with a particular device, the CPU places a device address on the address bus. Each interface attached to the common bus contains an address decoder that monitors the address lines. When the interface detects its own address, it activates the path between the bus lines and the device that it controls. All peripherals with addresses that do not correspond to the address on the bus are disabled by their interface. At the same time that the address is made available on the address bus, the CPU provides a function

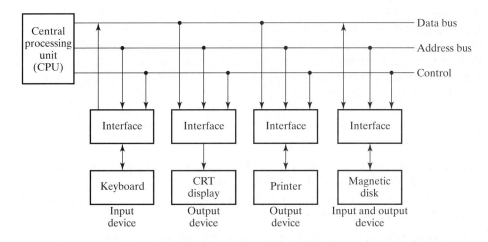

☐ **FIGURE 13-4**
Connection of I/O Devices to CPU

code on the control lines. The selected interface responds to the function code and proceeds to execute it. If data must be transferred, the interface communicates with both the device and the CPU data bus to synchronize the transfer.

In addition to communicating with the I/O devices, the CPU of a computer must communicate with the memory unit through an address and data bus. There are three ways that external computer buses communicate with memory and I/O. One method uses common data, address, and control buses for both memory and I/O. We have referred to this configuration as *memory-mapped I/O*. The common address space is shared between the interface units and memory words, each having distinct addresses. Computers that adopt the memory-mapped scheme read and write from interface units as if they were assigned memory addresses by using the same instructions that read from and write to memory.

The second alternative is to share a common address bus and data bus, but use different control lines for memory and I/O. Such computers have separate read and write lines for memory and I/O. To read or write from memory, the CPU activates the memory read or memory write control. To perform input to or output from an interface, the CPU activates the read I/O or write I/O control, using special instructions. In this way, the addresses assigned to memory and I/O interface units are independent from each other and are distinguished by separate control lines. This method is referred to as the *isolated I/O configuration*.

The third alternative is to have two independent sets of data, address, and control buses. This is possible in computers that include an *I/O processor* in the system in addition to the CPU. The memory communicates with both the CPU and I/O processor through a common memory bus. The I/O processor communicates with the input and output devices through separate address, data, and control lines. The purpose of the I/O processor is to provide an independent pathway for the transfer of information between external devices and internal memory. The I/O processor is sometimes called a *data channel*.

Example of I/O Interface

A typical I/O interface unit is shown in block diagram form in Figure 13-5. It consists of two data registers called *ports*, a control register, a status register, a bidirectional data bus, and timing and control circuits. The function of the interface is to translate the signals between the CPU buses and the I/O device and to provide the needed hardware to satisfy the two sets of timing constraints.

The I/O data from the device can be transferred into either port A or port B. The interface may operate with an output device, with an input device, or with a device that requires both input and output. If the interface is connected to a printer, it will only output data; if it services a scanner, it will only input data. A hard disk transfers data in both directions, but not at the same time; so the interface needs only one set of I/O bidirectional data lines.

The *control register* receives control information from the CPU. By loading appropriate bits into this register, the interface and the device can be placed in a variety of operating modes. For example, port A may be defined as an input port only. A magnetic tape unit may be instructed to rewind the tape or to start the tape moving in

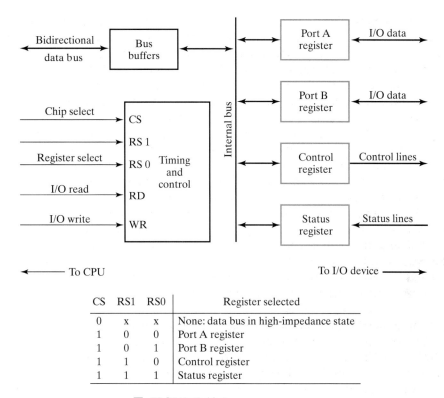

CS	RS1	RS0	Register selected
0	x	x	None: data bus in high-impedance state
1	0	0	Port A register
1	0	1	Port B register
1	1	0	Control register
1	1	1	Status register

□ **FIGURE 13-5**
Example of I/O Interface Unit

the forward direction. The bits in the status register are used for status conditions and for recording errors that may occur during data transfer. For example, a status bit may indicate that port A has received a new data item from the device, while another bit in the status register may indicate that a parity error has occurred during the transfer.

The interface registers communicate with the CPU through the bidirectional data bus. The address bus selects the interface unit through the chip select input and the two register select inputs. A circuit (usually a decoder or a gate) detects the address assigned to the interface registers. This circuit enables the chip select (*CS*) input when the interface is selected by the address bus. The two *register select inputs RS1* and *RS0* are usually connected to the two least significant lines of the address bus. These two inputs select one of the four registers in the interface, as specified in the table accompanying the diagram in Figure 13-5. The contents of the selected register are transferred into the CPU via the data bus when the I/O read signal is enabled. The CPU transfers binary information into the selected register via the data bus when the I/O write input is enabled.

The CPU, interface, and I/O device are likely to have different clocks that are not synchronized with each other. Thus, these units are said to be *asynchronous* with respect to each other. Asynchronous data transfer between two independent units requires that control signals be transmitted between the units to indicate the time at which data is being transmitted. In the case of CPU-to-interface communication, control signals must also indicate the time at which the address is valid. We will look at two methods for performing this timing: strobing, as it is called, and handshaking. Initially, we will consider generic cases in which no addresses are involved; subsequently, we will add addressing. The communicating units for the generic case will be referred to as the source unit and destination unit.

Strobing

Data transfers using *strobing* are shown in Figure 13-6. The data bus between the two units is assumed to be made bidirectional by the use of three-state buffers.

The transfer in Figure 13-6(a) is initiated by the destination unit. In the shaded area of the data signal, the data is invalid. Also, a change in Strobe at the tail of each arrow causes a change on the data bus at the head of the arrow. The destination unit changes the Strobe from 0 to 1. When the value 1 on Strobe reaches the source unit, the unit responds by placing the data on the data bus. The destination unit expects the data to be available, at worst, a fixed amount of time after Strobe goes to 1. At that time, the destination unit captures the data in a register and changes Strobe from 1 to 0. In response to the 0 value on Strobe, the source unit removes the data from the bus.

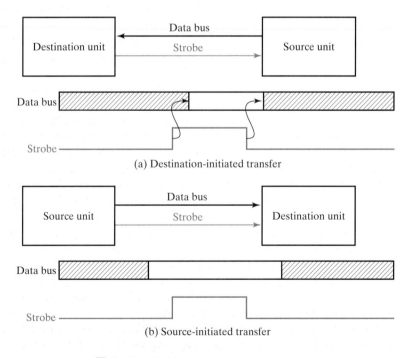

☐ **FIGURE 13-6**
Asynchronous Transfer Using Strobing

The transfer in Figure 13-6(b) is initiated by the source unit. In this case, the source unit places the data on the data bus. After the short time required for the data to settle on the bus, the source unit changes Strobe from 0 to 1. In response to Strobe equal to 1, the destination unit sets up the transfer to one of its registers. The source then changes Strobe from 1 to 0, which triggers the transfer into the register at the destination. Finally, after a short time required to ensure that the register transfer is done, the source removes the data from the data bus, completing the transfer.

Although simple, the strobe method of transferring data has several disadvantages. First, when the source unit initiates the transfer, there is no indication to it that the data was ever captured by the destination unit. It is possible, due to a hardware failure, that the destination unit did not receive the change in Strobe. Second, when the destination unit performs the transfer, there is no indication to it that the source has actually placed the data on the bus. Thus, the destination unit could be reading arbitrary values from the bus rather than actual data. Finally, the speeds at which the various units respond may vary. If there are multiple units, the unit initiating a transfer must wait for the delay of the slowest of the attached communicating units before changing Strobe to 0. Thus, the time taken for every transfer is determined by the slowest unit with which a given unit initiates transfers.

Handshaking

The *handshaking* method uses two control signals to deal with the timing of transfers. In addition to the signal from the unit initiating the transfer, there is a second control signal from the other unit involved in the transfer.

The basic principle of a two-signal handshaking procedure for data transfer is as follows. One control line from the initiating unit is used to request a response from the other unit. The second control line from the other unit is used to reply to the initiating unit that the response is occurring. In this way, each unit informs the other of its status, and the result is an orderly transfer through the bus.

Figure 13-7 shows data transfer procedures using handshaking. In Figure 13-7(a), the transfer is initiated by the destination unit. The two handshaking lines are called Request and Reply. The initial state is when both Request and Reply are disabled and in the 00 state. The subsequent states are 10, 11, and 01. The destination unit initiates the transfer by enabling Request. The source unit responds by placing the data on the bus. After a short time for settling of the data on the bus, the source unit activates Reply to signal the presence of the data. In response to Reply, the destination unit captures the data in a register and disables Request. The source unit then disables Reply and the system goes to the initial state. The destination unit may not make another request until the source unit has shown its readiness to provide new data by disabling Reply. Figure 13-7(b) represents handshaking for the source-initiated transfer. In this case, the source controls the interval between when the data is applied and when Request changes to 1 and between when Request changes to 0 and when the data is removed.

The handshaking scheme provides a high degree of flexibility and reliability, because the successful completion of a data transfer relies on active participation by both units. If one unit is faulty, the data transfer will not be completed. Such an

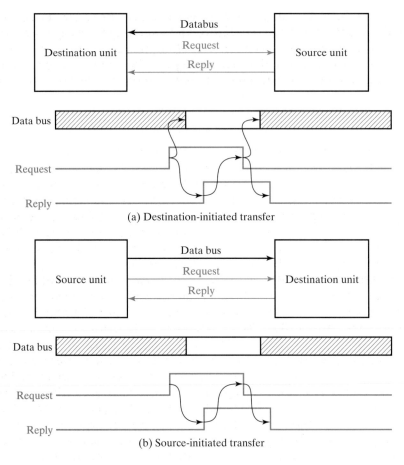

(a) Destination-initiated transfer

(b) Source-initiated transfer

☐ **FIGURE 13-7**
Asynchronous Transfer Using Handshaking

error can be detected by means of a time-out mechanism, which produces an alarm if the data transfer is not completed within a predetermined time interval. The time-out is implemented by means of an internal clock that starts counting time when the unit enables one of its handshaking control signals. If the return handshake does not occur within a given period, the unit assumes that an error occurred. The time-out signal can be used to interrupt the CPU and execute a service routine that takes appropriate error recovery action. Also, the timing is controlled by both units, not just the initiating unit. Within the time-out limits, the response of each unit to a change in the control signal of the other unit can take an arbitrary amount of time, and the transfer will still be successful.

The examples of transfers in Figure 13-6 and Figure 13-7 represent transfers between an interface and an I/O device and between a CPU and an interface. In the latter case, however, an address will be necessary to select the interface with which the CPU wishes to communicate and a register within the interface. In order to ensure that the CPU addresses the correct interface, the address must have set-tled on the address bus before the Strobe or Request signal changes from 0 to 1. Further, the address must remain stable until the change in the strobe or request from 1 to 0 has settled to 0 at the interface logic. If either of these conditions is vio-lated, another interface may be falsely activated, causing an incorrect data transfer.

13-4   SERIAL COMMUNICATION

The transfer of data between two units may be parallel or serial. In parallel data transfer, each bit of the message has its own path, and the entire message is trans-mitted at one time. This means that an n-bit message is transmitted in parallel through n separate conductor paths. In serial data transmission, each bit in the message is sent in sequence, one at a time. This method requires the use of one or two signal lines. Parallel transmission is faster, but requires many wires. It is used for short distances and when speed is important. Serial transmission is slower, but less expensive, since it requires only one conductor.

One way that computers and terminals that are remote from each other are connected is via telephone lines. Since telephone lines were originally designed for voice communication, but computers communicate in terms of digital signals, some form of conversion is needed. The devices that do the conversion are called *data sets* or *modems* (modulator-demodulators). A modem converts digital signals into audio tones to be transmitted over telephone lines and also converts audio tones from the line to digital signals for use by a computer. There are various modulation schemes, as well as several different grades of communication media and transmis-sion speeds. Serial data can be transmitted between two points in three different modes: simplex, half duplex, or full duplex. A *simplex* line carries information in one direction only. This mode is seldom used in data communication, because the receiver cannot communicate with the transmitter to indicate whether errors have occurred. Examples of simplex transmission are radio and television broadcasting.

A *half-duplex* transmission system is a system that is capable of transmitting in both directions, but in only one direction at a time. A pair of wires is needed for this mode. A common situation is for one modem to act as the transmitter and the other as the receiver. When transmission in one direction is completed, the roles of the modems are reversed to enable transmission in the opposite direction. The time required to switch a half-duplex line from one direction to the other is called the *turnaround time*.

A *full-duplex* transmission system can send and receive data in both direc-tions simultaneously. This can be achieved by means of a two-wire plus ground link, with a different wire dedicated to each direction of transmission. Alterna-tively, a single-wire circuit can support full-duplex communication if the frequency spectrum is subdivided into two nonoverlapping frequency bands to create sepa-rate receiving and transmitting channels in the same physical pair of wires.

The serial transmission of data can be synchronous or asynchronous. In *syn-chronous transmission*, the two units share a common clock frequency, and bits are transmitted continuously at that frequency. In long-distance serial transmission, the transmitter and receiver units are each driven by separate clocks of the same frequency. Synchronization signals are transmitted periodically between the two units to keep their clock frequencies in step with each other. In *asynchronous* transmission, binary information is sent only when it is available, and the line remains idle when there is no information to be transmitted. This is in contrast to synchronous transmission, in which bits must be transmitted continuously to keep the clock frequencies in both units synchronized.

Asynchronous Transmission

One of the most common applications of serial transmission is the communication of one computer with another via modems connected through the telephone system. Each character consists of an alphanumeric code of eight bits, with additional bits inserted at both ends of the code. In asynchronous serial transmission, each character consists of three parts: the start bit, the character bits, and the stop bits. The convention is for the transmitter to rest at the 1 state when no characters are transmitted. The first bit, called the start bit, is always 0 and is used to indicate the beginning of a character. An example of this format is shown in Figure 13-8.

A transmitted character can be detected by the receiver by applying the transmission rules. When a character is not being sent, the line is kept in the 1 state. The initiation of transmission is detected from the start bit, which is always 0. The character bits always follow the start bit. After the last bit of the character is transmitted, a stop bit is detected when the line returns to the 1 state for at least the time taken to transmit one bit. By means of these rules, the receiver can detect the start bit when the line goes from 1 to 0. By using a clock, the receiver examines the line at appropriate times to determine the bit values. The receiver knows the transfer rate of the bits and the number of character bits to accept.

After the character bits are transmitted, one or two stop bits are sent. The stop bits are always in the 1 state and frame the end of character to signify the idle or wait state. These bits allow both the transmitter and the receiver to resynchronize. The length of time that the line stays in the 1 state depends on the amount of time required for the equipment to resynchronize. Some older electromechanical terminals use two stop bits, but newer equipment often uses just one. The line remains in the 1 state until another character is transmitted. The stop time ensures that a new character will not follow for the time taken to transmit one or two bits.

As an illustration, consider serial transmission with a transfer rate of 10 characters per second. Suppose that each transmitted character consists of a start bit,

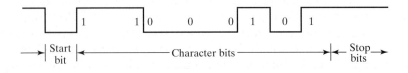

□ **FIGURE 13-8**
Format of Asynchronous Serial Transfer of Data

8 character bits, and 2 stop bits, for a total of 11 bits. If the bits are transmitted at a rate of 10 bits per second, then each bit takes 0.1 second for transfer. Since there are 11 bits to be transmitted, it follows that the *bit time* is 9.09 msec. The *baud rate* is defined as the maximum number of changes per second in the signal being transmitted. This is often, but not always, equivalent to the rate of data transfer in bits per second. Ten characters per second with an 11-bit format has a transfer rate of 110 baud.

Synchronous Transmission

Synchronous transmission does not use start or stop bits to frame characters. The modems employed in synchronous transmission have internal clocks that are set to the frequency at which bits are being transmitted. For proper operation, it is required that the clocks of the transmitter and receiver modems remain synchronized at all times. The communication line, however, carries only the data bits, from which information on the clock frequency must be extracted. Frequency synchronization is achieved by the receiving modem from the signal transitions that occur in the data that is received. Any frequency shift that may occur between the transmitter and receiver clocks is continuously adjusted by maintaining the receiver clock at the frequency of the incoming bit stream. In this way, the same rate is maintained in both the transmitter and the receiver.

Contrary to asynchronous transmission, in which each character can be sent separately with its own start and stop bits, synchronous transmission must send a continuous message in order to maintain synchronism. The message consists of a group of bits that form a block of data. The entire block is transmitted with special control bits at the beginning and the end, in order to frame the block into one unit of information.

The Keyboard Revisited

To this point, we have covered the basic nature of the I/O interface and serial transmission. With these two concepts available, we are now ready to continue with the example of the keyboard and its interface, as shown in Figure 13-9. The K-scan code produced by the keyboard microcontroller is to be transferred serially from the keyboard through the keyboard cable to the keyboard controller in the computer. The serial transfer on the Keyboard serial data line uses a format just like that shown for asynchronous transfer in Figure 13-8. In this case, however, a signal Keyboard clock is also sent through the cable. Thus, the transmission is synchronous with a transmitted clock signal, rather than asynchronous. These same signals are used to transmit control commands to the keyboard. In the keyboard controller, the microcontroller converts the K-scan code to a more standard *scan code*, which it then places in the Input register, at the same time sending an interrupt signal to the CPU indicating that a key has been pressed and a code is available. The interrupt-handling routine reads the scan code from the input register into a special area in memory. This area is manipulated by software stored in the Basic

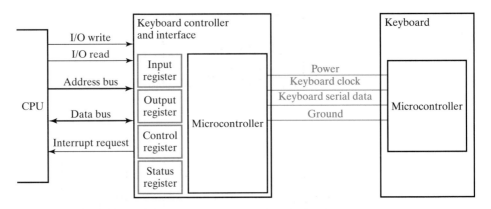

□ **FIGURE 13-9**
Keyboard Controller and Interface

Input/Output System (BIOS) that can translate the scan code into an ASCII character code for use by applications.

The Output register in the interface receives data from the CPU. The data can be passed on to control the keyboard—for example, setting the repetition rate when a key is held down. The Control register is used for commands to the keyboard controller. Finally, the Status register reports specific information on the status of the keyboard and the keyboard controller.

Perhaps one of the most interesting aspects of keyboard I/O is its high complexity. It involves two microcontrollers executing different programs, plus the main processor executing BIOS software (i.e., three different computers executing three distinct programs).

A Packet-Based Serial I/O Bus

Serial I/O, as described for the keyboard, uses a serial cable specifically dedicated to communicating between the computer and the keyboard. Whether parallel or serial, external I/O connections are typically dedicated. The use of these dedicated paths often requires that the computer case be opened and cards inserted with electronics and connectors specific to the particular I/O standard used for a given I/O device.

In contrast, packet-based serial I/O permits many different external I/O devices to use a shared communication structure that is attached to the computer through just one or two connectors. The types of devices supported include keyboards, mice, joysticks, printers, scanners, and speakers. The particular packet-based serial I/O we will describe here is the Universal Serial Bus (USB), which is becoming commonplace as the connection approach of choice for slow- to medium-speed I/O devices.

The interconnection of I/O devices by using USB is shown in Figure 13-10. The computer and attached devices can be classified as hubs, devices, or compound devices. A hub provides attachment points for USB devices and other hubs. A hub

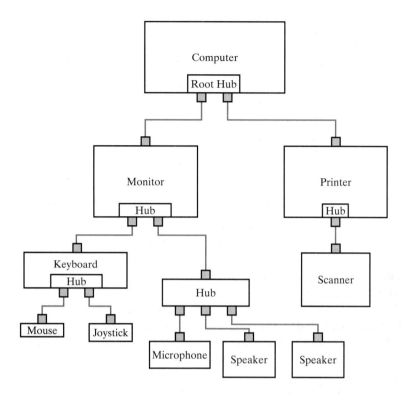

□ **FIGURE 13-10**
I/O Device Connection Using the Universal Serial Bus (USB)

contains a USB interface for control and status handling and a repeater for transferring information through the hub.

The computer contains a USB controller and the root hub. Additional hubs may be a part of the USB I/O structure. If a hub is combined with a device such as the keyboard shown in Figure 13-10, then the keyboard is referred to as a *compound device.* Aside from such compound devices, a USB device contains only one USB port to serve its function alone. The scanner is an example of a regular USB device. Without USB, the monitor, keyboard, mouse, joystick, microphone, speakers, printer, and scanner shown would all have separate I/O connections directly on the computer. The monitor, printer, scanner, microphone, and speakers might all require special cards to be inserted as discussed previously. With USB, only two connections are required.

The USB cables contains four wires: ground, power, and two data lines (D+ and D–) used for differential signaling. The power wire is used to provide small amounts of power to devices such as keyboards so that they do not need to have their own power supplies. To provide immunity to signal variation and noise, 0's and 1's are transmitted by using the difference in voltage between D+ and D–. If the voltage on D+ exceeds the voltage on D– by 200 millivolts or more, then the logic value is a 1. If the voltage on D– exceeds the voltage on D+ by 200 millivolts

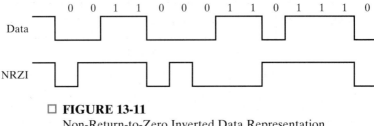

☐ **FIGURE 13-11**

Non-Return-to-Zero Inverted Data Representation

or more, the logic value is a 0. Other voltage relationships between D+ and D– are used as special signal states as well.

The logic values used for signalling are not the actual logic values of the information being transmitted. Instead, a Non-Return-to-Zero Inverted (NRZI) signalling convention is used. A zero in the data being transmitted is represented by a transition from 1 to 0 or 0 to 1 and a 1 is represented by a fixed value of 1 or 0. The relationship between the data being transmitted and the NRZI representation is illustrated in Figure 13-11. As is typical for I/O devices, there is no common clock serving both the computer and the device. NRZI encoding of the data provides edges that can be used to maintain synchronization between the arriving data and the time at which each bit is sampled at the receiver. If there are a large number of 1's in series in the data, there will be no transitions for some time in the NRZI encoding. To prevent loss of synchronization, a 0 is "stuffed" in before every seventh bit position in a string of 1's prior to NRZI encoding so that no more than six 1's appear in series. The receiver must be able to remove these extra zeros when converting NRZI data to normal data.

USB information is transmitted in packets. Each packet contains a specific set of fields depending on the packet type. Logical strings of packets are used to compose USB transactions. For example, an output transaction consists of an Out packet followed by a Data packet and a Handshake packet. The Out packet comes from the USB controller in the computer and notifies the device that it is to receive data. The computer then sends the Data packet. If the Data packet is received without error, then the device responds with the Acknowledge Handshake packet. Next, we detail the information contained in each of these packets.

Figure 13-12(a) shows a general format for USB packets and the formats for each of the three packets involved in an output transaction. Note that each packet begins with a synchronization pattern SYNC. This pattern is 00000001. Because of the sequence of zeros, the corresponding NRZI pattern contains seven edges, which provides a pattern to which the receiving clock can be synchronized. Since this pattern is preceded by a specific signal voltage state referred to as Idle, the pattern also signals the beginning of a new packet.

Following the SYNC, each of the packet formats contains 8 bits called the packet identifier (PID). In the PID, the packet type is specified by 4 bits, with an additional 4 bits that are complements of the first 4 to provide an error check on the type. A very large class of type errors will be detected by the repetition of the type as its complement. The type is optionally followed by information specific to the packet,

SYNC	PID	Packet Specific Data	CRC	EOP

(a) General packet format

SYNC 8 bits	Type 4 bits 1001	Check 4 bits 0110	Device Address 7 bits	Endpoint Address 4 bits	CRC	EOP

(b) Output packet

SYNC 8 bits	Type 4 bits 1100	Check 4 bits 0011	Data (Up to 1024 bytes)	CRC	EOP

(c) Data packet (Data0 type)

SYNC 8 bits	Type 4 bits 0100	Check 4 bits 1011	EOP

(d) Handshake packet (Acknowledge type)

☐ **FIGURE 13-12**
USB Packet Formats

which varies depending upon the packet type. Optionally, a CRC field appears next. The CRC pattern consisting of 5 or 16 bits is a Cyclic Redundancy Check pattern. This pattern is calculated at transmission of the packet from the packet-specific data. The same calculation is performed when the data is received. If the CRC pattern does not match the newly calculated pattern, then an error has been detected. In response to the error, the packet can be ignored and retransmitted. In the last field of the packet, an End of Packet (EOP) appears. This consists of D+ and D−, both low for two bit times, followed by the Idle state for a bit time. As its name indicates, this sequence of signal states identifies the end of the current packet. It should be noted that all fields are presented least significant bit first.

Referring to Figure 13-12(b), for the Output packet, the Type and Check fields are followed by a Device Address, an Endpoint Address, and a CRC pattern. The Device Address consists of seven bits and defines the device that is to input data. The Endpoint Address consists of four bits and defines which port of the device is to receive the information in the Data packet to follow. For example, there may be a port for data and one for control on a given device.

For the Data packet, the packet-specific data consists of 0 to 1024 data bytes. Due to the length of the packet, complex errors are more likely, so the CRC pattern is increased in length to 16 bits to improve its error detection capability.

In the Handshake packet, the packet-specific data is empty. The response to the receipt of the data packet is carried by the PID. PID 01001011 is an Acknowledge (ACK) indicating that the packet was received without any errors detected.

Absence of any HANDSHAKE packet when one would normally appear is an indication of an error. PID 01011010 is a No Acknowledge, indicating that the target is temporarily unable to accept or return data. PID 01111000 is a Stall (STALL), indicating that the target is unable to complete the transfer and that software intervention is required to recover from the stall condition.

The preceding concepts illustrate the general principles underlying a packet-based serial I/O bus and are specific to USB. USB supports other packet types and many different kinds of transactions. In addition, the attachment and detachment of devices is sensed and can trigger various software reactions. In general, there is substantial software in the computer to support the details of the control and operation of the Universal Serial Bus.

13-5 MODES OF TRANSFER

Binary information received from an external device is usually stored in memory for later processing. Information transferred from the central computer into an external device originates in the memory. The CPU merely executes the I/O instructions and may accept the data temporarily, but the ultimate source or destination is the memory. Data transfer between the central computer and I/O devices may be handled in a variety of modes, some of which use the CPU as an intermediate path, while others transfer the data directly to and from the memory. Data transfer to and from peripherals may be handled in one of four possible modes:

1. Data transfer under program control.
2. Interrupt-initiated data transfer.
3. Direct memory access transfer.
4. Transfer through an I/O processor.

Program-controlled operations are the result of I/O instructions written in the computer program. Each transfer of data is initiated by an instruction in the program. Usually, the transfer is to and from a CPU register and peripheral. Other instructions are needed to transfer the data to and from the CPU and memory. Transferring data under program control requires constant monitoring of the peripheral by the CPU. Once a data transfer is initiated, the CPU is required to monitor the interface to see when a transfer can again be made. It is up to the programmed instructions executed in the CPU to keep close tabs on everything that is taking place in the interface unit and the external device.

In the program-controlled transfer, the CPU stays in a program loop called a *busy-wait loop* until the I/O unit indicates that it is ready for data transfer. This is a time-consuming process, since it keeps the processor busy needlessly. The loop can be avoided by using the interrupt facility and special commands to inform the interface to issue an interrupt request signal when the data is available from the device. This allows the CPU to proceed to execute another program. The interface, meanwhile, keeps monitoring the device. When the interface determines that the device is ready for data transfer, it generates an interrupt request to the computer.

Upon detecting the external interrupt signal, the CPU momentarily stops the task it is performing, branches to a service program to process the data transfer, and then returns to the original task. This interrupt-initiated transfer is the type used for the keyboard controller shown in Figure 13-9.

Transferring of data under program control is performed through the I/O bus and between the CPU and a peripheral interface unit. In *direct memory access* (DMA), the interface unit transfers data into and out of the memory unit through the memory bus. The CPU initiates the transfer by supplying the interface with the starting address and the number of words needing to be transferred and then proceeds to execute other tasks. When the transfer is made, the interface requests memory cycles through the memory bus. When the request is granted by the memory controller, the interface transfers the data directly into memory. The CPU merely delays memory operations to allow the direct memory I/O transfer. Since the speed of a peripheral is usually slower than that of a processor, I/O memory transfers are infrequent compared with processor access to memory. DMA transfer is discussed in more detail in Section 13-7.

Many computers combine the interface logic with the requirements for DMA into one unit called an *I/O processor* (IOP). The IOP can handle many peripherals through a DMA-and-interrupt facility. In such a system, the computer is divided into three separate modules: the memory unit, the CPU, and the IOP. I/O processors are presented in Section 13-8.

Example of Program-Controlled Transfer

A simple example of data transfer from an I/O device through an interface into the CPU is shown in Figure 13-13. The device transfers bytes of data one at a time as they are available. When a byte is available, the device places it on the I/O bus and enables Ready. The interface accepts the byte into its data register and enables Acknowledge. The interface sets a bit in the status register, which we will refer to as a *flag*. The device can now disable Ready, but it will not transfer another byte until Acknowledge is disabled by the interface, according to the handshaking procedure established in Section 13-3.

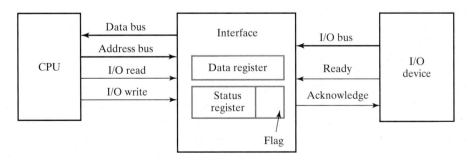

□ **FIGURE 13-13**
Data Transfer from I/O Device to CPU

Under program control, the CPU must check the flag to determine whether there is a new byte in the interface data register. This is done by reading the contents of the status register into a CPU register and checking the value of the flag. If the flag is equal to 1, the CPU reads the data from the data register. The flag is then cleared to 0 either by the CPU or the interface, depending on how the interface circuits are designed. Once the flag is cleared, the interface disables Acknowledge, and the device can transfer the next data byte.

A flowchart of the a program written for the preceding transfer is shown in Figure 13-14. The flowchart assumes that the device is sending a sequence of bytes that must be stored in memory. The program continually examines the status of the interface until the flag is set to 1. Each byte is brought into the CPU and transferred to memory until all of the data have been transferred.

The program-controlled data transfer is used only in systems that are dedicated to monitor a device continuously. The difference in information transfer rate between the CPU and the I/O device makes this type of transfer inefficient.

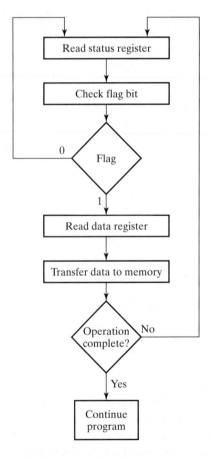

◻ **FIGURE 13-14**
Flowchart for CPU Program to Input Data

To see why, consider a typical computer that can execute the instructions to read the status register and check the flag in 100 ns. Assume that the input device transfers its data at an average rate of 100 bytes/s. This is equivalent to one byte every 10,000 μs, meaning that the CPU will check the flag 100,000 times between each transfer. Thus, the CPU is wasting time checking the flag instead of doing a useful processing task.

Interrupt-Initiated Transfer

An alternative to the CPU constantly monitoring the flag is to let the interface inform the computer when it is ready to transfer data. This mode of transfer uses the interrupt facility. While the CPU is running a program, it does not check the flag. However, when the flag is set, the computer is momentarily interrupted from proceeding with the current program and is informed of the fact that the flag has been set. The CPU drops what it is doing to take care of the input or output transfer. After the transfer is completed, the computer returns to the previous program to continue what it was doing before the interrupt. The CPU responds to the interrupt signal by storing the return address from the program counter into a memory stack or register, and then control branches to a service routine that processes the required I/O transfer. The way that the processor chooses the branch address of the service routine varies from one unit to another. In principle, there are two methods for accomplishing this: *vectored interrupt* and *nonvectored interrupt*. In a nonvectored interrupt, the branch address is assigned to a fixed location in memory. In a vectored interrupt, the source that interrupts supplies the branch address to the computer. This information is called the *vector address*. In some computers, the vector address is the first address of the service routine; in other computers, the vector address is an address that points to a location in memory where the first address of the service routine is stored. The vectored interrupt procedure was presented in Section 13-9 in conjunction with Figure 13-9.

13-6 PRIORITY INTERRUPT

A typical computer has a number of I/O devices attached to it that are able to originate an interrupt request. The first task of the interrupt system is to identify the source of the interrupt. There is also the possibility that several sources will request service simultaneously. In this case, the system must decide which device to service first.

A priority interrupt system establishes a priority over the various interrupt sources to determine which interrupt request to service first when two or more arrive simultaneously. The system may also determine which requests are permitted to interrupt the computer while another interrupt is being serviced. Higher levels of priority are assigned to requests that, if delayed or interrupted, could have serious consequences. Devices with high-speed transfers such as magnetic disks are given high priority, and slow devices such as keyboards receive the lowest priority.

When two devices interrupt the computer at the same time, the computer services the device with the higher priority first.

Establishing the priority of simultaneous interrupts can be done by software or hardware. Software uses a polling procedure to identify the interrupt source of highest priority. In this method, there is one common branch address for all interrupts. The program at the branch address takes care of interrupts by polling the interrupt sources in sequence. The priority of each interrupt source determines the order in which it is polled. The source with the highest priority is tested first, and if its interrupt signal is on, control branches to a routine which services that source. Otherwise, the source with the next lower priority is tested, and so on. Thus, the initial service routine for all interrupts consists of a program that tests the interrupt sources in sequence and branches to one of many other possible service routines. The particular service routine that is reached belongs to the highest priority device among all devices that interrupted the computer. The disadvantage of the software method is that if there are many interrupts, the time required to poll all the sources can exceed the time available to service the I/O device. In this situation, a hardware priority interrupt unit can be used to speed up the operation of the system.

A hardware priority interrupt unit functions as an overall manager in an interrupt system environment. The unit accepts interrupt requests from many sources, determines which of the incoming requests has the highest priority, and issues an interrupt request to the computer based on this determination. To speed up the operation, each interrupt source has its own interrupt vector address to access its own service routine directly. Thus, no polling is required, because all the decisions are made by the hardware priority interrupt unit. The hardware priority function can be established either by a serial or parallel connection of interrupt lines. The serial connection is also known as the daisy chain method.

Daisy Chain Priority

The *daisy chain* method of establishing priority consists of a serial connection of all devices that request an interrupt. The device with the highest priority is placed in the first position, followed by devices of priority in descending order, down to the device with the lowest priority, which is placed last in the chain. This method of connection between three devices and the CPU is shown in Figure 13-15. Interrupt request lines from all devices are ORed to form the interrupt line to the CPU. If any device has its Interrupt request at 1, the interrupt line goes to 1 and enables the interrupt input of the CPU. When no interrupts are pending, the interrupt line stays at 0, and no interrupts are recognized by the CPU. The CPU responds to an interrupt request by enabling Interrupt acknowledge. The signal that is produced is received by device 0 at its *PI* (priority in) input. The signal then passes on to the next device through the *PO* (priority out) output only if device 0 is not requesting an interrupt. If device 0 has a pending interrupt, it blocks the acknowledge signal from the next device by placing a 0 on the *PO* output and proceeds to insert its own interrupt vector address (*VAD*) onto the data bus for the CPU to use during the interrupt cycle.

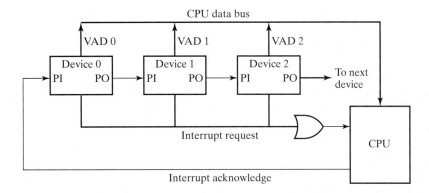

□ **FIGURE 13-15**
Daisy Chain Priority Interrupt

A device with a 0 on its *PI* input generates a 0 on its *PO* output to inform the device with next lower priority that the acknowledge signal has been blocked. A device that is requesting an interrupt and has a 1 on its *PI* input will intercept the acknowledge signal by placing a 0 on its *PO* output. If the device does not have pending interrupts, it transmits the acknowledge signal to the next device by placing a 1 on its *PO* output. Thus, the device with *PI* = 1 and *PO* = 0 is the one with the highest priority that is requesting an interrupt, and this device places its *VAD* on the data bus. The daisy chain arrangement gives the highest priority to the device that receives the Interrupt acknowledge signal from the CPU. The farther the device is from the first position, the lower is its priority.

Figure 13-16 shows the internal logic that must be included within each device connected in the daisy chain scheme. The device sets its *RF* latch when it is about to interrupt the CPU. The output of the latch functionally enters the OR that drives the interrupt line. If *PI* = 0, both *PO* and the enable line to *VAD* are equal to 0, irrespective of the value of *RF*. If *PI* = 1 and *RF* = 0, then *PO* = 1, the vector address is disabled, and the acknowledge signal passes to the next device through *PO*. The device is active when *PI* = 1 and *RF* = 1, which places a 0 on *PO* and enables the vector address onto the data bus. It is assumed that each device has its own distinct vector address. The *RF* latch is reset after a sufficient delay to ensure that the CPU has received the vector address.

Parallel Priority Hardware

The parallel priority interrupt method uses a register with bits set separately by the interrupt signal from each device. Priority is established according to the position of the bits in the register. In addition to the interrupt register, the circuit may include a mask register to control the status of each interrupt request. The mask register can be programmed to disable lower priority interrupts while a higher priority device is being serviced. It can also allow a high-priority device to interrupt the CPU while a lower priority device is being serviced.

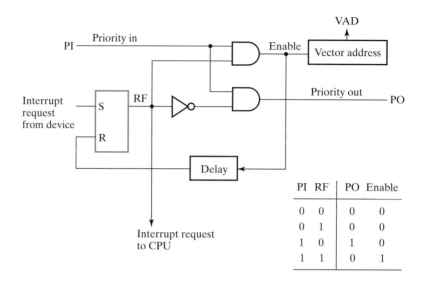

PI	RF	PO	Enable
0	0	0	0
0	1	0	0
1	0	1	0
1	1	0	1

☐ **FIGURE 13-16**
One Stage of the Daisy Chain Priority Arrangement

The priority logic for a system with four interrupt sources is shown in Figure 13-17. The logic consists of an interrupt register with individual bits set by external conditions and cleared by program instructions. Interrupt input 3 has the highest priority, input 0 the lowest. The mask register has the same number of bits as the interrupt register. By means of program instructions, it is possible to set or reset any bit in the mask register. Each interrupt bit and its corresponding mask bit are applied to an AND gate to produce the four inputs to a priority encoder. In this way, an interrupt is recognized only if its corresponding mask bit is set to 1 by the program. The priority encoder generates two bits of the vector address, which is transferred to the CPU via the data bus. Output V of the encoder is set to 1 if an interrupt request that is not masked has occurred. This provides the interrupt signal for the CPU.

The priority encoder is a circuit that implements the priority function. The logic of the priority encoder is such that, if two or more inputs are 1 at the same time, the input having the highest priority takes precedence. The circuit of a four-input priority encoder can be found in Section 4-4, and its truth table is listed in Table 4-5. Input D_3 has the highest priority so, regardless of the values of other inputs, when this input is 1, the output is $A_1 A_0 = 11$. D_2 has the next lower priority. The output is 10 if $D_2 = 1$, provided that $D_3 = 0$, regardless of the values of the other two lower priority inputs. The output is 01 when $D_1 = 1$, provided that the two higher priority inputs are equal to 0, and so on down the priority levels. The interrupt output labeled V is equal to 1 when one or more inputs are equal to 1. If all inputs are 0, V is 0, and the other two outputs of the encoder are not used. This is because the vector address is not transferred to the CPU when $V = 0$.

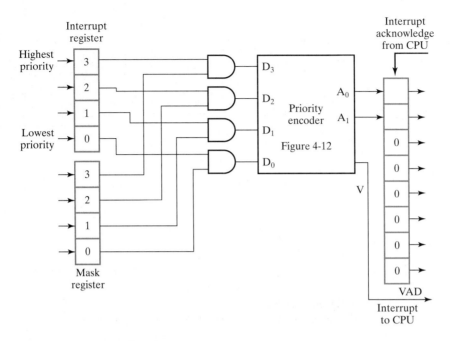

□ **FIGURE 13-17**
Parallel Priority Interrupt Hardware

The output of the priority encoder is used to form part of the vector address of the interrupt source. The other bits of the vector address can be assigned any values. For example, the vector address can be found by appending six zeros to the outputs of the encoder. With this choice, the interrupt vectors for the four I/O devices are assigned the 8-bit binary numbers equivalent to decimal 0, 1, 2, and 3.

13-7 DIRECT MEMORY ACCESS

The transfer of blocks of information between a fast storage device such as magnetic disk and the CPU can preoccupy the CPU and permit little, if any, other processing to be accomplished. Removing the CPU from the path and letting the peripheral device manage the memory buses directly will relieve the CPU from many I/O operations and allow it to proceed with other processing. In this transfer technique, called direct memory access (DMA), the DMA controller takes over the buses to manage the transfer directly between the I/O device and memory. As a consequence, the CPU is temporarily deprived of access to memory and control of the memory buses.

DMA may capture the buses in a number of ways. One common method extensively used in microprocessors is to disable the buses through special control signals. Figure 13-18 shows two control signals in a CPU that facilitate the DMA transfer. The bus request (*BR*) input is used by the DMA controller to request the

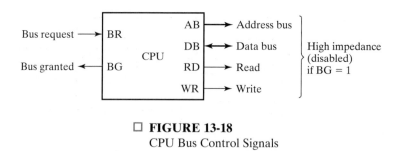

☐ **FIGURE 13-18**
CPU Bus Control Signals

CPU to relinquish control of the buses. When *BR* input is active, the CPU places the address bus, the data bus, and the read and write lines into a high-impedance state. After this is done, the CPU activates the bus granted (*BG*) output to inform the external DMA that it can take control of the buses. As long as the *BG* line is active, the CPU is unable to proceed with any operations requiring access to the buses. When the bus request input is disabled by the DMA, the CPU returns to its normal operation, disables the *BG* output, and takes control of the buses.

When the *BG* line is enabled, the external DMA controller takes control of the bus system in order to communicate directly with memory. The transfer can be made for an entire block of memory words, suspending operation of the CPU until the entire block is transferred, a process referred to as *burst transfer*. Or the transfer can be made one word at a time between executions of CPU instructions, a process called *single-cycle transfer* or *cycle stealing*. The CPU merely delays its bus operations for one memory cycle to allow the direct memory-I/O transfer to steal one memory cycle.

DMA Controller

The DMA controller needs the usual circuits of an interface to communicate with the CPU and the I/O device. In addition, it needs an address register, a word-count register, and a set of address lines. The address register and address lines are used for direct communication with memory. The word-count register specifies the number of words that must be transferred. The data transfer may be done directly between the device and memory under control of the DMA.

Figure 13-19 shows the block diagram of a typical DMA controller. The unit communicates with the CPU via the data bus and control lines. The registers in the DMA are selected by the CPU through the address bus by enabling the *DS* (DMA select) and *RS* (register select) inputs. The *RD* (read) and *WR* (write) inputs are bidirectional. When the *BG* (bus granted) input is 0, the CPU can communicate with the DMA registers through the data bus to read from or write to those registers. When *BG* = 1, the CPU has relinquished the buses, and the DMA can communicate directly with memory by specifying an address on the address bus and activating the *RD* or *WR* control. The DMA communicates with the external peripheral through the DMA request and DMA acknowledge lines by a prescribed handshaking procedure.

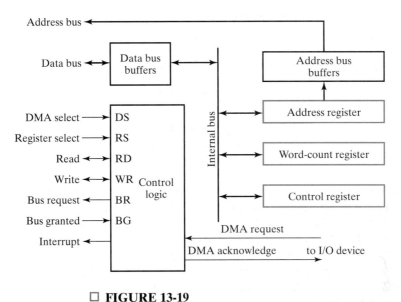

□ **FIGURE 13-19**
Block Diagram of a DMA Controller

The DMA controller has three registers: an address register, a word-count register, and a control register. The address register contains an address to specify the desired location of a word in memory. The address bits go through bus buffers onto the address bus. The address register is incremented after each word is transferred to memory. The word-count register holds the number of words to be transferred. This register is decremented by one after each word transfer and internally tested for zero. The control register specifies the mode of transfer. All registers in the DMA appear to the CPU as I/O interface registers. Thus, the CPU can read from or write to the DMA registers under program control via the data bus.

After initialization by the CPU, the DMA starts and continues to transfer data between memory and the peripheral unit until an entire block is transferred. The initialization process is essentially a program consisting of I/O instructions that include the address for selecting particular DMA registers. The CPU initializes the DMA by sending the following information through the data bus:

1. The starting address of the memory block in which data is available (for reading) or data is to be stored (for writing).

2. The word count, which is the number of words in the memory block.

3. A control bit to specify the mode of transfer, such as read or write.

4. A control bit to start the DMA transfer.

The starting address is stored in the address register, the word count in the word-count register, and the control information in the control register. Once the DMA is initialized, the CPU stops communicating with it unless the CPU receives an interrupt signal or needs to check how many words have been transferred.

DMA Transfer

The position of the DMA controller among the other components in a computer system is illustrated in Figure 13-20. The CPU communicates with the DMA through the address and data buses, as with any interface unit. The DMA has its own address, which activates the *DS* and *RS* lines. The CPU initializes the DMA through the data bus. Once the DMA receives the start control bit, it can begin transferring data between the peripheral device and memory. When the peripheral device sends a DMA request, the DMA controller activates the *BR* line, informing the CPU that it is to relinquish the buses. The CPU responds with its *BG* line, informing the DMA that the buses are disabled. The DMA then puts the current value of its address register onto the address bus, initiates the *RD* or *WR* signal, and sends a DMA acknowledge to the peripheral device.

When the peripheral device receives a DMA acknowledge, it puts a word on the data bus (for writing) or receives a word from the data bus (for reading). Thus, the DMA controls the read or write operation and supplies the address for memory. The peripheral unit can then communicate with memory through the data bus for a direct transfer of data between the two units while the CPU access to the data bus is momentarily disabled.

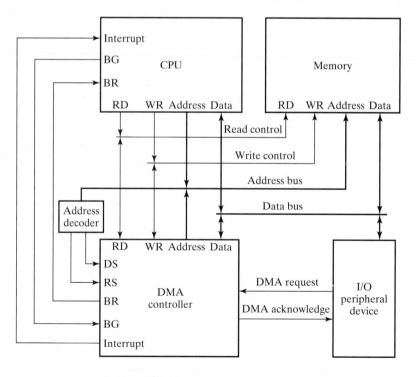

☐ **FIGURE 13-20**
DMA Transfer in a Computer System

For each word that is transferred, the DMA increments its address register and decrements its word-count register. If the word count has not reached zero, the DMA checks the request line coming from the peripheral. In a high-speed device, the line will be activated as soon as the previous transfer is completed. A second transfer is then initiated, and the process continues until the entire block is transferred. If the speed of the peripheral is slower, the DMA request line may be activated somewhat later. In this case, the DMA disables the bus request line so that the CPU can continue to execute its program. When the peripheral requests a transfer, the DMA requests the buses again.

If the word count reaches zero, the DMA stops any further transfer and removes its bus request. It also informs the CPU of the termination of the transfer by means of an interrupt. When the CPU responds to the interrupt, it reads the contents of the word-count register. A value of zero indicates that all the words were successfully transferred. The CPU can read the word-count register at any time, as well, to check the number of words already transferred.

A DMA controller may have more than one channel. In this case, each channel has a request and acknowledge pair of control signals that are connected to separate peripheral devices. Each channel also has its own address register and word-count register so that channels with high priority are serviced before channels with lower priority.

DMA transfer is very useful in many applications, including the fast transfer of information between magnetic disks and memory and between graphic displays and memory.

13-8 I/O PROCESSORS

Instead of having each interface communicate with the CPU, a computer may incorporate one or more external processors and assign them the task of communicating directly with all I/O devices. An input-output processor (IOP) may be classified as a processor with direct memory access capability that communicates with I/O devices. In this configuration, the computer system can be divided into a memory unit and a number of processors composed of the CPU and one or more IOPs. Each IOP takes care of input and output tasks, relieving the CPU of the "housekeeping" chores involved in I/O transfers. A processor that communicates with remote units over telephone and other communication media in a serial fashion is called a *data communication processor* (DCP). The benefit derived from using I/O processors is improved system performance, achieved through relieving the CPU of detailed tasks relating to I/O and assigning them to the appropriate I/O processors.

An IOP is similar to a CPU, except that it is designed to handle the details of I/O processing. Unlike the DMA controller, which must be set up entirely by the CPU, the IOP can fetch and execute its own instructions. IOP instructions are specifically designed to facilitate I/O transfers. In addition, the IOP can perform other processing tasks, such as arithmetic, logic, branching, and translation of code.

The block diagram of a computer with two processors is shown in Figure 13-21. The memory occupies a central position and can communicate with each processor by means of DMA. The CPU is responsible for processing data needed in the

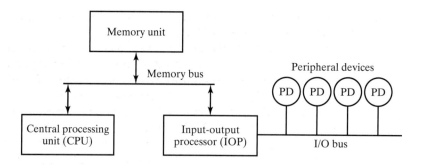

☐ **FIGURE 13-21**
Block Diagram of a Computer with I/O Processor

solution of computational tasks. The IOP provides a path for the transfer of data between various peripheral devices and the memory. The CPU is usually assigned the task of initiating the I/O program. From then on, the IOP operates independently of the CPU and continues to transfer data between external devices and memory. The data formats of peripheral devices often differ from those of memory and the CPU. The IOP must structure data words from many different sources. For example, it may be necessary to take four bytes from an input device and pack them into one 32-bit word before the transfer to memory. Data are gathered in the IOP at the device bit rate and bit capacity while the CPU is executing its own program. After assembly into a memory word, the data is transferred from the IOP directly into memory by stealing one memory cycle from the CPU. Similarly, an output word transferred from memory to the IOP is directed from the IOP to the output device at the device bit rate and bit capacity.

The communication between the IOP and the devices attached to it is similar to the program-controlled method of transfer. Communication with memory is similar to the DMA method. The way the CPU and IOP communicate with each other depends on the level of sophistication of the system. In very large-scale computers, each processor is independent of all the others, and any one processor can initiate an operation. In most computer systems, the CPU is the master, while the IOP is a slave processor. The CPU is assigned the task of initiating all operations, but I/O instructions are executed in the IOP. CPU instructions provide operations to start an I/O transfer and also to test I/O status conditions needed for making decisions on various I/O activities. The IOP, in turn, typically asks for attention from the CPU by means of an interrupt. It also responds to CPU requests by placing a status word in a prescribed location in memory, to be examined later by a CPU program. When an I/O operation is desired, the CPU informs the IOP where to find the I/O program and then leaves the details of the transfer to the IOP.

Instructions that are read from memory by an IOP are sometimes called *commands*, to distinguish them from instructions that are read by the CPU. An instruction and a command have similar functions. Commands are prepared by programmers and are stored in memory. The command words constitute the program for the IOP. The CPU informs the IOP where to find commands in memory when it is time to execute the I/O program.

Communication between the CPU and the IOP may take different forms, depending on the particular computer used. In most cases, the memory acts as a message center, where each processor leaves information for the other. To appreciate the operation of a typical IOP, we illustrate the method by which the CPU and IOP communicate with each other. This simplified example omits many operating details in order to provide an overview of basic concepts.

The sequence of operations may be carried out as shown in the flowchart of Figure 13-22. The CPU sends an instruction to test the IOP path. The IOP

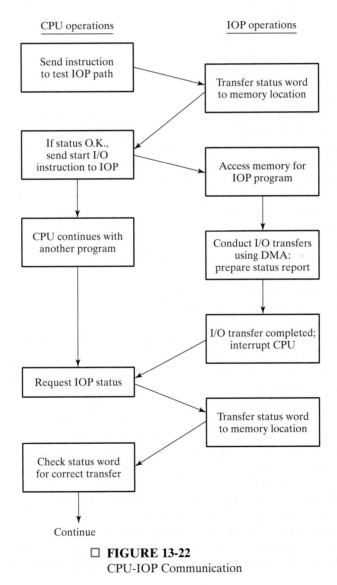

□ **FIGURE 13-22**
CPU-IOP Communication

responds by inserting a status word in memory for the CPU to check. The bits of the status word indicate the condition of the IOP and I/O device, such as "IOP overload condition," "device busy with another transfer," or "device ready for I/O transfer." The CPU refers to the status word in memory to decide what to do next. If all is in order, the CPU sends the instruction to start the I/O transfer. The memory address received with this instruction tells the IOP where to find its program.

The CPU can now continue with another program while the IOP is busy with the I/O program. Both programs refer to memory by means of DMA transfer. When the IOP terminates the execution of its program, it sends an interrupt request to the CPU. The CPU responds by issuing an instruction to read the status from the IOP. The IOP then responds by placing the contents of its status report into a specified memory location. The status word indicates whether the transfer has been completed or whether any errors occurred during the transfer. By inspecting the bits in the status word, the CPU determines whether the I/O operation was completed satisfactorily, without errors.

The IOP takes care of all data transfers between several I/O units and memory while the CPU is processing another program. The IOP and CPU compete for the use of memory, so the number of devices that can be in operation is limited by the access time of the memory. It is not possible for I/O devices to saturate the memory in most systems, as the speed of most devices is much slower than that of the CPU. However, multiple fast units, such as magnetic disks or graphics displays, can use an appreciable number of the available memory cycles. In that case, the speed of the CPU may deteriorate because the CPU often has to wait for the IOP to conduct memory transfers.

13-9 CHAPTER SUMMARY

In this chapter, we introduced I/O devices, typically called peripherals, and their associated digital support structures, including I/O buses, interfaces, and controllers. We studied the structure of a keyboard, a hard disk, and a graphics display. We looked at an example of a generic I/O interface and examined the interface and I/O controller for the keyboard. We introduced USB as an alternative solution to the attachment of many I/O devices. We considered timing problems between systems with different clocks and the parallel and serial transmission of information.

We also looked at modes of transferring information and saw how the more complex modes came about, principally to relieve the CPU from extensive, performance-robbing handling of I/O transfers. Interrupt-initiated transfers with multiple I/O interfaces lead to means of establishing priority between interrupt sources. Priority can be handled by software, serial daisy chain logic, or parallel interrupt-priority logic. Direct memory access accomplishes the transfer of data directly between an I/O interface and memory, with little CPU involvement. Finally, the I/O processor provides even greater independence of the CPU in handling I/O.

References

1. PATTERSON, D. A., and J. L. HENNESSY *Computer Organization and Design: The Hardware/Software Interface.* San Francisco, CA: Morgan Kaufmann, 1998.

2. VAN GILLUWE, F. *The Undocumented PC.* Reading, MA: Addison-Wesley, 1994.

3. MESSMER, H. P. *The Indispensable PC Hardware Book.* 2nd ed. Reading, MA: Addison-Wesley, 1995.

4. MindShare, Inc. (Don Anderson). *Universal Serial Bus System Architecture.* Reading, MA: Addison-Wesley Developers Press, 1997.

Problems

The plus (+) indicates a more advanced problem and the asterisk (*) indicates a solution is available on the Companion Website for the text.

13–1. *Find the formatted capacity of the hard disks described in the following table:

Disk	Heads	Cylinders	Sectors/ Track	Bytes/ Sector
A	1	1023	63	512
B	4	8191	63	512
C	16	16383	63	512

13–2. Estimate the time required to transfer a block of 1MB (2^{20} B) from disk to memory given the following disk parameters: seek time, 8.5 ms; rotational delay, 4.17 ms; controller time, negligible; transfer rate, 100 MB/s.

13–3. The addresses assigned to the four registers of the I/O interface of Figure 13-5 are equal to the binary equivalent of 240, 241, 242, and 243. Show the external circuit that must be connected between an 8-bit I/O address from the CPU and the CS, $RS0$, and $RS1$ inputs of the interface.

13–4. *How many I/O interface units of the type shown in Figure 13-5 can be addressed by using a 16-bit address, assuming

(a) that each of the chip select (CS) lines is attached to a different address line?

(b) that address bits are fully decoded to form the chip select inputs?

13–5. Six interface units of the type shown in Figure 13-5 are connected to a CPU that uses an I/O address of eight bits. Each one of the six chip select (CS) inputs is connected to a different address line. Specifically, address line 0 is connected to the CS input of the first interface unit, and address line 5 is connected to the CS input of the sixth interface unit. Address lines 7 and 6 are connected to the *RS1* and *RS0* inputs, respectively, of all six interface units. Determine the 8-bit address of each register in each interface (a total of 24 addresses).

13–6. *A different type of I/O interface does not have the *RS1* and *RS0* inputs. Up to two registers can be addressed by using a separate I/O read signal and I/O write signal for each address available. Assume that 50% of the registers at the interface with the CPU are read only, 25% of the registers are write only, and 25% of the registers are both read and write (bidirectional). How many registers can be addressed if the address contains four bits?

13–7. A commercial interface unit uses names different from those appearing in this text for the handshake lines associated with the transfer of data from the I/O device to the interface unit. The interface input handshake line is labeled *STB* (strobe), and the interface output handshake line is labeled *IBF* (input buffer full). A low-level signal on *STB* loads data from the I/O bus into the interface data register. A high-level signal on *IBF* indicates that the data has been accepted by the interface. *IBF* goes low after an I/O read signal from the CPU when it reads the contents of the data register.

   (a) Draw a block diagram showing the CPU, the interface, and the I/O device, along with the pertinent interconnections between the three units.

   (b) Draw a timing diagram for the handshaking transfer.

13–8. *Assume that the transfers with strobing shown in Figure 13-6 are between a CPU on the left and an I/O interface on the right. There is an address coming from the CPU for each of the transfers, both of which are initiated by the CPU.

   (a) Draw block diagrams showing the interconnections for the transfers.

   (b) Draw the timing diagrams for the two transfers, assuming that the address must be applied some time before the strobe becomes 1 and removed some time after the strobe becomes 0.

13–9. Assume that the transfers with handshaking shown in Figure 13-7 are between a CPU on the left and an I/O interface on the right. There is an address coming from the CPU for each of the transfers, both of which are initiated by the CPU.

   (a) Draw block diagrams, showing that interconnections for the transfers.

   (b) Draw the timing diagrams, assuming that the address must be applied some time before the request becomes 1 and removed some time after the request becomes 0.

13–10. *How many characters per second can be transmitted over a 57,600-baud line in each of the following modes? (Assume a character code of eight bits.)

   (a) Asynchronous serial transmission with two stop bits.

   (b) Asynchronous serial transmission with one stop bit.

   (c) Repeat a and b for a 115,200-baud line.

13–11. Sketch the timing diagram of the 11 bits (similar to Figure 13-8) that are transmitted over an asynchronous serial communication line when the ASCII letter E is transmitted with even parity. Assume that the ASCII character code is transmitted least significant bit first, with the parity bit following the character code.

13–12. What is the difference between the synchronous and the asynchronous serial transfer of information?

13–13. *Sketch the waveforms for the SYNC pattern used for USB and the corresponding NRZI waveform. Explain why the pattern selected is a good choice for achieving synchronization.

13–14. The following stream of data is to be transmitted by USB:
01111111001000001111110111111101
(a) Assuming bit stuffing is not used, sketch the NRZI waveform.
(b) Modify the stream by applying bit stuffing.
(c) Sketch the NRZI waveform for the result in b.

13–15. *The 8-bit ASCII word "Bye" is to be transmitted to a device address 39 and endpoint 2. List the Output and Data 0 packets and the Handshake packet for a Stall for this transmission prior to NRZI encoding.

13–16. Repeat problem 13-15 for the word "Hlo" and a Handshake packet of type No Acknowledge.

13–17. What is the basic advantage of using interrupt-initiated data transfer over transfer under program control without an interrupt?

13–18. *What happens in the daisy chain priority interrupt shown in Figure 13-15 when device 0 requests an interrupt after device 2 has sent an interrupt request to the CPU, but before the CPU responds with the interrupt acknowledge?

13–19. Consider a computer without priority interrupt hardware. Any one of many sources can interrupt the computer, and any interrupt request results in storing the return address and branching to a common interrupt routine. Explain how a priority can be established in the interrupt service program.

13–20. *What changes are needed in Figure 13-17 to make the four VAD values equal to the binary equivalent of 024, 025, 026, and 027?

13–21. Repeat problem 13-20 for VAD values 224, 225, 226 and 227.

13–22. *Design parallel priority interrupt hardware for a system with six interrupt sources.

13–23. A priority structure is to be designed that provides vector addresses.
(a) Obtain the condensed truth table of a 16×4 priority encoder.
(b) The four outputs w, x, y, z from the priority encoder are used to provide an 8-bit vector address in the form $10wxyz01$. List the 16 addresses, starting from the one with the highest priority.

13–24. *Why are the read and write control lines in a DMA controller bidirectional? Under what condition and for what purpose are they used as inputs? Under what condition and for what purpose are they used as outputs?

13–25. It is necessary to transfer 1024 words from a magnetic disk to a section of memory starting from address 2048. The transfer is by means of DMA as shown in Figure 13-20.

(a) Give the initial values that the CPU must transfer to the DMA controller.

(b) Give the step-by-step account of the actions taken during the input of the first two words.

14

MEMORY SYSTEMS

I n Chapter 9, we discussed basic RAM technology for implementing memory systems, including SRAMs and DRAMs. In the current chapter, we probe more deeply into what really constitutes a computer memory system. We begin with the premise that a fast, large memory is desirable and demonstrate that a straightforward implementation of such a memory for the typical computer is too costly and too slow. As a consequence, we study a more elegant solution in which most accesses to memory are fast (but some are slow) and the memory appears to be large. This solution employs two concepts: cache memory and virtual memory. A cache memory is a small, fast memory with special control hardware that permits it to handle a significant proportion of all accesses required by the CPU with an access time of the order of the CPU clock period. Virtual memory, implemented in software and hardware, using an intermediate-sized main memory (typically, DRAM), gives the appearance of a large main memory with access time similar to the main memory for the vast majority of accesses. The actual storage medium for most of the code and data in the virtual memory is a hard disk. Because there is a progression of components in the memory system having larger and larger storage capability, but slower and slower access (cache, main memory, and hard disk), the term *memory hierarchy* is applied.

In the generic computer at the beginning of Chapter 1, a number of components are heavily involved in the memory hierarchy. Within the processor, there is the memory management unit (MMU), which is hardware provided to support virtual memory. Also in the processor, the internal cache appears. Since this cache is too small to fully support the cache function, there is also an external cache attached to the CPU bus. Of course, the RAM is involved, and due to the presence of virtual memory, the hard disk, the bus interface, and the disk controller all have a role as parts of the memory system.

14-1 MEMORY HIERARCHY

Figure 14-1 shows a generic block diagram for a memory hierarchy. The lowest level of the hierarchy is a small, fast memory called a *cache*. For the hierarchy to function well, a very large proportion of the CPU instruction and operand fetches are expected to be from the cache. At the next level upward in the hierarchy is the *main memory*. The main memory serves directly most of the CPU instruction and operand fetches not satisfied by the cache. In addition, the cache fetches all of its data, some portion of which is passed on to the CPU, from the main memory. At the top level of the hierarchy is the *hard disk*, which is accessed only in the very infrequent cases in which a CPU instruction or operand fetch is not found in main memory.

With this memory hierarchy, since the CPU fetches most of the instructions and operands from the cache, it "sees" a fast memory, most of the time. Occasionally, when a word must come from main memory, a fetch takes somewhat longer. Very infrequently, when a word must be fetched from the hard disk, the fetch takes a very long time. In this last case, the CPU is likely to experience an interrupt that passes execution to a program which brings in a block of words from the hard disk. On balance, the situation is usually satisfactory, providing an average fetch time close to that of the cache. Moreover, the CPU sees a memory address space considerably larger than that of main memory.

With this general notion of a memory hierarchy kept in mind, we will proceed to consider an example that illustrates the potential power of such a hierarchy. However, there is one issue to be clarified first. In most instruction set architectures, the smallest of the objects that are addressed is a byte rather than a word. For a given load or store operation, whether a byte or word is affected is typically determined by the opcode. Addressing to bytes brings with it some assumptions and hardware details that are important, but, if used up to this point in the text, would have unnecessarily complicated much of the material covered. Consequently, for simplicity, we have assumed up to now that an addressed location contains a word. By contrast, in this chapter we will assume that addresses are defined

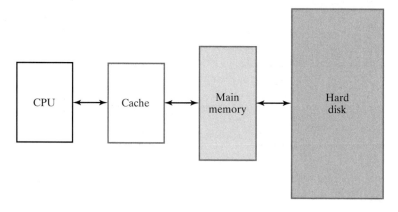

☐ **FIGURE 14-1**
Memory Hierarchy

for bytes, to match current practice. Nevertheless, we will still assume that data is moved around outside of the CPU as words or sets of words, to avoid messy explanations relating to the manipulation of bytes. This assumption simply hides some hardware details that would distract from the main focus of our discussion, but nevertheless must be handled by the hardware designer. To accomplish the simplification, if there are 2^b bytes per word, we will ignore the last b bits of the address. Since these bits are not needed to address a word, we show their values as 0's. For the examples we will present, b is always equal to 2, so two 0's are shown.

In Section 12-3, the pipelined CPU had a memory address with 32 bits and was able to access an instruction and data, if necessary, in each of the 1-ns clock cycles. Also, we assumed that the instruction and the data were, in effect, fetched from two different memories. To support this assumption in this chapter, we will suppose initially that the memory is divided in half—one-half for instructions and one-half for data. Each half of the memory must have an access time of 1 ns. In addition, if we utilize all the bits in the 32-bit address, then the memory can contain up to 2^{32} bytes, or 4 gigabytes (GB), of information. So the goal is to have two 2-GB memories, each with an access time of 1 ns.

Is such a memory realistic in terms of current (2003) computer technology? The typical memory is constructed of DRAM modules ranging in size from 16 to 64 Mbytes. The typical access time is about 10 ns. Thus, our two 2-GB memories would have an access time of somewhat more than 10 ns per word. This kind of memory both is too costly and operates at only one-tenth the desired speed. So our goal must be achieved another way, leading us to explore a memory hierarchy.

We begin by assuming a hierarchy with two caches, one for instructions and one for data, as shown in Figure 14-2. The use of these two caches permits one instruction and one operand to be fetched, or one instruction to be fetched and one result to be stored, in a single clock cycle if the caches are fast enough. In terms of the generic computer, we assume that the caches are internal, so that they can operate at speeds comparable to that of the CPU. Thus, fetches from the instruction cache and fetches from and stores to the data cache can be accomplished in

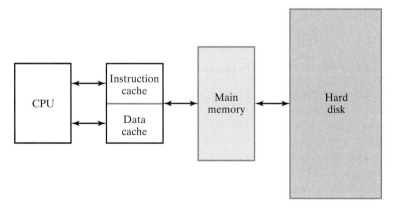

□ **FIGURE 14-2**
Example of Memory Hierarchy

2 ns. Hence, most of the fetches and stores for the CPU are from or to these caches and will take 2 CPU clock cycles. Suppose, then, that we are satisfied with most—say, 95%—of the memory accesses taking 2 ns. Suppose further that most of the remaining 5% of the memory accesses take 10 ns. Then the average access time is

$$0.95 \times 2 + 0.05 \times 10 = 2.4 \text{ ns}$$

This means that, on 19 out of every 20 memory accesses, the CPU operates at full speed, while the CPU will have to wait for 10 clock cycles for 1 out of every 20 memory accesses. This wait can be accomplished by stalling the CPU pipeline. Thus, we have accomplished our goal of "most" memory accesses taking 2 ns. But there is still the problem of the cost of the large memory.

Now suppose that, in addition to infrequently accepting a wait for a word from main memory that will take more than 10 ns, we are also willing to accept a very infrequent wait for a hard disk access taking 13 ms = 1.3×10^7 ns. Suppose that we have data indicating that about 95% of the fetches will be from a cache and about 4.999995% of the fetches will be from main memory. With this information, we can estimate the average access time as

$$0.95 \times 2 + 0.04999995 \times 10 + 5 \times 10^{-8} \times 1.3 \times 10^7 = 3.05 \text{ ns}$$

Thus, the average access time is about 3 times the 1 ns CPU clock period, but is about one-third of the 10 ns access time for main memory, again with 19 out of 20 of the accesses taking place in 2 ns. So we have achieved an average access time of about 3.05 ns for a memory structure with a capacity of 2^{32} bytes, not far from the original goal. Further, the cost of this memory hierarchy is tens of times smaller than the large, fast memory approach.

It therefore appears that the original goal of the appearance of a fast, large memory has been approached by the memory hierarchy at a reasonable cost. But along the way, we made some assumptions, namely, that 95% of the time the word desired would come from what we are now calling the cache and that 99.999995% of the time the words would come from either cache or main memory, with the remainder from hard disk. In the rest of this chapter, we will explore why assumptions similar to these usually hold, and we will examine the hardware and associated software components needed to achieve the goals of the memory hierarchy.

14-2 LOCALITY OF REFERENCE

In the previous section, we indicated that the success of the memory hierarchy is based on assumptions that are critical to achieving the appearance of a large, fast memory. We now deal with the foundation for making these assumptions, which is called *locality of reference*. Here "reference" means reference to memory for accessing instructions and for reading or writing operands. The term "locality" refers to the relative times at which instructions and operands are accessed (*temporal locality*) and the relative locations at which they reside in main memory (*spatial locality*).

Let us consider first the nature of the typical program. A program frequently contains many loops. In a loop, a sequence of instructions is executed many times before the program exits the loop and moves on to another loop or straight-line code not in a loop. In addition, loops are often nested in a hierarchy in which loops are contained in loops, and so on. Suppose we have a loop of eight instructions that is to be executed 100 times. Then for 800 executions, all instruction fetches will occur from just eight addresses in memory. Thus, each of the eight addresses is visited 100 times during the time the loop is executed. This is an example of temporal locality in the sense that an address which is accessed is likely to be accessed many times in the near future. Also, it is likely that the addresses of the instructions will be in sequential order. Thus, if an address is accessed for an instruction, nearby addresses are going to be addressed during the execution of the loop. This is an example of spatial locality.

In terms of accessing operands, similar temporal and spatial localities also occur. For example, in a computation on an array of numbers, there are multiple visits to the locations of many of the operands, giving temporal locality. Also, as the computation proceeds, when a particular address is accessed for a number, sequential addresses near to it are likely to be accessed for other numbers in the array, giving spatial locality.

From the prior discussion, we can conjecture that there is significant locality of reference in computer programs. To verify this decisively, it is necessary to study the patterns of execution of real programs. Such studies have demonstrated the presence of significant temporal and spatial locality of reference and play an important role in the design of caches and virtual memory systems.

The next question to answer is: What is the relation of locality of reference to the memory hierarchy? To examine this issue, we consider again the instruction fetch within a loop and look at the relationship between the cache and main memory. Initially, we assume that instructions are present only in main memory and that the cache is empty. When the CPU fetches the first instruction in a loop, it obtains the instruction from main memory. But the instruction and a portion of its address called the *address tag* are also placed in the cache. What then happens for the next 99 executions of this instruction? The answer is that the instruction can be fetched from the cache, which provides a much faster access. This is temporal locality at work: The instruction that was fetched once will tend to be used again and is now present in the cache for fast access.

Additionally, when the CPU fetches the instruction from main memory, the cache fetches nearby instructions into its SRAM. Now suppose that the nearby instructions include the entire loop of eight instructions presented in our example. Then all of the instructions are in the cache. By bringing in such a block of instructions, the cache is able to exploit spatial locality: It takes advantage of the fact that the execution of the first instruction implies the execution of instructions with nearby addresses by making the latter instructions available for fast access.

In our example, each of the instructions is fetched from main memory exactly once for the 100 executions of the loop. All other instruction fetches come from the cache. Thus, in this particular example, at least 99% of the instructions being executed are fetched from the cache, so that the rate of execution of instructions is governed almost completely by the cache access time and CPU speed, and very little by the main memory access time. Without temporal locality, many more accesses to main memory would occur, slowing down the system.

A relationship similar to that between cache and the main memory can exist between main memory and the hard disk. Again, both temporal and spatial locality of reference are of interest, except this time on a much larger scale. Programs and data are fetched from the hard disk, and data is written to the hard disk in blocks that range from kilowords to megawords. Ideally, once the code and initial data for a program reside in main memory, the hard disk need not be accessed except for storing final results of the program. But this can happen only if all of the code and data, including intermediate data used by the program, reside fully in main memory. If not, then it will be necessary to bring in code from the hard disk and to read and write data from and to the hard disk during program execution. Words are read from and written to the disk in blocks referred to as *pages*. If the movement of pages between main memory and hard disk is transparent to the programmer, then it will appear as if main memory is large enough to hold the entire program and all of the data. Hence, this automated arrangement is referred to as *virtual memory*. During the execution of the program, if an instruction to be executed is not in main memory, the CPU program flow is diverted to bring the page containing the instruction into main memory. Then the instruction can be read from main memory and executed. The details of this operation and the hardware and software actions required for it will be covered in Section 14-4.

In summary, locality of reference is absolutely key to the success of the concepts of cache memory and virtual memory. In the case of most programs, locality of reference is present to a fairly high degree. But occasionally, one does encounter a program that, for example, requires frequent access to a large body of data that cannot be accommodated in main memory. In such a case, the computer spends almost all of its time moving information between main memory and the hard disk and does little other computation. Continuous sounds emanating from the hard disk as the heads move from track to track is a telltale sign of this phenomenon, which is referred to as *thrashing*.

14-3 CACHE MEMORY

To illustrate the concept of cache memory, we assume a very small cache of eight 32-bit words and a small main memory with 1 KB (256 words), as shown in Figure 14-3. Both of these are too small to be realistic, but their size makes illustration of the concepts easier. The cache address contains 3 bits, the memory address 10. Out of the 256 words in main memory, only 8 at a time may lie in the cache. In order for the CPU to address a word in the cache, there must be information in the cache to identify the address of the word in main memory. If we consider the example of the loop in the last section, clearly, we find it desirable to contain the entire loop within the cache, so that all of the instructions can be fetched from the cache while the program is executing most of the passes through the loop. The instructions in the loop lie in consecutive word addresses.

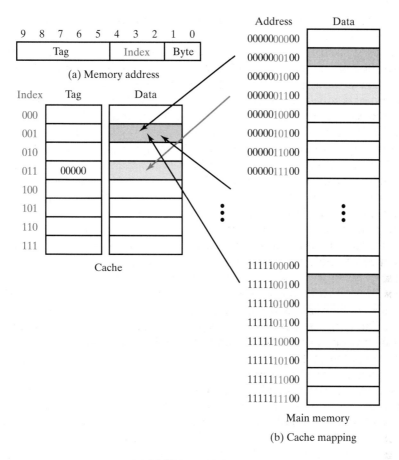

□ **FIGURE 14-3**
Direct Mapped Cache

Thus, it is desirable for the cache to have words from consecutive addresses in main memory present simultaneously. A simple way to facilitate this feature is to make bits 2 through 4 of the main memory address be the cache address. We refer to these bits as the *index*, as shown in Figure 14-3. Note that the data from address 0000001100 in main memory must be stored in cache address 011. The upper 5 bits of the main memory address, called the *tag*, are stored in the cache along with the data. Continuing the example, we find that for main memory address 0000001100, the tag is 00000. The tag combined with the index (or cache address) and 00 byte field identify an address in main memory.

Suppose that the CPU is to fetch an instruction from location 000001100 in main memory. This instruction may actually come from either the cache or main memory. The cache separates the tag 00000 from the cache address 011, internally fetches the tag and the stored word from location 011 in the cache memory, and compares the tag fetched with the tag portion of the address from the CPU. If the tag fetched is 00000, then the tags match, and the stored word fetched from cache memory is the desired instruction. Thus, the cache control places this word on the bus to the CPU, completing the fetch operation. This case in which the memory word is fetched from cache is called a *cache hit*. If the tag fetched from cache memory is not 00000, then there is a tag mismatch, and the cache control notifies main memory that it must provide the memory word, which is not available in the cache. This situation is called a *cache miss*. For a cache to be effective, the slower fetches from main memory must be avoided as much as possible, making considerably more cache hits than cache misses necessary.

When a cache miss occurs on a fetch, the word from main memory is not placed just on the bus for the CPU. The cache also captures the word and its tag and stores them for future access. In our example, the tag 00000 and the word from memory will be written in cache location 011 in anticipation of future accesses to the same memory address. The handling of writes to memory will be dealt with later in the chapter.

Cache Mappings

The example we just considered uses a particular association or mapping between the main memory address and the cache address; namely, the last three bits of the main memory word address are the cache address. Additionally, there is only one location in the cache for the 2^5 locations in main memory that have their last three bits in common. This mapping in Figure 14-3 in which only one specific location in the cache can contain the word from a particular main memory location is called *direct mapping*.

Direct mapping for a cache, however, does not always produce the most desirable situation. In our loop instruction fetch example, suppose that instructions and data are in the same cache and that data from location 1111101100 is frequently used. Then when the instruction in 0000001100 is fetched, location 011 in the cache is likely to contain the data from 1111101100 and tag 11111. A cache miss occurs and causes tag 11111 to be replaced in the cache with tag 00000 and the data to be replaced with the instruction. But the next time the data is needed, another cache miss occurs, since the location in the cache is now occupied by the instruction. Throughout the execution of the loop, both instruction fetch and data fetch cause many cache misses, significantly slowing CPU processing. To solve this problem, we explore alternative cache mappings.

In direct mapping, 2^5 addresses in main memory map to the single address in the cache that matches their last three bits. These locations are highlighted in gray in Figure 14-3 for index 001. As is illustrated, only one of the 2^5 addresses can have its word in cache address 001 at any time. In contrast, suppose that we let locations in main memory map into an arbitrary location in the cache. Then any location in memory can be mapped to any one of the eight addresses in the cache. This means that the tag will now be the full main memory word address. We examine the operation of such a cache having a *fully associative* mapping in Figure 14-4. Note that in this case there are two main memory addresses, 0000010000 and 1111110000, with bits 2 through 4 equal to 100 among the cache

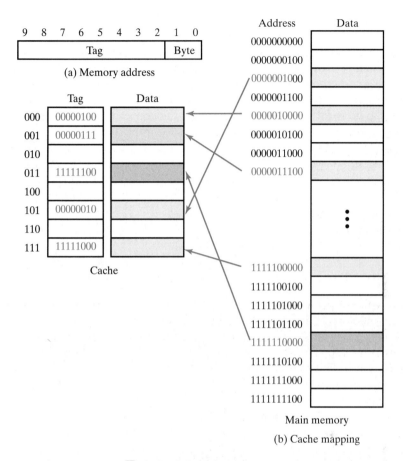

□ **FIGURE 14-4**
Fully Associative Cache

tags. These two addresses cannot be present simultaneously in the direct-mapped cache, as they would both occupy the cache address 100. Thus, a succession of cache misses due to alternate fetching of an instruction and data with the same index is avoided here, since both can be in the cache.

Now suppose that the CPU is to fetch an instruction from location 0000010000 in main memory. This instruction may actually be returned from either the cache or main memory. Since the instruction might lie in the cache, the cache must compare 00000100 to each of its eight tags. One way to do this is to successively read each tag and the associated word from the cache memory and compare the tag to 00000100. If a match occurs, as it will for the given address and cache location 000 in Figure 14-4, a cache hit occurs. The cache control then places the word on the bus to the CPU, completing the fetch operation. If the tag fetched from the cache is not 00000100, then there is a tag mismatch, and the cache control fetches the next successive tag and word. In the worst case, a match on the tag in cache address 111, eight fetches from the cache are required before the cache hit occurs. At 2 ns a fetch, this requires at least 16 ns, about half the time it would take to obtain the instruction from main memory. So successive reads of tags and words from the cache memory to find a match is not a very desirable approach. Instead, a structure called *associative memory* implements the tag portion of the cache memory.

Figure 14-5 shows an associative memory for a cache with 4-bit tags. The mechanism for writing tags into the memory uses a conventional write. Likewise, the tags can be read from the memory using the conventional memory read. Thus, the associative memory can use the bit slice model for RAM presented in

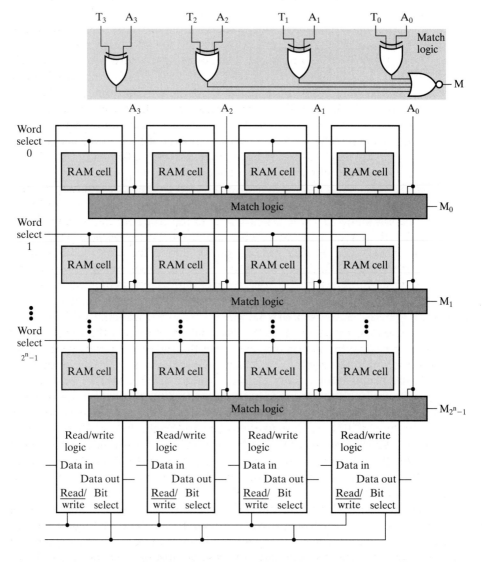

☐ **FIGURE 14-5**
Associative Memory for 4-bit Tags

Chapter 9. In addition, each tag storage row has match logic. The implementation of this logic and its connection to the RAM cells are shown in the figure. The match logic does an equality comparison or match between the tag T and the applied address A from the CPU. The match logic for each tag is composed of an exclusive-OR gate for each bit and a NOR gate that combines the outputs of the exclusive-ORs. If all of the bits of the tag and the address match, then the outputs of all the exclusive-ORs are 0 and the NOR output is a 1, indicating a match. If there is a mismatch between any of the bits in the tag and the address, then at least one exclusive-OR has a 1 output, which causes the output of the NOR gate to be 0, indicating a mismatch.

Since all tags are unique, only two situations can arise in the associative memory: there will be a match, with a 1 on the output of the match logic for one matching tag and a 0 on the remaining match logic outputs; or there will be no match, and all of the match logic outputs will be 0. With an associative memory holding the cache tags, the outputs of the match logic drive the word lines for the data memory words to be read. A signal must indicate whether a hit or a miss has occurred. If this signal is 1 for a hit and 0 for a miss, then it can be generated by using the OR of the match outputs. In the case of a hit, a 1 on Hit/$\overline{\text{miss}}$ places the word on the memory bus to the CPU; in the case of a miss, a 0 on Hit/$\overline{\text{miss}}$ tells the main memory that it is to provide the word addressed.

As in the case of the direct-mapped cache discussed earlier, the fully associative cache must capture the data word and its address tag and store them for future accesses. But now a new problem arises: Where in the cache are the tag and data to be placed? In addition to selecting a cache mapping, the cache designer must select a replacement approach that determines the location in the cache to be used for the incoming tag and data. One possibility is to select a *random replacement* location. The 3-bit address can be read from a simple hardware structure that generates a number which satisfies certain properties of random numbers. A somewhat more thoughtful approach is to use a first in, first out (*FIFO*) location. In this case, the location selected for replacement is the one that has occupied the cache for the longest time, based on the notion that the use of this oldest entry is likely to be finished. An approach that appears to attack the replacement problem even more directly is the *least recently used* (LRU) location approach. The goal of this approach is to replace the entry that has been unused for the longest time—hence the least recently used entry. The reason is that a cache entry that has not been used for the longest time is least likely to be used in the future. Thus, it can be replaced by a new cache entry. Although the LRU approach yields better results for caches, the difference between it and the other approaches is not large, and full implementation is costly. As a consequence, if used at all, the LRU approach is often only approximated.

There are also performance and cost issues surrounding the fully associative cache. Although such a cache provides maximum flexibility and good performance, it is not clear that the cost is justified. In fact, an alternative mapping that has better performance and eliminates the cost of most of the matching logic is a compromise between a direct-mapped cache and a fully associative cache. For such a mapping, lower order address bits act much as they do in direct mapping;

however, for each combination of lower order address bits, instead of having one location, there is a *set* of *s* locations. As with direct mapping, the tags and words are read from the cache memory locations addressed by the lower order address bits. For example, if the *set size s* equals two, then two tags and the two accompanying data words are read simultaneously. The tags are then simultaneously compared to the CPU-supplied address using just two matching logic structures. If one of the tags matches the address, then the associated word is returned to the CPU on the memory bus. If neither tag matches the address, then the two 0 matching values are used to send a miss signal to the CPU and main memory. Since there are sets of locations and associativity is used on sets, this technique is called *set-associative mapping*. Such a mapping with a set size *s* is an *s-way* set-associative mapping.

Figure 14-6 shows a two-way set-associative cache. There are eight cache locations arranged in four rows of two locations each. The rows are addressed by a 2-bit index and contain tags made up of the remaining six bits of the main memory address. The cache entry for a main memory address must lie in a specific row of the cache, but can be in either of the two columns. In the figure, the addresses are the same as they are in the fully associative cache in Figure 14-4. Note that no

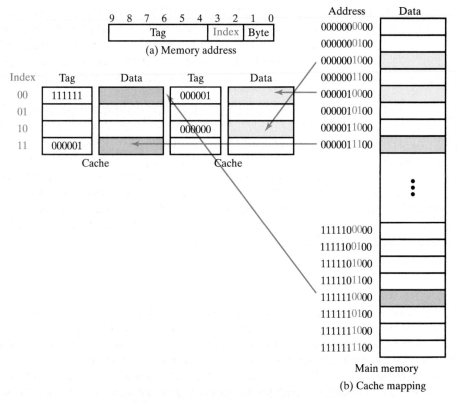

(b) Cache mapping

☐ **FIGURE 14-6**
Two-way Set-associative Cache

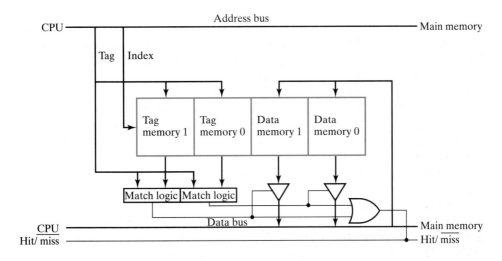

□ **FIGURE 14-7**
Partial Hardware Block Diagram for Set-associative Cache

mapping is shown for main memory address 1111100000, since the two cache cells in set 00 are already occupied by addresses 0000010000 and 1111110000. In order to accommodate 1111100000, the set size would need to be at least three. This example illustrates a case in which the reduced flexibility of a set-associative cache, compared to a fully associative cache, has an impact. The impact declines as the set size increases.

Figure 14-7 is a section of a hardware block diagram for the set-associative cache of Figure 14-6. The index is used to address each row of the cache memory. The two tags read from the tag memories are compared to the tag part of the address on the address bus from the CPU. If a match occurs, then the three-state buffer on the corresponding data memory output is activated, placing the data onto the data bus to the CPU. In addition, the match signal causes the output of the Hit/miss OR gate to become 1, indicating a hit. If a match does not occur, then Hit/miss is 0, informing the main memory that it must supply the word to the CPU and informing the CPU that the word will be delayed.

Line Size

To this point, we have assumed that each cache entry consists of a tag and a single memory word. In real caches, spatial locality is to be exploited, so additional words close to the one addressed are included in the cache entry. Then, rather than a single word being fetched from main memory when a cache miss occurs, a block of l words called a *line* is fetched. The number of words in a line is a power of two, and the words are aligned on address boundaries. For example, if four words are included in a line, then the addresses of the words in the line differ only in bits 2 and 3. The use of a block of words changes the makeup of the fields into which the cache divides the address. The new field structure is shown in Figure 14-8(a). Bits 2 and 3, the Word field, are used to address the word within the line. In this case, two bits are used, so there are four words per line. The next field, Index, identifies the set. Here there are two bits used, so there are four sets of tags and lines. The remainder of the address word is the Tag field, which contains the remaining four bits of the 10-bit memory address.

The resulting cache structure is shown in Figure 14-8(b). The tag memory has eight entries, two in each of the four sets. Corresponding to each of the tag entries is a line of four data words. To ensure fast operation, Index is applied to the tag memory to read two tags, one for each of the set entries, simultaneously. At the same time, Index and the Word address are applied to read out two words from the cache data memory that correspond to the two tags. Matching logic provided for each of the two set elements compares each tag to the CPU-supplied address. If a match occurs, then the associated cache data word already read is placed on the memory bus to the CPU. Otherwise, a cache miss is signaled, and the word addressed is returned from main memory to the CPU. The line containing the word

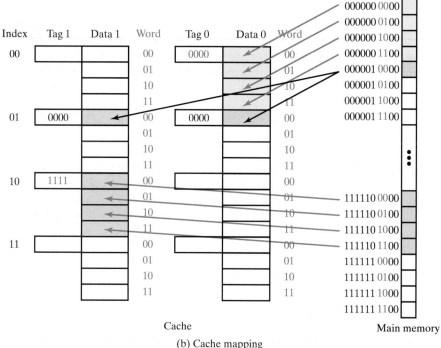

(a) Memory address

(b) Cache mapping

☐ **FIGURE 14-8**
Set-associative Cache with 4-word Lines

and its tag are also loaded into the cache. To facilitate loading the entire line of words, the width of the memory bus between main memory and the cache, as well as the cache load path, is made more than one word wide. Ideally, for our example the path is $4 \times 32 = 128$ bits wide. This allows the entire line to be placed in the cache in a single main memory read cycle. If the path is narrower, then a sequence of several reads from main memory is required.

An additional decision that the cache designer has to make is to determine the line size. A wide path to memory can affect both cost and performance, and a narrower path can slow transfer of the line to the cache. These features encourage a smaller cache line size, while spatial locality of reference encourages a larger line. In current systems, however, use of synchronous DRAM facilitates reading or writing large cache lines without the cost and performance issues associated with wide path. The rapid writing to and reading from memory of consecutive words achieved by using synchronous DRAM matches well the needs for transferring cache lines.

Cache Loading

Before any words and tags have been loaded into the cache, all locations contain invalid information. If a hit occurs on the cache at this time, then the word fetched and sent to the CPU cannot have come from main memory and is invalid. As lines are fetched from main memory into the cache, cache entries become valid, but there is no way to distinguish valid from invalid entries. To deal with this problem, in addition to the tag, a bit is added to each cache entry. This *valid bit* indicates that the associated cache line is valid (1) or invalid (0). It is read out of the cache along with the tag. If the valid bit is 0, then a cache miss occurs even if the tag matches the address from the CPU, requiring the addressed word to be taken from main memory.

Write Methods

We have focused so far on reading instructions and operands from the cache. What happens when a write occurs? Recall that, up to now, the words in a cache have been viewed simply as copies of words from main memory that are read from the cache to provide faster access. Now that we are considering writing results, this viewpoint changes somewhat. Following are three possible write actions from which we can select:

1. Write the result into main memory.
2. Write the result into the cache.
3. Write the result into both main memory and the cache.

Various realistic cache write methods employ one or more of these actions. Such methods fall into two main categories: write-through and write-back.

In *write-through*, the result is always written to main memory. This uses the main memory write time and can slow down processing. The slowdown can be partially avoided by using *write buffering*, a technique in which the address and word to be written are stored in special registers called write buffers by the CPU so that it can continue processing during the write to main memory. In most cache designs, the result is also written into the cache if the word is present there—that is, if there is a cache hit.

In the *write-back* method, also called *copy-back*, the CPU performs a write only to the cache in the case of a cache hit. If there is a miss, the CPU performs a write to main memory. There are two possible design choices for when a cache miss occurs. One is to read the line containing the word to be written from main memory into the cache, with the new word written into both the cache and main memory. This is referred to as *write-allocate*. It is done with the hope that there will be additional writes to the same block which will result in write hits and thus avoid writes to main memory. The other choice on a write miss is simply to write to main memory. In what follows, we will assume that write-allocate is used.

The goal of a write-back cache is to be able to write at the writing speed of the cache whenever there is a cache hit. This avoids having all writes performed at the slower writing speed of main memory. In addition, it reduces the number of accesses to main memory, making it more accessible to DMA, an I/O processor, or another CPU in the system. A disadvantage of write-back is that main memory entries corresponding to words in the cache that have been written are invalid. Unfortunately, this can cause a problem with respect to I/O processors or another CPU in the system accessing the same main memory, due to "stale" data in the memory.

The implementation of the write-back concept requires a write-back operation from the cache location to be used to store a new line being brought from main memory on a read miss. If the location in the cache contains a word that has been written into, then the entire line from the cache must be written back into main memory in order to release the location for the new line. This write-back requires additional time whenever a read miss occurs. To avoid a write-back on every read miss, an additional bit is added to each cache entry. This bit, called the *dirty bit*, is a 1 if the line in the cache has been written and a 0 if it has not been written. Write-back must be performed only if the dirty bit is a 1. With write-allocate used in a write-back cache, a write-back operation may also be required on a write miss.

Many other issues affect the choice of cache design parameters, particularly in the case of caches in a system in which the main memory may be read or written by a device other than the CPU for which the cache is provided.

Integration of Concepts

We now put together the basic concepts we have examined to determine the block diagram for a 256 KB, two-way set-associative cache with write-through. The memory address shown in Figure 14-9(a) contains 32 bits using byte addressing with line size $l = 16$ bytes. The index contains 13 bits. Since 4 bits are used for addressing words and bytes, and 13 bits are used for the index, the tag contains the remaining 15 bits of the 32-bit address. The cache contains 16,384 entries consisting of $2^{13} =$ 8192 sets. Each cache entry contains 16 bytes of data, a 15-bit tag, and a valid bit. The replacement strategy is random replacement.

Figure 14-9(b) gives the block diagram for the cache. There are two data memories and two tag memories, since the cache is two-way set associative. Each of these memories contains $2^{13} = 8192$ entries. Each entry in the data memory consists of 16 bytes. Since 32-bit words are assumed, there are four words in each data memory entry. Thus, each of the data memories consists of four 8192×32 memories in parallel with the index as their common address. In order to read a single word from these four memories on a cache hit, a 4-to-1 selector using three-state memory outputs selects the word, based on the two bits in the Word field of the address. The two tag memories are 8192×15; in addition to them, a valid bit is associated with each cache entry. These bits are stored in an 8192×2 memory and read out during a cache access with the data and tags. Note that the path between the cache and main memory is 128 bits wide. This allows us to assume that an entire cache line can be read from main memory in a single main memory cycle, an assumption that does not necessarily hold in practice. To understand the elements

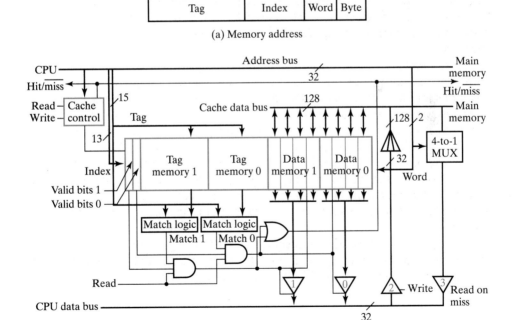

(a) Memory address

(b) Cache diagram

□ **FIGURE 14-9**
Detailed Block Diagram for 256K Cache

of the cache and how they work together, we will look at three possible cases of reading and writing. For each of these cases, we assume that the address from the CPU is $0F3F4024_{16}$. This gives Tag = 000011110011111_2 = $079F_{16}$, Index = 1010000000010_2 = 1402_{16}, and Word = 01_2.

First we assume a read hit—a read operation in which the data word lies in a cache entry, as in Figure 14-10. The cache uses the Index field to read out two tag entries from location 1402_{16} in Tag memory 1 and Tag memory 0. The match logic compares the tags of the entries, and in this case we assume that Tag 0 matches, causing Match 0 to be 1. This does not necessarily mean that we have a hit, since the cache entry may be invalid. Thus, the Valid 0 from location 1402_{16} bit is ANDed with Match 0. Also, the data can be placed on the CPU data bus only if the operation is a read. Thus, Read is ANDed with the Match 0 bit and the Valid 0 bit to form the control signal for three-state buffer 0. In this case, the control signal for the buffer 0 is 1. The data memories have used the Index field to read out eight words from location 1402_{16} at the same times the tags were read. The Word field selects the two of the eight words with word = 01_2 to place on the data buses going into the three-state buffers 1 and 0. Finally, with three-state buffer 0 turned on, the word addressed is placed on the CPU data bus. Also, the Hit/miss signal sends a 1 to the CPU and the main memory, notifying them of the hit.

In the second case, also shown in Figure 14-10, we assume a read miss—a read operation in which the data word is not in a cache entry. As before, the Index field address reads out the tag and valid entries, two tag comparisons are made, and two valid bits are checked. For both entries, a miss has occurred and is signaled

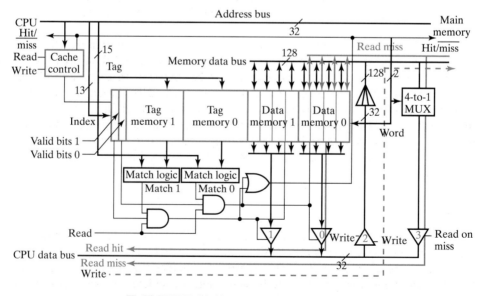

☐ **FIGURE 14-10**
256K Cache: Read and Write Operations

by Hit/miss at 0. This means that the word must be fetched from main memory. Accordingly, the cache control selects the cache entry to be replaced, and four words read from main memory are applied simultaneously by the memory data bus to the cache inputs and are written into the cache entry. At the same time, the 4-to-1 multiplexer selects the word addressed by the Word field and places it on the CPU data bus using the three-state buffer 3.

In the third case in Figure 14-10, we assume a write operation. The word from the CPU is fanned out to appear in all four of the word positions of the 128-bit memory data bus. The address to which the word is to be written is provided by the address bus to main memory for the write operation into the addressed word only. If the address causes a hit on the cache, the word addressed is also written into the cache.

Instruction and Data Caches

In most of the designs in previous chapters, we assumed that it was possible to fetch an instruction and to read an operand or write a result in the same clock cycle. To do this, however, we need a cache that can provide access to two distinct addresses in a single clock cycle. In response to this need, we discussed in a prior subsection an *instruction cache* and a *data cache*. In addition to easily providing multiple accesses per clock, the use of two caches permits caches that have different design parameters. The design parameters for each cache can be selected to fit the different characteristics of access for fetching instructions or reading and writing data. Because the demands on each of these caches are typically less than those on a single cache, a simpler design can be used. For example, a single cache may require a four-way set-association structure, whereas an instruction cache needs only direct mapping, and a data cache may need only a two-way set-associative structure.

In other instances, a single cache for both instructions and data may be used. Such a *unified cache* is typically as large as the instruction and data caches combined. The unified cache allows cache entries to be shared by instructions and data freely. Thus, at one time more entries can be occupied by instructions, and at another time more entries can be occupied by data. This flexibility has the potential for increasing the number of cache hits. This higher hit rate may be misleading, however, since the unified cache supports only one access at a time, and separate caches support two simultaneous accesses as long as one is for instructions and one is for data.

Multiple-Level Caches

It is possible to extend the depth of the memory hierarchy by adding additional levels of cache. Two levels of cache, often referred to as L1 and L2, with L1 closest to the CPU, are often used. In order to satisfy the demand of the CPU for instruction and operands, a very fast L1 cache is needed. To achieve the necessary speed, the delay that occurs when crossing IC boundaries is intolerable. Thus, the L1 cache is placed in the processor IC together with the CPU and is referred to as the *internal cache*, as in the generic computer processor. But the area in the IC is limited, so the L1 cache is typically small and inadequate if it is the only cache. Thus, a larger L2 cache is added outside of the processor IC.

The design of a two-level cache is more complex than that of a single-level cache. Two sets of parameters are specified. The L1 cache can be designed to specific CPU access needs including the possibility of separate instruction and data caches. Also, the constraint of external pins between the CPU and L1 cache is removed. In addition to permitting faster reads, the path between the CPU and the L1 cache can be quite wide, allowing, for example, multiple instructions to be fetched simultaneously. On the other hand, the L2 cache occupies the typical external cache environment. It differs, however, from the typical external cache in that, rather than providing instructions and operands to a CPU, it primarily provides instructions and operands to the first-level cache L1. Since the L2 cache is accessed only on L1 misses, the access pattern is considerably different than that for a CPU, and the design parameters are accordingly different.

14-4 VIRTUAL MEMORY

In our quest for a large, fast memory, we have achieved the appearance of a fast, medium-sized memory through the use of a cache. In order to have the appearance of a large memory, we now explore the relationship between main memory and hard disk. Because of the complexity of managing transfers between these two media, the control of such transfers involves the use of data structures and programs. Initially, we will discuss the most basic data structure used and the necessary hardware and software actions. Then we will deal with special hardware used to implement time-critical hardware actions.

With respect to large memory, not only do we want the entire virtual address space to appear to be main memory, but in most cases we would also like this complete space to appear to be available to each program that is executing. Thus, each program will "see" a memory the size of the virtual address space. Equally important to the programmer is the fact that real address space in main memory and real disk addresses are replaced by a single address space that has no restrictions on its use. With this arrangement, virtual memory can be used not only to provide the appearance of large main memory, but also to free up the programmer from having to consider the actual locations of the program and data in main memory and on the hard disk. The job of the software and hardware that implement virtual memory is to map each *virtual address* for each program into a *physical address* in the main memory. In addition, with a virtual address space for each program, it is possible for a virtual address from one program and a virtual address from another program to map to the same physical address. This allows code and data to be shared by multiple programs, thereby reducing the size of the main memory space and disk space required.

To permit the software to map virtual addresses to physical addresses, and to facilitate the transfer of information between main memory and hard disk, the virtual address space is divided into blocks of addresses, typically of a fixed size. These blocks, called *pages*, are larger than, but analogous to, lines in a cache. The physical address space in memory is divided into blocks called *page frames* that are the same size as the pages. When a page is present in the physical address space, it occupies a page frame. For purposes of illustration, we assume that a page consists of 4K bytes (1K words of 32 bits). Further, we assume that there are 32 address bits in the virtual address space. There are 2^{20} pages, maximum, in the virtual address space, and assuming a main memory of 16M bytes, there are 2^{12} page frames in main memory. Figure 14-11 shows the fields of virtual and physical addresses. The portion of the virtual address used to address words or bytes within a page is the *page offset*, which is the only part of the address that the virtual and physical addresses share. Note that words are assumed to be aligned in terms of their location with respect to their byte addresses such that each word address ends in

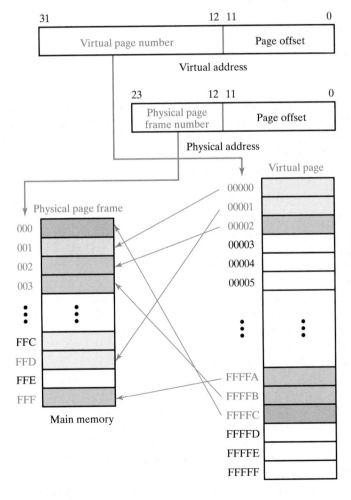

□ **FIGURE 14-11**
Virtual and Physical Address Fields and Mapping

binary 00. Likewise, pages are assumed to be aligned with respect to the byte addresses such that the page offset of the first byte in the page is 000_{16} and the page offset of the last byte in the page is FFF_{16}. The 20-bit portion of the virtual address used to select pages from the virtual address space is the *virtual page number*. The 12-bit portion of the physical address used to select pages in main memory is the *page frame number*. The figure shows a hypothetical mapping from the virtual address space into the physical address space. The virtual and physical page numbers are given in hexadecimal. A virtual page can be mapped to any physical page frame. Six mappings of pages from virtual memory to physical memory are shown. These pages constitute a total of 24K bytes. Note that there are no virtual pages mapped to physical page frames FFC and FFE. Thus, any data present in these pages is invalid.

Page Tables

In general, there may be a very large number of virtual pages, each of which must be mapped to either main memory or hard disk. The mappings are stored in a data structure called a *page table*. There are many ways to structure page tables and access them; we assume that page tables themselves are also kept in pages. Assuming that the representation of each mapping requires one word, 2^{10}, or 1K, mappings can be contained in a 4 KB page. Thus, the mappings for the entire address space for a program of 2^{22} bytes (4 MB) can be contained in one 4 KB page. A special table for each program called a *directory page* provides the mappings used to locate the 4 KB program page tables.

    A sample format for a page table entry is given in Figure 14-12. Twelve bits are used for the page frame number in which the page is located in main memory. In addition, there are three single bit fields: Valid, Dirty, and Used. If Valid is 1, then the page frame in memory is valid; if Valid is 0, the page frame in memory is invalid, meaning that it does not correspond to correct code or data. If Dirty is 1, then there has been a write to at least one byte in the page since it was placed in main memory. If Dirty is 0, there have been no writes to the page since it entered main memory. Note that the Valid and Dirty bits correspond exactly to those in a cache which uses write-back. When it is necessary for a page to be removed from main memory and the Dirty bit is 1, then the page is copied back to the hard disk. If the Dirty bit is 0, indicating that the page in main memory has not been written into, then the page coming into the same page frame is

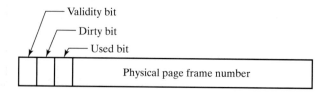

☐ **FIGURE 14-12**
Format for Page Table Entries

simply written over the present page. This can be done because the disk version of the present page is still correct. In order to use this feature, the software keeps a record of the location of the page on the disk elsewhere when it places the page in main memory. The Used bit is a simple mechanism for implementing a crude approximation to an LRU replacement scheme. Some additional bit positions in a page entry may be reserved for flags used by the computer operating system. For example, a few flags might represent the read and write protection status of a page and whether the page can be accessed in user mode or supervisor mode.

The page table structure we have just described is shown in Figure 14-13. The *directory page pointer* is a register that points to the location of the directory page in main memory. The directory page contains the locations of up to 1K page tables associated with the program that is executing. These page tables may be in main memory or on the hard disk. The page table to be accessed is derived from the most significant 10 bits of the virtual page number, which we call the *directory offset*. Assuming that the page table selected is in main memory, it can be accessed by the *page table page number*. The least significant 10 bits of the virtual page number, which we call the *page table offset*, can be used to access the entry for the page to be accessed. If the page is in main memory, the page offset is used to

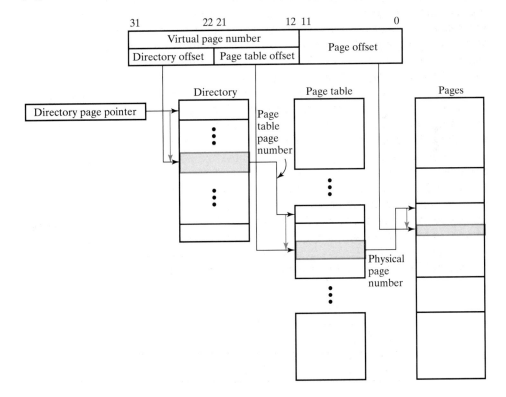

□ **FIGURE 14-13**
Example of Page Table Structure

locate the physical location of the byte or word to be accessed. If either the page table or the desired page is not in main memory, it must first be fetched by software from the hard disk to main memory before the word within it is accessed. Note that combining the offsets with register or table entries is done by simply setting the offset to the right of the page frame number, rather than adding the two together. This approach requires no delay, whereas addition would cause significant delay.

Translation Lookaside Buffer

From the preceding discussion, we note that virtual memory has a considerable performance penalty even in the best case, when the directory, the page table, and the page to be accessed are in main memory. For our assumed page table approach, three successive accesses to main memory occur in order to fetch a single operand or instruction:

1. Access for the directory entry.
2. Access for the page table entry.
3. Access for the operand or instruction.

Note that these accesses are performed automatically by hardware that is part of the MMU in the generic computer. Thus, to make virtual memory feasible, we need to drastically reduce accesses to main memory. If we have a cache, and if all of the entries are in the cache, then the time for each access is reduced. Nevertheless, three accesses are needed to the cache. To reduce the number of accesses, we will employ yet another cache for the purpose of translating the virtual address directly into a physical address. This new cache is called a *translation lookaside buffer* (TLB). It holds the locations of recently addressed pages to speed access to cache or main memory. Figure 14-14 gives an example of a TLB, which is typically fully associative or set associative, since it is necessary to compare the virtual page number from the CPU with a number of virtual page number tags. In addition to the latter, a cache entry includes the physical page number for those pages in main memory and a Valid bit. If the page is in main memory, the Dirty bit also appears. The Dirty bit serves the same function for a page in main memory as discussed previously for a line in a cache.

We now briefly look at a memory access using the TLB in Figure 14-14. The virtual page number is applied to the page number input to the cache. Within the cache, this page number is compared simultaneously with all of the virtual page number tags. If a match occurs and the Valid bit is a 1, then a TLB hit has occurred, and the physical page frame number appears on the page number output of the cache. This operation can be performed very quickly and produces the physical address required to access memory or a cache. On the other hand, if there is a TLB miss, then it is necessary to access main memory for the directory table entry and the page table entry. If there is a physical page in main memory, then the page table entry is brought into the TLB cache and replaces one of the entries there.

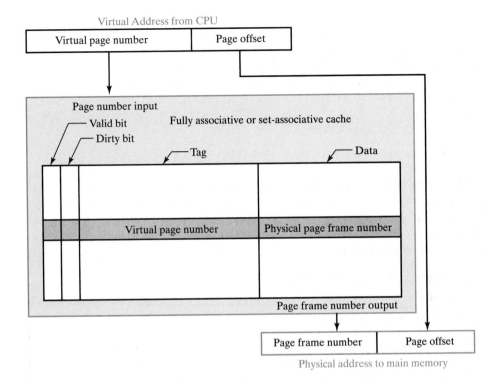

□ **FIGURE 14-14**
Example of Translation Lookaside Buffer

Overall, three memory accesses are required, including the one for the operand. If the physical page does not exist in main memory, then a *page fault* occurs. In this case, a software-implemented action fetches the page from its hard disk location to main memory. During the time required to complete this action, the CPU may execute a different program rather than waiting until the page has been placed in main memory.

Noting the prior hierarchy of actions based on the presentation of a virtual address, we see that the effectiveness of virtual memory depends on temporal and spatial locality. The fastest response is possible when the virtual page number is present in the TLB. If the hardware is fast enough and a hit also occurs on the cache, the operand can be available in as little as one or two CPU clock cycles. Such an event is likely to happen frequently if the same virtual pages tend to get accessed over time. Because of the size of the pages, if one operand is accessed from a page, then, due to spatial locality, it is likely that another operand will be accessed on the same page. With the limited capacity of the TLB, the next fastest action requires three accesses to main memory and slows processing considerably. In the worst of all situations, the page table and the page to be accessed are not in main memory. Then, lengthy transfers of two pages—the page table and the page from hard disk—are required.

Note that the basic hardware for implementing virtual memory, the TLB, and other optional features for memory access are included in the MMU in the generic computer. Among the other features is hardware support for an additional layer of virtual addressing called *segmentation* and for protection mechanisms to permit appropriate isolation and sharing of programs and data.

Virtual Memory and Cache

Although we have considered the cache and virtual memory separately, in an actual system they are both very likely to be present. In that case, the virtual address is converted to the physical address, and then the physical address is applied to the cache. Assuming that the TLB takes one clock cycle and the cache takes one clock cycle, in the best of cases fetching an instruction or operand requires two CPU clock cycles. As a consequence, in many pipelined CPU designs, two or more clock cycles are allowed for an operand fetch. Since instruction fetch addresses are more predictable, it is possible to modify the CPU pipeline and consider the TLB and cache to be a two-stage pipeline segment, so that an instruction fetch appears to require only one clock cycle.

14-5 CHAPTER SUMMARY

In this chapter, we examined the components of a memory hierarchy. Two concepts fundamental to the hierarchy are cache memory and virtual memory.

Based on the concept of locality of reference, a cache is a small, fast memory that, holds the operands and instructions most likely to be used by the CPU. Typically, a cache gives the appearance of a memory the size of main memory with a speed close to that of the cache. A cache operates by matching the tag portion of the CPU address with the tag portions of the addresses of the data in the cache. If a match occurs and other specific conditions are satisfied, a cache hit occurs, and the data can be obtained from the cache. If a cache miss occurs, the data must be obtained from the slower main memory. The cache designer must determine the values of a number of parameters, including the mapping of main memory addresses to cache addresses, the selection of the line of the cache to be replaced when a new line is added, the size of the cache, the size of the cache line, and the method for performing memory writes. There may be more than one cache in a memory hierarchy, and instructions and data may have separate caches.

Virtual memory is used to give the appearance of a large memory—much larger than the main memory—at a speed that is, on average, close to that of the main memory. Most of the virtual address space is actually on hard disk. To facilitate the movement of information between the memory and the hard disk, both are divided up in fixed size address blocks called page frames and pages, respectively. When a page is placed in main memory, its virtual address must be translated to a physical address. The translation is done using one or more page tables. In order to perform the translation on each memory access without a severe performance penalty, special hardware is employed. This hardware, called a translation lookaside buffer (TLB), is a special cache that is a part of the memory management unit (MMU) of the computer.

Together with main memory, the cache and the TLB give the illusion of a large, fast memory that is, in fact, a hierarchy of memories of different capacities, speeds, and technologies, with hardware and software performing automatic transfers between levels.

REFERENCES

1. MANO, M. M. *Computer Engineering: Hardware Design.* Englewood Cliffs, NJ: Prentice Hall, 1988.

2. HENNESSY, J. L., AND D. A. PATTERSON *Computer Architecture: A Quantitative Approach.* San Francisco, CA: Morgan Kaufmann, 1996.

3. BARON, R. J., AND L. HIGBIE *Computer Architecture.* Reading, MA: Addison-Wesley, 1992.

4. HANDY, J. *Cache Memory Book.* San Diego: Academic Press, 1993.

5. MANO, M. M. *Computer System Architecture,* 3rd Ed. Englewood Cliffs, NJ: Prentice Hall, 1993.

6. PATTERSON, D. A., AND J. L. HENNESSY *Computer Organization and Design: The Hardware/Software Interface.* San Francisco, CA: Morgan Kaufmann, 1998.

7. WYANT, G., AND T. HAMMERSTROM *How Microprocessors Work.* Emeryville, CA: Ziff-Davis Press, 1994.

8. MESSMER, H. P., *The Indispensable PC Hardware Book,* 2nd ed. Wokingham, U.K.: Addison-Wesley, 1995.

PROBLEMS

The plus (+) indicates a more advanced problem and the asterisk (*) indicates a solution is available on the Companion Website for the text.

14–1. *A CPU produces the following sequence of read addresses in hexadecimal:
54, 58, 104, 5C, 108, 60, F0, 64, 54, 58, 10C, 5C, 110, 60, F0, 64
Supposing that the cache is empty to begin with, and assuming an LRU replacement, determine whether each address produces a hit or a miss for each of the following caches: **(a)** direct mapped in Figure 14-3, **(b)** fully associative in Figure 14-4, and **(c)** two-way set associative in Figure 14-6.

14–2. Repeat Problem 14-1 for the following sequence of read addresses:
0, 4, 8, 12, 14, 1A, 1C, 26, 28, 2E, 30, 36, 38, 3E, 40, 46, 48, 4E, 50, 56, 58, 5E

14–3. Repeat problem 14-1 for the following sequence of read addresses in hexadecimal: 20, 04, 28, 60, 20, 04, 28, 4C, 10, 6C, 70, 10, 60, 70

14-4. *A computer has a 32-bit address and a direct-mapped cache. Addressing is to the byte level. The cache has a capacity of 1K bytes and uses lines that are 32 bytes. It uses write-through and so does not require a dirty bit.

    (a) How many bits are in the index for the cache?

    (b) How many bits are in the tag for the cache?

    (c) What is the total number of bits of storage in the cache, including the valid bits, the tags, and the cache lines?

14-5. A two-way set-associative cache in a system with 24-bit addresses has two 4-byte words per line and a capacity of 512K bytes. Addressing is to the byte level.

    (a) How many bits are there in the index and the tag?

    (b) Indicate the value of the index in hexadecimal for cache entries from the following main memory addresses in hexadecimal: 82AF82, 14AC89, 48CF0F and3ACF01.

    (c) Can all of the cache entries from part (b) be in the cache simultaneously?

14-6. *Discuss the advantages and disadvantages of:

    (a) separate instruction and data caches versus a unified cache for both.

    (b) a write-back cache versus a write-through cache.

14-7. Give an example of a sequence of program and data memory read addresses that will have a high hit rate for separate instruction and data caches and a low hit rate for a unified cache. Assume direct mapped caches with the parameters in Figure 14-3. Both the instructions and data are 32-bit words and the address resolution is to bytes.

14-8. *Give an example of a sequence of program and data memory read addresses that will have a high hit rate for a unified cache and a low hit rate for separate instruction and data caches. Assume that each of the instruction and data caches is two-way set associative with parameters as in Figure 14-6. Assume that the unified cache is four-way set associative with parameters as in Figure 14-6. Both the instructions and the data are 32-bit words and the address resolution is to bytes.

14-9. Explain why write-allocate is typically not used in a write-through cache.

14-10. A high-speed workstation has 64-bit words and 64-bit addresses with address resolution to the byte level.

    (a) How many words can be in the address space of the workstation?

    (b) Assuming a direct-mapped cache with 8192 32-byte lines, how many bits are in each of the following address fields for the cache: (1) Byte, (2) Index, and (3) Tag?

14-11. *A cache memory has an access time from the CPU of 4 ns, and the main memory has an access time from the CPU of 40 ns. What is the effective access time for the cache–main memory hierarchy if the hit ratio is: **(a)** 0.91, **(b)** 0.82, and **(c)** 0.96?

14–12. Redesign the cache in Figure 14-7 so that it is the same size, but is four-way set associative rather than two-way set associative.

14–13. +The cache in Figure 14-9 is to be redesigned to use write-back with write-allocate rather than write-through. Respond to the following requests, making sure to deal with all of the address and data issues involved in the write-back operation.

(a) Draw the new block diagram.

(b) Explain the sequence of actions you propose for a write miss and for a read miss.

14–14. *A virtual memory system uses 4K byte pages, 64-bit words, and a 48-bit virtual address. A particular program and its data require 4263 pages.

(a) What is the minimum number of page tables required?

(b) What is the minimum number of entries required in the directory page?

(c) Based on your answers to (a) and (b), how many entries are there in the last page table?

14–15. A small TLB has the following entries for a virtual page number of length 20 bits, a physical page number of 12 bits, and a page offset of 12 bits.

Valid bit	Dirty bit	Tag (Virtual Page Number)	Data (Physical Page Number)
1	1	01AF4	FFF
0	0	0E45F	E03
0	0	012FF	2F0
1	0	01A37	788
1	0	02BB4	45C
0	1	03CA0	657

The page numbers and offset are given in hexadecimal. For each of the virtual addresses listed, indicate whether a hit occurs and if it does, give the physical address: (a) 02BB4A65, (b) 0E45FB32, (c) 0D34E9DC, and (d) 03CA0777.

14–16. A computer can accommodate a maximum of 384M bytes of main memory. It has a 32-bit word and a 32-bit virtual address and uses 4K byte pages. The TLB contains only entries that include the Valid, Dirty, and Used bits, the virtual page number, and the physical page number. Assuming that the TLB is fully associative and has 32 entries, determine the following:

(a) How many bits of associative memory are required for the TLB?

(b) How many bits of SRAM are required for the TLB?

14–17. Four programs are concurrently executing in a multitasking computer with virtual memory pages having 4K bytes. Each page table entry is 32 bits.

What is the minimum numbers of bytes of main memory occupied by the directory pages and page tables for the four programs if the numbers of pages per program, in decimal, are as follows: 6321, 7777, 9602, and 3853.

14-18. *In caches, we use both write-through and write-back as potential writing approaches. But for virtual memory, only an approach that resembles write-back is used. Give a sound explanation of why this is so.

14-19. Explain clearly why both the cache memory concept and the virtual memory concept would be ineffective if locality of reference of memory-addressing patterns did not hold.

INDEX